Physics

Side Chain Liquid Crystal Polymers

To Maeve

SIDE CHAIN
LIQUID CRYSTAL POLYMERS

edited by
C.B. McARDLE
Loctite (Ireland) Ltd.
Dublin

Blackie
Glasgow and London

Published in the USA by
Chapman and Hall
New York

Blackie and Son Ltd
Bishopbriggs, Glasgow G64 2NZ

7 Leicester Place, London WC2H 7BP

Published in the USA by
Chapman and Hall
a division of Routledge, Chapman and Hall, Inc.
29 West 35th Street, New York, NY 10001-2291

British Library Cataloguing in Publication Data

Side chain liquid crystal polymers.
1. Polymers. Liquid crystals
I. McArdle, C.B.
547.7

ISBN 0-216-92503-7

For the USA, International Standard Book Number is
0-412-01761-X

Phototypesetting by Thomson Press (India) Limited, New Delhi
Printed in Great Britain by
Bell and Bain Ltd, Glasgow

Preface

This book represents the first state-of-the-art overview exclusively dedicated to the chemistry, physics and applications of thermotropic side chain liquid crystal polymers, and is made up of contributions from acknowledged experts worldwide.

Side chain liquid crystal polymers are an important class of electroactive polymers which have undergone extensive investigation, especially in the recent past, although their first concept and preparation dates back to the 1970's. The book describes the scope that these materials offer for molecular engineering, leading to the realisation of new mesomorphic species, the detailed macroscopic behaviour of which has been characterized by powerful physical techniques. From a technological standpoint, side chain liquid crystal polymers have generated much interest, originating primarily (but not exclusively) from the device physics community. A serious attempt has been made in the book to document progress made in all areas of application of these materials to date.

This text represents an extension to the earlier, more general reviews on LCP's, but exclusive treatment of the side chain type polymers in a single volume has presented the opportunity to provide what is hoped to be a continuous description of these materials, from theoretical concepts, through polymer formation and characterization, to their potential end uses. As such, this book should appeal to researchers working in pure and applied chemistry and physics.

The editor takes this opportunity to thank all contributors for their excellent cooperation throughout this project and to thank former colleague Professor M.G. Clark of GEC Research for permission to undertake the project in the first instance. The encouragement, understanding and practical assistance of my wife is warmly acknowledged.

C.B. McA.

Contents

Contributors

Dr G.S. Attard, Department of Chemistry, University of Southampton, Southampton SO9 5NH, UK.

Dr C. Boeffel, Max-Planck-Institut für Polymerforschung, P.O. Box 3148, D-6500, Mainz, West Germany.

Professor M.G. Clark, GEC Hirst Research Centre, East Lane, Wembley, Middlesex HA9 7PP, UK.

Dr J.C. Dubois, Laboratoire Central de Recherches, Thomson-CSF, Domain de Corbeville, 91400 Orsay Cedex, France

Professor H. Finkelmann, Institut für Makromolekulare Chemie, Universität Frieburg, Stefan-Meier Strasse 31, D-7800 Freiburg, West Germany.

Professor Ya. S. Freidzon, Department of Chemistry, Moscow State University, 119 899 Moscow, USSR.

Dr W. Gleim, Institut für Makromolekulare Chemie, Universität Frieburg, Stefan-Meier Strasse 31, D-7800 Freiburg, West Germany.

Professor G.W. Gray, F.R.S., School of Chemistry, University of Hull, Hull HU6 7RX, UK.

Professor W. Haase, Institut für Physikalisch Chemie, Technische Hochschule Darmstadt, D-6100 Darmstadt, West Germany.

Dr C.M. Haws, GEC Hirst Research Centre, East Lane, Wembley, Middlesex, HA9 7PP, UK.

Dr G.M. Janini, Department of Chemistry, University of Kuwait, Kuwait City, 13068 Kuwait.

Professor R.J. Laub, Department of Chemistry, San Diego State University, San Diego, California 92182, USA.

Dr P. Le Barny, Laboratoire Central de Recherches, Thomson-CSF, Domain de Corbeville, 91400 Orsay Cedex, France.

Dr C.B. McArdle, Loctite (Ireland) Ltd, Research Development Laboratories, Whitestown Industrial Estate, Tallaght, Dublin 24, Ireland.

Dr G.R. Möhlmann, Akzo Research Laboratories, Corporate Research, Applied Physics Department, P.O. Box 9300, 6800 SB Arnhem, Netherlands.

Dr C. Noël, École Supérieure de Physique et de Chemie Industrielles, Laboratoire de Physicochimie Structurale et Macromoléculare, 10 Rue Vauquelin, 75231 Paris Cedex 05, France.

Professor V. Percec, School of Macromolecular Science, Department of Engineering, Case Western Reserve University, Cleveland, Ohio 44106, USA.

Dr C. Pugh, School of Macromolecular Science, Department of Engineering, Case Western Reserve University, Cleveland, Ohio 44106, USA.

Professor J.H. Purnell, Department of Chemistry, University College of Swansea, Swansea, Wales SA2 8PP, UK.

Professor V.P. Shibaev, Department of Chemistry, Moscow State University, 119 899 Moscow, USSR.

Professor H.W. Spiess, Max-Planck-Institut für Polymerforschung, P.O. Box 3148, D-6500 Mainz, West Germany.

Dr O.S. Tyagi, Regional Research Laboratory, Council of Scientific and Industrial Research, Hyderabad 500 007, India.

Dr C.P.J.M. van der Vorst, Akzo Research Laboratories, Corporate Research, Applied Physics Department, P.O. Box 9300, 6800 SB Arnhem, Netherlands.

Dr M. Warner, Department of Physics, University of Cambridge, Cavendish Laboratory, Madingley Road, Cambridge CB3 0HE, UK.

1 Scope and potential for polymeric systems with mesogenic side chains

C.B. McARDLE,* GEC Research, Hirst Research Centre, East Lane, Wembley, Middlesex HA9 7PP, UK

1.1 Introduction

Liquid crystalline polymers (LCPs) are high-molecular-mass materials which exhibit mesomorphism. Traditionally two major classes of LCP have been identified: the so-called main chain and side chain types (MCLCPs and SCLCPs respectively). More recently other variants have appeared; these are combined LCPs (Reck and Ringsdorf, 1985, 1986) which are hybrid between MCLCPs and SCLCPs, and the rigid rod types described by Watanabe *et al.* (1987). A great wealth of literature already exists in the form of unified texts and reviews which detail both the major classes of LCPs (Blumstein, 1978*a, b*, 1985; Chapoy, 1985; Finkelmann and Rehage, 1984; Shibaev and Platé, 1984; Ciferi *et al.*, 1982). More recently bibliographic data has been compiled (Hinov, 1986) and reviews more or less specific to main chain (Cox, 1987) or comb polymer systems have appeared (Platé and Shibaev, 1988). The present text is dedicated specifically to developments in the chemistry, physics and applications of thermotropic side chain LCPs up to mid-1988. Lyotropic side chain LCPs have been completely omitted from this text, which has already far exceeded its predesignated length dealing with the thermotropics alone. The reader is referred to a recent review by Ringsdorf *et al.* (1988) (and references therein, particularly to the works of Finkelmann *et al.*) which covers lyotropic systems in some detail.

1.2 Structural progress in side chain LCPs

The concept of the side chain LCP is fairly self-explanatory from its name. The basic principles of these materials have already been covered at length in the literature cited above and hence will not be further elaborated upon. Figure 1.1, however, provides a schematic representation of how side chain LCPs have progressed from a structural chemistry standpoint over the last ten years. Progress in their physical chemistry and device physics has been equally great in this period, as is evident from the contents of Chapter 5 onwards.

In 1978, the side chain LCP was simply a hydrocarbon polymer with spaced-off simple mesogenic units. Today, almost a plethora of structural modifications exists for these materials, based upon numerous classes of polymers prepared by a variety of polymerization techniques. For example, relatively simple solid–liquid phase transfer catalysed synthetic procedures have been described for (specific) side chain LCPs

*Present address: Loctite (Ireland) Ltd, Research and Development Laboratories, Whitestown Industrial Estate, Tallaght, Dublin 24, Republic of Ireland.

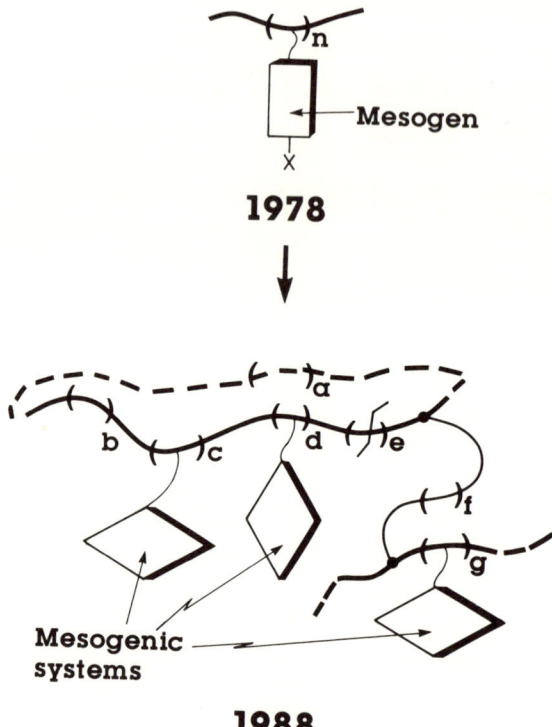

Figure 1.1 Schematic representation of the progress in structural development in side chain LCPs over the last decade.

(Keller, 1987). These techniques surely must be significant from a large-scale production standpoint, should this class of electroactive polymers fulfill their promises and become commercially important materials. Cyclic side chain LCPs now exist (Figure 1.1, $a \neq 0$) in addition to the more conventional linear materials which can be homo- or copolymeric. In the latter types, comonomers may be mesogenic or non-mesogenic (Figure 1.1, $b, c, d, e > 0$) and the non-mesogenic component can be functional (photochromic, photoconductive, hyperpolarizable) or non-functional (internal plasticizer or diluent).

The cores attached to spacers are now probably best referred to as 'mesogenic systems' rather than mere mesogens, since disc- and rod-shaped single and paired structures have been prepared and appended in different fashions. Today it is in fact difficult to conceive a shape appropriate to pictorially represent the broad class of mesogenic systems that have been prepared. Indeed, the creative organic chemist appears to be able to produce the most diverse structures and incorporate them into polymers which subsequently exhibit mesomorphism (Kreuder *et al.*, 1987). A firm understanding of the rôle of the spacer in side chain LCPs is now established, and interesting microphase separated LCPs have been studied (see Percec and Pugh, Chapter 3, and references therein) which have important ramifications from a device physics point of view. The chemistry and functionality of the spacer has also been modified; perfluorinated and ethylene oxide type spacers exist, and the latter can be organized by their terminal mesogens and function as open crown ethers. Inorganic siloxane spacers and crosslinks (Figure 1.1, $f \neq 0$) are well known. Use of the latter leads to various classes of liquid crystalline elastomer and lightly cross-linked networks which can be made to exhibit the interesting ferroelectric S_C^* phase (see Chapter 5, section 5.4.1). An in-depth discussion on side chain LCP formation is given in Chapters 3, 4 and 5.

1.3 Scope for applications of side chain LCPs

Given that the field of SCLCPs is now highly developed, it is perhaps more relevant than ever to stand back and ask the basic questions: why is research in this area generating so much interest at present, and what outlets might exist for these polymers? The materials are undoubtedly extremely interesting from an academic standpoint, and it is quite true that further fundamental research is required to underpin this branch of polymer and LC science. Fundamental experimentation on low-molar-mass LCs ultimately led to the realization of room-temperature nematic systems which had a major technological impact. With retrospective knowledge of low-molar-mass LCs, scientists working with side chain LCPs, in both industrial and academic research, are naturally intrigued by the potential scope of new materials which might combine the functionality of conventional LCs with the properties of macromolecules. It is all too easy, however, to elaborate on the last statement by immediately speculating as to the potentially facile processing technologies and novel embodiments which might be available through these unique materials. It is worth pointing out that the technology involved in processing side chain LCPs is little researched, even today, some 15 years after their discovery. Few examples exist where samples are prepared for study by conventional polymer thin film processing techniques. However, in the cases when techniques such as spin coating, polymer lamination and membrane preparation have been employed, very interesting results have been obtained, (see for example, Pinsl et al., 1987; Ueno et al., 1986; Loth et al., 1988). This area, then, is one yet to be studied, and is critically important for side chain and combined LCP product development. The importance of processability in electroactive main chain LCPs has already been recognized (Ulrich, 1987), and interesting processing techniques have been identified, for example, the control of shear fields by stagnation flow die geometries (Karasz, 1984) which presumably could be extended immediately to study nematic side chain or combined LCPs. Further aspects of processing lyotropic and thermotropic main chain LCPs are given by Zachariades and Porter (1987).

A variety of potential applications for side chain LCPs has been considered to date. Broadly speaking, applications fall into two classes: electro-thermo-optical effects and separation and complexation effects. Some examples from each class are briefly elaborated on below, but detail is given elsewhere in this book.

(1) *Electro-thermo-optical effects*

   (i) Optical data storage: digital, analogue and holographic optical storage have all been demonstrated (the latter two classes also with erasability) using cholesteric, nematic and smectic side chain LCPs. Write-once recording on main chain LCPs has also been demonstrated on polymers originally designed for resist work (Griffin *et al.*, 1988) (see Chapter 13).

(ii) Optical elements: selective wavelength, notch and bandpass filters may be made from cholesteric side chain (and main chain) LCPs. Erasable holographic optical elements, Fresnel zone plates and gratings have also been demonstrated (Chapter 13 and references therein; see also Shannon, 1986; Tsutsai *et al.*, 1980).

(iii) Optical claddings: optical fibre jacketing is a common application area for certain main chain LCPs. The side chain LCPs tend to be studied in integrated optics outlets, for example waveguides (see (iv), elastomers (viii) and Chapter 12).

(iv) Optically non-linear effects: side chain LCPs, loaded with appropriate dyes or with inherently coloured mesogenic systems are under investigation primarily for second-order effects such as second harmonic generation (SHG) and the electro-optic effect (see Chapter 12 and Le Grange *et al.*, 1987; Small *et al.*, 1986). Certain lyotropic main chain LCPs have been developed for second- and third-order non-linear effects (χ^3 effects), such as third harmonic generation, optical Kerr effects (Ulrich, 1987; Kuder, 1986).

(v) Photo effects: thermotropic and lyotropic side chain LCPs bearing photoconductive groups have been prepared and are currently under investigation (Lux *et al.*, 1987; Chapoy *et al.*, 1985). The aim in such work is to enhance electronic transport phenomena in well-ordered media. Photochromic side chain LCPs are referred to in Chapter 3. These may have applications in imaging science, but practical control of phase behaviour is required. Photocontrol of phase transformation in side chain LCPs may also be exploitable in imagery by analogy to low-molar-mass LC systems (see Chapter 13).

(vi) Electronic effects: today some speculation exists as to the possibility of producing conductive LCP systems (e.g., in the spinal columnar LCPs of Sirlin *et al.*, 1987). Discotic side chain LCPs containing oxidised triphenylene units may be potential candidates (Wenz, 1985).

(vii) Electro-optical displays: on their own, side chain LCPs are not practical display media, owing to their high viscosity. They have, however, been used as dopants to advantageously modify the elastic constants of conventional LCs (Coles, 1986, and references therein). Side chain LCPs have also been proposed as switchable alignment layers and polarizers in LCDs (Kruger and Rubner, 1984).

(viii) Elasto-optical effects: bulk samples of side chain LC elastomers have been prepared (1 cm thick), as have thin elastomeric membranes (Chapter 10) Applications may include modulators (cf. Maher *et al.*, 1976) and photoelastic coatings (cf. Zandman *et al.*, 1977) and media to unwind chiral S_C^* helices (cf. Chapter 5).

(ix) Piezo-pyro effects: see Chapter 5.

(2) *Separation and complexation effects*

Side chain LCPs with oxyethylene spacers can dissolve metal cations (Rodriguez-Parada and Percec, 1986). Interest in these and main chain LCPs with enhanced ionic conductivity may lead to applications in novel 'solid' polymer electrolytes (cf. Hall *et al.*, 1986, *a, b*; see also Ratner *et al.*, 1988). Control of permeation of gases and simple drugs has been reported recently using side chain LCP membranes (Tadayaki *et al.*, 1987; Loth *et al.*, 1988). For related work, see also Kajiyama *et al.* (1983); Terada *et al.*

(1982); Washizu *et al.* (1983) and Mariani *et al.*, 1987). Detailed treatment of one very well developed application of side chain LCPs as stationary phase in GC systems is presented in Chapter 14.

References

Blumstein, A. (ed.) (1978*a*) *Mesomorphic Order in Polymers and Polymerisation in Liquid Crystalline Media*, ACS Symp. Ser. 74, American Chemical Society, Washington DC.

Blumstein, A. (ed.) (1978*b*) *Liquid Crystalline Order in Polymers*. Academic Press, New York.

Blumstein, A. (ed.) (1985) *Polymeric Liquid Crystals*. Plenum, New York.

Chapoy, L. (ed.) (1985) *Recent Advances in Liquid Crystalline Polymers*. Elsevier Applied Science, London.

Chapoy, L., Munck, D., Rasmussen, K., Juul Diekmann, E. and Sethi, R. (1985) in *Recent Advances in Liquid Crystalline Polymers*, ed. Chapoy, L., Elsevier Applied Science, London.

Ciferi, A., Krigbaum, W. and Meyer, R. (eds.) (1982) *Polymer Liquid Crystals*. Academic Press, New York.

Coles, H. (1986) in *Fine Chemicals for the Electronics Industry*, ed. Bamfield, P. Vol. 60, Royal Society of Chemistry, London.

Cox, M. (1987) *Liquid Crystal Polymers*, ed. Meredith, R., RAPRA Report No. 4, Pergamon, Oxford.

Finkelmann, H. and Rehage, G. (1984) *Adv. Polym. Sci.* **60/61**, 99.

Griffin, A., Hall, C., Hoyle, C., Venataram, K. and McArdle, C. (1988) *Makromol. Chem. Rapid Commun.* **9**, 463.

Hall, P., Davies, G., Ward, I., McIntyre, J., Bannister, D. and LeBrocq, K. (1986*a*) *Polymer Commun.* **27**, 98.

Hall, P., Davies, G., Ward, I. and McIntyre, J. (1986*b*) *Polymer Commun.* **27**, 100.

Hinov, H. (1986) *Mol. Cryst. Liq. Cryst.* **136(2–4)**, 221.

Kajiyama, T., Nagata, Y., Washizu, S. and Takayanagi, M. (1983) *J. Membrane Sci.* **11**, 39.

Karasz, F. (1984) in *Ultrastructure Processing of Ceramics, Glasses and Composites*, eds. Hench, L. and Ulrich, D., Wiley Interscience, New York, Chapter 25.

Keller, P. (1987) *Macromolecules* **20**, 462.

Keuder, J. (1986) US Patent 4607095.

Kreuder, W., Ringsdorf, H., Herrmann-Schonherr, O. and Wendorff, J. (1987) *Angew. Chem. Int. Edn. Engl.* **26(12)**, 1249.

Kruger, H. and Rubner, R. (1984) US Patent 4469408.

Le Grange, J., Kuzyk, M. and Singer, K. (1987) *Mol. Cryst. Liq. Cryst.* **150b**, 567.

Loth, H. and Euschen, A. (1988) *Makromol. Chem. Rapid Commun.* **9**, 35.

Lux, M., Strohriegl, P. and Hocker, H. (1987) *Makromol. Chem.* **188**, 811.

Maher, J., Schank, R. and Pfister, G. (1976) *Appl. Phys. Letts.* **29(5)**, 293.

Mariani, R. and Abruna, H. (1987) *Electrochimica Acta* **32(2)**, 319.

Pinsl, J., Brauchle, Chr. and Kreuzer, F. (1987) *J. Molec. Electron.* **3**, 9.

Platé, N. and Shibaev, V. (1988) *Comb Shaped Polymers and Liquid Crystals*. Plenum, New York.

Ratner, M. and Shriver, D. (1988) *Chem. Rev.* **88**, 109.

Reck, B. and Ringsdorf, H. (1985) *Makromol. Chem. Rapid Commun.* **6**, 291.

Reck, B. and Ringsdorf, H. (1986) *Makromol. Chem. Rapid Commun.* **7**, 389.

Ringsdorf, H., Schlarb, B. and Venzmer, J. (1988) *Angew. Chem. Int. Edn. Engl.* **27**, 113.

Rodriguez-Parada, J. and Percec, V. (1986) *Polymer Preprints* **27(1)**, 360.

Shannon, P. (1986) US Patent 4614619.

Shibaev, V. and Platé, N. (1984) *Adv. Polym. Sci.* **60/61**, 173.

Sirlin, C., Bosio, L. and Simon, J. (1987) *J. Chem. Soc., Chem. Commun.* 379.

Small, R., Singer, K., Sohn, J., Kuzyk, M. and Lalama, S. (1986) *Proc. SPIE* **682**, 160.

Tadayuki, A., Shudo, Y. and Fumio, Y. (1987) JP62171706 [through *CA Selects*, Issue 3, 1988 entry 108: 39246h].

Terada, I., Kajiyama, T. and Takayanagi, M. (1982) *Rep. Progr. Polym. Phys. Jpn.* **25**, 303.

Tsutsui, T. and Tanaka, R. (1980) *Polymer Commun.* **21**, 1351.

Ueno, T., Nakamura, T. and Tani, C. (1986) *Proc. Japan Display'86*, 290.

Ulrich, D. (1987) *Polymer* **28**, 533.

Washizu, S., Terada, I., Kajiyama, T. and Takayanagi, M. (1983) *Rep. Progr. Polym. Phys. Jpn.* **26**, 235.

Watanabe, J., Goto, M. and Nagase, T. (1987) *Macromolecules* **20**, 298.

Wenz, G. (1985) *Makromol. Chem. Rapid Commun.* **6**, 577.

Zachariades, A. and Porter, R. (1987) *High Modulus Polymers*, Marcel Dekker, New York.

Zandman, F., Redner, S. and Dally, J. (1977) *Photoelastic Coatings.* Iowa State University Press and the Society for Experimental Stress Analysis (Iowa and Connecticut).

2 The physical principles of side chain polymer liquid crystals

MARK WARNER, Cavendish Laboratory, Cambridge, UK

2.1 Introduction

Polymer entropy is antagonistic to nematic order. A flexible polymer is expelled from a nematic solvent, since a chain stretched out to be parallel to a nematic field loses a lot of entropy; equally the nematic field, distorting to be locally parallel to a random polymer chain, would develop a large nematic elastic energy.

The antagonism displayed in these examples underlies all the novel properties of polymer liquid crystals: molecular architecture links nematic and flexible elements together in one component, and the competition between polymer and liquid crystal cannot be resolved by phase separation. Initial speculation that polymer and mesogenic properties could be combined to dramatic effect comes apparently from de Gennes (1969); we return to this in discussing networks (de Gennes, 1975). Some new phases resulting from this competition were first suggested by Vasilenko et al. (1985).

In this chapter we shall present some qualitative ideas describing the observed and proposed new nematic phases of comb-like polymers and their complex phase diagrams, smectic phases and elastomers. We shall conclude by summarizing and by making some speculations about dielectric response. Other chapters in this book will make clear the complexities of the molecular possibilities in making comb polymers and the delicacy of phases when confronted by changes in chemical structure. Here we shall look at model molecules only to illustrate how competition can arise and how it can be resolved.

At the same time as discussing phases we shall discuss chain shape. Indeed, chain configurations are a good way of microscopically understanding the new comb phases.

Mesomorphic character in general has been given to polymers by incorporating stiff elements (mesogens) in many ways: concatenating stiff elements to give worm-like molecules as in main chain LCPs, attaching rods or discs to flexible backbones via the so called 'spacer' or 'hinge', as in side chain LCPs, and combining the stiffness in both the backbone and pendants, as in the combined LCPs (cf. Reck and Ringsdorf, 1985, 1986). Various mesophases result, and a considerable degree of molecular tailoring is possible, as is made abundantly clear by Percec and Pugh (Chapter 3). Full characterization of these molecularly engineered systems is all-important if we are to understand, further refine and ultimately exploit these materials. A number of techniques for macroscopic evaluation of side chain LCPs have been described in this book, such as dielectric and nuclear magnetic resonance spectroscopy (Chapters 7 and 8). Model calculations, with a range of parameters selected, reveal the qualitative aspects of ordering and the delicate interplay of phases possible by engineering new molecules.

In referring to particular side chain polymers as combs, we shall mean this in a literal sense rather than simply polymers branched with the topology of a comb. The pendant entities, the teeth of the comb, are rigid elements as in a real comb and impart mesogenic character to the molecule. In general our comb-like polymers are the 'combined' type, where the backbone (sometimes referred to as the main chain) can also be stiff.

To anticipate our results, we see in Figure 2.1 three nematic phases termed N_I, N_{II}, and N_{III} by Wang and Warner (1987). Since we have two nematic elements, backbone and teeth, we can have two order parameters, each of which can be positive or negative. The directions in the molecule needed to define order parameters are the tangent vector for the backbone and the rod axis for the teeth (see Figure 2.1). Given two order parameters, we can conceive of three phases, the case where both order parameters are negative being unstable. A detailed analysis of what types of transitions are possible between the various phases is given by Renz and Warner (1988). Vasilenko et al. (1985) treat the comb components in a more detailed manner in order to construct a lattice theory. The spacer connecting a tooth to the main chain is given an order parameter as well, but their phases have similarities with those below, and will be discussed.

In the nematic-forming phases, a coupling between teeth and polymer backbone was first seen by Koch et al. (1985) who looked at the stress-optical coefficients of isotropic and nematic networks of comb polymers and found negative values. Clearly the chain shape is affected. (The effect of nematic ordering on the chain is examined in section 2.2.) If the teeth order most strongly and the spacer is such that the chain is compelled to be perpendicular to them, then it will be flattened down from the spherical shape of greatest randomness toward being an oblate spheroid. N_I can then be called an oblate phase. Indeed, the shape of a polymer chain is a mirror of its configurations and seems to be a good microscopic tool for understanding the new comb phases. N_{II} and N_{III} are both prolate phases: in N_{II} the chain clearly dominates (it has the positive order) and the spacer forces the teeth to be negatively ordered; in N_{III} it is not clear a priori which is stronger, but in any case the nematic cross-coupling, driving main chain and teeth parallel, overcomes the spacer. We discuss experiments revealing chain shape and the symmetry of phases further in section 2.4, but much fuller and more explicit discussions

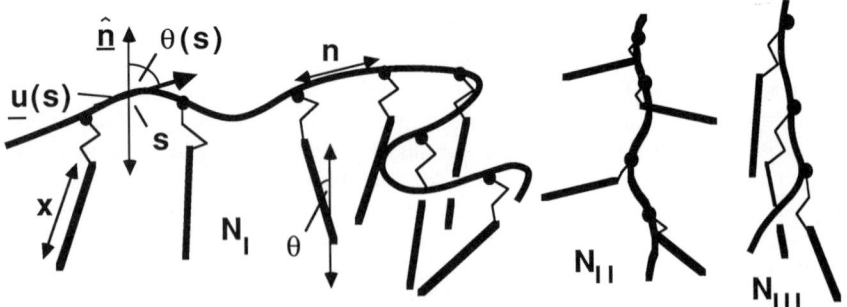

Figure 2.1 Three possible phases for comb nematics. If (S_A, S_B) are the order parameters of side chains (A) and main chain (B) then we have N_1 with $(+, -)$, N_{II} with $(-, +)$ and N_{III} $(+, +)$. The length of teeth is x and their spacing along the backbone is n in units of l, the chain diameter. The angle of the chain tangent vector at arc position s along the chain is $\theta(s)$ with respect to the director n and the angle of a tooth is θ.

are given in Chapters 8 and 10. Chain shape will also be very important in describing the microscopic origin of mesogenic network elasticity (section 2.6) and the behaviour of smectics (section 2.5).

2.2 Polymer entropy, nematic ordering and nematic worms

Before discussing the ordering of side chain polymers we must review the three elements essential to their understanding—polymer entropy, nematic orientational ordering and the combination of these two effects in the backbone part of the problem (nematic worms). We accordingly explain why polymers wish to be chaotic, describe Maier–Saupe-like models for nematic rods, and give a model for nematic worms (semi-flexible polymers with no pendant groups).

2.2.1 *Polymers*

A crude but qualitatively accurate model for the entropy of a polymer is to represent the bonds of a polymer as bonds (steps) on a lattice (see Figure 2.2). If the coordination number of the lattice is z, then there are $z - 1$ choices of where to put each bond, slightly fewer than this if one considers excluded volume correlations. The coordination z might be 6 in three dimensions and 4 in two dimensions. The total number of ways of putting down a polymer of N bonds is $\sim (z - 1)^N$ times unimportant powers of N. Taking the logarithm of this number, the entropy S of a chain of N steps in 3-D is $S_3 \sim k_B N \ln (5)$, in 2-D is $S_2 \sim k_B N \ln (3)$ and in 1-D is $S_1 \sim k_B N \ln (1) = 0$. Interesting cases are:

(i) The reduction in entropy if a chain is confined toward a plane (an extreme case of an oblate chain is chosen for illustration) is $\Delta S = S_2 - S_3 \sim - k_B N \ln (5/3)$, giving a substantial rise in the entropic contribution $- TS$ to the free energy, $\Delta F \sim k_B T N \ln (5/3)$. This rise in F resists the nematic ordering of the teeth, affecting phase equilibria and transitions. Confinement toward a plane was seen in neutron scattering from comb smectics (Keller *et al.*, 1985). In section 2.5 we discuss how the entropy penalty of confinement is relieved by hopping between layers, that is, by molecular defects.

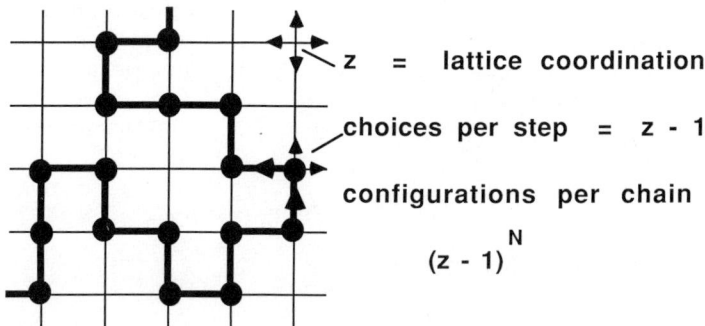

Figure 2.2 A lattice model for the entropy of a polymer. Restrictions (nematic, smectic or mechanical) on chain entropy are fiercely resisted.

(ii) A polymer in a prolate phase is stretched out toward being a rod by the nematic field. In this crude model, the number of choices per step is $z - 1 \sim 1$ and hence $S_1 \sim 0$. The rise in free energy is then even greater, $\Delta F \sim k_B T N \ln(5)$. De Gennes (1982) has proposed 'hairpin' defects as a mechanism for entropy recovery, an interesting subject that would take us away from comb polymers.

The above entropy arguments give a rise in F proportional to T. This is a well-known result, for instance in polymer networks. The resistance to mechanical stretching, the modulus, of a rubber is given by the product of the absolute temperature and the number of chains cross-linked. For comb nematics, an elastic contribution resisting the applied oblate or prolate nematic field ((i) or (ii) above) thence couples the chain to the nematic part of the free energy. Transitions in nematic polymers are perturbed by the competition between these two energies. The picture of entropically elastic chains under tension from the nematic field is a good one to keep in mind when thinking of the differences between conventional and polymer liquid crystals. The extent to which, in practice, chains are flattened or extended, varies enormously according to the chemical structure (primarily the spacers coupling the chain to its teeth) and according to phase, smectic effects being very great. The idea of elastic chains is applied literally in section 2.6 where the elasticity of nematic networks is modelled.

2.2.2 Simple nematics

The ordering influences in conventional nematics are well understood and we merely sketch them here in order to introduce our notation. Two rods prefer to be parallel because of steric and anisotropic van der Waals' forces between them. Such an interaction, like the nematic ordering itself, does not distinguish between up and down (both are of quadrupolar symmetry). Thus in a mean sense a given rod sees a potential $U = -v_A S_A P_2(\cos \theta)$ where v_A is a coupling constant, S_A is the mean order, the average $\langle \cdots \rangle$ of P_2, and where P_2, the second Legendre polynomial, is $P_2(x) = (3/2)x^2 - \frac{1}{2}$. θ is the angle made by the rod to the z axis, the mean ordering direction (see Figure 2.1 where a tooth is used to illustrate such a rod). The potential is minimal for $\theta = 0°$ or $180°$. This is the Maier–Saupe (M–S) picture (see Chandrasekhar, 1977). Possibly more important than the van der Waals' forces are the steric requirements that molecular rods do not overlap. Rods densely enough packed have more translational freedom (entropy) if they are aligned (nematic) rather than randomly directed. This is the Onsager–Flory picture. The effective potential felt by a rod would now be temperature-dependent and more complicated than P_2, but it has the same up-down symmetry and P_2 is the first term in its expansion. We shall take the M–S form, but understand that it partially covers other effects.

The order is simply calculated by averaging P_2 over the Boltzmann weight of the potential:

$$S_A = \frac{\int_{-1}^{1} \mathrm{d}(\cos \theta) \exp\{-U(\cos \theta)/k_B T\} P_2(\cos \theta)}{\int_{-1}^{1} \mathrm{d}(\cos \theta) \exp\{-U(\cos \theta)/k_B T\}} \tag{2.1}$$

The denominator is actually the partition function Z and normalizes the probability distribution function. If we put $\cos \theta = x$ and substitute $U = -v_A S_A P_2(x)$, then we get explicitly

$$S_A = \frac{\int_{-1}^{1} dx \exp\{v_A S_A P_2(x)/k_B T\} P_2(x)}{Z(v_A S_A/k_B T)} \equiv \frac{k_B T}{v_A} \frac{\partial}{\partial S_A} \ln[Z(S_A)] \qquad (2.2)$$

and we see that, because U dependend on the mean order, S_A appears in both sides. Solving eqn. (2.2) for S_A makes the theory self-consistent. Maier and Saupe found, on reducing temperature, a first-order (discontinuous) transition to a nematic state ($S_A \neq 0$) at $T = T_{NI} = 0.22 v_A/k_B$ where $S_{NI} = 0.43$. The free energy, F, also emerges, since $F(S_A)/k_B T = -\ln Z + (1/2) v_A S_A^2/k_B T$ (the last term correcting overcounting in the mean field approach). If we minimize $F(S_A)$ with respect to S_A, that is, we take $\partial F/\partial S_A = 0$, we recover precisely the M–S condition, eqn (2.2), since the right-hand side is proportional to $\partial \ln Z/\partial S_A$. We can also rearrange variables, $y_A = v_A S_A/k_B T$, and see that F becomes

$$\frac{F(y_A)}{k_B T} = -\ln Z(y_A) + \frac{1}{2}\left(\frac{k_B T}{v_A}\right) y_A^2 \qquad (2.3)$$

It is the shape of $F(y_A)$ that gives the desired M–S result, that is, a first-order phase transition. The function in eqn (2.2) is sigmoid. One of its intersections with the straight line S_A appears discontinuously at a finite value of S_A as T is decreased.

2.2.3 Worm nematics (main chain LCP)

The general, 'combined' comb polymers will have stiff, nematogenic elements along their backbones. Before describing combs, we must describe the interplay of competitive polymeric and nematic tendencies in the backbone. This takes us a little afield, since the backbone, stripped for the moment of its teeth, is in effect a main chain LCP. Its drive to order or disorder is, however, an important part of the description of the comb as a whole, and we accordingly develop a model of main chain LCPs here.

The main chain will have limited flexibility because of mesogens along its length, and because bulky side units (the teeth) impede bend. A first model is to consider it a worm, that is, a curve where bend, the rate of change of unit tangent vector, is penalized. The total bend energy along its length L is

$$U_{bend} = \frac{1}{2}\int_0^L ds \varepsilon \left(\frac{\partial u}{\partial s}\right)^2 \qquad (2.4)$$

where ε is a bend modulus, s in the arc position and $u(s)$ is the tangent vector, see Figure 2.1. The tangent vector is a unit vector for a worm, since a worm is locally length-preserving. It is important to use a worm: over short arc lengths, chains can appear quite stiff, and worms are rod-like for lengths of the order of their natural persistent length $l_p = 2\varepsilon/k_B T$. Worms long compared with l_p are Gaussian. The entropy scale of section 2.2.1 is set by comparing l_p with the monomer dimension l. Worms are necessary for describing nematic properties, since such forces couple to the tangent vector, the local direction of the chain. The concept of a tangent vector is absent for a real Gaussian.

The nematic elements want to order under a mean potential depending on the angle the worm makes with the z axis. Calling $u \cdot z \equiv u_z = \cos\theta(s)$, with $\theta(s)$ being the local angle the chain makes with z, see Figure 2.1, we have a total potential

$$U_B = \int_0^L ds \left\{ \frac{\varepsilon}{2} \left(\frac{\partial u}{\partial s} \right)^2 - v_B S_B P_2[\cos \theta(s)] \right\} \tag{2.5}$$

where S_B is the mean order of the backbone and v_B is the mean coupling. One can follow through the M–S procedure again, albeit with a little more complication since one is not summing over the angle θ of a single rod, but over the configurations of an entire worm. We write this sum symbolically as $\int \delta u(s)$, and get the partition function Z_B:

$$Z_B(y_B) = \int \delta u(s) \exp \left\{ - U_B[u(s)]/k_B T \right\} \tag{2.6}$$

This model was first introduced by Jähnig (1979, 1981) in the context of membranes (where there are additional influences acting), and for nematic polymers by ten Bosch et al. (1983a, b, c), Rusakov and Schliomis (1985), Warner et al. (1985), and Wang and Warner (1986a). The problem of Z_B is reducible to the study of the spheroidal wave equation, and the free energy of a long chain is simply the ground-state eigenvalue. The free energy and also other properties such as the configurations can be derived from this approach, either perturbatively, numerically or, for some low-temperature properties like hairpins, only by asymptotic analysis.

All that is important to qualitatively understand eqn (2.6) is (i) that the free energy F_B is a function only of the combination, y_B, of bend, nematic potential and temperature: $y_B = (\varepsilon v_B S_B)^{1/2}/k_B T$, and (ii) that the M–S condition for the equilibrium order S_B is again just the minimization of $F(y_B)$. Since order is quadrupolar, the shape of $F(y_B)$ is qualitatively the same as in the conventional case, that is, sigmoid. The significance of this shape is explained in the context of combs by Renz and Warner (1988). One again gets a first-order transition to a nematic state at

$$T_{NI} = 0.388(v_B \varepsilon)^{1/2}/k_B \tag{2.7}$$

$$S_{NI} = 0.356 \tag{2.8}$$

The nematic transition of a worm is very similar to that of a conventional nematic, the important difference being that the latent entropy is predicted to be much higher, $\Delta S = 1.69k_B$ per persistent unit of backbone (Wang and Warner, 1986a). This is expected from a polymer system and seems to be in qualitative accord with experiments reviewed by Luckhurst (1985), although there are also systematic quantitative alternations in ΔS with spacer length within the backbone. Details such as the alternation with spacer length cannot be described with a worm model, as the flexibility is smeared out.

The M–S and worm results represent a gross simplification of the nematic problem. Order parameters at the transition vary from the universal values given. There is much debate about the reasons for this: fluctuations away from the constant mean field, or the interplay with steric effects are two of the suggested causes. Nevertheless, a qualitative picture is given by these approaches and we proceed in this spirit by putting conventional and worm nematics together in a model for combs.

2.3 Competing nematic tendencies in comb polymer liquid crystals

Having discussed the elements of a comb, we must assemble them and consider their mutual interactions (Wang and Warner, 1987).

For the nematic part of the comb problem, we must consider the two component parts, backbone and teeth, and their mutual coupling. The tooth–tooth interaction (v_A) and the backbone–backbone interaction (v_B) have been discussed above in sections on simple and worm nematics, and remain as before. The tooth–backbone interaction has two parts: since there are in general stiff elements in both the teeth and the backbone, there is a drive to parallelism, a standard nematic requirement. The spacer connecting the teeth to the backbone will have a limited flexibility and, according to its precise structure (for instance the number of CH_2 units), may drive the tooth toward being perpendicular to the backbone, that is, oppose the nematic tendency. We sometimes refer to this attachment spacer as a 'hinge' to avoid confusion with spacers along the backbone. Percec and others suggest even a microphase separation if hinges are long and A and B are grossly incompatible (see Chapter 3). These competing effects are together represented by v_m. Positive v_m shows nematic dominance over the hinges, negative, vice versa (see further discussion in section 2.4). We continue to denote the order parameters of the teeth and backbones by $S_A = \langle P_2(\cos\theta_A)\rangle$ and $S_B = \langle P_2(\cos\theta(s))\rangle$ respectively. The mean field potentials U seen by teeth and backbone are now extended because of cross-coupling:

$$U_A = -\{\chi v_A S_A + (1-\chi)v_m S_B\}P_2(\cos\theta_A) \tag{2.9}$$

$$U_B = \int_0^L ds\left[\frac{\varepsilon}{2}\left(\frac{\partial u}{\partial s}\right)^2 - \{(1-\chi)v_B S_B + \chi v_m S_A\}P_2(\cos\theta(s))\right] \tag{2.10}$$

The bend energy has been added to U_B as in eqn (2.5). χ is the volume fraction of teeth and $(1-\chi)$ is that of backbone, $\chi = x/(x+n)$, with x the tooth length and n their spacing along the backbone as defined in Figure 2.1. The rod length x is intended to both represent the tooth volume in the volume fraction χ and to scale the nematic interaction—see the discussion of v_m below. If the hinge occupies a significant volume compared with the tooth, then the expression for χ must be slightly modified. For simplicity, we ignore this distinction between rod and side chain lengths here, that is, we ignore the volume of the hinge. The volume fraction factors give the probabilities of seeing the appropriate species.

In effect we have in eqns (2.9) and (2.10) two, coupled Maier–Saupe-like systems. The nematic ordering field felt by the teeth depends not only on their own order through $\chi v_A S_A$, but also on the order S_B of the backbones to which they are coupled via $(1-\chi)v_m S_B$ in U_A. There is likewise an effect of the teeth on the backbone energy through the S_A term in U_B. The transitions and order of one phase are hence linked to those of the other phase. There is a strong similarity between this model for combs and that of ten Bosch et al. (1983b) for polymers in nematic solvents. Cser (1984) and Cser et al. (1985), cite an extensive assembly of experimental data, and also view comb polymers as a system of many components linked together in one molecule. The full solution of the free energy and resultant phases is not difficult. It is reducible to the simultaneous solution of the coupled M–S and worm systems, either perturbatively and numerically (Wang and Warner, 1987), or by graphical and analytical analysis (Renz and Warner, 1988). However, a qualitative understanding of the possible behaviour can be achieved by discussing the relative sizes of the three couplings v_A, v_B and v_m in the context of specific phase diagrams. We shall find that, depending on the choice of coupling constants, one or other component will dominate the ordering and that the other component will follow in a manner determined by the relative magnitude of the self-coupling to the mutual coupling v_m.

We have sought to understand the competitive influences involved in comb ordering by emphasizing concepts such as chain shape and conformation, and their implications for the entropically inspired elastic resistance to distortion by the nematic field. We thereby make contact with notions familiar to polymer physics. One can also adopt a lattice model of polymer chains extended beyond our crude illustration of section 2.2.1 to account for chain stiffness. Equally, instead of simply insisting on the P_2 symmetry of the nematic part of the problem (section 2.2.2), one can explicitly put in the steric contribution to the potential $U(\cos \theta)$ (eqn 2.1 and the paragraph before where the rôle of rod shape is discussed). The Flory lattice model, or a refinement thereof, can be employed to model the packing of rods. These lattice approaches to the two parts of the problem have been utilized by Vasilenko et al. (1985). They only consider the effect of volume fraction on packing, and not the effect of temperature on nematic order, that is, they consider lyotropic systems (see the end of section 2.4.2).

2.4 Competition between new nematic phases

We see that there are three nematic ordering influences. Rather than trying to estimate the magnitudes of the coupling constants and of the bend ε directly, we shall assume some values in order to illustrate the complex possibilities for ordering.

2.4.1 Competition

The analysis of simple nematics and worm nematics shows that if A and B were decoupled ($v_m = 0$) at some volume fraction χ_0 say, then the component teeth and backbones would separately order at temperatures $k_B T_A = 0.22 \chi v_A$ and $k_B T_B = 0.388\{(1 - \chi)v_B \varepsilon / l\}^{1/2}$, the former being the result of M–S theory, the latter the solution of the polymer liquid crystal problem (see section 2.3.3). We have put a factor of l into the ε as defined before, so that ε/l appears henceforth. This follows the convention of Renz and Warner (1988). The component dominating the ordering is that with the greater transition temperature T_A or T_B. Thus the ratio $T_A : T_B = \chi v_A / \sqrt{\{(1 - \chi)v_B \varepsilon / l\}}$ sets the character of the phases available. The sign of v_m will then determine which comb phase is exhibited. The mutual coupling v_m between A and B is thus very important, as is much emphasized in Chapter 3, and has been studied by NMR (Chapter 8). It has two components: a nematic coupling v_C, driving A and B parallel, and the hinge coupling v_f. A positive v_f makes A and B perpendicular, a negative v_f adds to the parallel influence of v_C. These influences combine to give $v_m = v_C - v_f / x(1 - \chi)$. Inserting this back into eqns (2.9) and (2.10) shows that teeth order under v_C according to their length x, and that there is one hinge effect, v_f, counted per tooth or per backbone repeat length n. As χ is varied at fixed side chain length x (by varying n), so the mutual coupling changes, thereby altering the phase competition. In fact, it can be shown (Renz and Warner, 1988) that this model depends on only three potential terms (and not χ), and hence that the Us are really very simple. Thus the whole problem has a simple structure, but in this section we explicitly exhibit the way v_m varies to give an idea of the molecular structural modifications that are needed to effect phase behaviour via v_m. The point χ_0 where v_m changes sign is now seen to be $\chi_0 = 1 - v_f / x v_C$. To illustrate the competition, a series of schematic phase diagrams based on ones calculated in this model by Wang and Warner (1987) are analysed below. Temperature is reduced by T_B, the transition temperature of the pure B worms (volume fraction χ of teeth is zero). For more details, consult the original paper.

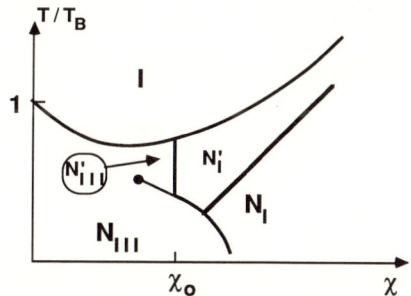

Figure 2.3 Phase diagram for side chains of length $x = 3$. Temperature is reduced by the value at the N–i transition for pure main chain, B. Volume fraction χ of side chains is changed by varying the spacing n of teeth along the backbone (see Figure 2.1). The cross-coupling v_m accordingly varies from $v_m > 0$ for $\chi < \chi_0$ to $v_m < 0$ for $\chi > \chi_0$. Coupling constants are in the ratio $v_A:v_B:v_C:v_f:\varepsilon/l = 3:2:2:3:2$.

2.4.2 Illustrative model phase competitions

If v_m is > 0, the two components want to be parallel and, since individually their ordering is naturally positive, there is no conflict and N_{III} results. This is so for $\chi < \chi_0$ in Figure 2.3.

If v_m is negative (χ greater than χ_0), then A and B wish to be perpendicular, unless their own ordering is overwhelming. What actually happens depends on the relative sizes of the individual ordering temperatures for the two components A and B at the prevailing χ, $T_A(\chi)$ and $T_B(\chi)$, and their comparison with the prevailing temperature T. Let us take the example of Figure 2.3. At $\chi = \chi_0$ where the components are effectively decoupled we have $T_A(\chi) > T_B(\chi)$; the teeth order before the backbone. Hence lowering T at constant $\chi(> \chi_0)$ yields the isotropic to N_I transition at T around $T_A(\chi_0)$—see Figure 2.1 to confirm that N_I is a tooth-dominated phase. It can be proved (Renz and Warner, 1988) that the N_{III}/N_I phase line is exactly vertical at $\chi = \chi_0$. The N'_{III} and N'_I around χ_0 are so denoted with primes because the B component, which naturally orders at lower temperatures, is only weakly ordered with $|S_B| \leqslant 0.17$. The teeth have already ordered (since $T < T_A(\chi_0)$) and, via the cross-coupling, impose an order (of the same sign as v_m) on B. It is as if the teeth were an external field inducing a para-nematic phase. In fact, polymer liquid crystals in an external field are qualitatively the same as conventional nematics (Wang and Warner, 1986b). This analogy can be used to explore the region around the critical point (·) in the phase diagram—see the analysis of Renz and Warner (1988).

As temperature is further lowered so that both components would individually be ordered in the absence of cross-coupling, $T < T_B(\chi_0) \ll T_A(\chi_0)$, then the N_I phase for $\chi \geqslant \chi_0$ is supplanted by the N_{III} phase—now the ordering of the backbone takes place and overcomes the negative effect of the hinge, whence $S_A > 0$ and $S_B > 0$.

This interplay between the three couplings $v_A/(v_B\varepsilon)$ and v_m is further illustrated in Figure 2.4. At χ_0, where $v_m = 0$, the components are effectively decoupled and the self-coupling are such that B dominates instead, that is $T_B(\chi_0) > T_A(\chi_0)$. This means that the N_{III} phase from the $v_m > 0$ region meets an N_{II} phase from the $v_m < 0$ region where $T_B(\chi_0) > T > T_A(\chi_0)$. The N_{II} phase is where the backbone B orders first and the hinges drive the A component to be perpendicular with negative order parameter. Contrast

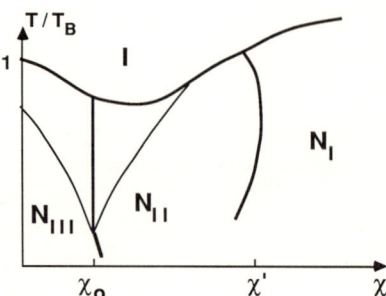

Figure 2.4 Phase diagram for side chains of length $x = 3$. As for Figure 2.3, but the coupling constants are in the ratio $v_A : v_B : v_C : v_f : \varepsilon/l = 1 : 2 : 1.2 : 3 : 2$; note the re-entrancy.

this with Figure 2.3 where the opposite is true; teeth dominate and N_I results. As χ is increased, volume fraction terms in U_A and U_B cause the rôle of the side chains to increase in importance. Eventually they order as strongly as the main chain, $T_A(\chi) \sim T_B(\chi)$, which occurs at χ', say. Then the side chains can play a dominant role and the backbone must follow—N_I is the result. Because of quantitative differences between Maier–Saupe theory and that of nematic worms, in particular the difference in the temperature dependence of ordering, the N_{II}/N_I coexistence is predicted to be re-entrant. At χ fixed around χ', variation with T is predicted to yield the sequence $N_{II}/N_I/N_{II}/I$.

By varying the A–B couplings, the mutual interaction can be made anti-nematic (perpendicular) across the whole range of χ. An example is given in Figure 2.5a where v_C and v_f are such that $v_m = 0$ at $\chi = 0$. It is the system of Figure 2.3, but with v_C halved. The hinges dominate over the entire range of χ. N_{III} has now been eliminated, except perhaps in the low T, χ corner, in favour of the 'perpendicular' phases N_{II} and N_I. For a wide range of values of χ it is predicted that, at fixed χ, varying T yields the phase sequence $N_{II}/N_I/I$. This involves the exchange in sign of the order parameters and is

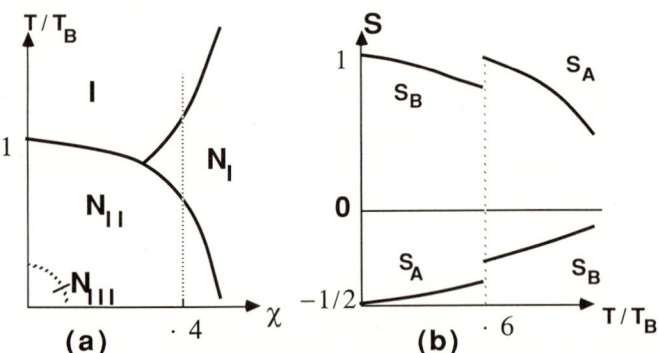

Figure 2.5 Negative main chain–side chain coupling. (a) Combs as in Figure 2.3, but nematic component of the A–B coupling, v_C, has been halved. v_m is negative for all χ, thus effectively eliminating N_{III}. (b) Temperature variation of a phase with χ fixed at 0.4 (broken line in (a)) shows A and B exchanging roles. This is seen from the abrupt change in sign of their order parameters when plotted against T.

plotted in Figure 2.5b. If the dominant optical anisotropy were to occur in the side chains A, we would measure a negative apparent optical anisotropy (assuming a positive anisotropy of the side chains A). This becomes less negative with increasing temperature until it discontinuously becomes large and positive. This is an apparently anomalous sequence of orderings.

The lattice model (Vasilenko *et al.*, 1985) also finds phases delicately balanced against each other. Detailed comparison with this work is difficult since different influences in the problem are allowed to vary. One case where the attachment spacers are eliminated but where the main and side chains order strongly (and would yield N_{III} here) gives a phase with the backbone kinked between each tooth, a possibility accessible only in the lattice approach. Other cases yield the equivalents of our N_I and N_{II} phases with critical points also.

2.4.3 Experimental suggestions for new nematic phases

In the introduction (section 2.1) to what phases might be expected, the work on elastomers of Koch *et al.* (1985) has already been mentioned. In the isotropic phase of networks of comb polymers, a negative-stress optical effect grew larger around the phase transition as pretransitional nematic fluctuations increased. This suggested that, when chains were stretched out by the applied stress, the teeth (where most of the optical anisotropy lay) were thereby induced toward the plane perpendicular to the elongation. Thus for these molecules a negative v_m is suggested, and hence most likely either the N_I or N_{II} phases when the system becomes nematic. These phases differ in that N_{II} remains uniaxial when stress is applied along the director but N_I becomes biaxial, as was indeed observed. Figure 2.1 helps to visualize this. The difficulty of relating hinge effect to chemical structure is revealed by later experiments of a similar kind (Schätzle and Finkelmann, 1987). In a family of comb molecules, teeth were induced to be parallel or perpendicular to the applied stress, and thus presumably to the chain backbone, alternately as the number of carbon atoms in the spacer was increased. Further discussions of what one can deduce about phases and ordering tendencies from network studies, optically, by IR dichroism, and by NMR are given in Chapter 10 and briefly in Chapter 8.

Chain shape is a good monitor of the polymer entropy reservoir that distinguishes polymer liquid crystals from their conventional low-molar-mass counterparts. Direct observation of the shape of individual chains in a melt of their own kind is only possible by neutron scattering from individual deuteron-labelled chains. Experiments so far also point to a perpendicular coupling and in particular to an N_I phase. Kirste and Ohm (1985) and Keller *et al.* (1985) observed the radius of gyration (proportional to the root mean square size) of comb polymers in the isotropic phase. They found a value which then diminished parallel and increased perpendicular to the ordering direction when measured in the nematic phase. Both these groups used very long chains for which an anisotropic Gaussian interpretation seems safe—see section 2.6.3 for more on such random walks. But in general care is needed in interpreting such experiments: mostly one observes polymers which are sufficiently long compared with their local persistence that they are essentially Gaussian random walks. Stiff units in the backbone and bulky side groups (including the teeth) give a greater effective stiffness and a longer persistence length, so that for a given molecular mass the condition to be Gaussian is harder to fulfil. The theoretical models thus far employed are all based on worms and are hence general to chains of all lengths. The precise way in which nematic worm shape and order differs with length is given by Wang and Warner (1986a), but the particular examples of phase diagrams given above are for long chains. Short-chain cases can easily be calculated. Certainly the small degree of anisotropy in the cases of the nematic phases observed by neutrons suggest that v_m is only weakly negative for these systems.

A comb polymer in a nematic solvent can offer sufficient contrast that its shape can be seen by small-angle X-ray scattering, obviating the need for deuteration. Although not a probe of pure comb phase behaviour, Mattoussi et al. (1986) exploited the observed difference between the mean sizes of the polymer perpendicular and parallel to the nematic ordering direction to deduce the nematic coupling to the main chain. The parallel dimension was greater than the perpendicular, suggesting a parallelism between solvent, teeth and main chain and a possible predisposition to N_{III} phases in the melt. One difference between Mattoussi et al.'s molecules and those of the neutron experiments is the use of a siloxane backbone, known for its great flexibility.

Another, very sensitive, technique (Chapter 8) is to measure the order parameter of a specific part of the chain directly by NMR. Local geometric factors have to be known before one translates these measurements into a gross order parameter characteristic of the oblate or prolate spheroidal shape of the chain as a whole, but they immediately reveal the subtle rôle played by the connecting spacer and by the backbone repeat unit. Some polymer species have a much stronger coupling than others, even though the chemical differences are not extreme (see Chapter 8).

2.5 Smectic combs

Many side chain polymers have teeth that are conventional smectics when un-connected to polymer backbones. The corresponding polymers often display smectic phases, most simply the smectic A phase (Sm_A) to which we shall limit ourselves here. (Other smectic phases have very exciting properties, for instance in ferroelectricity, Sm_C^*, and new types of critical behaviour, Sm_C.) Ideas of entropy help one understand such phases, and a model can be constructed which attempts to describe the effect of smectic ordering of teeth transmitted via cross-coupling to the backbone.

2.5.1 Smectic phases and their restriction of polymer freedom

Conventional smectic ordering is where molecules, say rods, in addition to being orientationally ordered (nematic) are spatially organized into 'layers'. In the simplest case, Sm_A, the layers are perpendicular to the director. The smectic structure is weak. X-ray studies show that a more accurate description than sharp layers is given by a periodic modulation in density $\rho(z)$, where

$$\rho(z) = \rho_0 + \rho_1 \cos(kz) \tag{2.11}$$

with the amplitude ρ_1 much less than the mean ρ_0. This is indicated by the absence of higher-order reflections; only the 001 reflection is observed. The wave vector k of the density wave is $2\pi/d$, where d is the layer spacing.

The neutron observation of nematic chain shape by Keller et al. (1985) was continued to lower temperatures into the smectic A phase where a large anisotropy in radius of gyration parallel and perpendicular to z was observed. This suggests that, in addition to the nematic coupling driving the *direction* (tangent vector) of the chain toward the perpendicular plane, there is a strong coupling driving the *position* of the chain to be in a smectic layer.

Since the teeth are confined toward being in layers, the backbone that is firmly attached to them will be driven toward layers as well. Polymer entropy is much reduced if a Gaussian chain is limited to a plane instead of being allowed to explore three dimensions freely—see the crude model for chain freedom in section 2.1.1. There is undoubtedly an entropic penalty against forming a smectic state, and this would shift the nematic (or isotropic) to smectic phase transition. Rieger (1988, unpubl.) points out that if the walk is sharply enough confined so as to be very strongly two dimensional, then there is an excluded volume contribution to chain statistics and free energy. Although in three-dimensional melts self-avoidance is screened out, this is not so in two dimensions. (There is a fundamental reason for this: an ideal polymer has a fractal dimension of 2 and hence in spatial dimension 2 (a plane) it can be expected to eventually fill this space.) If a smectic can rigidly confine the polymer to a layer, the excluded volume effect would force chains to jump between layers. Various single-chain models of random walks have been adopted to model these effects. Against these polymeric penalties is the effect of backbones inducing additional spatial correlation between teeth adjacent in arc position, thereby in some cases promoting smectic correlation. All these factors, and those developed in the model below, have an as yet unknown effect on the order of the transition or, as appropriate, on the width of the critical (Ginsburg) region. We content ourselves with proposing a model that allows some entropy recovery in the face of smectic restrictions and predicts an unusual form for chain shape in strongly ordered systems.

2.5.2 A model for chains in comb smectics

The density wave eqn (2.11) is a consequence of a smectic potential acting on the teeth. When these are attached to a backbone, as in the comb polymer case, the chain also experiences a spatial potential, say for simplicity of the mean field type $b\sigma \cos(kz(s))$, with b a coupling constant and where $z(s)$ is the z coordinate of the chain at arc position s along its length. The smectic order is $\sigma = \langle \cos kz(s) \rangle$ which we shall henceforth assume to be close to the value $\sigma = 1$ and no longer changing significantly. We then absorb σ into b. This means that we are describing chains trapped near the bottom of one or more troughs of a deeply corrugated potential. The same total nematic potential $\int \{\cdots\} P_2$ acting on the chain in eqn (2.10) from the nematic teeth and from the chain itself is acting here too. We shall not further change χ. We shall assume that the nematic order of the teeth is (i) saturated at $S_A = 1$ and does not change further, and (ii) we are describing an N_I phase where, predominantly, the teeth are driving the backbone. All this allows us to subsume the various couplings into a single coupling c to yield a mean field potential $cP_2(u_z(s))$. The strongly positive coupling c gives the chain a negative-order parameter tending to $-\frac{1}{2}$. The overall potential U is then

$$U = \int_0^L ds \left\{ \frac{\varepsilon}{2} \left(\frac{\partial u(s)}{\partial s} \right)^2 + cP_2(u_z(s)) + b \cos(kz(s)) \right\} \quad (2.12)$$

The problem is now complicated, since part of the energy relates to the tangent vector u (the nematic field cP_2 depends on its direction, the bend $\varepsilon(\partial u/\partial s)$ on its rate of change), and part of it to the position of the chain (the smectic energy $b \cos [kz(s)]$). It is helpful to draw a diagram of the smectic and nematic parts, as in Figure 2.6.

B

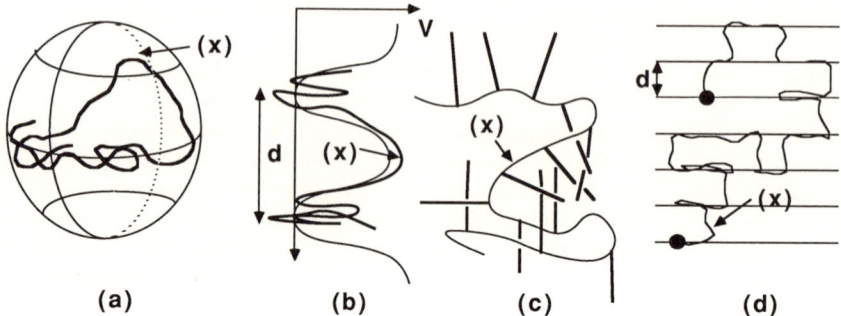

Figure 2.6 Nematic and smectic energy of layer hops. (a) The tangent sphere of the chain with a trajectory of the point representing the end of $u(s)$ traced out as s goes from 0 to L. The polar caps are regions of high nematic potential. (b) The smectic potential V with the trajectory of a chain position shown without the attached teeth. (c) A section of a smectic comb, with teeth shown. Layer transitions (hops) are denoted in all figures by (x). (d) The resultant picture of a one-dimensional random walk up and down the layers.

In the N_I phase, the tangent vector is confined to the tropics by a repulsive nematic potential centred on each pole. The nematic potential penalises temperate excursions such as that incurred in a layer hop (x) in (a), seen also in real space in (c) where the teeth attendant to that stretch of chain involved in the hop are in nematically unfavourable configurations. This influence demands the chain adopt an oblique as possible angle during a layer hop. However a transition at a gentle, more favourable angle in (c), that is an excursion in (a) not so close to the polar regions, means that longer is spent in the smectically unfavourable regions around (x) in (b), that is between layers. Smectic energy is thereby increased. With these two competing influences there is a clear case for the existence of a solitary defect representing the minimization of the total smectic plus nematic energies. This defect of a hop starting in one well and ending in the next was proposed by Renz and Warner (1986) and is akin to the hairpin defects in nematic ordering proposed by de Gennes (1982) for main chain nematics.

The energy can be worked out for an infinitely long chain at zero temperature with just one hop in it. This assumes that the effects of successive hops do not overlap. Ignoring bend and discarding the irrelevant constant part of P_2 one must minimize

$$U = \int_{-\infty}^{+\infty} ds \left\{ \frac{3c}{2} \left(\frac{\partial z}{\partial s} \right)^2 + b \cos (kz(s)) \right\} \tag{2.13}$$

The minimal energy trajectory from one well to the next is an easy problem:

$$\tan \left(\frac{\pi z(s)}{2d} \right) = \exp \left\{ \frac{s}{s_0} \right\} \tag{2.14}$$

the characteristic width of a hop being $s_0 = (d/2\pi) \cdot \sqrt{(b/3c)}$ and the energy E_L of a layer hop being

$$E_L = 4d(3bc)^{1/2}/\pi \tag{2.15}$$

To evaluate the thermal properties of a chain, one simply puts the energy E_L in a Boltzmann statistical weight and obtains the probability per unit length of polymer of achieving a layer hop, p_h:

$$p_h = \frac{1}{l_h} \exp\left\{- E_L/k_B T\right\} \tag{2.16}$$

The unit length l_h which makes this a 'frequency' cannot be obtained from our simple method. The full problem of a nematic worm in a sinusoidal potential is solved by Renz and Warner (1986) where it is seen *post hoc* that the neglect of bend was justified, as was the method of evaluating E_L at zero temperature and putting it in the exponential for finite temperature results. Deviations from the simplifications used to arrive at eqn (2.16) come in the pre-exponential structure of p_h, that is into the length l_h. We return to this below.

The picture is now that of the chain making random jumps up and down a one-dimensional ladder, the corrugated potential (see Figure 2.6d). The probability of a jump tells us that the number of jumps per chain is $n_h = (L/l_h)\cdot\exp\left\{- E_L/k_B T\right\}$, which can be interpreted as the chain attempting to recover some of the entropy lost to the smectic confinement toward layers: there is an entropy of (L/l_h) associated with the number of places along the chain where the defect can be put, the second factor coming from the energy penalty associated with the creation of each hop.

For long chains with many hops we can use the 1-D random walk formula for the mean square extent of the chain in the z direction taking n_h steps of length d:

$$\langle z^2 \rangle \equiv \langle \{z(L) - z(0)\}^2 \rangle = d^2 n_h = \frac{d^2 L}{l_h} \exp\left\{- E_L/k_B T\right\} \tag{2.17}$$

This is a Gaussian result, $\langle z^2 \rangle \sim L \sim N$ (the number of monomers in the chain), but suppressed by the activation factor. The effective step length for the 1-D walk is $l_{eff} = (d^2/l_h)\cdot\exp\left\{- E_L/k_B T\right\}$. In fact l_h has a curious structure, with the result that l_{eff} is not proportional to d^2 as one naively expects (see Renz and Warner, 1986, for the detailed results). Further experiments (Moussa *et al.*, 1987, unpubl.) suggest that there is indeed an activated form to the strongly suppressed z component of the radius of gyration.

2.6 Elastomers of comb polymer liquid crystals

De Gennes (1975) suggested at an early stage that nematic polymers might be cross-linked to produce networks of quite remarkable properties, for instance non-linear or discontinuous stress–strain relations, spontaneous shape changes and coupling between mechanical and optical and electric fields.

His proposal was quite general to all types of nematic polymers and was based on powerful and simple symmetry arguments independent of any microscopic model. We shall review his idea and its consequences and then tie this via a molecular model to the ideas of sections 2.2–2.4. We conclude with a discussion of experiments and their various difficulties and with the outlook for future science and technology.

2.6.1 *The Landau–de Gennes picture of nematic elastomers*

The basic idea was that an isotropic phase responds to a weak applied external field by developing weak small nematic order. As the field is increased, a discontinuous jump to a high state of nematic order occurs provided that the temperature is not above some

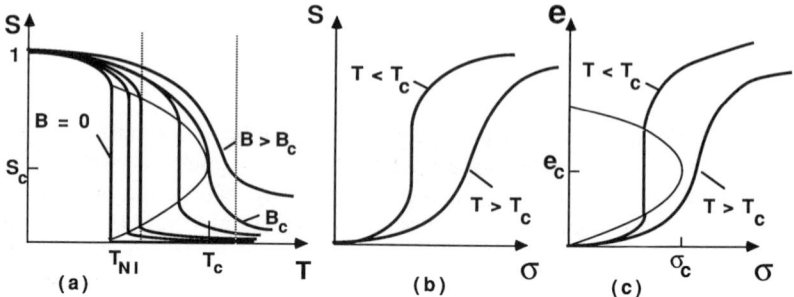

Figure 2.7 A nematic in an external field. (*a*) Order parameter *S* against temperature *T* for several values of field, *B*. (*b*) Variation of *S* against stress σ at the constant temperatures indicated by the two broken lines in Figure 2.7*a*. (*c*) Strain *e* against stress σ at two temperatures.

critical value. Above this critical value the transition is washed out, however large the field. Normally one envisages a conventional nematic with a magnetic field *B* applied. Figure 2.7*a* summarizes by plotting the order parameter *S* against *T* for several values of B^2 (electric fields are equivalent, the nematic coupling to the square of *E* as is appropriate to a liquid of quadrupolar symmetry). Note the critical point (S_c, T_c) or equivalently (B_c, T_c) above which there is no discontinuity of *S*.

To nematic elastomer, one is also able to apply a mechanical field (stress). The shape of chains reflects their nematic order, but also the macroscopic strain of the sample. Since applied stress couples to strain, it also then couples to the nematic order. Thus the stress σ fulfils the role of the square of the fields, B^2 or E^2. It can replace *B* in Figure 2.7*a*, and σ is used in Figures 2.7*b* and *c*.

For simplicity we describe the case of uniaxial nematic order and incompressible strain, both in the principal frame. The order parameter and strain tensors are diagonal and traceless and can each be characterized by a number, in the first case the order parameter *S* and in the second case the strain e ($\equiv e_{zz}$) in the *z* direction. Small extensions λ, the ratio of the new to the old lineal dimensions, are related to strain by $\lambda_z = e_{zz} - 1$ and are sketched in Figure 2.8.

The free energy expansion that takes a simple form

$$F(S) = F_0(S) - USe + \tfrac{1}{2}\mu e^2 - \sigma e \qquad (2.18)$$

where $F_0(S)$ is the nematic energy in the absence of any network effects and the coupling of strain to nematic order is $-USe$. The elastic strain energy is $\mu e^2/2$ and σ is a stress (force per unit area) applied in the *z* direction only. The coupling *U* and the modulus μ (both proportional to cross-link density) have been redefined by simple numbers from those used by de Gennes (1975) where the full tensorial forms $Q_{\alpha\beta}e_{\alpha\beta}$ and $e_{\alpha\beta}\sigma_{\alpha\beta}$ are to be found, and the coefficients, including *U* are discussed (see also a trivial erratum pointed out by de Gennes, 1982).

With no external field, $\sigma = 0$, a spontaneous distortion can happen, since the minimum of *F* is, from $\partial F/\partial e = 0$, at $e_m = US/\mu$. Inserting this *e* and the form for F_0 into *F*, one obtains a Landau–de Gennes expansion for $F(S)$:

$$F(S) = \tfrac{1}{2}AS^2 + \tfrac{1}{3}BS^3 + \tfrac{1}{4}CS^4 + \cdots - \frac{U^2S^2}{2\mu} \qquad (2.19)$$

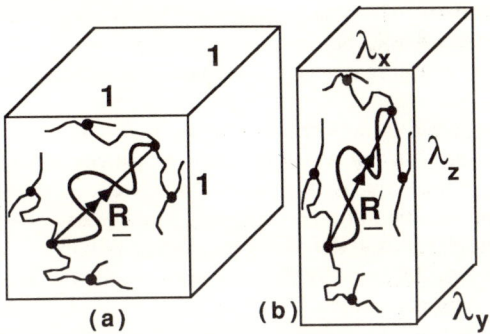

Figure 2.8 Isotropic and nematic phases of the same network. A strand (bold) is shown connecting two cross-links (denoted by ●) in a section of a network: (a) in the initial isotropic state with a vector separation R and (b) in the nematic state with a separation R'. A block of rubber has dimensions $1 \times 1 \times 1$ initially which becomes $\lambda_x \times \lambda_y \times \lambda_z$ on distortion on going to the nematic state.

The transition temperature T_{NI} of a nematic is given partly in terms of the point where the coefficient $A \equiv A_0 \cdot (T - T^*)$ vanishes. T_{NI} is slightly higher than T^*. The $-U^2/2\mu$ addition to A represents a shift in T^* and hence an elastically inspired shift upwards in T_{NI}.

With an applied stress, the minimum condition $\partial F/\partial e = 0$ yields the strain $e_m = (\sigma + US\hat{\jmath}/\mu$, whereupon, neglecting terms in σ^2, the shift in F away from F_0 becomes

$$\Delta F = -\frac{1}{2}\frac{U^2}{\mu}S^2 - \sigma SU/\mu \qquad (2.20)$$

The first term is again a shift in the transition temperature, the second is exactly that of a conventional nematic with an external field B^2 or $E^2 \equiv \sigma$ and with U/μ a susceptibility.

One can now obtain the stress–strain relation. Firstly, the relation between S and σ (Figure 2.7b), is obtained by varying the applied field (σ replacing B^2 in Figure 2.7a) at fixed T. Note that varying σ at $T < T_c$ eventually yields a jump in S. In the equilibrium relation $e = (\sigma + US(\sigma))/\mu$, putting in $S(\sigma)$ yields e versus σ in Figure 2.7c. The stress–strain relation has the slope $1/\mu_{eff}$, the reciprocal of the effective modulus of the network taking into consideration any nematic or incipient nematic tendencies. Initially the slope is finite, but when the discontinuity is induced by the applied field the slope is infinite, and μ_{eff} is zero over a range of strains between the light curve. The discontinuous stress–strain relations are lost beyond de Gennes' mechanical critical point.

2.6.2 A molecular model for nematic network elasticity

Schwarz (1986) was concerned about the connection at the molecular level between structural detail (primarily the length of spacer connecting teeth to the main chain) and the apparently differing tendencies of comb elastomers under stress and under temperature change. He concluded that positive or negative stress-optical behaviour depends on the detailed functioning of the hinge, a view we have reinforced throughout this chapter. In this section we wish to sketch how, under certain circumstances, the well-tested classical theory of rubber elasticity can be simply applied to comb networks and, given the symmetry (N_I, N_{II} or N_{III}) of the uncross-linked state, the ideas of entropy and nematic coupling can give a molecular understanding of the de Gennes proposal.

The important aspect of a network is that we have a system of chains sufficiently linked together that they can convey stress. The new element is that now the chains may naturally wish to be anisotropic in their shape. We have seen that this may be because of the anisotropy in the orientation of their teeth in the case of combs or because of mesogenic backbone elements in the purely main chain polymers. We sketch a network of comb chains in Figure 2.8, but with the teeth suppressed, so that this could equally be a model for main chain elastomers.

For the classical picture of networks it is important that the chain length between cross-links is long compared with the persistence length of the chain. For the first time in this section, this assumption is critical. We now sketch the model of Warner et al. (1988) based on this Gaussian-like distribution of chains in the melt from which the network is formed. The probability $p(R')$ of finding two ends of a strand connecting two cross-links a distance R' apart is

$$p(R') = \left(\frac{3}{(2\pi)^3 l_\parallel l_\perp^2 L^3} \right)^{1/2} \exp \left\{ -\frac{3R_\parallel'^2}{2l_\parallel L} - \frac{3R_\perp'^2}{2l_\perp L} \right\} \qquad (2.21)$$

The mean square sizes of the chain parallel and perpendicular to the ordering direction characterise the Gaussian and are

$$\langle R_\parallel^2 \rangle = l_\parallel L \qquad (2.22a)$$

$$\langle R_\perp^2 \rangle = l_\perp L \qquad (2.22b)$$

where L is the arc length of the strand and $l_\parallel$ and $l_\perp$ are the effective step lengths for the random walks in the two directions. Equations (2.22) define these two lengths which have been measured for some systems by neutron scattering. For instance, for the N_I phase where the parallel direction is flattened and the perpendicular extended, we have

$$l_\parallel(S_B) < l_0 < l_\perp(S_B) \qquad (2.23)$$

where S_B (< 0) is the backbone order (we shall simply call it S for brevity now) and l_0 the persistence length (the effective step) in the isotropic phases. In the N_{II} and N_{III} phases, the inequality (2.23) is reversed. The free energy of a strand with length R' in Figure 2.8b is then the logarithm of p, that is $F(R') = -k_B T \ln Z(R')$. We make the usual assumption of classical elasticity, namely that the cross-links (indicated by dots ($\cdot$) in Figure 2.8) deform affinely (in geometric proportion) with the bulk deformation λ. Fluctuations away from this are discussed by Warner et al. (1988). This means that we have R' coming from an R in the original isotropic network (a) where $R' = \lambda \cdot R$. On Figure 2.8 is indicated λ in its principal (diagonal) frame. We henceforth identify z with $\parallel$ and x and y with $\perp$. The free energy is $F(R') = -k_B T \ln Z(\lambda \cdot R)$ which we require be averaged over initial choices R of the strand size at cross-linking; one uses for this averaging

$$p_0(R) = \left(\frac{3}{2\pi l_0 L} \right)^{3/2} \exp \left\{ -\frac{3R^2}{2l_0 L} \right\} \qquad (2.24)$$

where now all three components of the walk are identical. Putting $R' = \lambda \cdot R$, taking the logarithm of eqn (2.21) and averaging over $p_0(R)$ one obtains the elastic component of the free energy

$$F_N^{(\text{el})}(S, \lambda_\parallel, \lambda_\perp) = \frac{k_B T}{2} \left[\lambda_\parallel^2 \frac{l_0}{l_\parallel} + 2\lambda_\perp^2 \frac{l_0}{l_\perp} + \ln\left(\frac{l_\parallel l_\perp^2}{l_0^3}\right) \right] \tag{2.25}$$

which is a function of the order parameter through the effective step lengths $l_\parallel$ and $l_\perp$.

A spontaneous distortion λ will occur to minimize $F_N(\lambda)$. Volume changes in all rubbers cost much more energy that that involved in elastic network distortions, which are therefore treated as occurring without change of volume. If the solid of Figure 2.8b is to have the same volume as the original in (a), then clearly $\lambda_\parallel \cdot \lambda_\perp^2 = 1$. Denoting $\lambda_\parallel$ by λ, say, and hence $\lambda_\perp$ by $1/\sqrt{\lambda}$ in eqn (2.25) gives us an $F(\lambda)$ which we minimize, $\partial F/\partial\lambda = 0$, to yield the minimum λ_m, the spontaneous distortion the network suffers when it turns nematic:

$$\lambda_m = (l_\parallel / l_\perp)^{1/3} \tag{2.26}$$

Check that this conforms to our expectations: if the chains whish to become N_I with $l_\parallel < l_\perp$ then $\lambda_m < 1$ and the $\parallel$ direction shrinks and the $\perp$ directions expand. The reverse is true for the N_{II} and N_{III} phases and for main chain polymer liquid crystals (all prolate), and this is what is sketched in Figure 2.8b. Inserting λ_m into eqn (2.25) yields

$$F_N^{(\text{el})}(S) = \frac{k_B T}{2} \left[3\left(\frac{l_\parallel l_\perp^2}{l_0^3}\right)^{1/3} + \ln\left(\frac{l_\parallel l_\perp^2}{l_0^3}\right) \right] \tag{2.27}$$

We emphasize that this is the energy of a strand; to get the free energy per unit volume we must multiply by the number of strands per unit volume, a simple multiple of the number density of cross-links if we ignore wasted links.

The free energy is given by $k_B T$ times a factor $[\cdots]$ which depends on T through $l_\parallel \cdot l_\perp^2$ only. The $k_B T$ factor indicates, as in classical networks, that the entropy of the strands is very important (the classical result for $F(\lambda)$ is $1/2 k_B T[\lambda^2 + 2/\lambda]$). The physical principle behind distortions away from the anisotropic distribution, eqn (2.21), is that, although chains would ideally like to change their shape to this which is natural in the melt nematic state, they cannot: to do so they would need $\lambda_\parallel^2 = l_\parallel/l_0$ and $\lambda_\perp^2 = l_\perp/l_0$ at which point they would recover the same elastic energy they had in the isotropic network. The balance between isotropic and nematic phases would then not be influenced by elastic factors arising from cross-linking. However, these values of λ are not compatible with the incompressibility constraint, and the compromise λ_m is reached. The chains are not at their ideal nematic melt shape, and free energy is raised by an amount F_N, depending on the state of nematic order. At $S = 0$ we have $F_I = (k_B T/2) \cdot 3$ since $l_\parallel = l_\perp = l_0$.

Subtracting F_I from $F_N(S)$ gives us the free energy difference required by de Gennes. The step lengths $l_\parallel(S)$ and $l_\perp(S)$ can be calculated (Warner et al., 1988), but here we can proceed more simply: returning to the nematic melt (non-cross-linked) it is clear that $l_\parallel$ and $l_\perp$ must have the form (with coefficients a and b depending on the chain structure)

$$l_\parallel = l_0(1 + aS + b_\parallel S^2 + \cdots) \tag{2.28a}$$

$$l_\perp = l_0\left(1 - \frac{a}{2}S + b_\perp S^2 + \cdots\right) \tag{2.28b}$$

since distorting the original Gaussian spherical chain to either prolate ($S > 0$) or oblate

$(S < 0)$ spheroidal always increases the chain radius of gyration and hence the mean square size above the isotropic value $(l_0 L)$

$$\langle R^2 \rangle = (l_\parallel + l_\perp)L = l_0 L(1 + (b_\parallel + 2b_\perp)S^2 + \cdots) \tag{2.29}$$

(an increase irrespective of the sign of S). The sign of a is positive, so that $l_\parallel > l_0$ for $S > 0$ and vice versa. The free energy $F_N(S)$ depends only on the combination W of $l_\parallel, l_\perp$ and l_0: $W = (l_0^3/l_\parallel \cdot l_\perp^2)^{1/3}$. The structure of F_N is now easily seen since $W = 1 + cS^2 + \cdots$ (with $c = 3a^2/4 - b_\parallel - 2b_\perp > 0$) and $F(W)$ is minimal around $W = 1$; $F(W) = 3(W-1)^2$ whence

$$F(S) \approx \tfrac{3}{4}c^2 S^4 \tag{2.30}$$

Warner *et al.* (1988) give the full structure, and also show that if the network is cross-linked when the order was $S_{(x)}$ (no longer zero), a memory of the nematic state is chemically locked in and $F(S)$ becomes

$$F(S) = \tfrac{3}{4}c^2(S^2 - S_{(x)}^2)^2 \approx -\tfrac{3}{2}c^2 S_{(x)}^2 S^2 + \tfrac{3}{4}c^2 S^4 \tag{2.31}$$

(neglecting the constant $S_{(x)}^4$ term common to both the nematic and isotropic phases of this network).

The above elastic terms should be added to the Landau free energy of the nematic part of the problem, as was done by de Gennes (1975). The result is a shift in the nematic transitions from those of the un-cross-linked equivalent. For a network cross-linked in the isotropic state $(S_{(x)} = 0)$, the addition (eqn 2.27) to F is $\sim S^4$, which lowers the transition temperature. When the cross-linking is done in the nematic state, $(S_{(x)} \neq 0)$, then there is the $-S^2$ term discussed after eqn (2.19) which causes a rise in T_{NI}. The strength of this term is $S_{(x)}^2$. This accords with intuition; if nematic order is linked into the network at formation, it will always favour the nematic state and allow it to persist to higher temperatures than the non-cross-linked system.

The spontaneous distortion is first order in S; $\lambda_m \sim aS/2$. Questions of the stress–strain relation can likewise be modelled by the classical picture—see Warner *et al.* (1988) for further details.

2.6.3 *Elastomer experiments and future outlook*

The predictions of the foregoing phenomenological and molecular models of networks, namely a spontaneous distortion and a shifting of the nematic–isotropic transition according to the cross-linking history, can be clouded by many factors in practice. We discuss some of these here in so far as they relate to the theoretical picture developed, leaving a thorough discussion of networks to Chapter 10.

Liquid crystals seldom have a single ordering direction, but rather one that varies spatially because of fluctuations, defects and domain formation. The above models were for monodomains, perhaps realizable for a sample cross linked in a nematic state subject to an aligning field. Mismatch in domain direction causes elastic strains. Figure 2.7 shows that the transition of a nematic in a field is shifted, and thus a range of fields of differing magnitudes will give a blurred transition over a finite temperature interval for a bulk sample. This is sometimes observed, though there may be other reasons for this (see below).

In order that theory confronts experiment, the cross-linking process should lead to a well-determined number of cross-links of known functionality and with a degree of wastage that can be estimated. Elastically inspired shifts in T_{NI} are proportional to actual cross-link density, not necessarily to the amount of cross-linking agent added. Cross-links are assumed to simply link chains together, and are not of sufficient number and volume that they act as solvent, causing biphasic fluctuations and thence blurring out transitions, as well as shifting transitions by simple dilution, thence masking elastic effects.

Networks should not be too highly cross-linked if a connection with a Gaussian (classical) theory of rubber is to be sought. If the length of strands between links is too short, then they are not flexible on this length scale and must be treated by other means.

Clearly networks offer many new scientific and technological possibilities. One can link in types of structure other than nematic, for instance twist (cholesteric networks), poled dipole fields and smectic order. Calculations of spontaneous elastic distortions and moduli already point to very interesting effects. Certain smectics, for instance the ferroelectric chiral smectic C*, would have a greatly enhance device potential from the new mechanical possibilities afforded by polymerization and cross-linking (for instance free thin film formation with mechanical integrity). The coupling between mechanical state and nematic order (and thence optical properties) has been the main theme of this chapter. Other external fields, such as electric ones, can also be simultaneously applied to enrich the possibilities. The reader can well imagine future possibilities such as acousto-optical couplers and strain–temperature devices.

2.7 Conclusions and outlook

The theme of this chapter has been that of competition between polymer entropy and mesogenic order, either nematic or smectic. The polymer backbone of a comb nematic has a strong requirement, if it has any flexibility at all, to be as random as possible in its configurations and shape. This conflicts with any requirement of orientational order that nematic elements, present as attached teeth or in the backbone itself, might have. This conflict is already present in worm nematics, but an additional element enters for comb polymers—hinges couple the teeth to the backbone in such a way that the teeth and backbone might have to compete to dominate the nematic ordering. This gives the possibility for at least three different nematic phases which then compete with each other and the isotropic state and produce complex phase diagrams. We have discussed several of these in order to elucidate, by way of examples, the competitions arising. Confinement to planes in the case of smectic combs also represents a reduction in entropy of chains and is resisted. The partial recovery of chain entropy is discussed in the section on smectic combs. Finally, the theme of entropy, chain shape and nematic order is applied to solids (nematic elastomers). The phenomenological and classical molecular theories of elastomer phases spontaneous distortions and non-linear stress–strain relations are presented. Beginning from a picture of the polymeric nematic state of chains under elastic strain imposed by the nematic ordering fields, one has arrived at nematic chains that are, in addition, put under actual mechanical stress.

There is a plethora of problems for the future. On the theoretical side, the full range of phases has not been exhausted by the simple three discussed here. More complex but still uniaxial nematic phases can be envisaged. In addition, questions of biaxiality have not been discussed at all. With teeth or backbone perpendicular to the ordering direction and themselves capable of ordering, at low enough temperatures two of the phases discussed will be unstable with respect to biaxial ordering. The smectic A phase still demands an understanding of its transitions, both to the three nematic phases and to other rearrangements within the Sm_A phases analogous to the I, II, and III phases of the nematic. The other smectic phases have rich properties and it is not yet clear what differences evolve from their polymerization into combs. Work on smectic networks proceeds, but other network effects remain unsolved, for instance anomalous solvent absorption, domain structure and the interplay with dipolar effects.

Many experimental problems continue to be of vital importance to establish the validity or otherwise of theoretical models. The neutron investigation, requiring deuteration, of chain shape in many different nematic comb melts would help establish the relation, if any, between variation in hinge structure and nematic main chain and side chain coupling and the resultant phase symmetry. For networks, many difficulties are to be overcome (section 2.6) if experiment is to confront the theory that has been developed. These are largely questions of synthesis, since the physical techniques used for investigation are highly developed and of great power.

With the advent of a wider range of syntheses, many more proposals and speculations for new types of behaviour and effects can be made. Most exciting in my opinion are speculations involving cooperative dielectric behaviour offering giant dipolar response and possibly non-linear optical efficiencies hundreds of times greater than those currently available in conventional nematic media. For this, subtle synthesis with well-defined tacticities will be required, and will demand the close collaboration of chemists and physicists.

Acknowledgements

I have had many collaborators, especially Dr J.M.F. Gunn with whom the work on worms and dielectrics was initiated, and with whom all other aspects of this work were discussed; Dr X.-J. Wang, with whom the model of combs and their phases was conceived; and Dr W. Renz with whom the model of smectic layer hopping and the analytic approach to comb phases were developed. Dr S. Picken (Akzo Corporate Research NL) critically read this review and made useful suggestions. I am grateful to Unilever plc for financial support.

References

Bosch, A. ten, Maissa, P. and Sixou, P. (1983a) *Phys. Letts.* **94A**, 298.
Bosch, A. ten, Maissa, P. and Sixou, P. (1983b) *J. Chem. Phys.* **79**, 3462.
Bosch, A. ten, Maissa, P. and Sixou, P. (1983c) *J. Phys. (Paris) Letts.* **44**, L105.
Chandrasekhar, S. (1977) *Liquid Crystals.* Cambridge University Press, Cambridge.
Cser, F. (1984) in *Liquid Crystals and Ordered Fluids*, Vol. 4, ed. Griffin, A.C. and Johnson, J.F. Plenum, New York.
Cser, F., Hovárth, J., Nyitrai, K. and Hardy, Gy. (1985) *Israel J. Chem.* **25**, 252.
Gennes, P.-G. de (1969) *Phys. Lett.* **28A**, 11.
Gennes, P.-G. de (1975) *C.R. Acad. Sci. Paris* **281**, 101
Gennes, P.-G. de (1982) in *Polymer Liquid Crystals*, eds. Ciferri, A., Krigbaum, W.R. and Meyer, R.B., Academic Press, New York.
Halperin, A. (1986) *J. Chem. Phys.* **85**, 1081.
Jähnig, F. (1979) *J. Chem. Phys.* **70**, 3279.
Jähnig, F. (1981) *Mol. Cryst. Liq. Cryst.* **63**, 157.
Keller, P., Carvalho, B., Cotton, J.P., Lambert, M., Moussa, F. and Pepy, G. (1985) *J. Phys. (Paris) Letts.* **46**, L1065.
Kirste, R.G. and Ohm, H.G. (1985) *Makromol. Chem. Rapid Commun.* **6**, 179.
Koch, H.J., Finkelmann, H., Gleim, W. and Rehage, G. (1985) in *Polymeric Liquid Crystals*, ed. Blumstein, A., Plenum, London.

Luckhurst, G.R. (1985) in *Recent Advances in Liquid Crystalline Polymers*, ed. Chapoy, L. Elsevier, London, 105.

Mattoussi, H., Ober, R., Veyssie, M. and Finkelmann, H. (1986) *Europhysics Letts.* **2**, 233.

Reck, B. and Ringsdorf, H. (1985) *Makromol. Chem. Rapid Commun.* **6**, 291.

Reck, B. and Ringsdorf, H. (1986) *Makromol. Chem. Rapid Commun.* **7**, 389.

Renz, W. and Warner, M. (1986) *Phys. Rev. Letts.* **56**, 1268.

Renz, W. and Warner, M. (1988) *Proc. Roy. Soc.* **A417**, 213.

Rusakov, V.V. and Schliomis, M.I. (1985) *J. Phys. (Paris) Lett.* **46**, L1065.

Schätzle, J. and Finkelmann, H. (1987) *Mol. Cryst. Liq. Cryst.* **142**, 85.

Schwarz, J. (1986) *Makromol. Chem. Rapid Comm.* **7**, 21.

Vasilenko, S.V., Shibaev, V.P. and Khokhlov, A.R. (1985) *Makromol. Chem.* **186**, 1951.

Wang, X.-J. and Warner, M. (1986a) *J. Phys.* **A19**, 2215.

Wang, X.-J. and Warner, M. (1986b) *Phys. Letts.* **A119**, 181.

Wang, X.-J. and Warner, M. (1987) *J. Phys.* **A20**, 713.

Warner, M., Gunn, J.M.F. and Baumgärtner, A. (1985) *J. Phys.* **A18**, 3007.

Warner, M., Gelling, K.P. and Vilgis, T.A. (1988) *J. Chem. Phys.* **88**, 4008.

3 Molecular engineering of predominantly hydrocarbon-based LCPs

VIRGIL PERCEC and COLEEN PUGH, Department of Macromolecular Science, Case Western Reserve University, Cleveland, Ohio 44106, USA

3.1 Introduction

Systematic investigations of the synthesis, characterization and applications of side chain (SC) liquid crystalline polymers (LCP) began only after Ringsdorf and co-workers (Finkelmann et al., 1978, 1978a, b) proposed that a flexible spacer should be inserted between the polymeric main chain and the mesogenic side groups to decouple the motions of the main chain and side groups in the liquid crystalline state. Although numerous, previous investigations concerning the synthesis of polymers with low-molar-mass liquid crystalline compounds attached to their backbones did not lead to a synthetic method which could systematically produce side chain LCPs, since most involved polymers with the mesogen attached directly to the backbone. The research prior to the introduction of the spacer concept in 1978 was repeatedly reviewed (Shibaev and Platé, 1977; Blumstein, 1978; Blumstein and Hsu, 1978; Elias, 1977; Wendorff, 1978). Only selected examples will be discussed here.

The research performed after 1978 has been even more comprehensively reviewed (Attard and Williams, 1986; Bata, 1981; Blumstein, 1985; Chapoy, 1985; Engel et al., 1985; Finkelmann, 1982, 1983, 1987a, b; Finkelmann and Rehage, 1984; Griffin and Johnson, 1982; Le Barny, 1987; Lipatov et al., 1984; Magagnini, 1981; Platé et al., 1984a, b, 1985, 1987; Platé and Shibaev, 1984, 1987; Saeva, 1981; Shibaev and Platé, 1984, 1985; Talroze et al., 1983; Tsukruk et al., 1986). Particularly helpful are the reviews by Finkelmann and Rehage (1984), and Shibaev and Platé (1984), which tabulated and organized most of the side chain liquid crystalline polymers and their phase transitions available at that time.

Based on the spacer model, a large number of side chain liquid crystalline polymers and copolymers were synthesized and details are currently available in the literature. Different smectic, chiral smectic, nematic and cholesteric mesophases are exhibited by these polymers based on different mesogenic groups and polymer backbones. Nevertheless, even the most elementary synthetic and structural principles of this field are not completely elucidated. Answering textbook questions related to the influence of polymer molecular mass, polydispersity and tacticity, the nature of the spacer and its length, and the nature of the mesogenic group and polymer backbone on phase transitions, and on the nature of the mesophase or the extent of decoupling, is not a trivial task. This is largely because there are no well-defined synthetic methods to prepare liquid crystalline polymers with predictable molecular masses and narrow polydispersities, as are available for most other classes of polymers. In addition, much of the literature's experimental data does not provide the polymer molecular mass, polydispersity and purity. Because most reported molecular masses were determined by either gel permeation chromatography or viscosimetry, it must be mentioned that the most reliable results are obtained by light-scattering and vapour-pressure osmometry since they are the only direct methods of measuring the molecular masses of polymers. However, it has been shown that side chain LCPs aggregate in solution, and care must therefore be taken to perform measurements below the polymer's critical concentration of aggregation (Springer and Weigelt, 1983a, b; 1985; Duran and Strazielle, 1987).

In addition, differential scanning calorimetric (DSC) characterization demonstrated that the thermal behavior of several SC–LCPs is apparently dependent on thermal history (Springer and Weigelt, 1985). Therefore, we must still question the thermal transitions and their thermodynamic parameters reported in the literature as to whether or not the experiments were performed under equilibrium conditions. Additional research must be performed before a molecular engineering approach to designing side chain liquid crystalline polymers can be taken.

The present chapter is not intended to be an updated comprehensive collection of SC–LCP data. Rather, it will emphasize two major problems of the field. First, the synthetic and structural principles which are accepted by investigators active in the field will be critically discussed, underlining additional experiments necessary for further elucidation. Second, a number of trends which seem to be emerging as basic principles but are not yet sufficiently supported by experimental data will be discussed, with the hope of raising additional research interest in this area. Given the intent, content and length of this chapter, many contributors to this field should not be surprised at not seeing some of their work cited in this discussion.

3.2 Principles of side chain LCPs

3.2.1 Structural principles

The structural principles currently used to synthesize side chain LCPs are outlined in Schemes 3.1–3.3. Scheme 3.1 presents different types of polymers already synthesized from either linear or ring-shaped flexible backbones containing rod-like mesogenic groups exhibiting calamitic phases. Rod-like mesogens are based on either rigid rod-like molecules or flexible rod-like molecules which undergo conformational isomerism. The rod-like mesogens are usually selected from the literature on low-molar-mass liquid crystals (Kelker and Hatz, 1980; Demus and Zaschke, 1984).

The structural principles of low molar mass liquid crystals based on *rigid rod-like mesogens* have been repeatedly discussed and reviewed (Gray, 1962, 1979, 1982, 1983, 1985; Gray and Winsor, 1974; Luckhurst and Gray, 1979; Osman, 1983a, b; Toyne, 1987), and will therefore be mentioned only briefly here. The traditional synthetic pathway to low-molar-mass liquid crystals based on rigid rod-like mesogens implies the interconnection of two rigid cyclic units, which may be similar or dissimilar. The interconnecting group should cause the resulting compound to have a linear and consequently planar conformation. Therefore, aromatic, *trans*-cyclohexane, bicyclo-octane, and *trans*-2, 5-disubstituted-1, 3-dioxane, 1, 3-dithiane, or 1, 3-dioxathiane groups are preferred. Linking units containing multiple bonds such as $-(C \equiv C)_n-$, $-CH = N-, -N = N-, -(CH = CH)_n-, -CH = N-N = CH-$, etc. are also frequently employed since they restrict the freedom of rotation. Such groups can conjugate with phenylene rings, enhancing the anisotropic polarizability. This increases the mol-

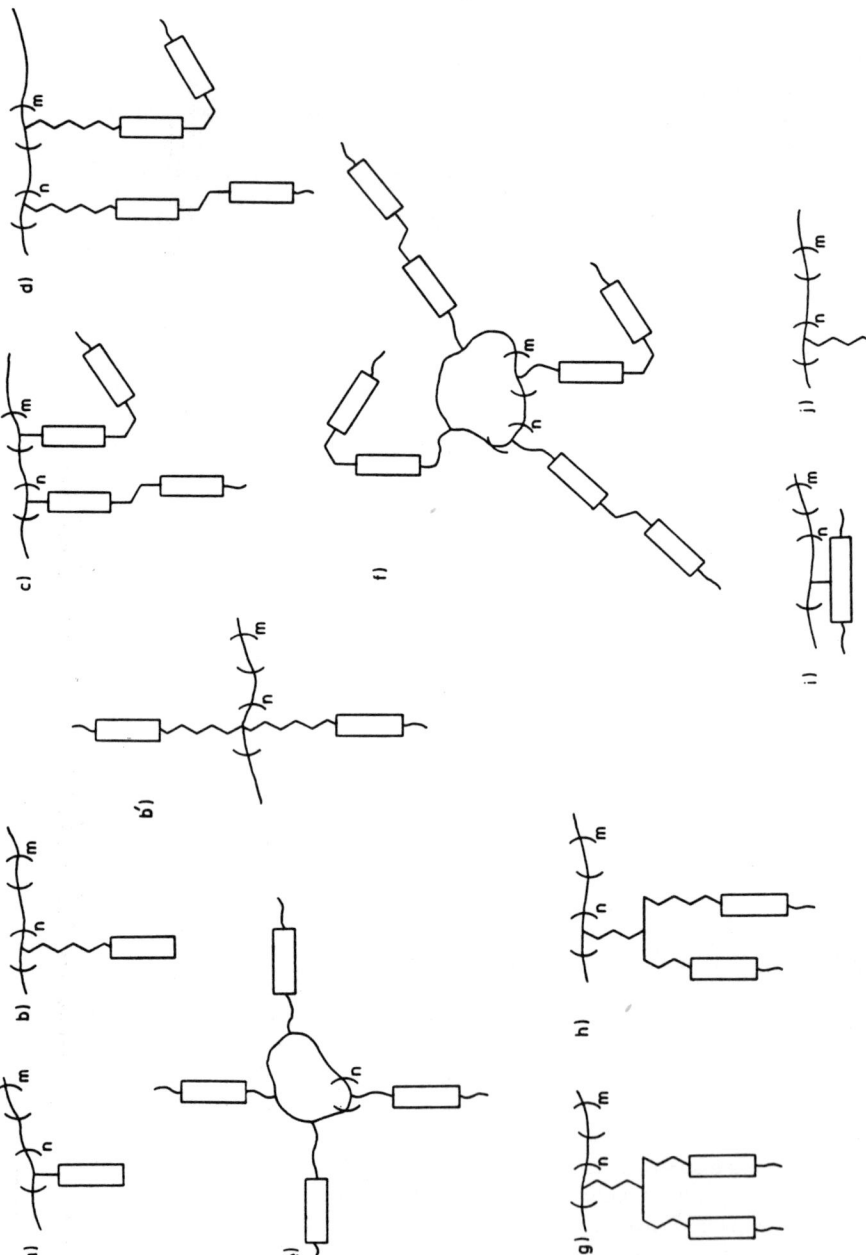

Scheme 3.1 Linear and cyclic side chain liquid crystalline oligomers and polymers containing flexible backbones and polymers containing flexible backbones and rod-like mesogens attached either longitudinal or parallel to the polymer backbone

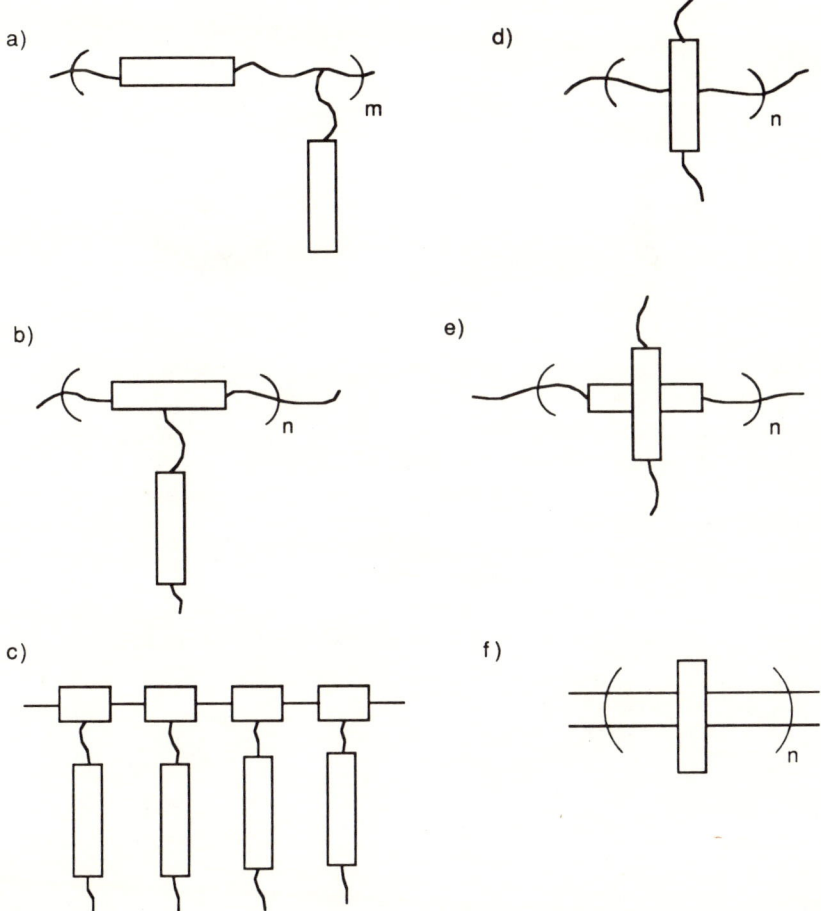

Scheme 3.2 Combined liquid crystalline polymers containing mesogenic units in the main chain and as side groups.

ecular length and maintains rigidity. The ester group is also effective since resonance interactions confer double-bond character at the $\overset{O}{\overset{\|}{C}}\!-\!O$ link, thereby restricting rotation from the *trans* to *cis* conformer. Therefore, in all these examples, the extended conformation of a certain molecule is accomplished and maintained through the rigidity and linearity of its constituents, i.e. its rigid rod-like character.

Flexible rod-like mesogens are obtained when the rigid link is replaced with flexible ethane or methyleneoxy units. Although ethane and methyleneoxy units can adopt an extended conformation which is similar to the *trans* conformation provided by an ester unit, they are flexible and undergo free rotation, leading to a number of different conformational isomers which are in dynamic equilibrium. Nevertheless, the *anti* and

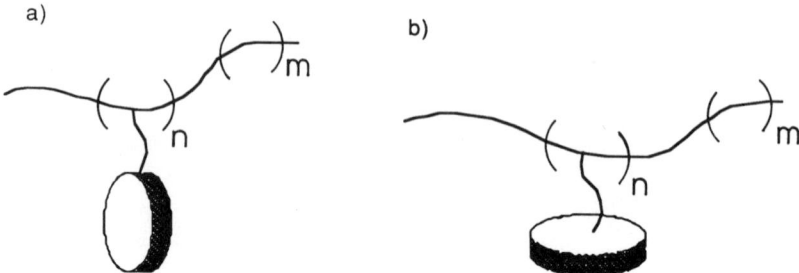

Scheme 3.3 Side chain liquid crystalline polymers containing disc-like mesogens.

gauche isomers predominate since they are the lowest-energy isomers (Figure 3.1). These two isomers are in dynamic equilibrium; the *anti/gauche* ratio in an isotropic solution is usually 3/2 (Eidenschink, 1985). The *anti* conformer presents an extended linear conformation which is identical to the rod-like conformation of (for example) a phenyl benzoate derivative. The *gauche* conformer, on the other hand, is similar to a kinked molecule. Consequently, a polymer containing these flexible rod-like mesogenic side groups behaves as a copolymer containing rod-like and kinked side groups; their ratio and therefore copolymer composition is dynamic. In contrast to mesogenic units, where the linearity and planarity of the molecules is realized and maintained through their rigid rod-like character, the rod-like character of flexible mesogens is realized

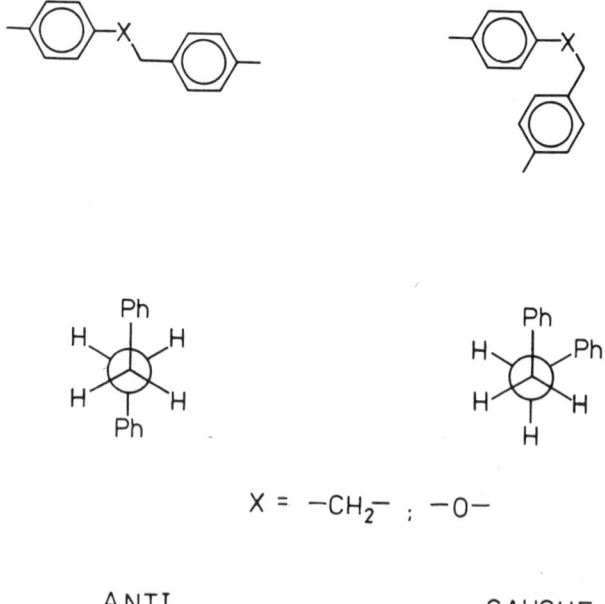

Figure 3.1 The lowest energy *anti* and *gauche* conformational isomers of 1, 2-diphenylethane- and benzyl-ether-based mesogens.

through the conformational isomerism and is maintained through the dynamic equilibrium between different conformers. We therefore use in this case the name *rod-like mesogenic units based on conformational isomerism*, or *flexible rod-like mesogenic units*. Within this chapter, the term 'flexible' refers to conformational freedom which can change the shape of a molecule from elongated to kinked, and not to the conformational isomerism of different extended conformers, nor to rotational freedom. Although rotational freedom is often possible in conjugated units like biphenyl or phenyl benzoate groups, it does not affect the elongated shape of the molecule. The idea of replacing ester groups with methyleneoxy or ethane groups was only recently proposed by organic chemists working with low-molar-mass liquid crystals (Abdullah *et al.*, 1985; Carr and Gray, 1985; Carr *et al.*, 1985; Eidenschink, 1985; Gray and McDonnell, 1979; Gray, 1981; Kelly and Schad, 1985; Osman, 1982*a, b*; 1983*a*; Takatsu *et al.*, 1984; Tinh *et al.*, 1985). In general, replacement of an ester group with a flexible methyleneoxy group depresses the thermal transition temperatures, thereby facilitating physical investigations.

Additional rod-like mesogenic units which exhibit conformational isomerism are based on *trans*-2, 5-disubstituted-1, 3-dioxane, -1, 3-dithiane, -1, 3-dioxathiane, 2, 5-disubstituted-1, 3-dioxa-2-borinane and *trans*-1, 4-disubstituted cyclohexane. The *trans* chair isomers adopt fully extended conformations independent of whether the substituents in the 2 and 5 positions are in an equatorial or in an axial placement as shown in Figure 3.2. Therefore, although the two *trans* chair conformers are in dynamic equilibrium, their ratio does not significantly affect their mesomorphic properties, at least in low-molar-mass liquid crystals, since both isomers exhibit extended conformations. Consequently, although this class of mesogens exhibits conformational isomerism, they can be classified as rigid rod-like mesogens. Low-molar-mass liquid crystals containing heterocycloalkanediyl units have only recently become of interest (Haramoto and Kamogawa, 1983, 1985*a, b*; Haramoto *et al.*, 1984*a, b*, 1986; Seto *et al.*, 1985*a, b*; Sucrow *et al.*, 1985; Zaschke *et al.*, 1984).

Both rigid and flexible rod-like mesogens have been attached to flexible linear polymer backbones in a 'normal,' i.e. longitudinal way, either directly (Scheme 3.1, *a, c*) or through flexible spacers (Scheme 3.1, *b, d*). Rigid and flexible rod-like mesogens have also been attached to cyclic oligomers through flexible spacers (Scheme 3.1, *e, f*). Each polymer structural unit can carry either one (Scheme 3.1, *a − f*) or two (Scheme 3.1, *b'*) 'single' mesogenic groups; i.e. each flexible spacer contains a single mesogenic unit (Scheme 3.1, *b'*). Since polymers containing 'single' rod-like mesogens represent the traditional class of SC–LCP, we will refer to them in the rest of this chapter as polymers containing rod-like mesogens. In contrast to 'single' mesogens, paired rigid rod-like mesogens (either similar or dissimilar) attached normal to the single spacer through either similar (Scheme 3.1, *g*) or dissimilar (Scheme 3.1, *h*) spacers have also been used in the synthesis of side chain liquid crystalline polymers. Alternatively, rigid rod-like mesogens can be attached to the polymer backbone in a parallel manner, either directly (Scheme 3.1, *i*) or through flexible spacers (Scheme 3.1, *j*). This last approach was inspired by the work of Demus *et al.* on low-molecular-mass rod-like liquid crystals containing lateral long-chain substituents (Weissflog and Demus, 1983, 1984, 1985; Takenaka *et al.*, 1986; Demus *et al.*, 1985).

A second class of side chain LCPs are the 'combined' liquid crystalline polymers, in which the mesogen is attached in a normal way through flexible spacers to either semi-flexible backbones (i.e. thermotropic main chain LCPs containing mesogenic units and

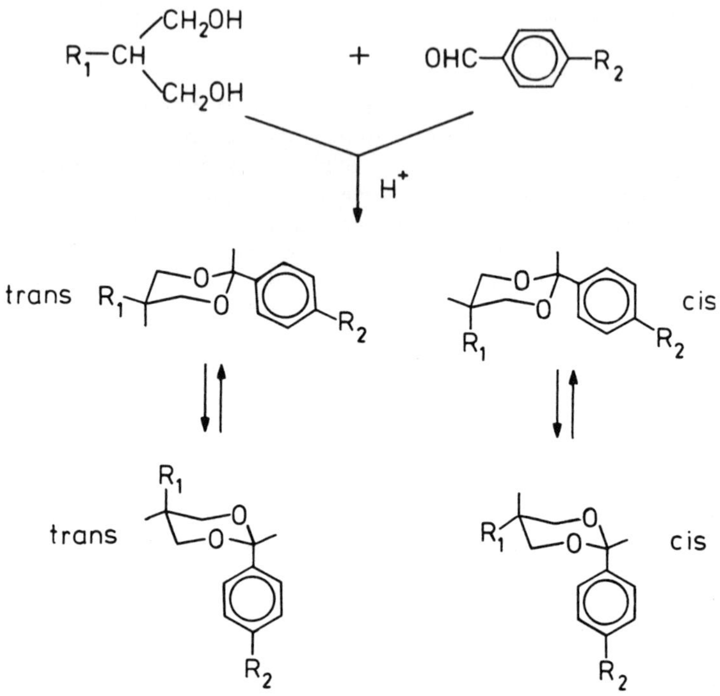

R1	R2
alkyl	alkyl
alkyl	alkoxy
alkyl	acyloxy, halogen, CN
alkyl, alkoxy } 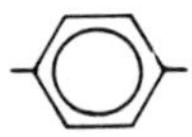	CN

Figure 3.2 Synthesis of *cis*- and *trans*- 2, 5-disubstituted-1, 3-dioxane mesogens and the conformational isomerism exhibited by both isomers.

flexible spacers in the main chain) or to rigid backbones (i.e. rigid thermotropic main chain liquid crystalline polymers without flexible spacers). In the first case, the mesogenic side group can be attached to either the flexible spacer of the main chain (Scheme 3.2, *a*) or to the mesogenic unit of the main chain (Scheme 3.2, *b*). As shown in Scheme 3.2, *c*, there is only one possible attachment of the mesogen to a rigid backbone.

Two additional variations of liquid crystalline polymers are to insert the mesogenic unit perpendicular to a flexible main chain (Scheme 3.2, *d*), or to insert two cross-shaped mesogens within a flexible main chain (Scheme 3.2, *e*).

The second class of mesogens which have already been used in the synthesis of side chain LCPs is based on disc-like molecules. Low-molar-mass disc-like liquid crystals were discovered almost simultaneously by several research groups (Billard et al., 1978; Chandrasekhar et al., 1977; Dubois, 1978; Destrade et al., 1979; Huu et al., 1978), and have been reviewed (Chandrasekhar, 1982, 1983; Destrade et al., 1983; Dubois and Billard, 1984; Strzelecka et al., 1983). Disc-like flat molecules containing a rigid core of aromatic or alicyclic units form either nematic, cholesteric or columnar mesophases when they are fitted with (most frequently) three, four, six or eight side chains. The only SC–LC polymeric structure synthesized to date was accomplished by normal attachment of the disc-like molecule to a flexible polymeric backbone through flexible spacers (Scheme 3.3, a). However, a parallel attachment is also probable (Scheme 3.3, b). Examples of these polymers will be presented in the following sections.

Although the polymeric structures outlined in Schemes 3.1–3.3 do not exhaust the entire family of low-molar-mass liquid crystals generally considered in the synthesis of SC–LCPs, apparently no other mesogenic units were used in their preparation at the time this review was written.

3.2.2 Partial decoupling by the spacer concept

Attainment of a thermotropic liquid crystalline mesophase from a flexible backbone containing mesogenic side groups requires reconciliation of the main chain's tendency to form a statistical random coil conformation and the side groups' tendency to arrange anisotropically. This is shown schematically in Figure 3.3. The principle that the motions of both the main chain and the side groups are coupled when the mesogenic groups are directly attached to the flexible backbone is well accepted. In this case, the conformation of the main chain is disturbed when the side groups adopt an anisotropic arrangement.

In order to balance the competition between the backbone's random-coil conformation and the side groups alignment, Ringsdorf and co-workers predicted that the motion of the polymer main chain must be decoupled from that of the anisotropically orientable mesogenic side groups in the fluid state (Finkelmann et al., 1978a, b, c). As demonstrated schematically in Figure 3.3 and explained in Figure 3.4, decoupling should be achieved when a flexible spacer is inserted between the polymer backbone and the mesogenic side groups. Based on the spacer model, the side groups should be able to self-orient into an anisotropic mesophase even when the main chain adopts a random-coil statistical conformation. The spacer concept therefore assumes that the main chain should do little to hinder the orientation of the mesogenic side groups; that is, for a given mesogen and spacer, the nature of the polymer backbone should theoretically not affect the type of mesophase formed and its thermal stability. However, the polymer backbone does in fact affect both aspects of the mesomorphic behaviour of SC–LCPs, as will be discussed further in section 3.4.3.7.

Since the spacer concept was proposed, side chain liquid crystalline polymers have been routinely synthesized. Subsequent experiments performed by Ringsdorf and Spiess (Boeffel et al., 1983, 1986; Blumich et al., 1987; Geib et al., 1982; Wassmer et al., 1985; Spiess, 1985) demonstrated that although a spacer helps to decouple the mesogenic groups from the main chain, and that decoupling becomes more effective with increasing spacer length, decoupling is nevertheless incomplete (see also Chapter 8). At least part of the spacer is anisotropically oriented with the mesogen.

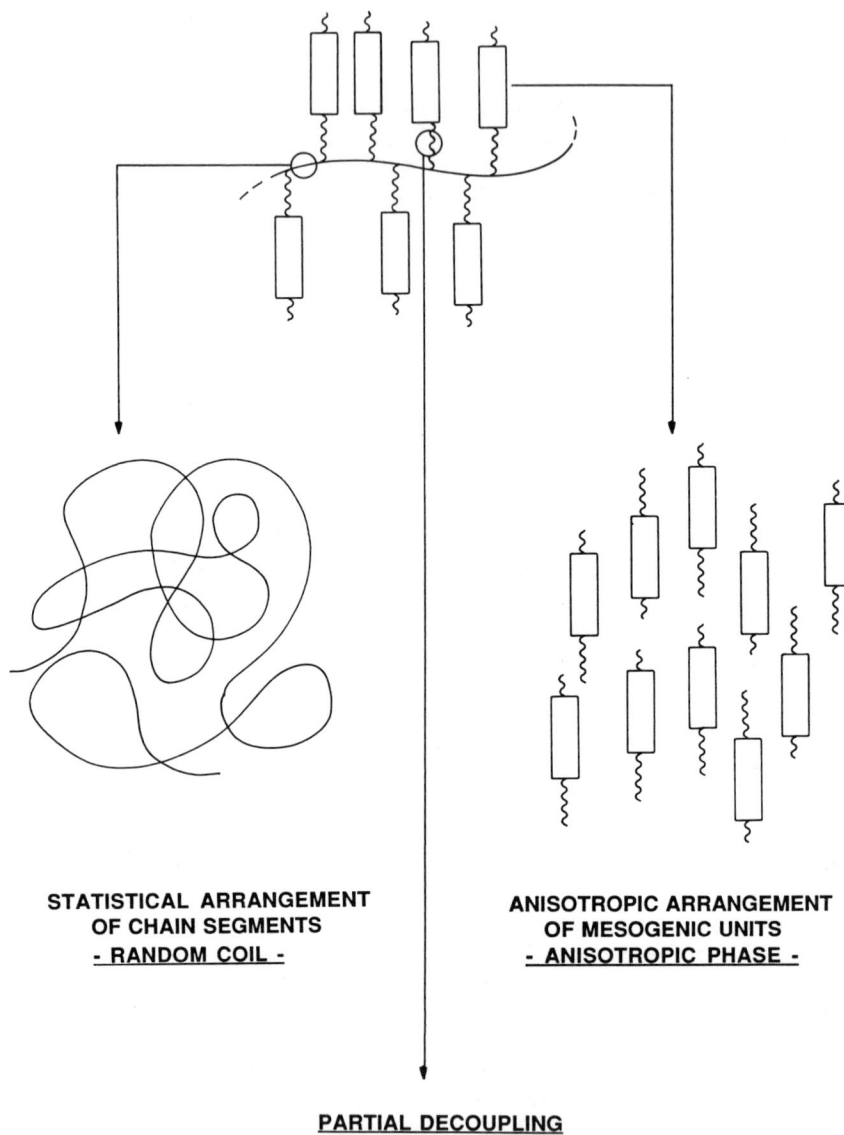

**STATISTICAL ARRANGEMENT
OF CHAIN SEGMENTS
- RANDOM COIL -**

**ANISOTROPIC ARRANGEMENT
OF MESOGENIC UNITS
- ANISOTROPIC PHASE -**

**PARTIAL DECOUPLING
BY FLEXIBLE SPACERS**

Figure 3.3 Schematic structure of liquid crystalline side chain polymers showing the necessity of decoupling the mesogenic groups and the polymer main chain through flexible spacers (Spiess, 1985).

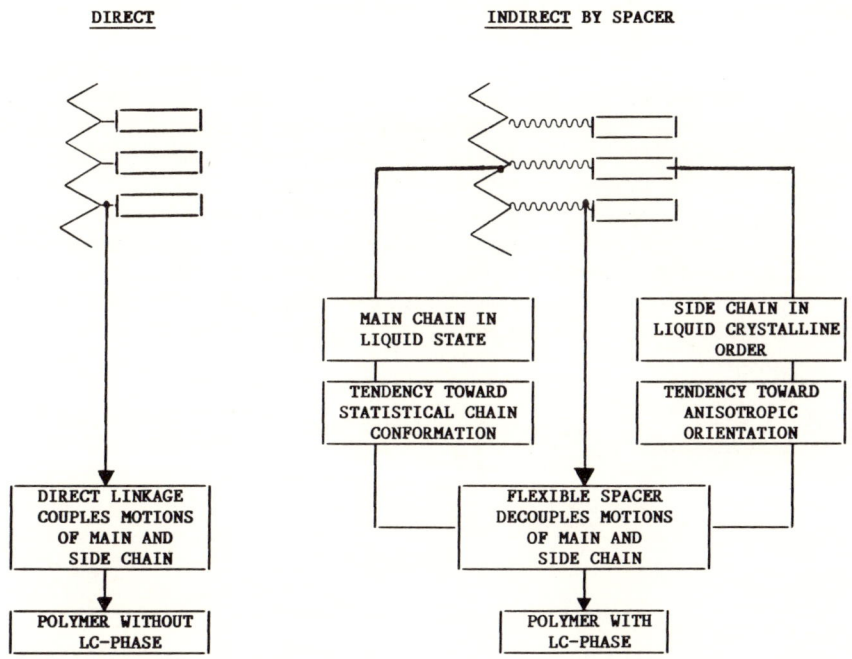

Figure 3.4 Spacer model considerations for the preparation of side chain liquid crystalline polymers (Finkelmann *et al.*, 1978a).

Therefore, the spacer also causes a dynamic coupling of the motions of the main chain and side groups. Consequently, the most accurate visualization of a side chain liquid crystalline polymer is the interaction of two semi-independent thermodynamic subsystems through the flexible spacer; ie. a section of the spacer plasticizes the main chain, while another section of the spacer stabilizes the mesogenic units (Engel *et al.*, 1985; Spiess, 1985; Cser, 1984).

Additional support for only partial decoupling was obtained by neutron scattering experiments performed on deuterated LCP backbones. These experiments demonstrated that the radius of gyration in the nematic phase has an anisotropy of approximately 25% (Kirste and Ohm, 1985). In contrast, the backbone is highly anisotropic in the aligned smectic phase, and is confined within a smectic layer (Keller *et al.*, 1985). However, the system's conformational freedom is increased by the spacer when compared to the model proposed for polymers containing no flexible spacer (Strzelecki and Liebert, 1973). Most interestingly, the motions of the mesogen are frozen at temperatures below the glass transition temperature, with the exception of 180° rotational jumps of the phenyl ring about the mesogen's molecular axis (Geib *et al.*, 1982; Chapter 8).

The spacer concept has been successfully transplanted to the preparation of synthetic polymeric lipids (Elbert *et al.*, 1985; Kunitake *et al.*, 1984; Laschewsky *et al.*, 1987), which form lyotropic liquid crystalline phases. As with thermotropic liquid crystals, the mobilities of different structural elements within the polymeric lipids are vastly different. Although the lipid chains are almost as mobile as the corresponding monomeric membranes well above the transition to a liquid crystalline phase (Spiess, 1985), the motions of those groups closest to the polymer chain are prevented.

Hydrophilic spacers have also been introduced into polymeric lipids (Elbert *et al.*, 1985; Kunitake *et al.*, 1984; see also Ringsdorf *et al.*, 1988). As shown in Figure 3.5, the hydrophilic spacer can be either part of the side group, part of the main chain, or a combination of both main chain and side chain spacer. The latter alternative is obtained by preparing copolymers in which the main chain spacer is formed from easily accessible unsaturated amphiliphiles and hydrophilic comonomers (Laschewsky *et al.*, 1987). As outlined in Figure 3.5, the three types of spacers are homologues of, respectively, (A') thermotropic homopolymers containing spacers; (B'), copolymers of nonmesogenic monomers and monomers containing mesogens directly attached to the

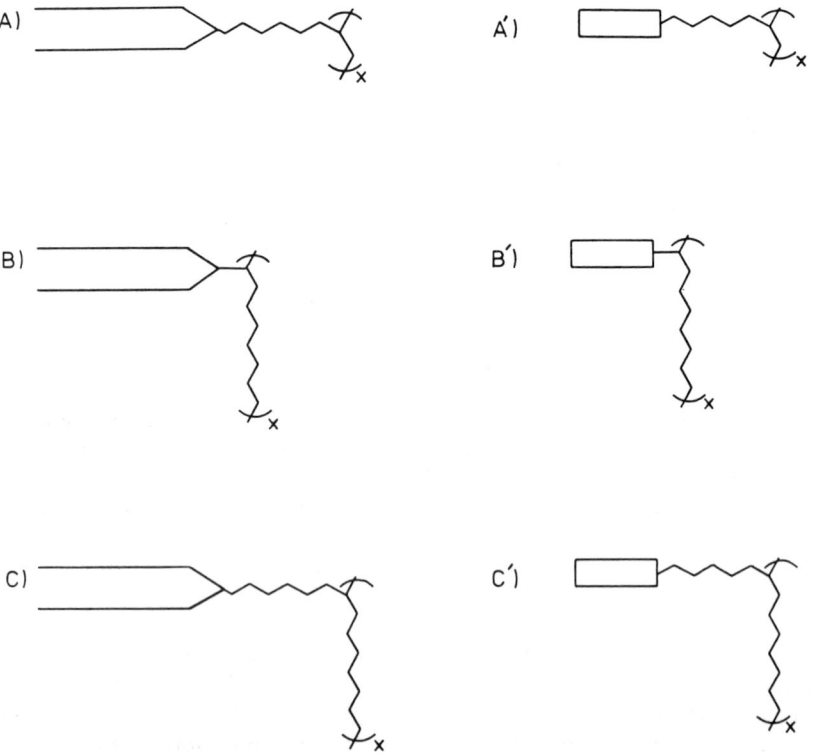

Figure 3.5 Schematic representation of polymeric lipids containing hydrophilic spacer groups: (*A*) side group spacer; (*B*) main chain spacer; (*C*) main chain and side chain spacers; and of the corresponding thermotropic side chain LCPs (*A'*) side group spacer; (*B'*) main chain spacer; (*C'*) main chain and side chain spacers (Laschewsky *et al.*, 1987).

polymerizable group; and (C′) copolymers of non-mesogenic monomers and mono-mers containing mesogens attached to the polymerizable group through a flexible spacer. An example of a SC-LC polymalonate containing flexible spacers in both the main chain and side groups was recently synthesized by Ringsdorf *et al.* (1987c). Because of a lack of sufficient systematic investigations, it is still a question as to whether or not this analogy is enlightening. Presently, we believe that completely or at least highly decoupled side chain liquid crystalline polymers are possible by the combination of main chain and side chain spacers as in Figure 3.5, C and C′. This is already supported by some experimental results, as will be discussed speculatively in the following section.

3.2.3 *Complete decoupling by the spacer concept?*

One general conclusion seems to emerge from the short literature survey in section 3.4.3.7 concerning the influence of the polymer backbone on the mesophase formation of side chain LCPs. That is, in most cases the nature of the least ordered mesophase exhibited by a polymer with a given mesogen is dictated primarily by the spacer length, with only the thermal stability of the mesophases being influenced by the nature of the polymer backbone. The broadest range of mesophase thermal stability is exhibited by polymers containing the most flexible backbone, namely polysiloxanes. Nevertheless, compared to more rigid polymers, polymers containing extremely flexible backbones usually exhibit additional mesophases or even side chain crystalline phases. Perhaps then, the polymer backbone is more important in obtaining a highly decoupled side chain LCP than previously considered.

The enhanced mesophase thermal stability and the increased isotropization temperature of polymers with flexible backbones can be associated with a high degree of freedom of the side groups. Certainly the conformation of a more flexible backbone can be easily disturbed and subsequently the side groups can adopt a higher degree of order compared to those of a more rigid backbone. However, as we have recently stated (Hsu and Percec, 1987a, b, c; Hahn and Percec, 1987; Percec, 1988b), this behavior may also be related to the miscibility of the polymer backbone and the side chains. A simplistic comparison between graft copolymers and side chain LCPs should clarify this. When the graft and backbone segments of graft copolymers are miscible, the graft copolymer exhibits a single phase morphology, and the overall properties are approximately mass averaged. On the other hand, when the graft and the backbone segments are immiscible, the graft copolymer exhibits a microphase separated biphasic morphology, and presents synergistic properties contributed by both the individual graft and backbone segments (Sperling, 1986).

In our opinion, the existence of such a single phase morphology in side chain liquid crystalline polymers has little chance of more than a partial decoupling of the side chain and main chain motions. Rather, we feel that a microphase separated system in analogy to graft copolymers may be the key to obtaining a highly decoupled side chain LCP. This is because a microphase separated side chain LCP should contain microdomains exclusively of polymer backbones, and microdomains exclusively of side groups. Such a system would not only resemble a fine dispersion of an immiscible low-molar-mass liquid crystal in an amorphous polymer, but should also exhibit two glass transition temperatures (T_g) if the side groups do not crystallize; one due to the independent motion of the main chain, and the other due to the cooperative motion of the side groups. The cooperative motion of the side groups would therefore be independent of the main chain.

The next question is what structural requirements should be considered in designing such a system. One must certainly consider the same parameters governing the immiscibility of polymer blends and/or graft copolymers such as immiscibility of the two polymeric segments, the segment molecular masses, and the weight ratio of these segments. The microdomain size in a phase separated system is dictated by the mass ratio of the two subsystems. This is because at least a very small mass fraction of any polymer can be blended with any other polymer to form a miscible one-phase system. Therefore, in a microphase separated system, large domain sizes are expected at mass ratios of about 50/50. Considering this parameter in relation to the three structural models depicted in Figure 3.5, one finds that the homopolymer architecture depicted as A' containing only a flexible spacer is the least likely to form a microphase separated system, since the mass percent of the polymer's side groups is usually 80–90% of the overall mass of the polymer. The situation is more favourable in case B', but is the most favourable for model C'.

Before speculating further, we should first consider the possibility of independent motions of main chain and side groups in non-liquid crystalline comb-like polymers. Although this approach may seem strange, Cowie et al. published a paper in 1979 entitled 'Independent relaxation of alkyl side-chains in poly(di-n-heptyl itaconate): evidence for a double glass transition,' and concluded that the two changes in heat capacity observed on the DSC thermograms are due to the independent motion of the main chain, and the cooperative but independent motion of the alkyl side groups. The same authors offered additional support for this concept (Cowie et al., 1981), and explained that since these polymers have an extremely high side chain density, and since the side chains are less polar than the main backbone, these polyitaconates may be regarded as a comb-polymer, or even copolymer, in which the side chains are not significantly entangled with the main chain because of their effective shielding. The ease of detection of side chain motion is not surprising considering that these polymers behave as two phases in the glass. Additional examples of independent motions were reported for poly(α-amino acid)s containing side chains (Pezzin et al., 1975; Poliks et al., 1988; Sugai et al., 1966; Tsutsumi et al., 1973; Yamashita et al., 1975) and copolymers of α-olefins with maleic anhydride (Rim, 1985, 1986).

Lipatov et al. (1983, 1984), Tsukruk et al. (1985, 1986) and Krigbaum (1985) first suggested that smectic side chain LCPs might have a microphase structure with clusters of mesogenic side groups embedded in a continuous phase formed by the polymeric backbone.

The first examples of side chain liquid crystalline copolysiloxanes containing combined main chain and side chain spacers according to Figure 3.5 (C'), were synthesized by Ringsdorf and Schneller (1981, 1982) by the hydrosilation of poly(methyl-co-dimethyl siloxane)s. In this case, the main chain spacer is an oligodimethyl siloxane segment, while the side chain spacer is paraffinic. Although these authors obtained several polymers exhibiting two glass transition temperatures, the two T_g values were assigned to the two blocky segments (oligomethyl siloxane containing mesogenic side groups and oligodimethyl siloxane) of the 'block' copolymers.

Recently, we have synthesized a large variety of copolysiloxanes with different molecular masses and compositions containing mesogenic side groups and long spacers which exhibit two glass transition temperatures by DSC (Hsu and Percec, 1987a,b,c; Hahn and Percec, 1987; Percec 1988b). We assigned the lower glass transition temperatures to the independent motions of the main chain, and the higher glass transition temperatures to the cooperative (but independent from the main chain) motions of the side groups.

There are, however, two other possible explanations for these two changes in the heat capacity observed on DSC thermograms. First, as proposed by Ringsdorf and Schneller (1981, 1982), the two glass transitions may be a consequence of the blockiness of the copolymer. However, if this was the case, the T_g due to the oligodimethylsiloxane segment should be independent of the nature of the side groups attached to the methylsiloxane segment. On the contrary, we have shown that in both cases T_g depends on the nature of the side groups (Hsu and Percec, 1987b,c; Hahn and Percec, 1987, 1988; Percec, 1988b). In fact, the higher temperature T_g is proportional to the mass fraction of the side groups, as expected for side chain polymers with independent motions of the side chains.

The second alternative is that only one of the changes in the heat capacity may be due to a glass transition, while the other is the result of a secondary β transition. The W.L.F. equation (Ferry, 1980) is frequently used to discriminate between a secondary β transition and a glass transition (Monnerie, 1985). We recently represented the two transitions associated with the changes in the heat capacity in a 'relaxation map' ($\log f$ v. $1/T$). This experiment suggested that they are both glass transitions (Hahn et al., 1988).

Although the complete development of microphase separated side chain liquid crystalline polymers is in a very early stage of development, it certainly deserves further consideration since it offers a viable solution to the synthesis of highly or even completely decoupled liquid crystalline polymers. For instance, real polymer homologues of low-molar-mass liquid crystals, i.e. a fluid in a rigid matrix, should be obtained with a biphasic LCP system in which the glass transition of the backbone is above room temperature, while that of the side groups is below room temperature. If the system is not biphasic, the motions of the mesogens would simply be frozen at temperatures below the T_g of the polymer. Such a biphasic system has potentially very important practical applications.

3.3 Synthesis

3.3.1 Chain polymerization reactions

Monomers containing mesogenic groups are most frequently synthesized as methacrylates, acrylates, acrylamides, chloroacrylates (Zentel and Ringsdorf, 1984), itaconates (Hisgen and Ringsdorf, 1983) and styrene derivatives (Percec et al., 1987a) (Figure 3.6). Therefore, the most convenient method to polymerize mesogenic monomers is by radical initiation. However, in order to obtain different tacticities and molecular masses, and narrow molecular mass distributions, methacrylates, acrylates and styrene derivatives could be suitably polymerized anionically, while methacrylates, acrylates and acrylamides could be polymerized by group transfer polymerization. Although both anionic (Frosini et al., 1981; Hahn et al., 1981) and group transfer polymerizations (Kreuder et al., 1986; Pugh and Percec, 1985) have already been used to polymerize liquid crystalline monomers, living polymerizations were not obtained experimentally.

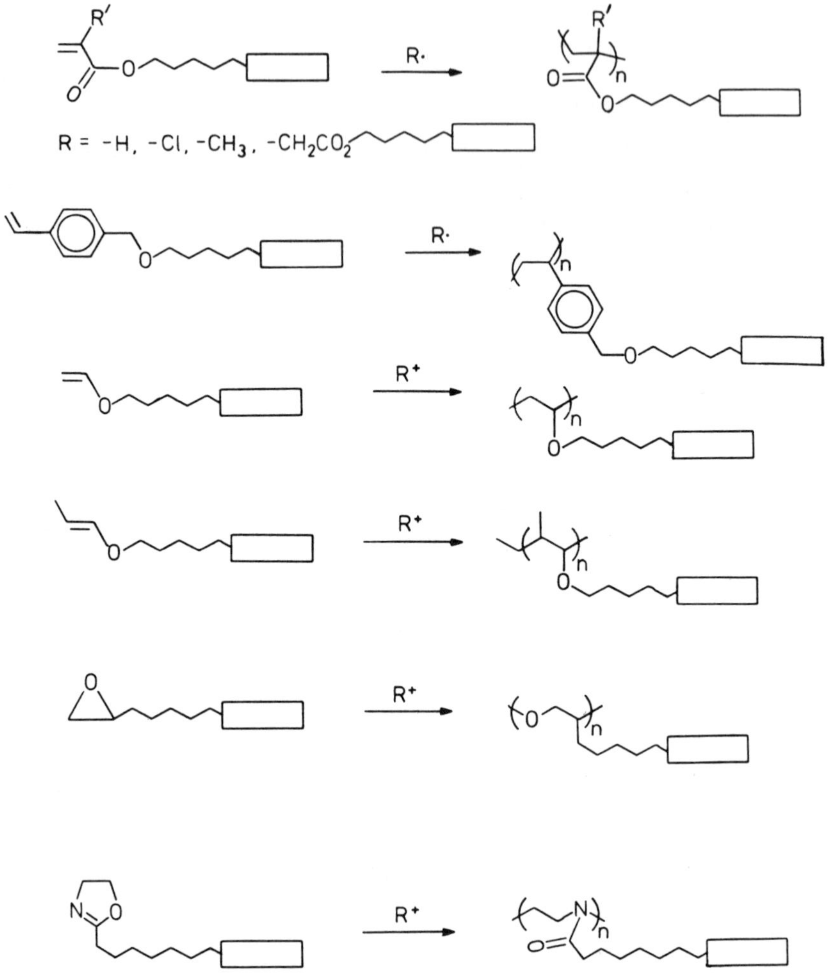

Figure 3.6 Polymerization of some representative classes of monomers containing mesogenic groups.

This is because the monomers contained benzoate esters as either part of the mesogen or at the direct attachment of the mesogen to the polymerizable group. Benzoate esters are too reactive towards nucleophilic displacement at the electrophilic carbonyl to be present in a monomer polymerized anionically or by group transfer polymerization. Although related problems must be considered in cationic polymerizations, a suitable selection of the functional groups in the mesogenic monomer should overcome the difficulties of living polymerizations.

The cationic polymerization of nucleophilic olefins and heterocyclic monomers offers additional synthetic avenues to vary the nature of the polymer backbone and to eventually achieve living polymerizations. Liquid crystalline polymers were recently prepared by the cationic polymerization of monomers containing vinyl and propenyl ethers (Rodriguez-Parada and Percec, 1986; Percec and Tomazos, 1987), cyclic imino ethers (Rodriguez-Parada and Percec, 1987), and oxiranes (Cser *et al.*, 1985) as polymerizable groups (Figure 3.6). However, living polymerizations were not achieved with these systems either. In addition, since cyclic oligomers are also obtained in the cationic polymerization of oxiranes, the cyclic compounds should be carefully monitored and considered in the characterization of polyoxiranes.

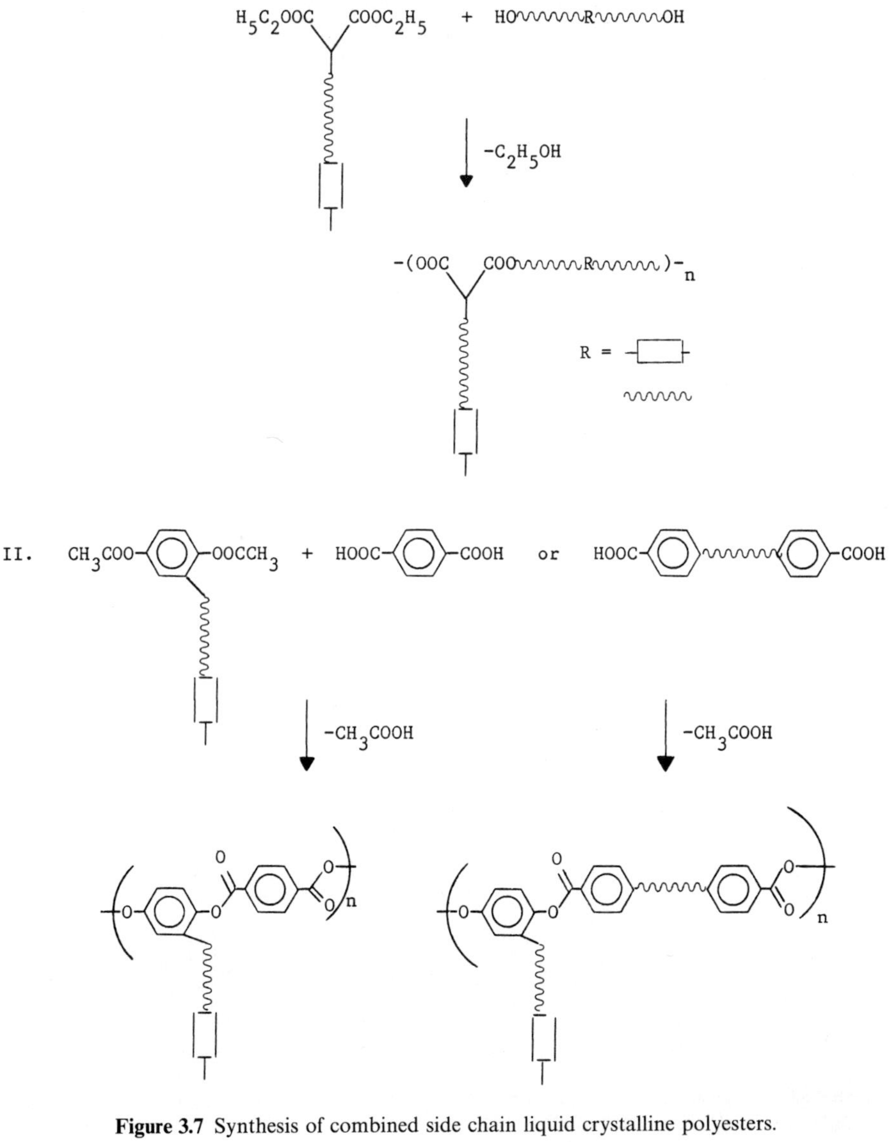

Figure 3.7 Synthesis of combined side chain liquid crystalline polyesters.

3.3.2 Step polymerization reactions

A large variety of side chain liquid crystalline polyesters containing mesogenic side groups were recently synthesized by Ringsdorf and co-workers (Reck and Ringsdorf, 1985; Berg *et al.*, 1986). The architecture of the resulting polymers can be designed to correspond to structures *a*, *b* and *j* in Scheme 3.1, and to any of the structures shown in Scheme 3.2. One synthetic route is to attach the mesogenic side group to a monomer such as a substituted malonate derivative that gives rise to the flexible spacer, and then polyesterifying it with a diol or with a diol containing a mesogenic unit. Alternatively, the mesogenic side groups can be attached to hydroquinone, and then condensed with either a flexible or a rigid diacarboxylic acid. These two routes allow insertion of the pendant mesogenic side groups into either a flexible or a rigid structural unit (Figure 3.7).

A number of synthetic examples based on these systems will be described in section 3.9.

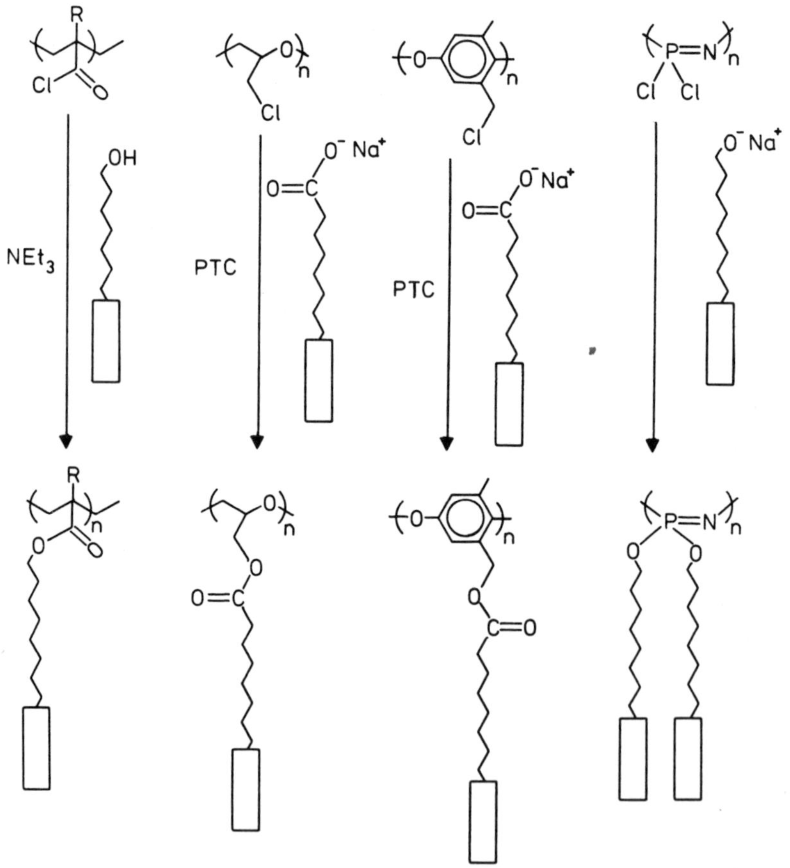

Figure 3.8 Synthesis of some liquid crystalline polymers by polymer homologous reactions (R = H or CH$_3$, PTC = phase transfer catalyst).

3.3.3 *Polymer homologous reactions*

In principle, the reactivity of a functional group should not be altered by attachment to a polymer (Flory, 1939). However, certain effects can either activate or inhibit the reactivity of functional groups (Morawetz, 1987). These special effects include neighbouring group effects, stereoisomerism, different local polarity of the polymer domains from that of the bulk solvent, and the attraction or repulsion of a reagent to/from the polymer. Since such effects often limit the conversion of polymer homologous reactions, they should be considered in attempts to prepare LCPs by such

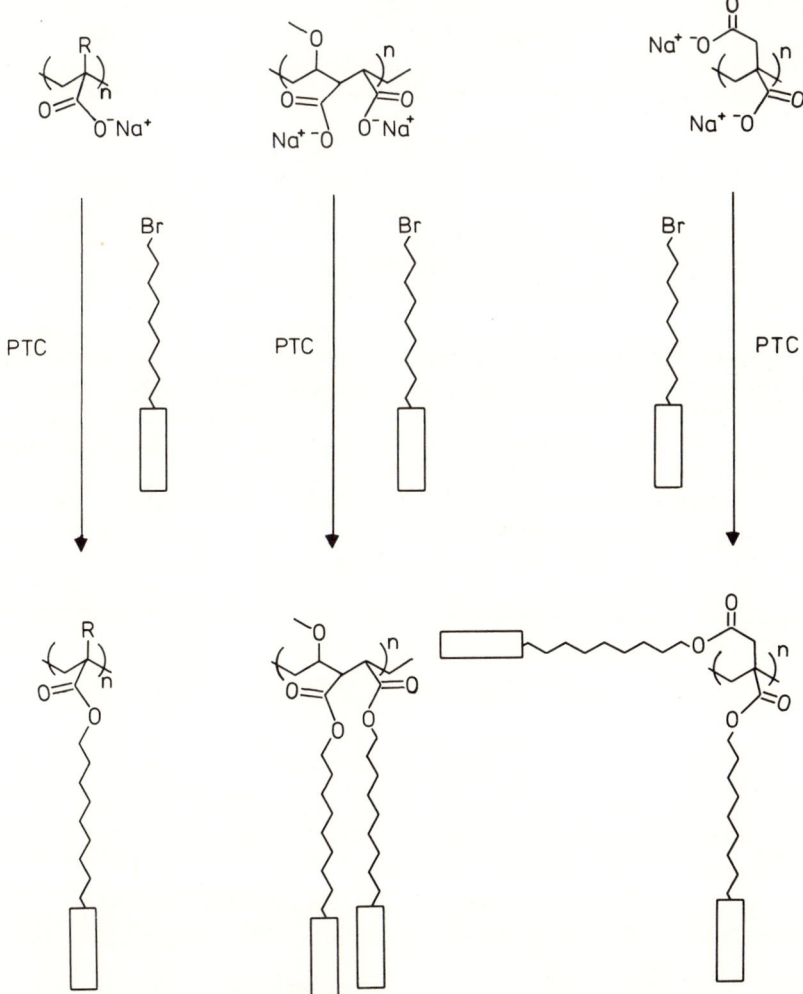

Figure 3.9 Synthesis of polyacrylates (R = H), polymethacrylates (R = CH₃), poly(methylvinylether-co-maleate)s and polyitaconates by phase transfer catalysed (PTC) esterification.

'quantitative' polymer homologous reactions. The two organic reactions frequently used to synthesize side chain liquid crystalline polymers by polymer homologous reactions are nucleophilic displacement and hydrosilation.

Nucleophilic displacement reactions can be further divided into two categories. The first consists of reactions performed on a polymer backbone containing electrophilic side groups which are displaced by mesogenic units containing a nucleophile at the opposite end of the flexible spacer. As shown schematically in Figure 3.8, this includes the displacement of halides from polyacryloyl- or polymethacryloylchloride (Paleos *et al.*, 1981, 1982), poly(epichlorohydrin) (Pugh and Percec, 1986*a*, 1987), radically brominated poly(2,6,-dimethyl-1,4-phenylene oxide) (Pugh and Percec, 1986*b*,1987) and poly(dichlorophosphazene) (Singler *et al.*, 1987; Kim and Allcock 1987). Alternatively, polyacrylates (Keller, 1984), polymethacrylates (Keller, 1985*b*), poly(methylvinylether-co-maleate)s (Keller, 1985*a*), and polyitaconates (Keller, 1985*c*) can be prepared by displacing bromine from bromoalkyl containing mesogens with polymeric nucleophiles as illustrated schematically in Figure 3.9.

As suggested by Keller, these reactions can be performed by either liquid–liquid (Keller, 1984, 1985*a*, *b*, *c*) or solid–liquid (Keller, 1987) phase transfer catalysed reactions. The liquid–liquid (organic solvent—aqueous solution of the sodium salt of the polyacid) phase transfer catalysed reactions described in Figure 3.9 are, however, susceptible to side reactions, particularly when bromoalkyl esters as used in Keller's experiments are present in the bromoalkyl mesogen. Since polyacrylic, polymethacrylic and polyitaconic acids are weak acids, the aqueous solution of their sodium salts are basic enough to cleave an aromatic–aliphatic ester unit and give rise to a number of side reactions such as that illustrated in Figure 3.10 (Chen and Maa, 1988). In this case, the aromatic–aliphatic ester is activated by the aromatic–aromatic ester in the *p*-position, as is typical of many mesogens. Under less than proper reaction conditions, this side reaction may become the main reaction. Therefore, when polymer homologous phase transfer catalysed esterifications are used, we recommend that solid–liquid (sodium salt–organic solvent) reaction conditions be used, and that the mesogen does not contain ester groups (Pugh and Percec, 1986*a*, *b*, 1987). Additional discussions of

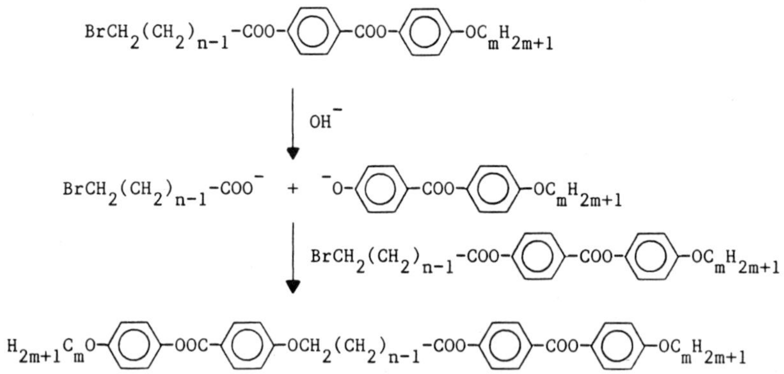

Figure 3.10 Side reactions resulting from the hydrolysis of an ester-based mesogen by hydroxide present in liquid–liquid PTC reactions (Chen and Maa, 1988).

the principles of phase transfer catalysed reactions should also be consulted by experimentalists willing to use this synthetic procedure (Mathias and Carraher, 1984; Percec, 1987).

Lastly, one of the most frequently used polymer homologous reaction is hydrosilation. Hydrosilations are difficult to perform to completion, and are accompanied by a number of side reactions (Auman et al., 1987a, b; Gray et al., 1986; Nestor et al., 1987). Typical procedures used in several laboratories that confront these problems are detailed in the literature (Gray et al., 1986; Hsu and Percec, 1987c; Hahn and Percec, 1987; Nestor et al., 1987). The detailed synthesis of liquid crystalline polysiloxanes and copolysiloxanes by hydrosilation reactions is treated in Chapter 4 of this book.

Nevertheless, it must be mentioned that in most cases, the purities of the polymers reported in the literature are not determined by gel permeation chromatography (GPC), high pressure liquid chromatography (HPLC), or thin layer chromatography (TLC). Since none of the chain polymerization reactions or polymer homologous reactions discussed here is quantitative, a double precipitation of the polymer in a non-solvent such as methanol leaves, in many cases, a large amount of unreacted monomer in the polymer sample (Nestor et al., 1987). At the same time, condensation polymerizations and cationic ring opening polymerizations of oxiranes produce cyclic compounds (Percec and Pugh, 1987) which cannot easily be removed at the end of the polymerization. Therefore, we recommend that both the purity of the final products and at least their relative molecular masses be determined by HPLC or TLC, and GPC respectively.

3.4 Polymers containing rigid rod-like mesogens

3.4.1 Linear polymers without flexible spacers

In order for a polymer containing mesogenic side groups attached to it without a flexible spacer to form a mesophase, the polymer backbone must be significantly distorted from its normal random-coil conformation. At the same time, the backbone sterically hinders the packing of the mesogens. Because of this, most polymers with the mesogenic groups directly attached to the backbone are amorphous (Shibaev and Platé, 1977, 1984). Nevertheless, when the side chain ordering is sufficiently strong to overcome the normal barriers associated with the random-coil conformation of the backbone, the polymer should exhibit liquid crystalline mesomorphism.

Table 3.1 presents some of the most representative examples of LC polymers containing biphenyl side groups. These examples are very unusual, since the non-substituted biphenyl unit itself exhibits very weak mesogenic character (Gray, 1962, 1979, 1982). In addition, it is remarkable that only small changes in the architecture of the polymer containing biphenyl side groups can dramatically affect its properties. While poly(4-biphenyl acrylate), poly(4-cyclohexylphenyl acrylate) and poly[N-(4-biphenyl)acrylamide] exhibit mesomorphic properties, changes from poly(4-biphenyl acrylate) to poly(4-biphenyl methacrylate) or to poly(vinyl-4-phenyl benzoate) lead to amorphous polymers. Even the insertion of a flexible spacer does not lead to a LC polymer. It seems that the addition of a methyl group to each structural unit of the polyacrylate backbone creates a higher conformational barrier to mesogen packing. However, the fact that poly(4-biphenyl acrylate) is liquid crystalline, while the corresponding poly(vinyl-4-phenyl benzoate) is amorphous, has not yet been ex-

Table 3.1 Representative polymers containing biphenyl side groups

Polymer	Phase transitions* (°C)		Reference
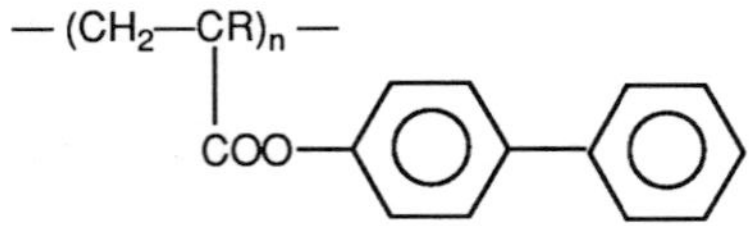 $-(CH_2-CR)_n-$ COO- [biphenyl]	R = H	g 110 S 280 i	Bresci et al. (1980) Baccaredda et al. (1971)
$-(CH_2-CH)_n-$ COO- [cyclohexyl-phenyl]	R = CH₃	g 170 i	Magagnini et al. (1974)
		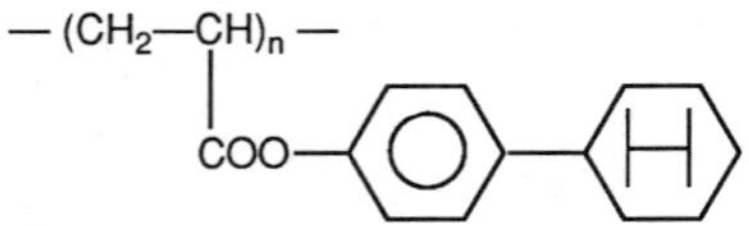 g 101 S 205 i	Magagnini et al. (1974)
$-(CH_2-CH)_n-$ CONH- [biphenyl]		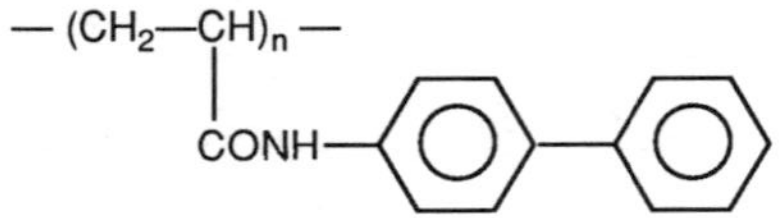 S 325 i	Lupinacci et al. (1980)
$-(CH_2-CH)_n-$ OOC- [biphenyl]		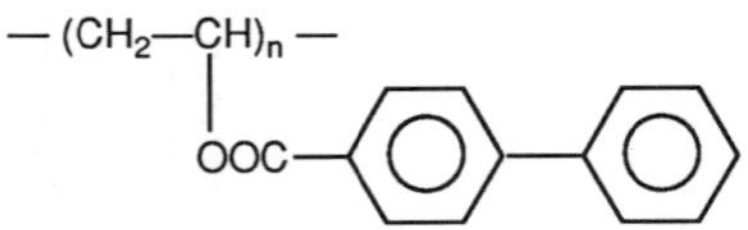 g 114 i	Magagnini et al. (1974)
$-(CH_2-CH)_n-$ COOCH₂- [biphenyl]		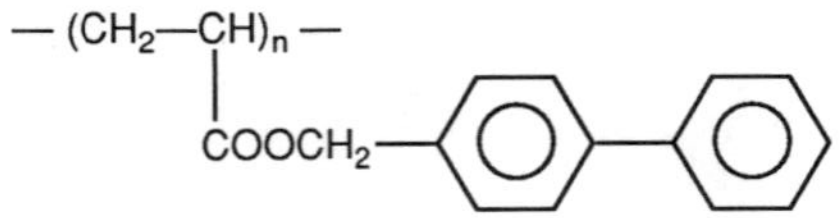 g 50 i	Magagnini et al. (1974)
$-(CH_2-CR)_n-$ COO(CH₂)ₘ-O- [biphenyl]	R = H, m = 2 R = H, m = 6 R = CH₃, m = 6	g 80 i g 48 K 57 i g 50 i	Bresci et al. (1980) Bresci et al. (1980) Finkelmann et al. (1978)

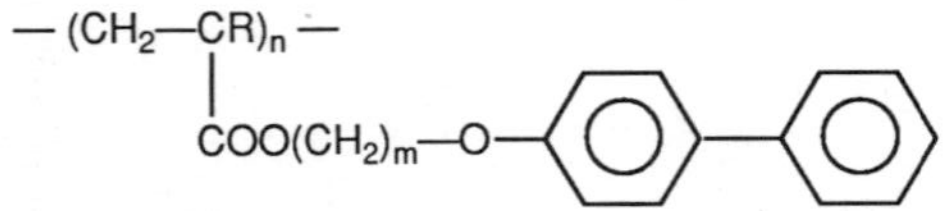

*Abbreviations: g, glass; S, smectic; i, isotropic; K, crystal

plained. We speculate that the different placement of the ester group attached to the 4-position of the biphenyl changes the size of the mesogen as outlined below:

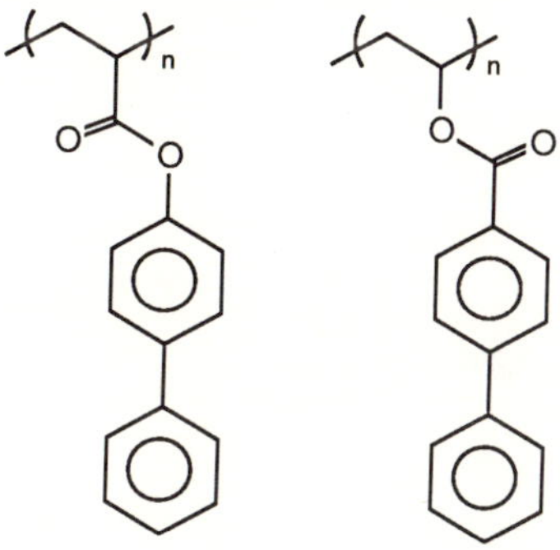

The phenolate arrangement probably induces double-bond character to the $\overset{O}{\overset{\|}{C}}-O$ linkage, forming an additional quasi-phenyl or quasi-2, 6-disubstituted naphthalene unit and thus increasing the length of the mesogen and restricting the rotation of the polyacrylate about its $\overset{O}{\overset{\|}{C}}-O$ bond. In contrast, there is free rotation about the polyvinylbenzoate $\overset{O}{\overset{\|}{C}}-O$ bond, and the elongated character of this mesogenic group is not enlarged. This idea is somewhat supported by the similar difference in the phase transitions of main chain thermotropic polyesters containing the ester group in the same two different positions (Ober et al., 1984); we believe the same explanation may be applicable here. Consequently, although all the mesogenic units presented in Table 3.1 at first glance appear identical, they actually are not.

Table 3.2 presents several additional examples of LCPs containing mesogenic groups attached directly to the backbone without flexible spacers. All, however, contain more effective mesogens than those of the polymers in Table 3.1, and subsequently, most polymers with the same mesogen exhibit mesomorphic properties indifferent of the nature of the backbone. X-ray diffraction studies of these polymers suggested that the polymer backbones lie in essentially parallel planes within perpendicular smectic-like layers of the directly attached and rigid anisotropic side chain substituents (Alimoglu et al., 1984). These examples demonstrate that a proper combination of side groups and main chain lead to LCPs without using a flexible spacer.

The influence of polymer tacticity on the phase transitions of poly(4-biphenyl acrylate) is consistent with the idea that the normal conformation of a polymer backbone must be distorted to achieve liquid crystallinity (Frosini et al., 1981; Magagnini, 1981). The results in Table 3.3 demonstrate that the mesophase of the atactic polymer has a larger thermal stability than that of the isotactic polymer. Therefore, although both polymers give rise to the same type of mesophase, a stereoregular (e.g. isotactic) backbone decreases the thermal stability of the mesophase.

C

Table 3.2 Representative polymers with direct attachment of the mesogen to the polymer backbone

Polymer	Phase transitions (°C)			Reference
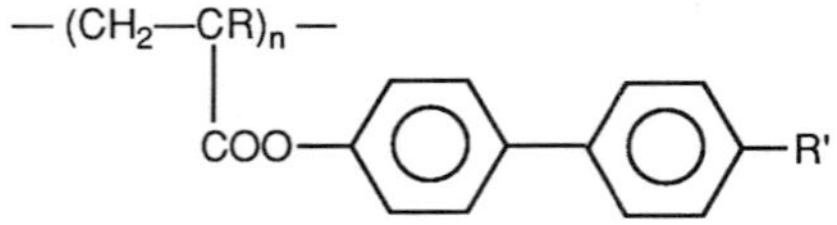	$R = H$,	$R' = CN$	S 270 i	Alimoglu et al. (1984)
	$R = CH_3$,	$R' = CN$	S 240 i	Alimoglu et al. (1984)
	$R = H$,	$R' = C_5H_{11}$	S 303 i	Alimoglu et al. (1984)
	$R = CH_3$,	$R' = C_5H_{11}$	S 232 i	Alimoglu et al. (1984)
	$R = CH_3$,	$R' = OCH_3$	S 255 i	Duran et al. (1987a)
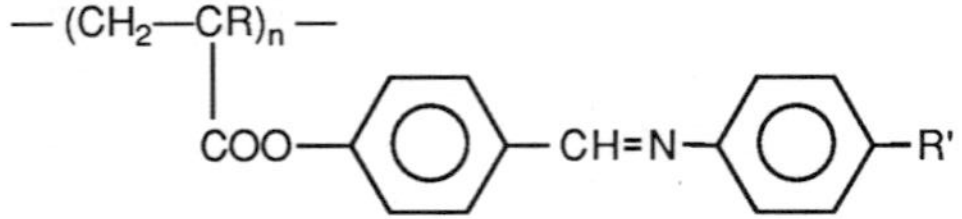	$R = H$,	$R' = H$	g 93 S 258 i	Frosini et al. (1981)
	$R = CH_3$,	$R' = H$	g 163 i	Frosini et al. (1981)
	$R = H$,	$R' = COOH$	K 201 S 226 i	Clough et al. (1977)
	$R = CH_3$,	$R' = COOH$	K_1 182 K_2 201 S 205 i	Blumstein et al. (1975a)
	$R = H$,	$R' = OC_2H_5$	g 88 S 268 i	Frosini et al. (1981)
	$R = CH_3$,	$R' = OC_2H_5$	g 198 S 284 i	Frosini et al. (1981)
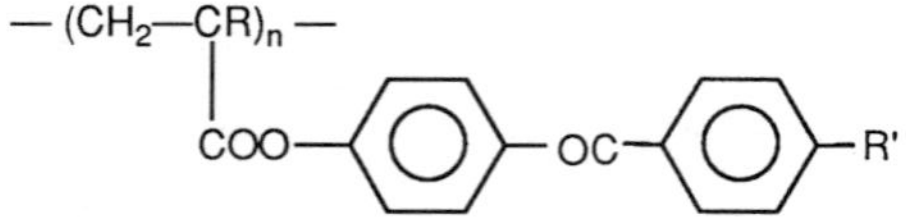	$R = CH_3$	$R' = OC_4H_9$	g 180 S 254 i	Konstantinov et al. (1983)
	$R = CH_3$,	$R' = OC_9H_{19}$	g 140 S 209 i	Konstantinov et al. (1983)
	$R = CH_3$,	$R' = OC_{12}H_{25}$	g 130 S 208 i	Konstantinov et al. (1983)
	$R = H$,	$R' = OC_{16}H_{33}$	g 100 S 200 i	Konstantinov et al. (1983)
	$R = CH_3$	$R' = OC_{16}H_{33}$	g 160 S 213 i	Konstantinov et al. (1983)
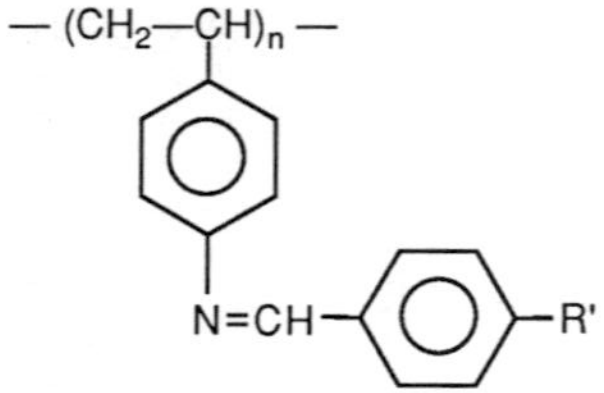		$R' = OC_4H_9$	K 88 N 121	Blumstein et al. (1975a)
		$R' = OC_6H_{13}$	K 94 S 98 N 116 i	Blumstein et al. (1975a)
		$R' = CN$	K 114 N 140 i	Clough et al. (1977)

Table 3.3 The influence of tacticity on the phase transitions of poly(4-biphenylacrylate) (Frosini *et al.*, 1981; Magagnini, 1981).

$$— (CH_2—CH)_n —$$
$$|$$
$$COO—⬡—⬡$$

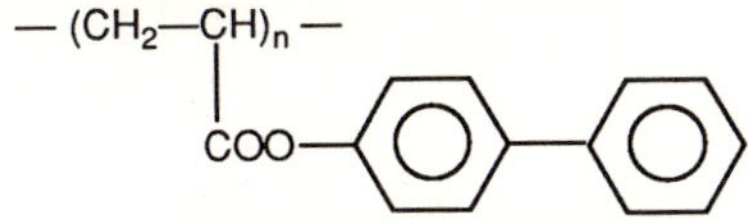

Tacticity	Phase transitions (°C)
Atactic	g 110 S 280 i
Isotactic	g 110 S 233 i

Table 3.4 The influence of molecular mass on the phase transitions of atactic poly(4-biphenylacrylate) (Frosini *et al.*, 1981)

$\overline{M}_n$	$\overline{DP}$	T_i(°C)	ΔH_i(J/g)
12 500	56	235	9.0
24 800	111	260	18.1
91 200	407	271	24.4

Only one experiment is available in the literature concerning the influence of molecular mass on the mesomorphic behavior of a polymer with the mesogens directly attached to it (Table 3.4) (Frosini *et al.*, 1981; Magagnini, 1981). Although this experiment was performed with only three different molecular masses, it clearly shows that there is a sharp increase in both the isotropization temperature and the change in enthalpy of isotropization up to a polymerization degree (*n*) of about 100. Above this degree of polymerization, the increase of both values is much less.

3.4.2 *Linear polymers with mesogens attached laterally without flexible spacers*

The previous section discussed the behavior of side chain LCPs containing mesogenic units attached to the polymer backbone in a longitudinal, or normal way. Alternatively, the mesogen, can be attached parallel to the backbone as in Figure 3.11, in which case substitution of the mesogen must be lateral. Figure 3.11 also shows the single example of a polymer with the mesogen attached laterally through one methylenic unit; it can be considered to be a polymer without a spacer (Zhou *et al.*, 1987). This polymer exhibits a nematic mesophase, and as speculated, has the potential of exhibiting a fully extended conformation due to the proposed 'mesogen-jacket' surrounding it. Although this model is interesting, further work is required to support it.

3.4.3 *Linear polymers with paraffinic flexible spacers*

3.4.3.1 *Influence of molecular mass on phase transitions.* It is difficult to systematically discuss the influence of the architecture of a polymer on its phase transitions before

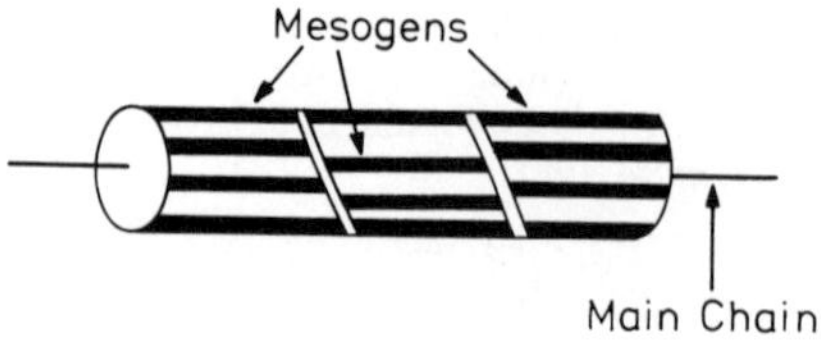

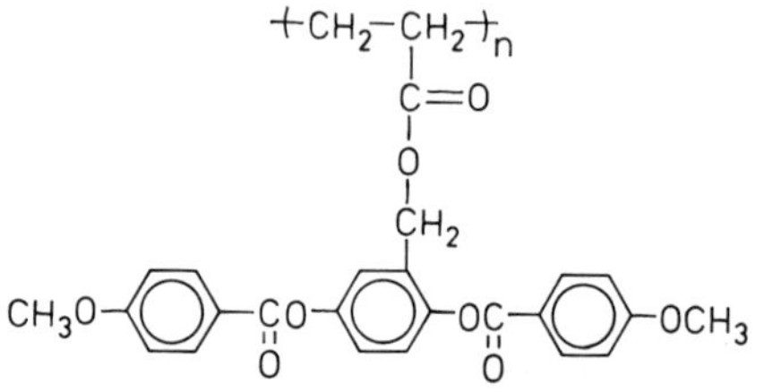

Figure 3.11 Speculative model showing a 'mesogen-jacket' around a polymer main chain (top) (Zhou et al., 1987).

having sufficient knowledge of the influence of molecular weight on phase transitions. A molecular mass study is presented in Table 3.5 for a nematic polyacrylate (Portugall et al., 1982a). Glass transition, isotropization temperature and the change in enthalpy at isotropization all increase up to a degree of polymerization (n) of 114, which is the highest value available. Unfortunately, this seems to be the only report on the influence of molecular mass on the thermodynamic parameters of SC-LCPs with a flexible spacer, and sufficient experimental evidence is not available to confirm the observed trend in the isotropization enthalpy.

A more comprehensive set of molecular mass results is illustrated in Figure 3.12 (Kostromin et al., 1982). However, with the exception of one data point, the thermal transitions are plotted only as a function of intrinsic viscosity (a relative indication v. direct measurement of molecular mass), and the general trend of different polymer

Table 3.5 Influence of molecular mass on the phase transitions of a LC polyacrylate (Portugall et al., 1982)

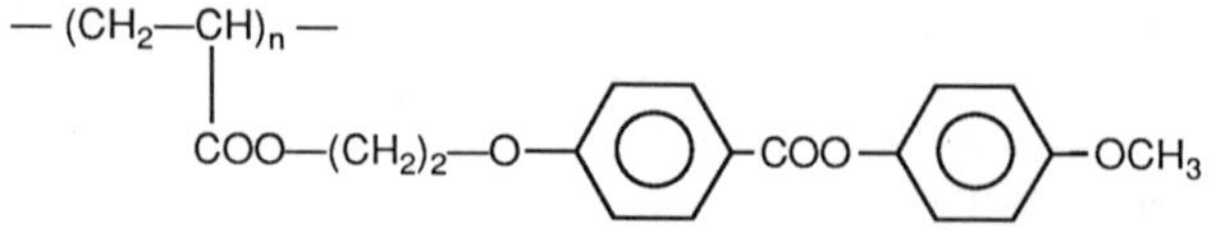

M_n	n	Phase transitions (°C)	ΔH_i(kJ/mol)
4500	13	g 53 N 100 i	0.6
14 000	41	g 59 N 114 i	0.6
39 000	114	g 62 N 116 i	0.9

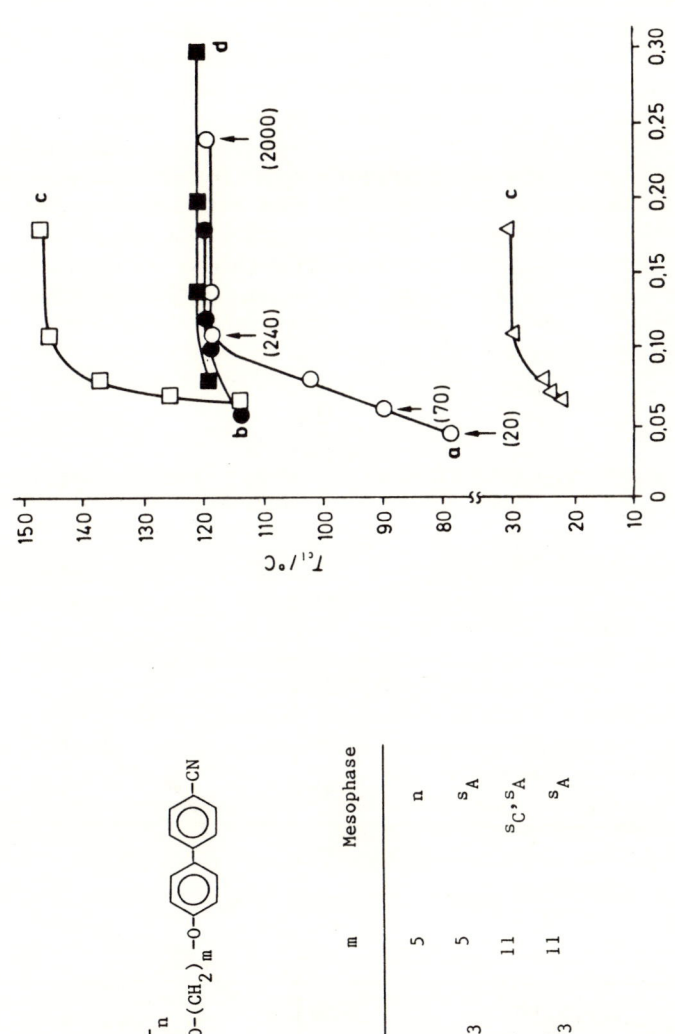

Figure 3.12 Isotropization (○ ● □) and $s_C \to s_A$ (△) transition temperatures of 4-cyanobiphenyl-based mesogenic acrylates and methacrylates as a function of intrinsic viscosity (Kostromin *et al.*, 1982*b*). (Numbers in parentheses are the weight average degrees of polymerization.)

backbones can therefore not be quantitatively compared. There seems to be a sharp increase in isotropization temperatures up to mass average degrees of polymerization of about 240. This value is surprisingly high when compared with any other class of oligomers (Percec and Pugh, 1987), especially for polymers with narrow molecular mass distribution. Above this degree of polymerization, all phase transitions are independent of molecular mass. Apparently polyacrylates, which in this case are either nematic or smectic, exhibit a greater dependence on molecular mass than the corresponding polymethacrylates. In the case of the polymers containing eleven methylenic units in the flexible spacer, the polyacrylate isotropization temperatures are higher than those of the corresponding polymethacrylate; the isotropization transitions of the two classes of polymers are quite similar when the flexible spacer contains only five methylene units. Unfortunately, no thermodynamic data were reported.

The most comprehensive study to date concerning the influence of molecular mass on phase transitions was performed on fractionated samples of polysiloxanes (Stevens et al., 1984). Figures 3.13 and 3.14 present two representative sets of results. The

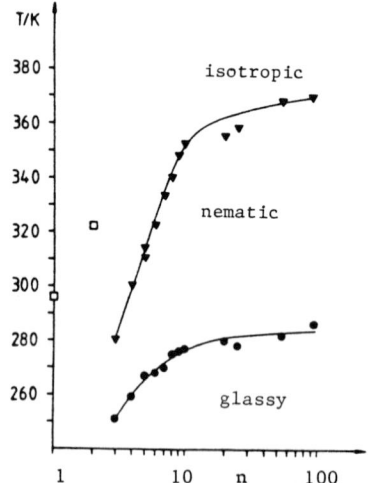

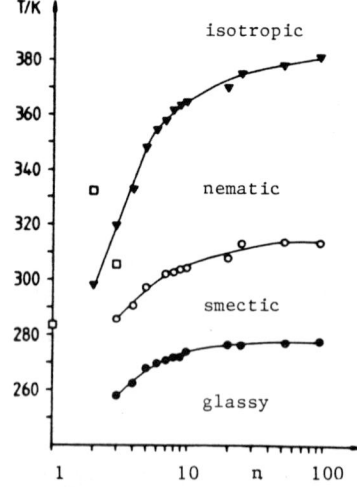

Figure 3.13 Transition temperatures of a mesogenic polysiloxane with four methylenic units in the spacer as a function of molecular weight (□ crystalline to isotropic transition) (Stevens *et al.*, 1984).

Figure 3.14 Transition temperatures of a mesogenic polysiloxane with six methylenic units in the spacer as a function of molecular weight (□ crystalline to isotropic transiton) (Stevens *et al.*, 1984).

tendency towards side chain crystallization apparently decreases in going from monomer to dimer, to trimer and so on, and vanishes for higher-molecular-mass oligomers and polymers. Although there are not enough experiments to definitively support this trend, the decrease in side chain melting temperature with increasing polymer molecular mass has been theoretically predicted (Rim, 1986). In contrast to side chain melting transitions, there is in all cases an initial sharp increase in the isotropization, smectic–nematic and glass transition temperatures. The increase drastically decelerates at approximately $n = 10$. The generally accepted explanation for this strong dependence is that an increase in polymer molecular mass creates a denser packing of the mesogens and therefore decreases the specific volume at the phase transition. Nevertheless, there is no definitive concept to date as to whether the degree of polymerization should effect the phase transition enthalpy.

In conclusion, the phase transitions of different polymers are always influenced by molecular mass, and this dependence must therefore first be established for each polymer under consideration. Although many polymers in the literature are of unknown molecular mass, we will try in the following sections to select for discussion only those polymers which we feel have high enough molecular mass that the phase transitions are somewhat independent of it.

3.4.3.2 *Influence of tacticity on phase transitions.* Table 3.6 presents the only data available in the literature concerning the influence of polymer tacticity on the phase transitions of SC–LCPs with a flexible spacer (Hahn *et al.*, 1981). In contrast to poly(4-biphenyl acrylate) with no spacer, the melting temperature of the isotactic polymer is higher than that of the atactic polymer, while the isotropization temperature is only slightly higher for the isotactic polymer. However, consistent with poly(4-biphenyl acrylate) and as expected, the thermal stability of the isotactic polymer's mesophase is therefore depressed. Evidently, polymer tacticity influences the crystalline phase more than liquid crystalline phases. This is an important result, since it was recently demonstrated that the tacticity of side chain LCPs obtained by radical polymerization is dependent on the spacer length (Duran and Gramain, 1987).

3.4.3.3 *Influence of spacer length on phase transitions.* For the same mesogenic unit and polymer backbone, the length of the flexible spacer influences both the nature of the

Table 3.6 Influence of tacticity on the phase transitions of poly[6-(4′-methoxy-4-biphenyloxy)hexyl]methacrylate (Hahn *et al.*, 1981)

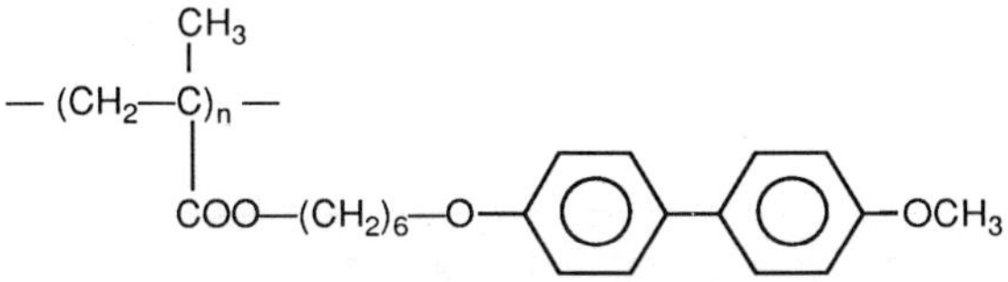

Tacticity	Phase transitions (°C)
Atactic	K 117 S 127–131 i
Isotactic	K 131 S 135 i

Table 3.7 Influence of spacer length on the phase transitions of poly[ω-(4'-methoxy-4-biphenyloxy)alkanoyl] methacrylates

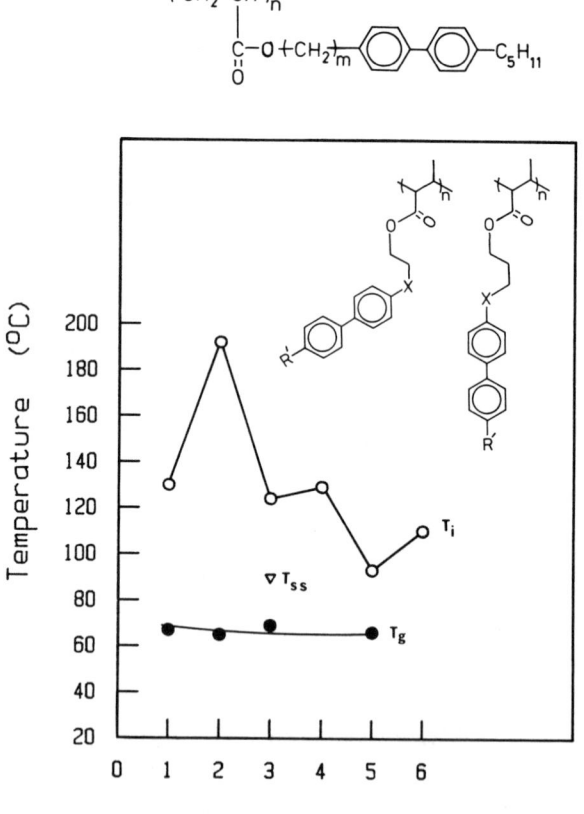

Spacer length m	Phase transitions (°C)	Reference
0	S 255 i	Duran *et al.* (1987*b*)
2	g 120 N 152 i	Finkelmann *et al.* (1978*a*)
6	K 119 S 136 i	Finkelmann *et al.* (1978*a*)
11	g 54 S$_c$ 87 S$_A$ 142 i	Hsu *et al.* (1987*a*)

Figure 3.15 Transition temperatures of poly[ω-(4'-*n*-pentyl-4-biphenyl)alkanoyl] acrylates as a function of spacer length m ($\nabla T_{s \to s_A}$) (data from Gemmell *et al.*, 1985*b*).

mesophase(s) and the thermal transitions. Table 3.7 summarizes selected examples of polymethacrylates containing 4-methoxy biphenyl as the mesogen and different spacer lengths. The polymer with no spacer exhibits a smectic mesophase, that with the shortest spacer displays a nematic mesophase, and the polymers with longer spacers again exhibit smectic mesophases. As is true for low-molar-mass LCs, the tendency towards smectic mesomorphism increases with increasing spacer length. At the same time, most of the polymers without spacers exhibit smectic mesomorphism (Tables 3.1, 3.2, 3.7). With the insertion of a flexible spacer, the isotropization temperature first decreases, and then increases with long spacers.

Figure 3.15 presents an additional set of data illustrating the influence of spacer length on phase transitions (Gemmell *et al.*, 1985*b*). Although the glass transition is essentially constant as a function of spacer length in this case (probably as a result of similar molecular masses regardless of spacer length), the isotropization temperatures clearly show an odd–even effect. The odd–even effect can be explained if we consider the placement of the mesogenic groups in relation to the polymer backbone, as shown schematically in Figure 3.15. When the interconnecting group between the spacer and mesogen is considered part of the flexible spacer, a spacer containing an even number of atoms places the mesogen essentially perpendicular to the polymer chain in the case of

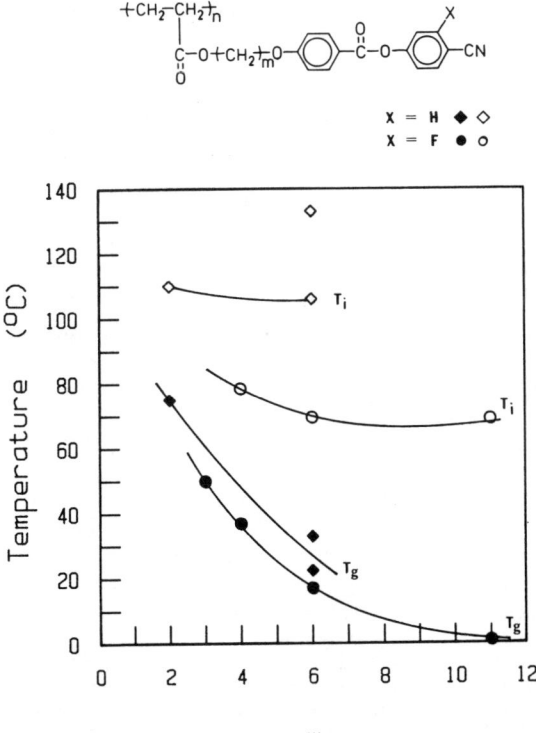

Figure 3.16 Transition temperatures of a laterally substituted liquid crystalline mesogenic polyacrylate as a function of spacer length *m* (data from Portugall *et al.*, 1982; Le Barny *et al.*, 1986*b*).

polymethacrylates and polyacrylates. Therefore, the formation of a smectic layer is easier when the spacer contains an even number of atoms than an odd number of atoms. This is because the mesogen is tilted in relation to the backbone in the latter situation. Consequently, an even number of atoms in the spacer should give higher transition temperatures. This situation can be reversed depending on the interconnecting group. Eventually the odd–even effect should vanish with increasing spacer length.

Figure 3.16 summarizes a series of polymers which demonstrate the influence of spacer length on both the glass transition and isotropization temperatures, and the influence of different side groups attached to the mesogen on the polymer's phase transitions. The glass transition temperature decreases continually with increasing spacer length, while the isotropization temperature first decreases then increases. The insertion of a side group into the mesogenic unit decreases the thermal stability of the mesophase.

The phase transitions of a series of polyacrylates and polymethacrylates containing different spacer lengths and different alkoxy groups in the *para* position of the mesogen are collected in Table 3.8. Unfortunately, the molecular masses of many of these

Table 3.8 Influence of spacer length and para-substituent on the phase transitions of mesogenic polyacrylates, polymethacrylates and poly(chloroacrylates)

$$- (CH_2-CR)_n -$$
$$COO-(CH_2)_m-O-\langle O \rangle-COO-\langle O \rangle-R'$$

m	R	R'	Phase transitions (°c)	Reference
0	H	H	g 80 K 180 S 228 i	Horvath *et al.* (1985b)
0	H	OCH$_3$	g 110 K 180 S 296 i	Horvath *et al.* (1985b)
0	H	OC$_4$H$_9$	g 120 K 180 S 321 i	Horvath *et al.* (1985b)
2	H	H	g 10 K 45 S 72 i	Horvath *et al.* (1985b)
2	H	OCH$_3$	g 25 K 55 S 116 i	Horvath *et al.* (1985b)
2	H	OCH$_3$	g 62 N 116 i	Portugall *et al.* (1982)
2	CH$_3$	OCH$_3$	g 101 N 121 i	Finkelmann *et al.* (1978c)
2	CH$_3$	OC$_3$H$_7$	g 120 S 129 i	Finkelmann *et al.* (1978c)
2	H	OC$_4$H$_9$	g 30 K 64 S 119 S 154 i	Horvath *et al.* (1985b)
2	CH$_3$	OC$_4$H$_9$	g 137 S$_A$ 163 i	Zentel and Ringsdorf (1984)
2	Cl	OC$_4$H$_9$	g 132 S 175 i	Zentel and Ringsdorf (1984)
2	H	OC$_6$H$_{13}$	g 78 S 188 i	Portugall *et al.* (1982)
2	CH$_3$	OC$_6$H$_{13}$	g 100 S 140 i	Finkelmann *et al.* (1987c)
6	H	H	g −8 K 4 S 23 s 39 i	Horvath *et al.* (1985b)
6	H	OCH$_3$	g 5 K 20 S 86 S 104–118 i	Horvath *et al.* (1985b)
6	H	OCH$_3$	g 35 S 97 N 123 i	Portugall *et al.* (1982)
6	CH$_3$	OCH$_3$	g 42 N 107 i	Kreuder *et al.* (1986)
6	CH$_3$	OCH$_3$	g 95 N 105 i	Finkelmann *et al.* (1978c)
6	H	OC$_4$H$_9$	g 5 K 30 S 103 N 114 i	Horvath *et al.* (1985a)
6	CH$_3$	OC$_4$H$_9$	g 39 S$_A$ 109 N 114 i	Zentel and Ringsdorf (1984)
6	Cl	OC$_4$H$_9$	g 39 S$_C$ 110 N 117 i	Zentel and Ringsdorf (1984)
6	H	OC$_6$H$_{13}$	g 28 S 130 i	Portugall *et al.* (1982)
6	CH$_3$	OC$_6$H$_{13}$	g 60 S 115 i	Finkelmann *et al.* (1978c)

Table 3.9 Influence of mesogen length on the phase transitions of mesogenic polysiloxanes (Apfel *et al.*, 1985)

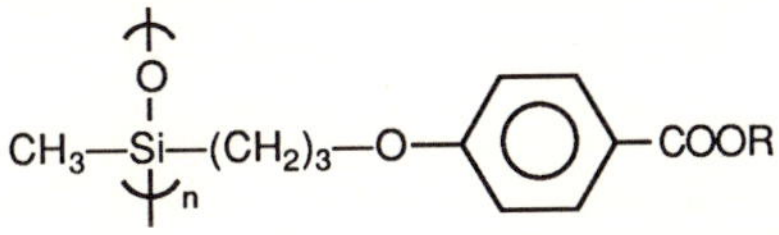

R	Phase transitions (°C)
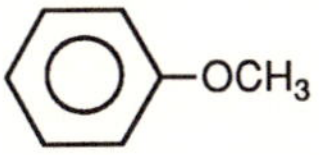 —OCH$_3$	g 15 N 61 i
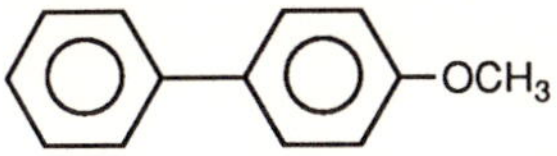 —OCH$_3$	K 139 N 319 i
—OCH$_3$	K 200 N 360 i
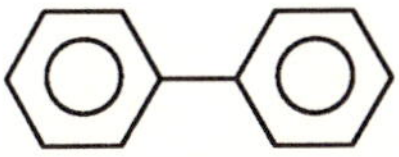	K 95 N 125 i
	K 200 N 360 i

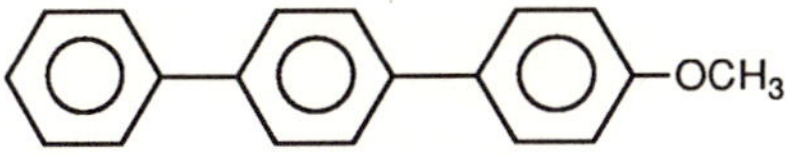

polymers were not determined, and the trends in the different phase transitions of the various backbones can therefore not be quantitatively compared.

3.4.3.4 *Influence of the mesogen length on phase transitions.* As presented in Table 3.9 for two different homologous series of mesogens, an increase in the mesogen length leads to an enhancement of the thermal stability of the mesophase (Apfel *et al.*, 1985; see also Chapter 14). At the same time, even if the highest-temperature mesophase is the same, more ordered phases (either crystalline or liquid crystalline) occur with increasing mesogen length.

An alternative situation concerning the influence of mesogen length on phase transitions is illustrated in Scheme 3.4 (Gemmell *et al.*, 1985*b*). Here the mesogenic unit contains an alkyl group in the 4′-position of the biphenyl unit. As in the case of low-molecular-mass LCs, the odd or even number of atoms in the 4′-substituent should cause the phase transitions to exhibit an odd–even behaviour. As illustrated in Scheme 3.4, when the methyl group at the end of the substituent is not counted, an 'even' number of atoms places the terminal methyl group 'on-axis,' enhancing the anisotropy of the molecular polarizability and the linear character of the mesogen. Therefore, the phase transitions of the even-number substituent should be higher in both temperature and order than those of the compound containing an 'odd' number of atoms. The latter arrangement in which the terminal group is 'off-axis' and therefore non-linear with the mesogen is more difficult to pack in a parallel arrangement needed

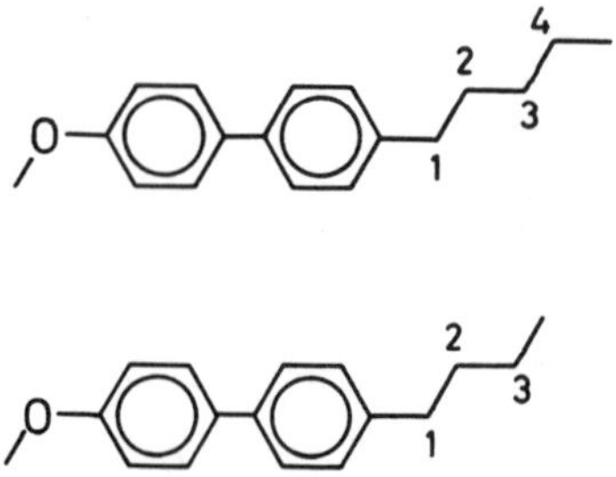

Scheme 3.4 Conformation of odd and even spacers in 4, 4′-disubstituted biphenyl compounds.

to form a LC phase and, at the same time, exhibits a decreased anisotropy of the molecular polarizability.

3.4.3.5 *Influence of the constitutional isomerism of the mesogen on phase transitions.* Figure 3.17 presents the phase transitions of two polysiloxane series containing different spacer lengths and opposite placements of the ester unit in the

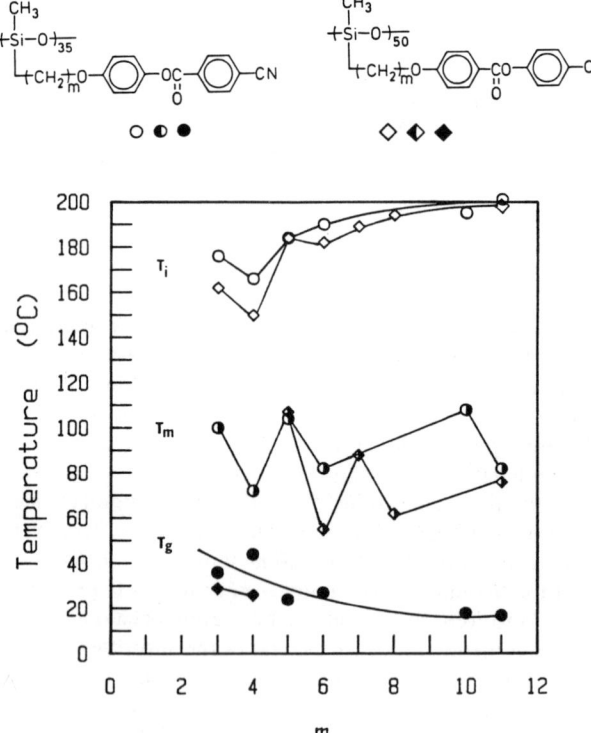

Figure 3.17 Glass transition, melting and isotropization temperatures of polysiloxanes containing isomeric phenyl-benzoate-based mesogens as a function of spacer length m (data from Mauzac *et al.*, 1986; Gemmell *et al.*, 1985a).

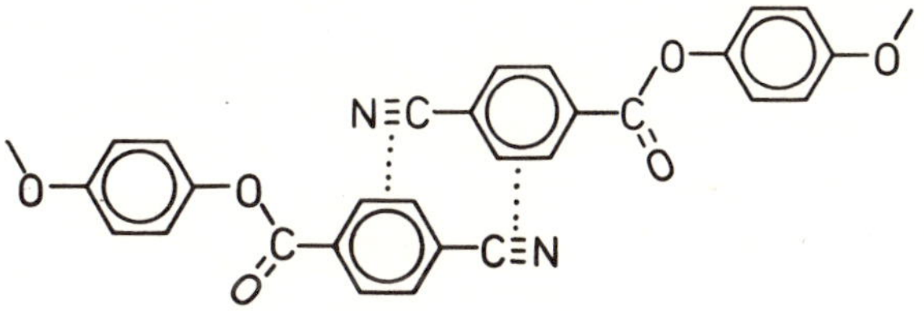

I

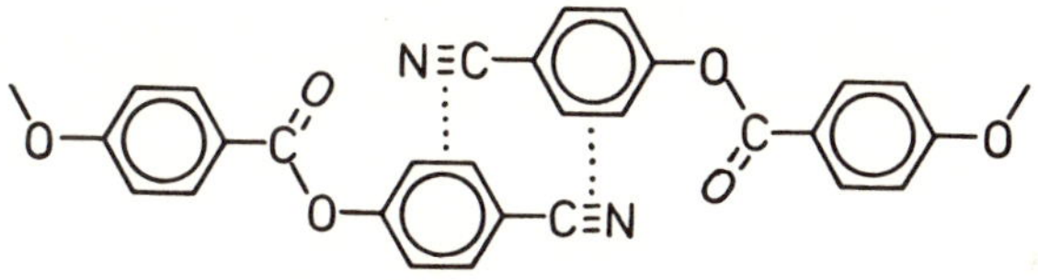

II

Scheme 3.5 Interaction and interdigitation of cyanobenzoyl ester- and cyanophenyl ester-based mesogens.

phenylbenzoate-based mesogen (Mauzac *et al.*, 1986; Gemmell *et al.*, 1985*a*). **I** designates the polymer that contains the cyano benzoyl ester unit, and **II** designates the polymer that contains the cyano phenyl ester. Although the polymerization degrees of the two polymers are different, they seem to be high enough that the phase transitions are no longer molecular-mass dependent. All of polymer **I**'s phase transitions are higher than those of polymer **II**'s. In addition, polymer **I** gives rise to side chain crystallization, while polymer **II** does not. Both series of polymers, however, exhibit smectic mesomorphism, indifferent of their spacer length. Although no conclusive explanation of this behavior was advanced by the original authors, we can speculate from the drawing in Scheme 3.5 that the mesogens in **I** are less sterically restricted in forming an interdigitated bilayer arrangement than the mesogens in **II**, since steric interactions between the —C≡N and —C=O groups in **II** probably limit their ability to interdigitate. In addition, the electron withdrawing character of the cyano phenyl group will decrease the electron density and therefore double-bond character of the ester —C—O— bond in **II**, but should have little effect on this linkage in **I**. Therefore, the mesogen in polymer **II** should be less effective than that in **I** at forming more ordered mesophases and crystalline phases.

As found earlier with other systems, the data in Figure 3.17 shows the glass transition temperatures decrease continuously with increasing spacer length, while the melting and isotropization transitions at first decrease, and then increase in an odd–even manner.

3.4.3.6 *Influence of the interconnecting group between the mesogen and the spacer and between the spacer and the backbone.* The nature of the interconnecting groups between the mesogen and spacer, and the spacer and backbone, has important implications on both the type of mesophase formed and the transition temperatures. That is, the

Table 3.10 Influence of the interconnecting group between the spacer and the 4-cyanobiphenyl mesogen of polyacrylates on their phase behaviour

Polymer	Phase transitions (°C)	Reference
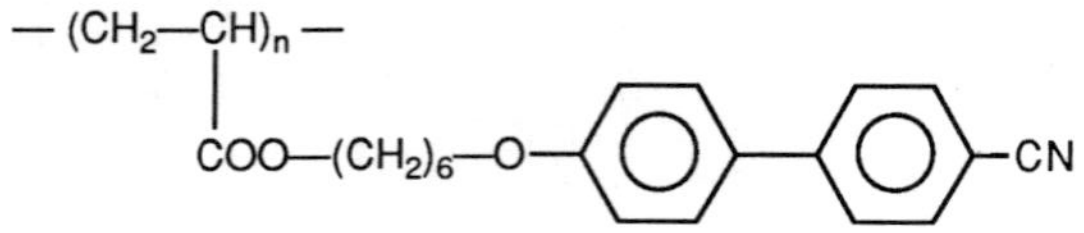	g 32 N$_{Re}$ 80 S$_A$ 124 N 132 i	Le Barny *et al.* (1986)
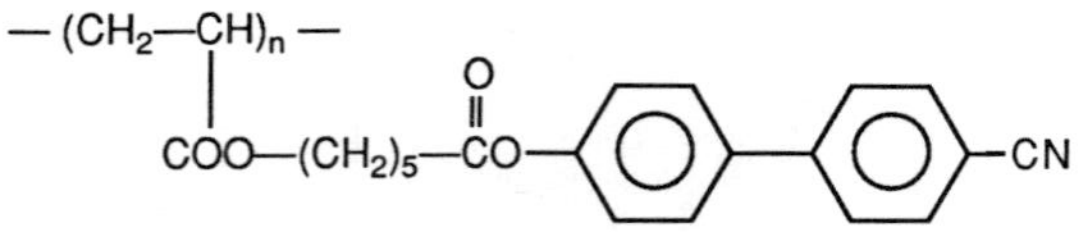	g 30 N 115 i	Le Barny *et al.* (1986)
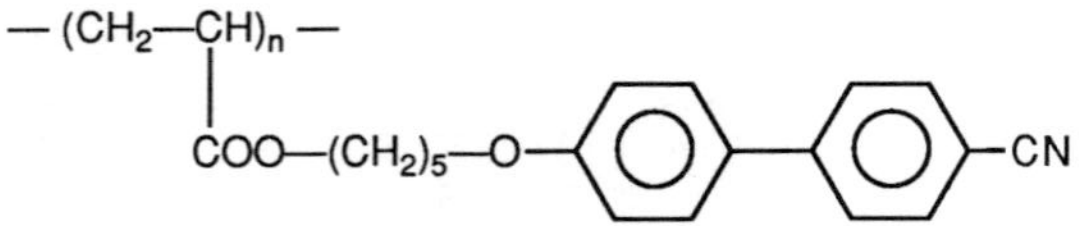	g 40 N 119 i	Kostromin *et al.* (1982*b*)

flexibility of both interconnecting groups and also the polarizability of the group connecting the spacer and mesogen should affect the phase behavior of the polymer in the same way as they affect the mesophase of low-molar mass liquid crystals. Although there is very little in the literature on this topic, we will try to discuss it using three examples.

Table 3.10 presents three polyacrylates containing medium-length spacers and identical mesogenic units, but different interconnecting groups between the spacer and mesogen. It is best to compare the polymer containing six methylenic units in the spacer with the one containing five methylenic units and an ester unit. The mesomorphic behaviour of these two is completely different, with the polymer containing an ester group having a lower isotropization temperature.

Table 3.11 presents another set of polymers in which the same mesogenic group is attached to the polymer backbone, but this time through long spacers. In this case, the nature of the interconnecting group does not affect the type of mesophase formed, but does affect the isotropization temperature; that is, the isotropization temperature is decreased by the presence of an ester group when considering an identical number of atoms in the side chain.

Figure 3.18 illustrates the phase transitions of two series of polymers containing identical mesogens interconnected to the spacer through an ether bond, and interconnected to the backbone through either an ester or an ether unit. The structural units of the backbones are different, but are actually constitutional isomers. Although the molecular masses of the poly(propenyl ether)s are lower than those of the

Table 3.11 Influence of the interconnecting group between long spacers and the 4-cyanobiphenyl mesogen of polymethacrylates on their phase behaviour (Shibaev et al., 1982)

Polymer	Phase transitions (°C)
$-(CH_2-\underset{\underset{COO-(CH_2)_{11}-O-\bigcirc-\bigcirc-CN}{|}}{\overset{\overset{CH_3}{|}}{C}})-_n$	g 40 S_A 121 i
$-(CH_2-\underset{\underset{COO-(CH_2)_{10}-CH_2-\bigcirc-\bigcirc-CN}{|}}{\overset{\overset{CH_3}{|}}{C}})-_n$	g 30 S_A 81 i
$-(CH_2-\underset{\underset{COO-(CH_2)_{10}-COO-\bigcirc-\bigcirc-CN}{|}}{\overset{\overset{CH_3}{|}}{C}})-_n$	g 45 S_A 93 i

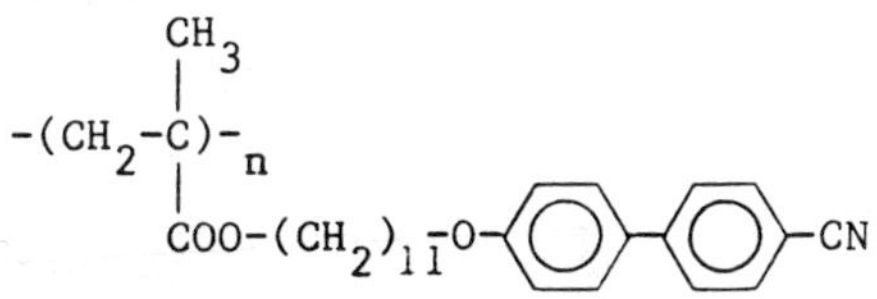

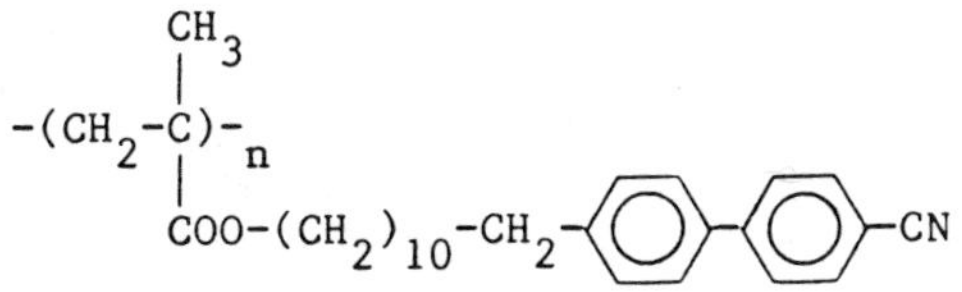

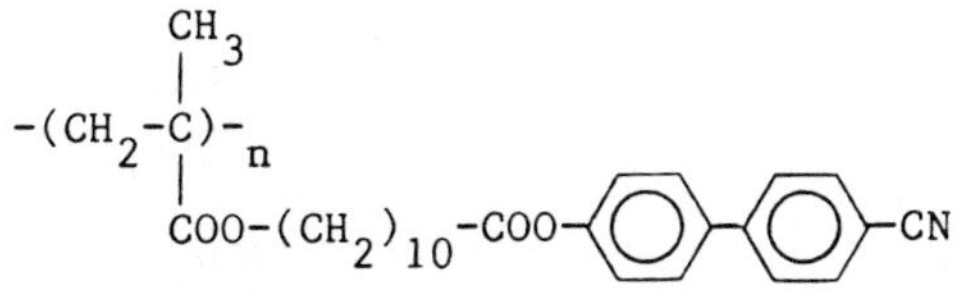

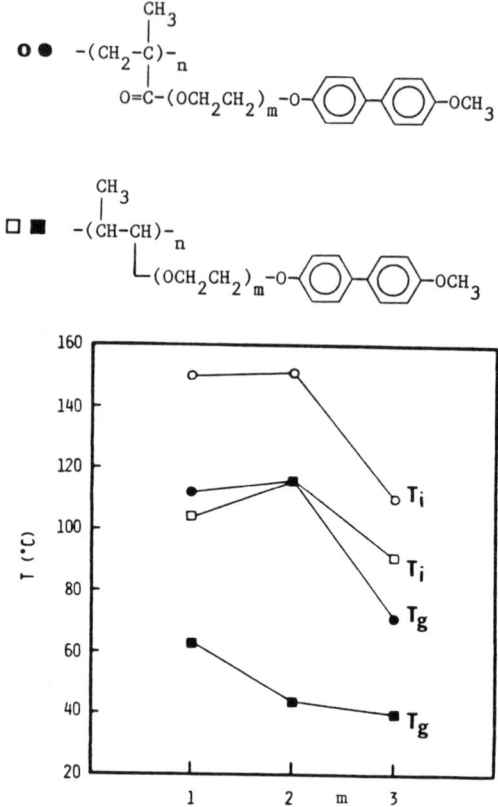

Figure 3.18 Transition temperatures of methoxybiphenyl-based liquid crystalline polymethacrylates and poly(propenyl ether)s as a function of the number m of oligooxyethylene units in the spacer (Rodriguez-Parada and Percec, 1986).

corresponding polymethacrylates (Rodriguez-Parada and Percec, 1986), a qualitative conclusion can be made regarding the influence of the interconnecting unit between the backbone and spacer on phase transitions. Since for the poly(propenyl ether)s, T_g values are much lower than those of the corresponding polymethacrylates, and the isotropization temperatures only slightly lower, a broader range of mesomorphism is observed for polymers containing interconnecting ether units. Additionally, all the poly(propenyl ether)s present higher-order mesophases than do the polymethacrylates. This is most evident with the two polymers containing one oligoethylene unit in the spacer ($m = 1$); the polymethacrylate is nematic, whereas the poly(propenyl ether) is smectic.

3.4.3.7 *Influence of the polymer backbone on phase transitions.* A discussion of the influence of the nature of the polymer backbone on phase transitions requires knowledge of the influence of the molecular mass of each polymer on its phase transitions. Since both requirements have not been met in the literature, we have tried to select the most accurate literature examples for such a comparison.

Table 3.12 Influence of the polymer backbone on the phase transitions of polymers containing

$$R = -(CH_2)_2-O-\langle\bigcirc\rangle-COO-\langle\bigcirc\rangle-OCH_3$$

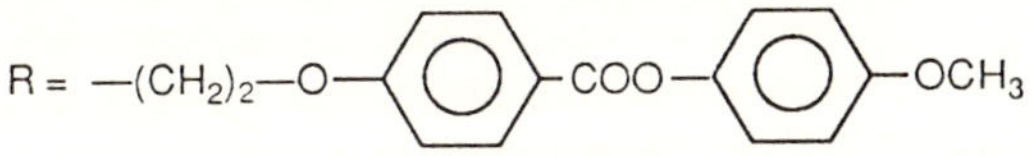

as mesogen (Finkelmann *et al.*, 1984)

Polymer backbone	Phase transitions (°C)	ΔT
$-(CH_2-\underset{\underset{COOR}{\mid}}{\overset{\overset{CH_3}{\mid}}{C}})_n-$	g 96 N 121 i	25
$-(CH_2-\underset{\underset{COOR}{\mid}}{\overset{\overset{H}{\mid}}{C}})_n-$	g 47 N 77 i	30
$-(O-\underset{\underset{CH_2R}{\mid}}{\overset{\overset{CH_3}{\mid}}{Si}})_n-$	g 15 N 61 i	46

Table 3.13 Influence of the polymer backbone on the phase transitions of polymers containing $R = -(CH_2)_6-O-\langle\bigcirc\rangle-COO-\langle\bigcirc\rangle-OC_4H_9$ as mesogen

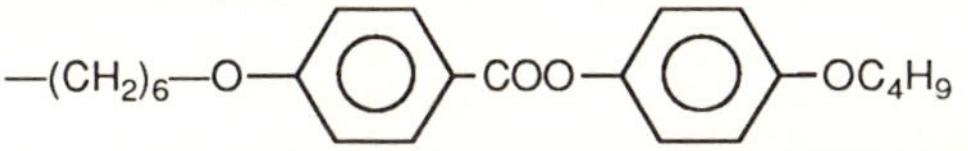

Polymer backbone	Phase transitions (°C)	Reference
$-(CH_2-\underset{\underset{COOR}{\mid}}{\overset{\overset{CH_3}{\mid}}{C}})_n-$	g 40–45 S 110 N 115 i g 48–52 S 92–95 N 108–110 i g 39 S$_A$ 109 N 114 i	Keller *et al.* (1985a) Davidson *et al.* (1985) Zentel and Ringsdorf (1984)
$-(CH_2-\underset{\underset{COOR}{\mid}}{CH})_n-$	g 95 K 30 lc 103 lc 114 i	Horvath *et al.* (1985b)
$-(CH_2-\underset{\underset{COOR}{\mid}}{\overset{\overset{Cl}{\mid}}{C}})_n-$	g 39 S$_C$ 110 N 117 i	Zentel and Ringsdorf (1984)

The first is shown in Table 3.12, in which the same mesogenic unit is attached to three different backbones (Finkelmann *et al.*, 1984). No molecular masses were reported. Although both the glass transition and isotropization temperatures decrease with increasingly flexible backbones, the overall thermal stability range of the nematic mesophase increases. However, the isotropization temperature generally increases with increasing polymer backbone flexibility. For example, Tables 3.13 and 3.14 both show two different polymer backbones with the same (but different in the two tables) side chains, and seem to indicate that the isotropization temperature increases with increasing polymer backbone flexibility. Again, however, the molecular masses were not reported, and there is some contradiction in the trends in glass transition temperature.

A comprehensive compilation of data from several different laboratories is presented in Table 3.15, and includes the use of mesogenic 4-cyanobiphenyl attached to polymethacrylate, polyacrylate and polysiloxane backbones through flexible spacers of two to eleven methylenic units. Although there are some contradictory results, the general trends further illustrated in Figure 3.19 include steadily decreasing glass transition temperatures with increasing spacer length and with increasing polymer backbone flexibility. Isotropization temperatures first decrease with increasing spacer length and then increase. In addition, the highest isotropization temperatures are always obtained with the most flexible backbones.

3.4.4 *Linear polymers with oligooxyethylene and oligosiloxane spacers*

In order to enhance the degree of decoupling through a more flexible spacer and to prepare liquid crystalline pseudo-crown ethers, several research groups have synthesized side chain LCPs containing oligooxyethylene or oligosiloxane spacers (Engel *et al.*, 1985; Rodriguez-Parada and Percec, 1986; Percec *et al.*, 1987a; Duran *et al.*,

Table 3.14 Influence of the polymer backbone on the phase transitions of polymers containing

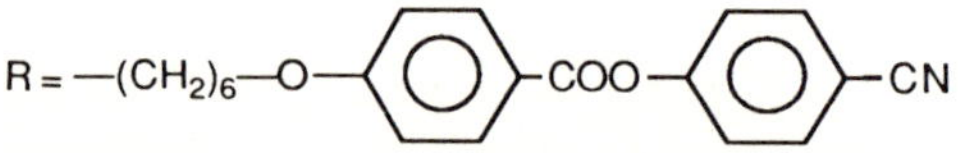

$$R = —(CH_2)_6—O—\bigcirc—COO—\bigcirc—CN \text{ as mesogen}$$

Polymer	Phase transitions (°C)	Reference
$—(CH_2—\overset{CH_3}{\underset{COOR}{C}})_n—$	g 55–60 N 107–110 i	Davidson *et al.* (1985)
$—(CH_2—\overset{H}{\underset{COOR}{C}})_n—$	g 33 N 133 i	Le Barny *et al.* (1986b)
$—(O—\overset{CH_3}{\underset{R}{Si}})_n—$	g 55 S 185 i	Gemmell *et al.* (1985a)

Table 3.15 Influence of the polymer backbone and spacer length on the phase transitions of polymers containing 4-cyanobiphenyl based mesogenic groups

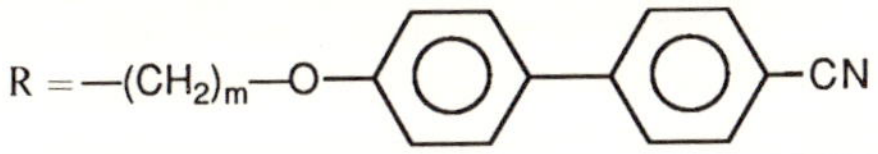

$$R = -(CH_2)_m-O-\langle\bigcirc\rangle-\langle\bigcirc\rangle-CN$$

Polymer	m	R'	Phase transitions	Reference
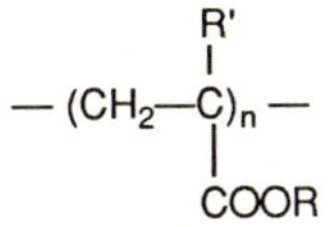 $-(CH_2-\overset{\overset{R'}{\mid}}{\underset{\underset{COOR}{\mid}}{C}})_n-$	0	CH_3	S 240 i	Alimoglu et al. (1984)
	0	H	S 270 i	Alimoglu et al. (1984)
	2	CH_3	g 95 i	Kostromin et al. (1984); Shibaev et al. (1982)
	2	H	g 50 N 112 i	Kostromin et al. (1984); Shibaev et al. (1982)
	2	H	g 84 K 114 i	Dubois et al. (1986)
	3	H	g 54 S_A 82 i	Dubois et al. (1986)
	4	H	g 42 N 229 i	Dubois et al. (1986)
	5	CH_3	g 60 S_A 120 i	Kostromin et al. (1984); Shibaev et al. (1982)
	5	H	g 40 N 119 i	Kostromin et al. (1984); Shibaev et al. (1982)
	5	H	g 35 S_A 120 N 124 i	Dubois et al. (1986)
	6	CH_3	g 55 S 100 i	Shibaev and Platé, (1985)
	6	H	g 32 N_{Re} 80 S_A 124 N 132 i	Dubois et al. (1986)
	6	H	g 32 N_{Re} 80 S_A 124 N 132 i	Le Barny et al. (1986)
	11	CH_3	g 40 S_A 121 i	Kostromin et al. (1984) Shibaev et al. (1982)
	11	H	g 25 S_C 30 S_A 145 i	Kostromin et al. (1982a, 1984)
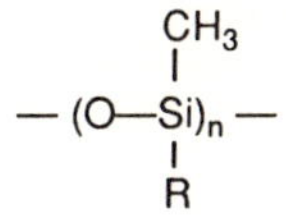 $-(O-\overset{\overset{CH_3}{\mid}}{\underset{\underset{R}{\mid}}{Si}})_n-$	3		g 40 S_A 152 i	Gemmell et al. (1985a)
	3		g 32 S 117 i	Ringsdorf and Schneller (1982)
	4		g 28 S_A 132 i	Gemmell et al. (1985a)
	5		g 14 S_A 170 i	Gemmell et al. (1985a)
	5		g 16 S 152 i	Ringsdorf and Schneller (1982)
	6		g 14 S_A 166 i	Gemmell et al. (1985a)
	11		g − 1 S_C 48 S_A 157 i	Hsu et al. (1987b); Hsu and Percec (1987b)

1986; Duran and Gramain, 1987; Kostromin et al., 1987; Kim and Allcock, 1987). It is well known from the literature on low-molar-mass liquid crystals that the insertion of oligooxyethylene segments in mesogenic derivatives destabilizes the thermal stability of the mesophase (Dietrich and Steiger, 1972). Contradictory results are obtained with polymers, however. When the mesogenic unit contains cyanophenyl end groups, the resulting polymers do not exhibit liquid crystallinity (Tables 3.16, 3.17) (Engel et al., 1985; Kostromin et al., 1987). Even when mesomorphic behavior is observed, the thermal stability of the mesophase is strongly depressed in comparison to that of the homologous polymer containing a paraffinic spacer (Table 3.17). However, all the polymers synthesized so far with p-methoxyphenyl mesogen exhibit liquid crystallinity, as for example in Figure 3.18 and Table 3.16 (Engel et al., 1985; Rodriguez-Parada and Percec, 1986; Duran et al., 1986; Percec et al., 1987a; Duran and Gramain, 1987).

We can speculate that the electron-donating or basic character of the oligooxyethylenic segment enables the spacer to interact with the acidic or electron-accepting cyano

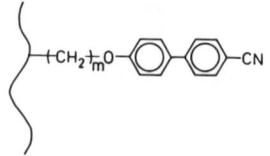

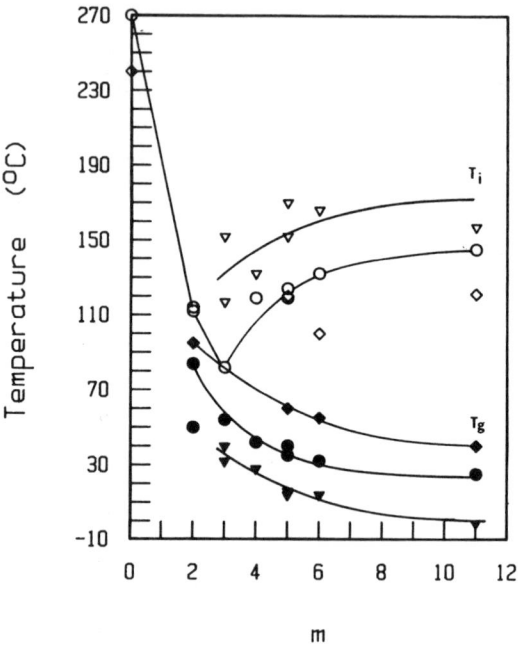

Figure 3.19 Glass transition (filled symbols) and isotropization (open symbols) temperatures of 4-cyanobiphenyl-based liquid crystalline polymethacrylates ($\diamond$), polyacrylates ($\bigcirc$) and polysiloxanes ($\triangledown$) as a function of spacer length m (data from Table 3.15).

groups. Since the donor character of an oligooxyethylenic segment is higher than that of the cyanophenyl unit, this interaction dominates, decreasing the ability of the cyanophenyl unit to interdigitate and form smectic layers. This is not the case with electron-donating p-methoxyphenyl units and therefore mesophase formation is not disturbed by an oligooxyethylenic spacer.

A series of polymers containing an oligosiloxane segment as the center of a flexible spacer is presented in Table 3.18. Unlike paraffinic spacers, the oligosiloxane spacer requires a longer mesogenic unit to induce liquid crystallinity. This led to the conclusion that a flexible spacer should be divided into at least two units: one which stabilizes the mesophase and the other which provides the real decoupling (Engel *et al.*, 1985). This idea is outlined in Table 3.18.

Table 3.16 Comparison of mesogenic monomers and side-chain polymers with oligooxyethylene units in the spacer

Polymer backbone and spacer	Interconnecting group and mesogen	m	Phase transitions (°C)		Reference
			Monomer	Polymer	
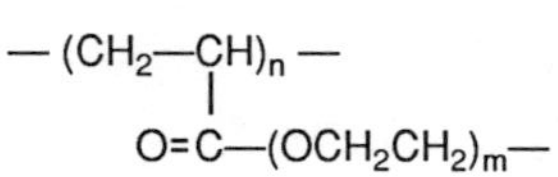 $-(CH_2-CH)_n-$ $O=C-(OCH_2CH_2)_m-$	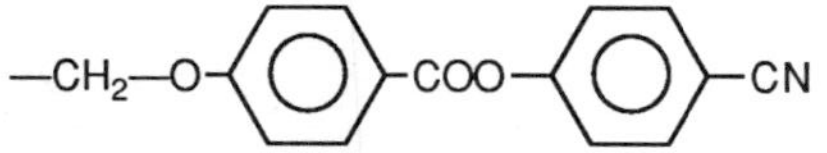 —CH₂—O—⬡—COO—⬡—CN	1	K 87 i	g 24 N 67 i	Engel et al. (1985)
	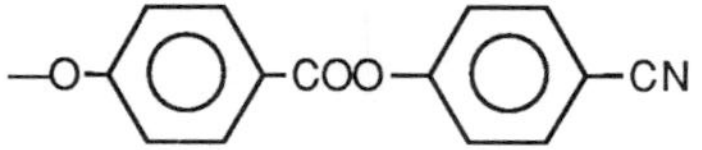 —⬡—COO—⬡—CN	2	K 87 i	g 46 i	Engel et al. (1985)
		3	K 3 i	g — 5 i	Engel et al. (1985)
	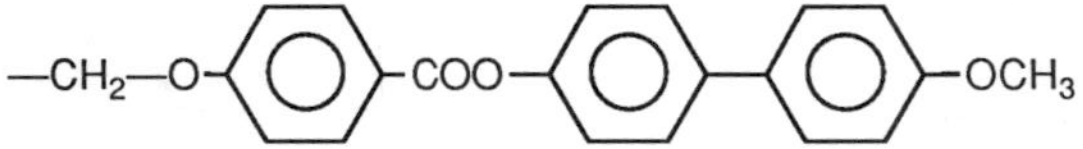 —CH₂—O—⬡—COO—⬡—⬡—OCH₃	1	K 102 N 202 i	g 67 K 87 S 127 N 290 i	Engel et al. (1985)
	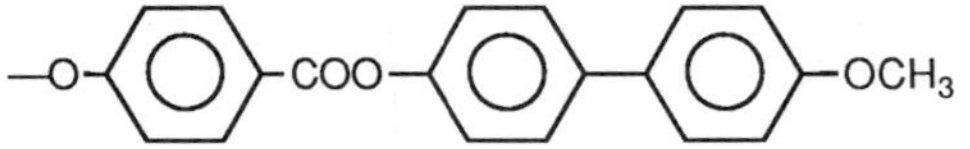 —⬡—COO—⬡—⬡—OCH₃	2	K 108 N 174 i	K 137 N 277 i	Engel et al. (1985)
	—O—⬡—⬡—OCH₃	3	K 76 N 134 i	K 115 N 237 i	Engel et al. (1985)
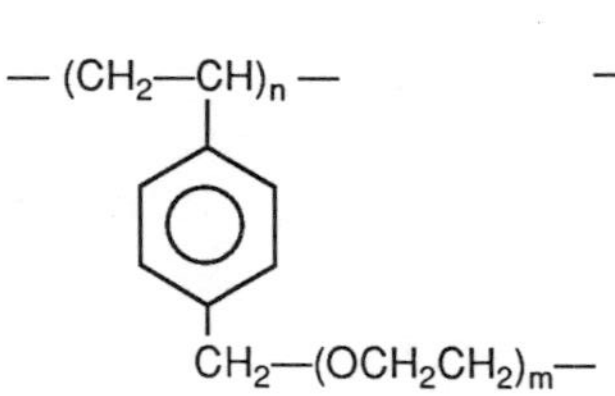 $-(CH_2-CH)_n-$ ⬡ $CH_2-(OCH_2CH_2)_m-$	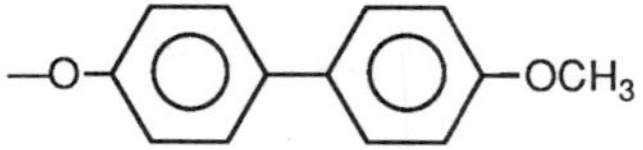 —O—⬡—N=N—⬡—OCH₃	1	K 130–131 i	S 142 i	Percec et al. (1987a)
		2	K 80–95 i	g 57 S 104 i	Percec et al. (1987a)
		3	K 60–63 i	g 30 S 86 i	Percec et al. (1987a)
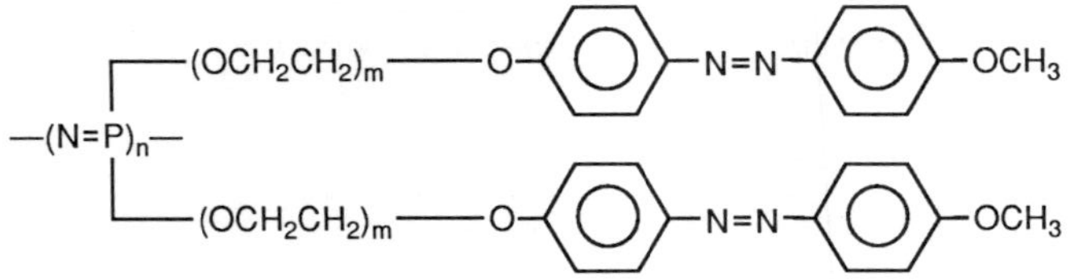 $-(N=P)_n-$ $(OCH_2CH_2)_m$	—O—⬡—N=N—⬡—OCH₃	3	—	K 118 N 127 i	Kim and Allcock (1987)

Table 3.17 Mesogenic polymers with oligooxyethylene spacers

Polymer	Phase transitions (°C)	Reference
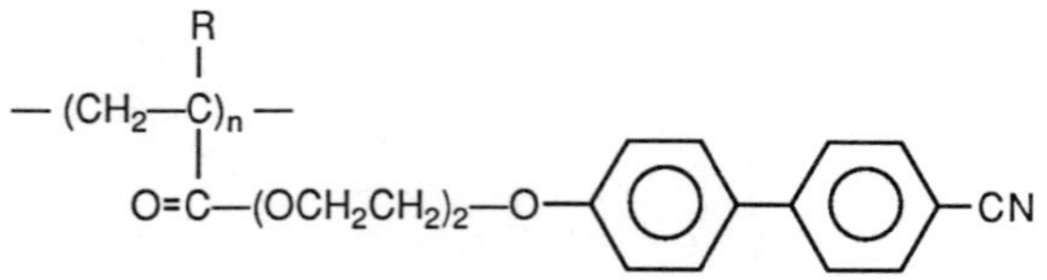	$R = H$ g 37 i $R = CH_3$ g 65 i	Kostromin et al. (1987) Kostromin et al. (1987)
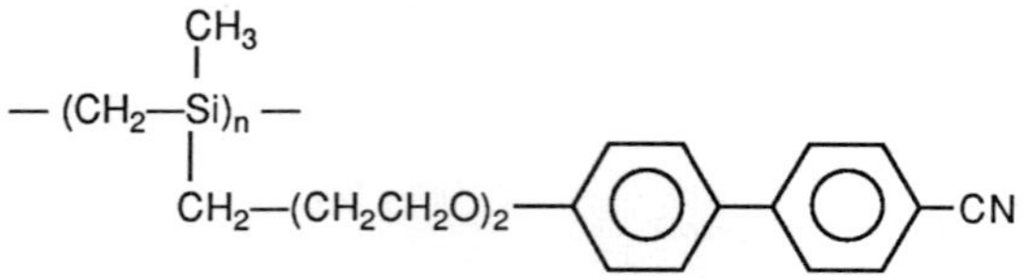	g 5 K 17 S$_A$ 102 i	Kostromin et al. (1987)
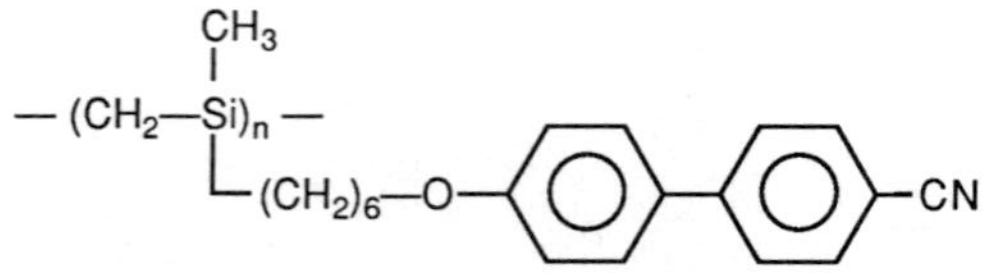	g 14 S$_A$ 166 i	Gemmell et al. (1985a)

Table 3.18 Mesogenic monomers and side chain polymers with siloxane units in the spacer (Engel et al., 1985)

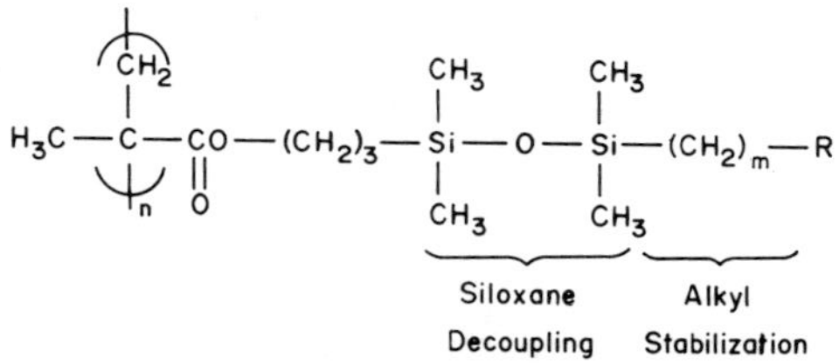

R		m	Phase transitions (°C)	
			Monomer	Polymer
$-O-$⬡$-COO-$⬡$-CN$		3	K 6 i	g 4 i
		5	K 1 i	g 0 i
		11	K − 4 N 54 i	g − 20 N 37 i
$-O-$⬡$-COO-$⬡$-$⬡$-C_3H_7$		3	K 27 N 73 i	g 27 S 102 i
		5	K 16 S 38 N 54 i	g 0 S 131 i
		11	K 30 S 42 N 102 i	g 11 K 54 S 177 i

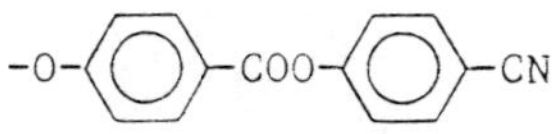

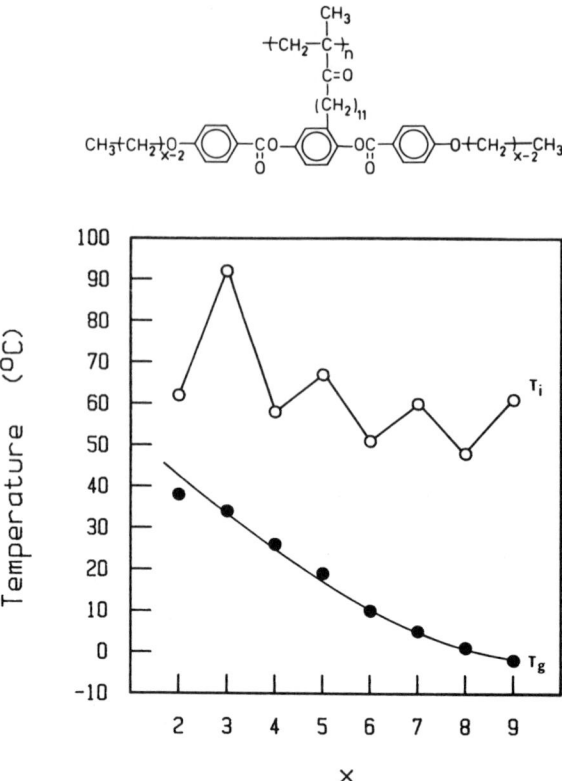

Figure 3.20 Glass transition (●) and isotropization (○) temperatures of a laterally attached liquid crystalline polymethacrylate as a function of the number x of atoms in the terminal substituents (Hessel *et al.*, 1987).

3.4.5 *Linear polymers with flexible spacers and laterally attached mesogens*

Semi-flexible polymers containing mesogens attached laterally through a flexible spacer were recently synthesized (Hessel and Finkelmann, 1985a, b; Hessel *et al.*, 1987). Representative examples are presented in Figure 3.20, together with the dependence of the glass transition and isotropization temperatures on the number of atoms in the terminal group. Interestingly, none of these polymers exhibits smectic mesomorphism; all of them exhibit biaxial nematic mesophases indifferent of their spacer or terminal group lengths. The isotropization temperature follows the expected odd–even trend, while the glass transition temperature decreases with the increasing number of atoms in the terminal group of the mesogenic unit.

3.5 Polymers containing flexible rod-like mesogens

3.5.1 *Linear polymers containing flexible rod-like mesogens without flexible spacers*

These polymers were prepared following the principles of conformational isomerism of benzyl ether units outlined in Figure 3.1. Table 3.19 presents two examples of

Table 3.19 Polyvinylbenzyl ethers containing biphenyl and 4-methoxybiphenyl units (Percec *et al.*, 1987*a*)

	R	Phase transitions (°C)
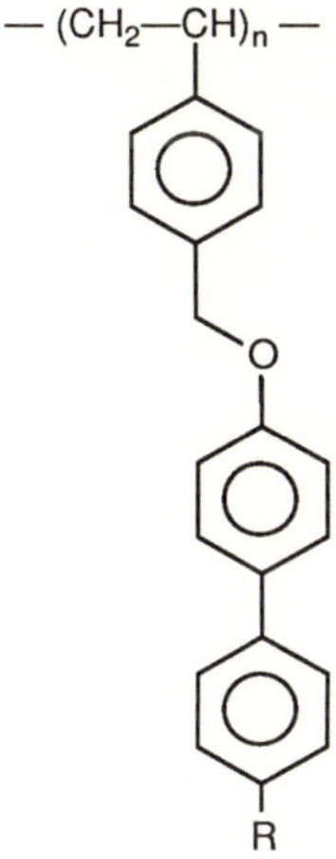	H	g 97 N 182 i
	OCH$_3$	K 183 N 233 i

$- (CH_2{-}CH)_n -$ (structure with vinylbenzyl ether linked to biphenyl bearing R substituent)

poly(vinylbenzyl ether)s containing biphenyl and 4-methoxybiphenyl units, respectively (Percec *et al.*, 1987*a*). Considering the discussion of the conformational isomerism detailed in sections 3.2.1 and 3.5.2.1, the two structures outlined in Table 3.19 should be considered to be polyethylenes containing the benzyl ether of 4-hydroxybiphenyl or 4-hydroxy-4'-methoxybiphenyl as the mesogenic unit, and not as *p*-substituted polystyrenes (for side chain LC polystyrenes see Crivello *et al.*, 1988).

3.5.2 *Linear polymers containing flexible rod-like mesogens and flexible spacers*

3.5.2.1 *Mesogens based on 1,2-diphenylethane and benzyl ether units.* As previously discussed in section 3.2.1 and outlined in Figure 3.1, the *anti* conformers of 1,2-diphenylethane and benzyl-ether based molecules exhibit extended conformations and therefore should also display liquid crystallinity. Attachment of any of these two types of mesogens to polymer backbones should lead to liquid crystalline 'copolymers' containing the two conformational isomers, i.e. *anti* and *gauche*, of the same side group as structural units. The composition of these conformationally isomeric copolymers is dynamic, since the two conformers are in dynamic equilibrium. Although most low-molar-mass liquid crystals containing either 1,2-diphenylethane or benzyl ether units as part of the rigid core exhibit virtual or monotropic mesophases, their homologous polymers present enantiotropic mesophases (Hsu and Percec, 1987*d*; Hsu and Percec, 1988). This exemplifies enhancement of the tendency to form mesophases by attachment to a polymer backbone, which is sometimes referred to as the 'polymer effect.' However, we disagree with the commonly used meaning of the 'polymer effect.' The polymer backbone does enhance the tendency of a low-molar-mass compound towards mesomorphism, but does so by depressing or even cancelling the ability of low-molar-mass mesogenic-like compounds to crystallize after polymerization. Subsequently, virtual or monotropic mesophases become enantiotropic. This is

Table 3.20 Phase behaviour of LCPs containing 1,2-diphenylethane- and benzylether-based mesogenic groups

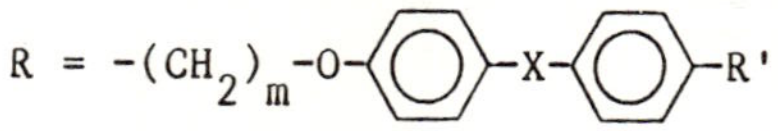

$$R = -(CH_2)_m -O-\bigcirc-O-X-\bigcirc-R'$$

Polymer	m	X	R'	Phase transitions (°C)	Reference
$-(CH_2-\overset{CH_3}{\underset{COOR}{C}})_{50}-$	11	$-CH_2O-$	$-OCH_3$	g 35 S 61 S 83 i	Hsu and Percec (1988a)
	11	$-CH_2O-$	$-CN$	g 19 K 64 S 68 i	Hsu and Percec (1988a)
	11	$-CH_2CH_2-$	$-CN$	g 3 S 56 S 68 i	Hsu and Percec (1988a)
	6	$-CH_2CH_2-$	$-CN$	g 29 N 49 i	Hsu and Percec (1988a)
$-(CH_2-\underset{COOR}{CH})_{30}-$	11	$-CH_2O-$	$-OCH_3$	g 36 K 63 S 85 S 92 i	Hsu and Percec (1988a)
	11	$-CH_2O-$	$-CN$	g 19 K 72 S 77 S 84 i	Hsu and Percec (1988a)
	11	$-CH_2CH_2-$	$-CN$	g 16 S 28 S 34 S 72 i	Hsu and Percec (1988a)
	6	$-CH_2CH_2-$	$-CN$	g 20 S 60 i	Hsu and Percec (1988a)
$-(O-\overset{CH_3}{\underset{R}{Si}})_{80}-$	11	$-CH_2O-$	$-OCH_3$	g 44 K 82 S 99 i	Hsu and Percec (1987d)
	11	$-CH_2O-$	$-CN$	g 31 K 92 S 119 i	Hsu and Percec (1987d)
	11	$-CH_2CH_2-$	$-CN$	g -16 S 23 S 28 S 61 S 76 i	Hsu and Percec (1987d)
	6	$-CH_2CH_2-$	$-CN$	g -10 S 41 i	Hsu and Percec (1987d)
	11	$-CH_2O-$	$-PhOCH_3$	g 63 S 146 S 182 S 228 S 235 i	Hsu and Percec (1987d)
	11	$-CH_2O-$	$-PhCN$	g 15 S 217 i	Hsu and Percec (1987d)
	11	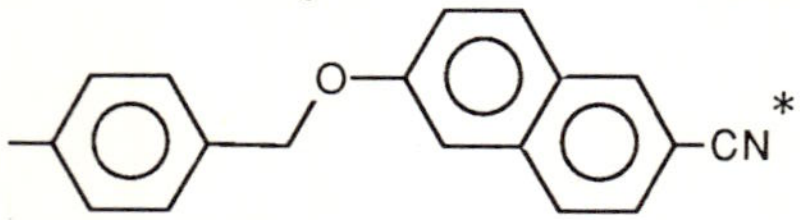 *		g 19 S 164 i	Hsu and Percec (1987d)

*corresponds to the entire mesogen attached to $-(CH_2)_m-O-$

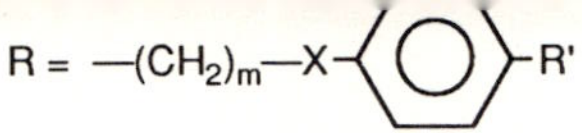

$$R = -(CH_2)_m - X - \langle benzene \rangle - R'$$

Polymer	m	X	R'	Phase transitions (°C)	Reference
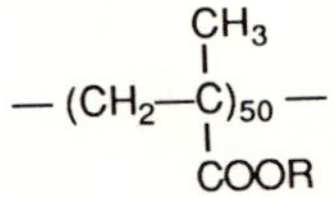 CH_3 / $-(CH_2-C)_{50}-$ / $COOR$	11	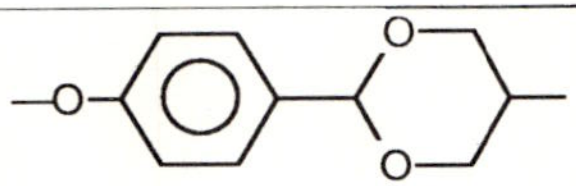 (dioxane–phenyl)	$-OCH_3$	g 44 S 127 S 190 i	Hsu et al. (1987a)
	11	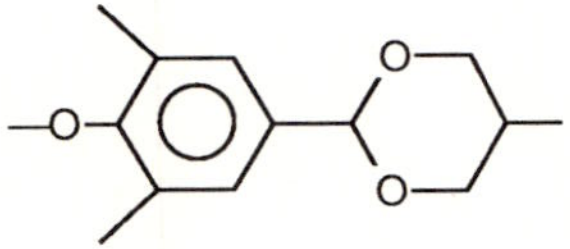 (dioxane–dimethylphenyl)	$-OCH_3$	g 21 N 107 i	Hsu et al. (1988b)
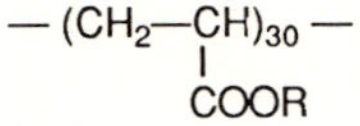 $-(CH_2-CH)_{30}-$ / $COOR$	11	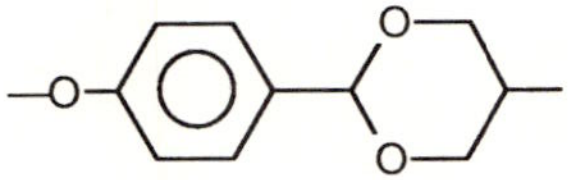 (dioxane–phenyl)	$-OCH_3$	g 32 S 59 S 116 S 177 i	Hsu et al. (1987a)
	11	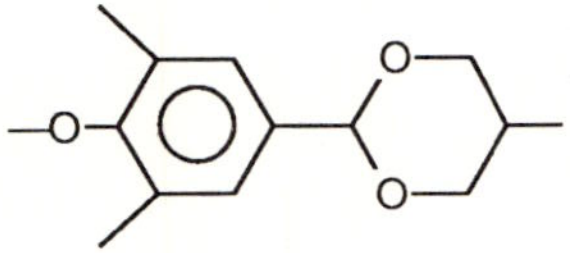 (dioxane–dimethylphenyl)	$-OCH_3$	g 10 N 67 i	Hsu and Percec (1988b)
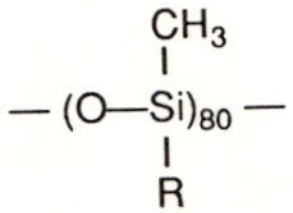	11	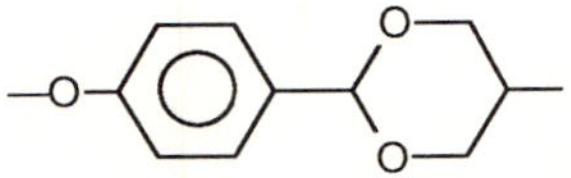 (dioxane–phenyl)	$-OCH_3$	g 44 S 76 S 158 i	Hsu et al. (1987d)
CH_3 / $-(O-Si)_{80}-$ / R	11	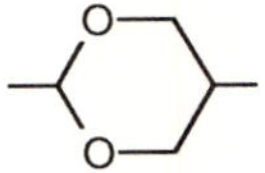 (dioxane)	$-OCH_3$	g −17 S 3 S 63 i	Hsu and Percec (1987a)
	10	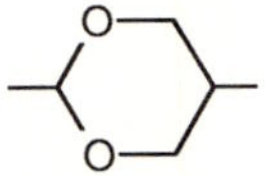 (dioxane)	$-OCH_3$	g −7 S 84 S 93 i	Hsu et al. (1987d)
	11	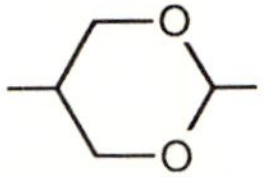 (dioxane)	$-CN$	g −10 S_A 149 i	Hsu and Percec (1987c)

nicely illustrated by the influence of polymer-molecular-mass on phase transitions as depicted by the examples described in Figures 3.13 and 3.14. Monomers, dimers and even trimers crystallize and melt at higher temperatures than the isotropization transitions. Above these degrees of polymerization, the tendency towards side chain crystallization decreases and the polymers exhibit only LC mesophases. Additional work is required to demonstrate this polymer effect. Nevertheless, a completely decoupled LCP should exhibit thermotropic behavior similar to the parent low-molar-mass structural units.

Table 3.20 summarizes some representative examples of polymethacrylates, polyacrylates and polysiloxanes containing 1, 2-diphenylethane and benzyl ether based mesogens. Several comments should be made on the behavior of these polymers. As demonstrated in previous sections, the isotropization temperatures are highest for the most flexible backbones in almost all cases. On the other hand, the isotropization temperatures are lower than even those of the homologous polymers containing a 4-substituted biphenyl moiety with the same substituent in the 4-position. This is supported by comparison of any of the cyano substituted polymers from Table 3.20 with the homologous 4-cyanobiphenyl-containing polymers from Table 3.21. In addition, Table 3.20 demonstrates that a shortening of the flexible spacer from eleven to six methylenic units transforms the mesophase from a smectic to a nematic one in the case of the polymethacrylate backbone with 1-phenyl-2-cyanophenyl ethane based mesogens, but leaves it unchanged for the polyacrylate and polysiloxane backbones. Interestingly, none of these polymers give rise to side chain crystallization, even when the spacer contains eleven methylenic units. This is understandable, since a mixture of conformational isomers should have a lower tendency towards co-crystallization than a single mesogenic unit towards crystallization.

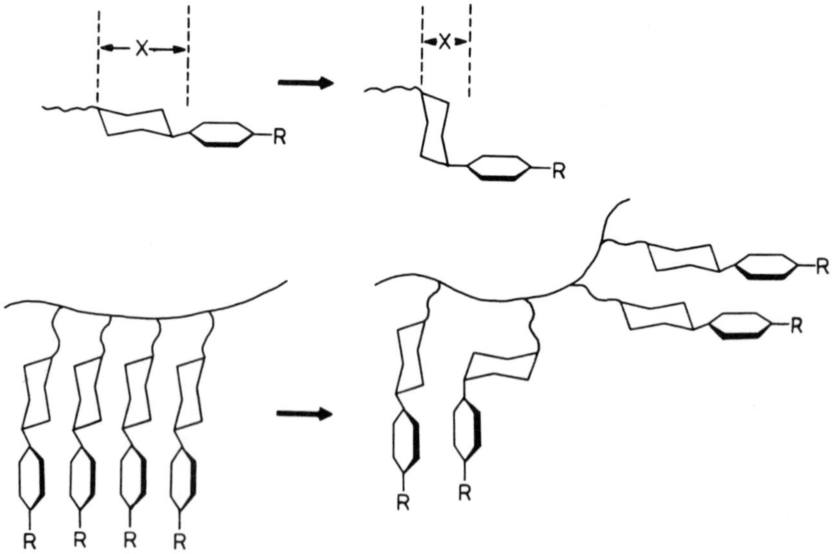

Figure 3.21 Model depicting the effect of equatorial to axial isomerism of *trans*-2, 5-disubstituted-1, 3-dioxane based mesogens (Hahn and Percec, 1988).

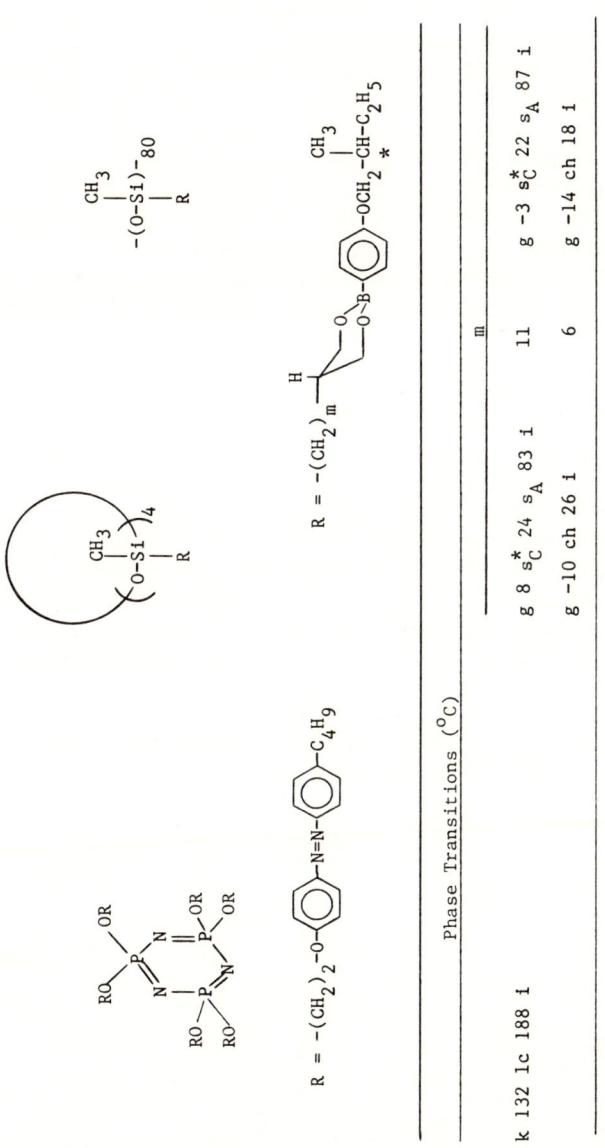

Figure 3.22 Phase transitions of liquid crystalline cyclic triphosphazene, tetramethylcyclotetrasiloxane and the corresponding linear polysiloxane.

3.5.2.2 *Mesogens based on trans-2, 5-disubstituted-1, 3-dioxane.* Figure 3.2 outlines the synthesis of *trans*-2, 5-disubstituted-1, 3-dioxane derivatives and the conformational isomerism exhibited by the *trans* isomer. It was hypothesized that the dynamic equilibrium between the axial and equatorial arrangement of substituents might decrease the tendency towards side chain crystallization in polymers containing long flexible spacers. In the case of low-molar-mass liquid crystals, there is little difference between the *trans* isomers having substituents in the axial versus the equatorial position. As depicted in Figure 3.21, once these mesogenic units are attached to polymeric systems as side groups, the isomeric mixture does indeed depress the side chain crystallization (Table 3.21). However, the difficulty in achieving complete hydrosilation when bulky substituents are involved prevents accurate comparison of the transition temperatures of these polymers. Interestingly, by substituting a phenyl ring of the mesogenic group with two methyl groups, nematic mesophases are obtained even with polymers containing eleven methylenic units in the flexible spacer (Hsu and Percec, 1988*a*).

3.6 Cyclic oligomers containing rod-like mesogens

At present, there are only two examples of cyclic oligomers containing rod-like mesogens in the side chains (Singler *et al.*, 1987; Hahn and Percec, 1987, 1988, and unpublished data). In the first example (Figure 3.22), the chlorine groups in hexachlorotriphosphazene were displaced with a nucleophilic 'side chain' containing the mesogen. Unfortunately, the mesophase of the resulting compound was not assigned and its phase behavior cannot be compared with that of its linear homologous polymer since the 'polymer' was obtained only as a copolymer (Singler *et al.*, 1987).

The second mesogenic cyclic oligomer series is based on tetramethylcyclotetra-siloxane as shown in Figure 3.23 (Hahn and Percec, 1987, 1988, and unpublished data). These oligomers resemble a crown-like conformation and and are of particular interest since theoretical calculations have predicted that they will form disc-like mesophases at low temperatures (Everit *et al.*, 1988). Figure 3.22 presents the phase transitions of both

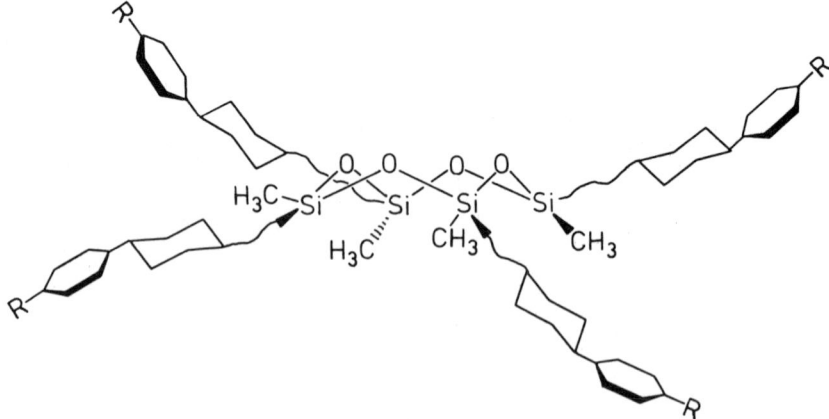

Figure 3.23 Crown-like conformation of tetramethylcyclo-tetrasiloxane-containing mesogenic groups (Hahn and Percec, 1988).

sets of cyclic oligomers and the homologous linear siloxane polymers (Hahn and Percec, 1987, 1988, and unpublished data). The tabulated data show that the linear and cyclic compounds exhibit identical thermal transitions at similar temperatures. The transition temperatures of the cyclic structures are shifted only slightly from those of the linear structures. This is quite unexpected, considering the difference between the degree of polymerizations of the cyclic ($n = 4$) and linear ($n = 80$) polymers. Additional work is required to elucidate their behavior.

3.7 Polymers containing disc-like mesogens

Of the two published examples of side chain liquid crystalline polymers containing disc-like mesogens, one is a homopolysiloxane and the other is a copolysiloxane (Figure 3.24) (Kreuder and Ringsdorf, 1983). Both polymers exhibit liquid crystallinity although their discotic mesophases have not yet been identified. Recently, polysiloxanes containing monofunctionalized myo- and scyllo-inositol as discs were synthesized (Kohne et al., 1988). As expected from the behavior of the parent low-molar-mass discotic compounds, only the polymers containing scyllo-inositol side groups exhibit discotic mesophases.

3.8 Polymers and copolymers with paired mesogens

All the liquid crystalline polymers discussed up to this point contain one mesogenic unit attached to one flexible spacer. As already mentioned, the thermal stability of the polymer's mesophase is higher than that of the corresponding monomer, and in many circumstances liquid crystalline polymers are obtained from crystalline monomers. The

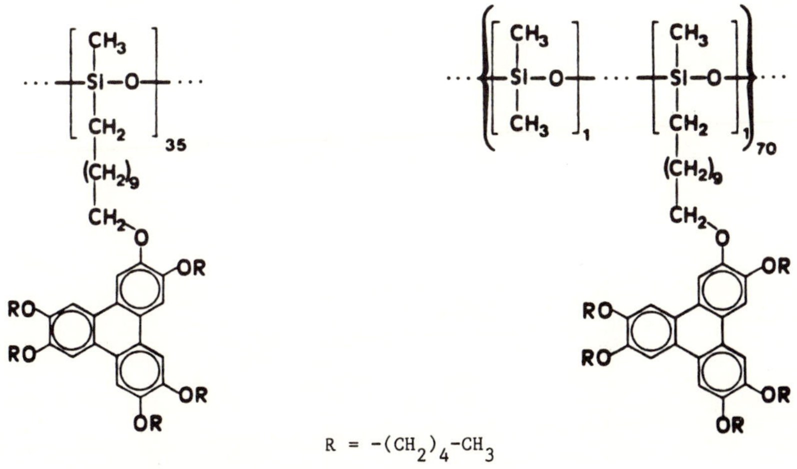

Figure 3.24 Liquid crystalline side chain polymers with disc-like mesogens (Kreuder and Ringsdorf, 1983).

Table 3.22 Polysiloxanes containing paired mesogens

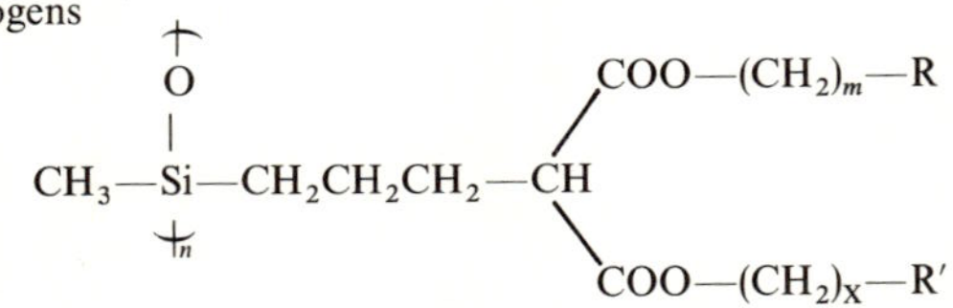

$$
\left[O-Si-CH_2CH_2CH_2-CH \underset{COO-(CH_2)_x-R'}{\overset{COO-(CH_2)_m-R}{<}} \right]_n
$$

R	R'	m	X	Phase transitions (°C)	Reference
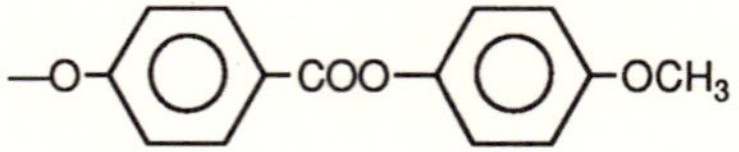 —O—⟨◯⟩—COO—⟨◯⟩—OCH₃	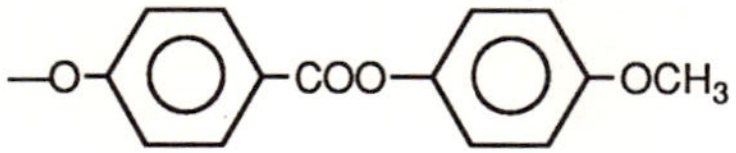 —⟨◯⟩—COO—⟨◯⟩—OCH₃	2	2	g 35 K 94 S 98 N 103 i	Engel et al. (1985)
		2	6	g 18 S 100 N 113 i	Engel et al. (1985)
		6	6	g 4 S 130 i	Engel et al. (1985)
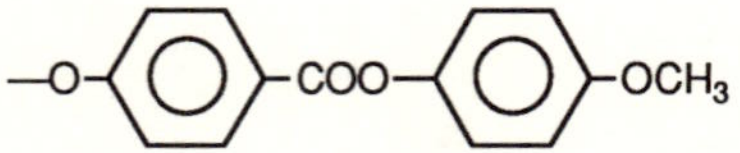 —O—⟨◯⟩—COO—⟨◯⟩—OCH₃	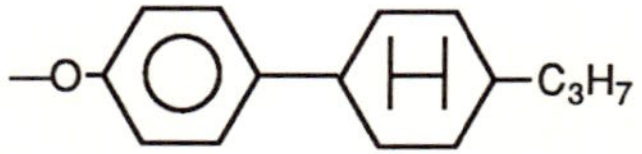 —O—⟨◯⟩—COO—⟨H⟩—C₃H₇	2	2	g 13 S 73 S$_A$ 104 i	Diele et al. (1986)
		2	6	g 25 K 70 S 72 N 80 i	Diele et al. (1986)

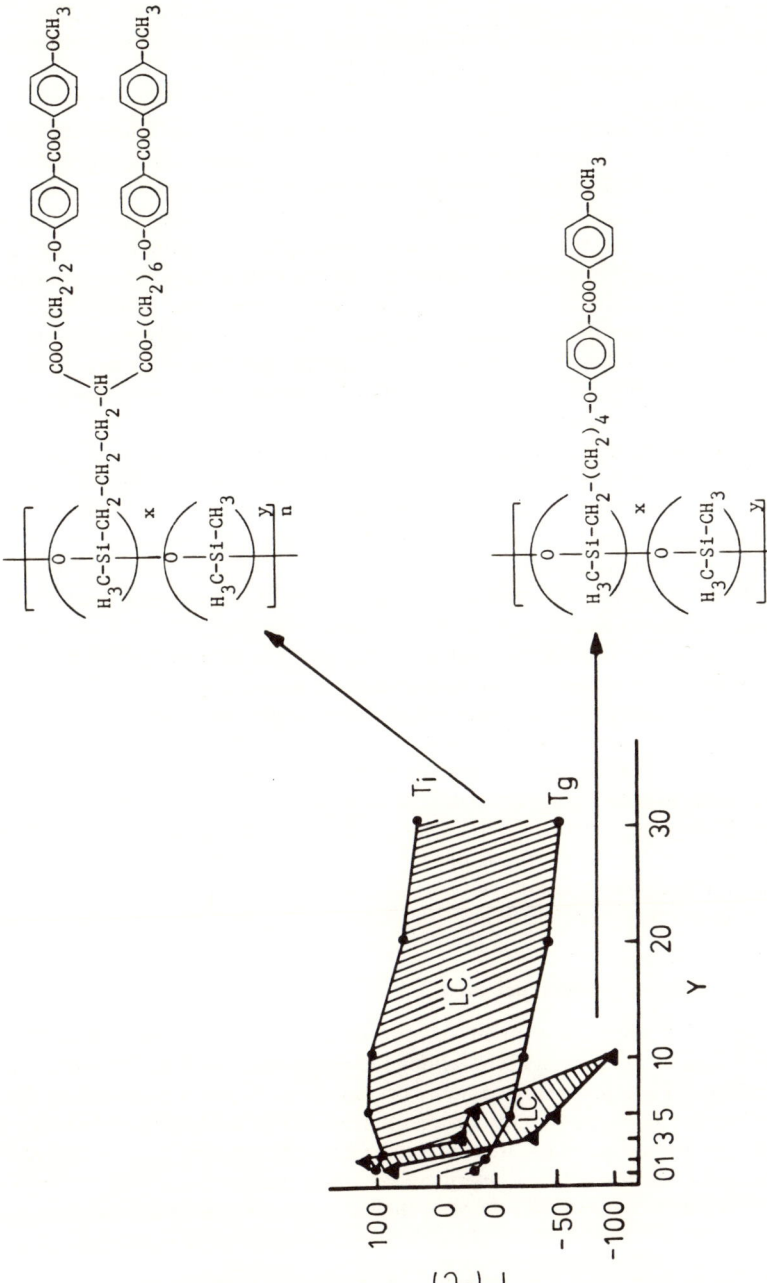

Figure 3.25 Comparison of side chain liquid crystalline polysiloxanes and copolysiloxanes containing single and paired mesogenic side groups (Engel *et al.*, 1985).

D

formation of a mesophase is even more favoured, i.e., the crystallization tendency is disfavoured, when each flexible spacer carries more than one mesogenic side group (Engel *et al.*, 1985; Hisgen and Ringsdorf, 1983; Diele *et al.*, 1986, 1987).

Table 3.22 presents several examples of side chain LCPs containing paired mesogens. Paired mesogens are of particular interest in the synthesis of side chain liquid crystalline copolymers containing mesomorphic and nonmesomorphic structural units since they represent LCPs containing main chain and side chain spacers (Figures 3.5, C′). A copolymer based on nonmesogenic structural units and structural units containing a single mesogen is usually non-mesomorphic at a molar ratio of non-mesogenic to mesogenic structural units of 10:1. At the same time, a copolymer based on structural units containing paired-mesogenic and non-mesogenic structural units is mesomorphic and exhibits the same mesophase thermal stability as the parent mesomorphic homopolymer, even when the molar ratio of non-mesogenic to mesogenic units is as high as 30:1. Two series of such copolymers covering a large range of compositions are presented in Figure 3.25.

It has been suggested that the liquid crystalline phase obtained at as low as 10% mesogenic units in the copolymer is due to the strong lateral interactions of the paired mesogenic units, thus leading to a close packing of the mesogens in layers, and rejection of the polysiloxane backbone from this array (Diele *et al.*, 1987). However, due to the flexibility of the siloxane chains, it seems that the polymer chains are able to accommodate the order of the mesogenic side groups. As a consequence, a separation of the structure into two sublayers must be considered, one of which consists of the mesogenic side chains ordered parallel, and one which is formed by the dimethyl-siloxane segments. This model is schematically illustrated in Figure 3.26 (Diele *et al.*, 1987), and completely meets the requirements of a microphase separated model of highly decoupled systems as suggested in section 3.2.3. Recent research from Ringsdorf's group has definitively demonstrated that copolysiloxanes containing paired mesogens as side groups are indeed microphase separated systems (Westphal *et al.*, 1988; Diele *et al.*, 1988). Therefore, copolymers containing paired mesogens are suitable candidates for the synthesis of fully decoupled side chain liquid crystalline polymers, and the investigation of polymers containing two or more mesogens attached to the same spacer deserves expansion.

3.9 Combined liquid crystalline polymers

As outlined in Scheme 3.2 and briefly mentioned in section 3.2.1, combined liquid crystalline polymers are polymers containing mesogenic units in both the main chain and side chains. The term 'combined liquid crystalline polymers' was coined by Reck and Ringsdorf to emphasize that the structural principles of both main chain and side chain liquid crystalline polymers are combined within the same molecule (Engel *et al.*, 1985; Reck and Ringsdorf, 1985). The principles of their synthesis are briefly outlined in Figure 3.7. We will mention only some of their characteristic properties here.

3.9.1 *Combined LCPs containing the side groups attached from the flexible spacer of the main chain*

These polymers are most frequently synthesized as polymalonates, as outlined in Figure 3.7 and Table 3.23 (Engel *et al.*, 1985; Reck and Ringsdorf, 1985; Zentel *et al.*,

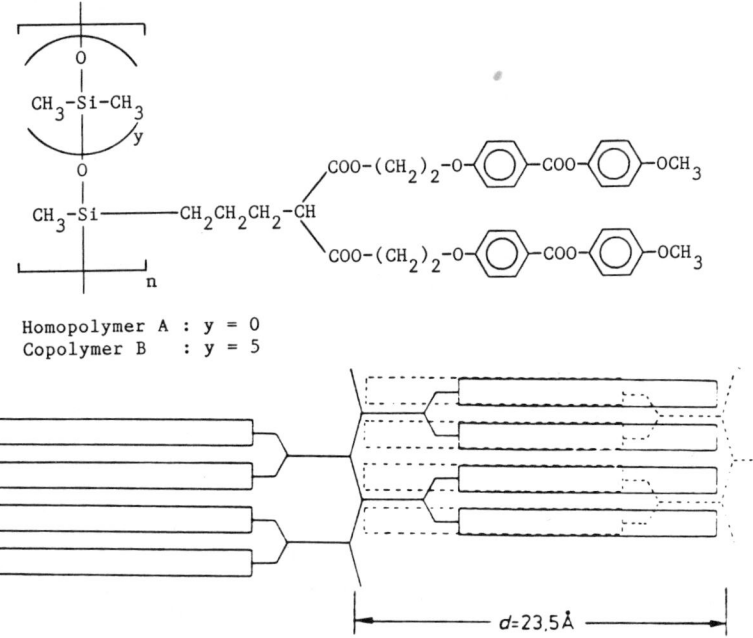

Homopolymer A : y = 0
Copolymer B   : y = 5

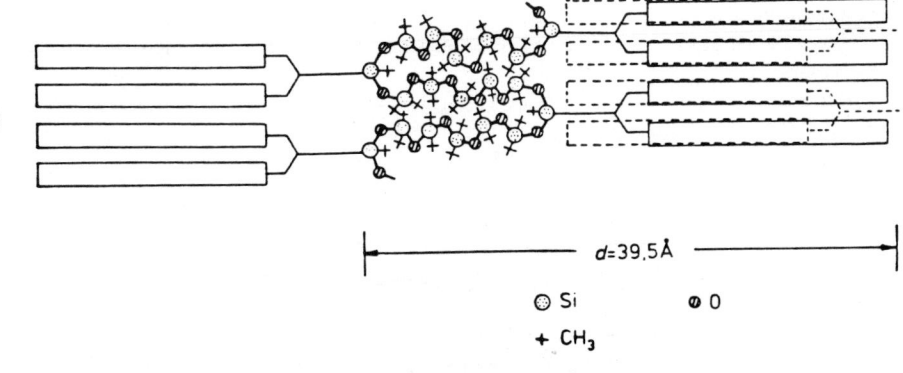

Diele et al., 1987

Figure 3.26 Schematic model of (A) the structure of the homopolymer (y = 0), and (B) the layer of a copolymer (y = 5) with an assumed configuration of the dimethylsiloxane segments between the Si atoms linked to the dimesogenic side groups (Diele *et al.*, 1987). The dashed lines are side groups positioned above the plane of the drawing.

Table 3.23 Combined main chain and side chain liquid crystalline polyesters (Reck and Ringsdorf, 1985)

$$-[-O-\langle\bigcirc\rangle-\langle\bigcirc\rangle-O-(CH_2)_6-OOC-\underset{\underset{R}{|}}{CH}-COO-(CH_2)_6-]_n-$$

R		m	Phase transitions (°C)
H		—	$g\,97\,S_1\,126\,S_C\,135\,S_A\,147\,i$
$-(CH_2)_m-O-\langle\bigcirc\rangle-N=N-\langle\bigcirc\rangle-OCH_3$		2	$K_1\,109\,K_2\,120\,N\,153\,i$
		6	$K_1\,108\,K_2\,112\,S_C\,131\,S_A\,136\,N\,155\,i$
		10	$g\,72\,S_1\,120\,S_C\,139\,S_A\,154\,i$
$-(CH_2)_m-O-\langle\bigcirc\rangle-N=N-\langle\bigcirc\rangle-CN$		6	$g\,61\,S_1\,112\,S_2\,184\,i$

1987*a*). The two most characteristic properties of these polymers are their enhanced solubility compared to the parent main chain LCP, and their enhanced mesophase thermal stability compared to both the corresponding main chain and side chain homologues. Table 3.23 presents some selected examples (Engel *et al.*, 1985; Reck and Ringsdorf, 1985). These polymers exhibit both smectic and nematic mesophases. Figure 3.27 explains schematically their ease in forming smectic mesophases (Reck and Ringsdorf, 1986; Endres *et al.*, 1987; Eich *et al.*, 1987). Bidirectionally oriented fibres are obtained from them (Engel *et al.*, 1985; Reck and Ringsdorf, 1985; Zentel *et al.*, 1987*b*).

Since these polymers behave more like side chain LCPs than main chain LCPs from the point of view of the mesophase they form, they may induce different kinds of mesomorphism in combined liquid crystalline polymers. Recently, chiral smectic C (C*) combined LCPs were synthesized (Zentel *et al.*, 1987*a*; Kapitza and Zentel, 1988). Table 3.24 summarizes some representative examples of combined smectic C* polymers.

Alternatively, the synthetic principle outlined in the first equation of Figure 3.7 can be used to prepare side chain liquid crystalline polymalonates without mesogenic units in their main chain (Engel *et al.*, 1985; Eich *et al.*, 1987), considerably broadening the

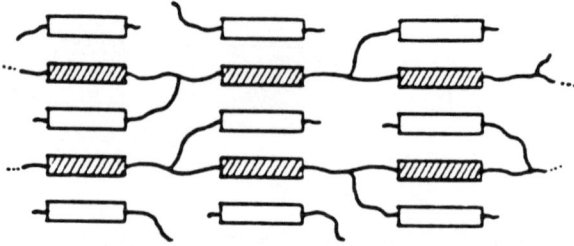

Figure 3.27 Smectic mesophase of a combined liquid crystalline polymer containing the side group attached from the flexible spacer of the main chain (Reck and Ringsdorf, 1986).

Table 3.24 Combined main chain and side chain liquid crystalline polyesters with chiral smectic C phases (S_C^*) (Zentel *et al.*, 1987)

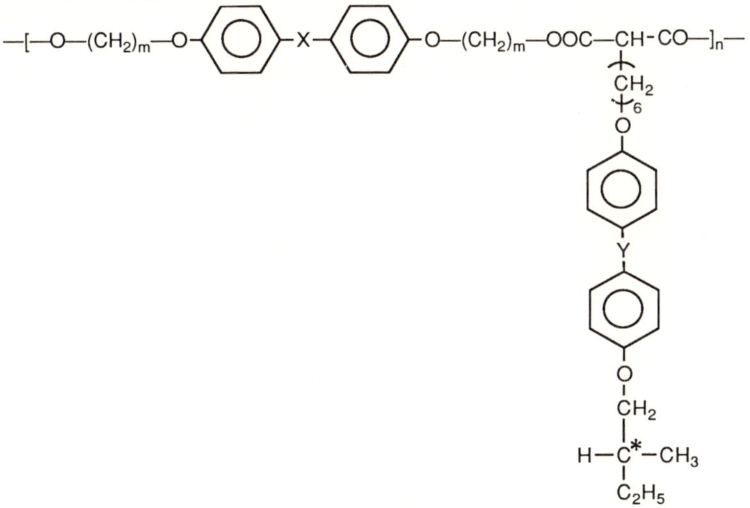

X	*m*	Y	Molecular mass (GPC)	Phase transition (°C)
—	6	—	56 000	K_1 151 K_2 154 i
—N=N—	6	—	3 200	K 108 S_C^* 115 N* 128 i
—N=N—	2	—	21 000	K 150 i[+]
—	6	—N=N—	30 000	K 130 S_C^* 136 i
—N=N—	6	—N=N—	36 000	K 107 S_C^* 111 N* 137 i

[+]Monotropic S_A or S_C phase not identified further.

synthetic methods available to tailor-make side chain LCPs. Figure 3.28 presents two examples of side chain polymalonates. Their phase transitions demonstrate that both smectic and nematic polymers can be prepared by this technique. It is particularly interesting that the proper selection of the diol used in the polyesterification reaction provides a simple and convenient tool to tailor T_g for LCPs.

3.9.2 Combined LCPs containing side groups attached from the mesogenic group of the main chain

Combined LCPs in which the side groups are attached to the mesogenic groups of the main chain can be subdivided into the two subclasses shown in Scheme 3.2 and Figure 3.7. The first one refers to combined LCPs in which the polymer main chain contains flexible spacers, while the second refers to LCPs with rigid rod-like polymer main chains, i.e. main chains without flexible spacers. Figure 3.29 presents a few examples from both categories, together with their phase transitions (Reck and Ringsdorf, 1986). In contrast to the previous class of combined LCPs, these polymers exhibit broad nematic mesophases and are not able to crystallize, although they do

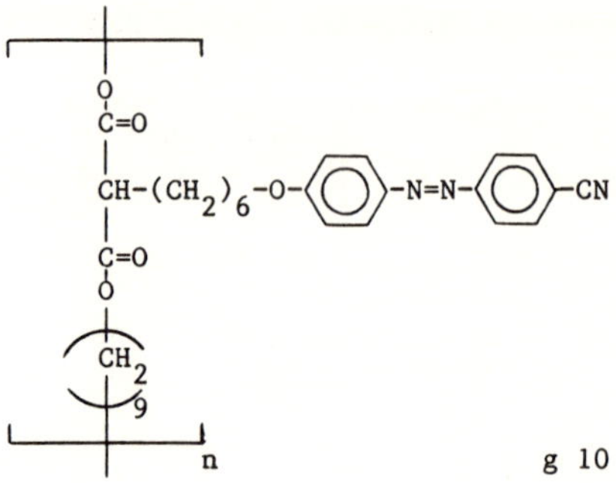

Figure 3.28 Side chain liquid crystalline polymalonates with flexible (Engel *et al.*, 1985) and semi-flexible (Eich *et al.*, 1987) backbones.

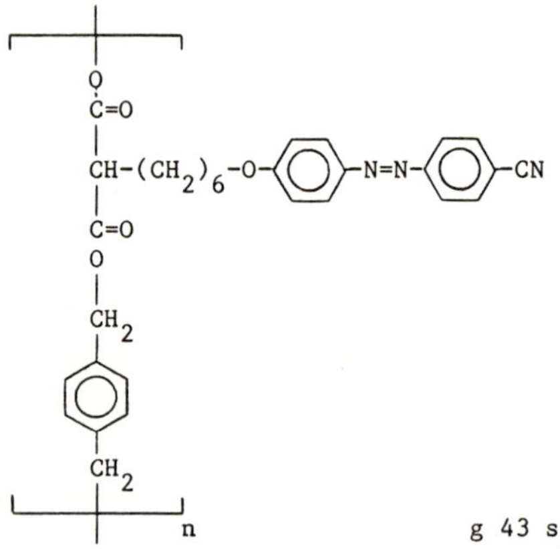

exhibit some additional layered non-classical smectic phases at lower temperatures (Reck and Ringsdorf, 1988, pers. comm.). At the same time, they present very broad mesophase thermal stabilities. X-ray diffraction of the melt-drawn fibres obtained from these polymers demonstrate that their mesogens are arranged in a parallel manner as shown schematically in Figure 3.30 (Reck and Ringsdorf, 1986). This schematic arrangement of the mesogens is in contrast to that shown in Figure 3.27, and can easily explain the differences in the mesomorphic behavior of these two classes of combined LCPs, especially the fact that this last LCP class forms only nematic mesophases.

Combined LCPs containing rigid rod-like main chains present much lower melting phase transitions than the parent non-substituted backbones, which in most cases decompose before melting (Ober *et al.*, 1984).

3.9.3 *Combined LCPs containing cross-shaped mesogens*

As outlined schematically in Scheme 3.2, the most recent addition to combined LCPs are the polymers containing cross-shaped mesogens (Berg *et al.*, 1986). These can be

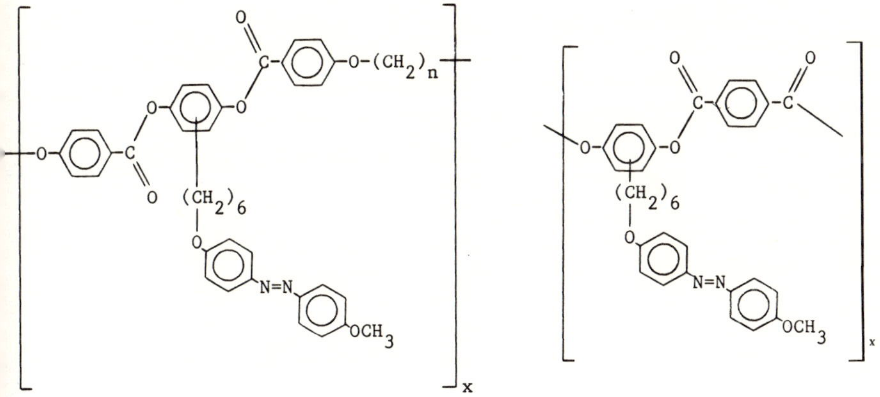

1a   n=9

b   n=6

c   n=2

Polymer	Phase Transitions (°C)
1a	g 45 n 221 i
1b	g 56 n 268 i
1c	g 90 n 310 i (dec)
2	g 67 k 221 n>350 (dec)

Figure 3.29 Combined liquid crystalline polymers containing mesogenic side groups attached from the mesogenic group of the main chain (Reck and Ringsdorf, 1986).

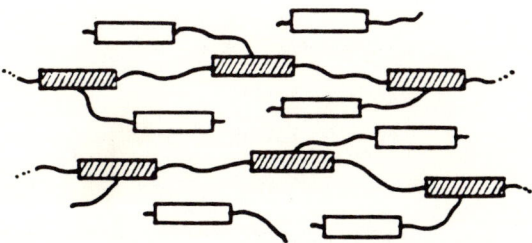

Figure 3.30 Nematic mesophase displayed by combined liquid crystalline polymers containing mesogenic side groups attached from the mesogenic group of the main chain (Reck and Ringsdorf, 1986).

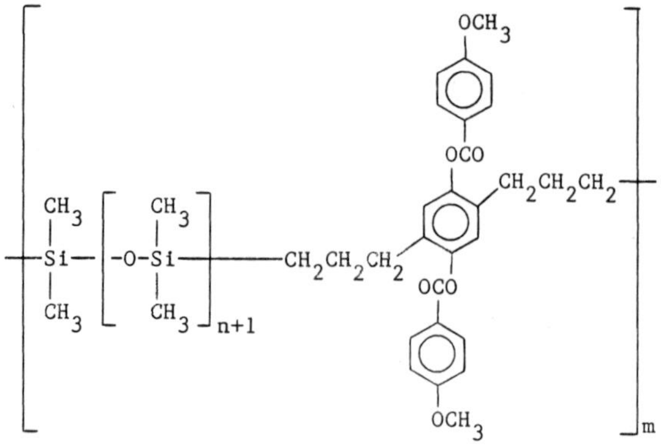

1a n=0
b n=5.5

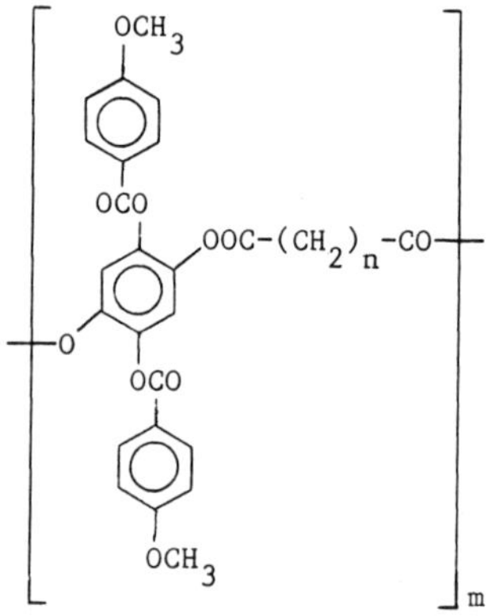

2a n=10 ($\overline{M}n$)
b n=10 ($\overline{M}n$)
c n=12
d n=14
e n=20

Polymer	n	Phase Transitions (°C)	$\overline{M}n$
1a	0	k 53 i	2100
1b	5.5	s 10.5 i	3400
2a	10	k 162 i	2500
2b	10	g 30 k_1 114 k_2 146 i	10000
2c	12	g 27 lc 88 k_1 121 k_2 140 i	3700
2d	14	g 22 lc 73 k_1 85 k_2 105 i	4600
2e	20	g 17 lc 54 k 94 i	3400

Figure 3.31 Combined liquid crystalline polymers containing mesogenic groups inserted perpendicular to the contour of flexible polysiloxane and polyester backbones (Berg *et al.*, 1986).

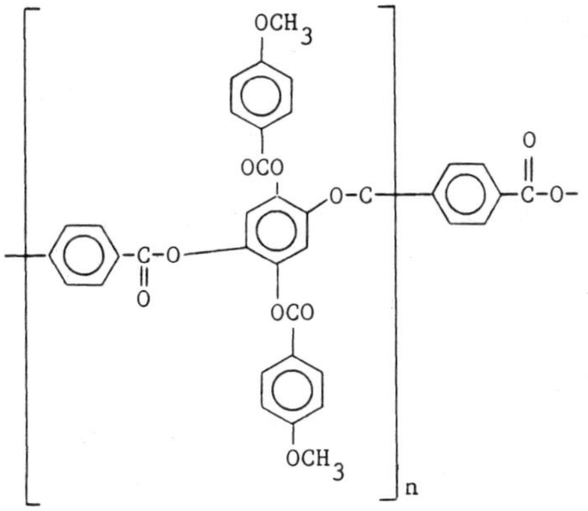

a

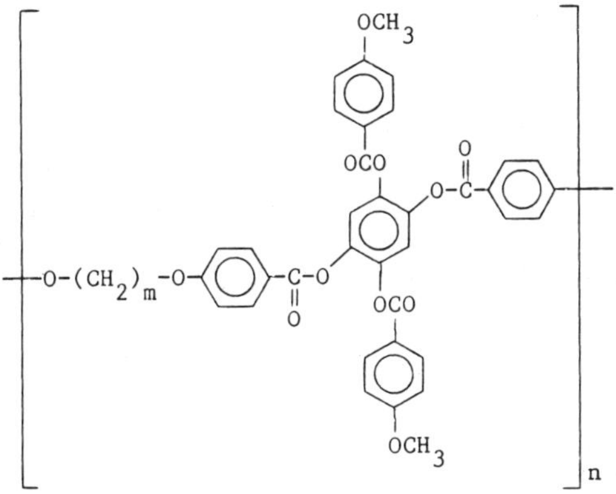

b m=2

c m=6

d m=9

	m	Phase Transitions (°C)	$\overline{Mn}$
a	–	g 162 k 186 n 265 i*	–
b	2	g 203 k 253 n 295 i*	–
c	6	g 60 n 217 i*	6000
d	9	g 54 n 214 i*	10000

*decomposes

Figure 3.32 Combined liquid crystalline polymers containing cross-shaped mesogens within rigid (a) and flexible (b–d) polymer backbone portions (Berg *et al.*, 1986).

subdivided into polymers containing the mesogens inserted perpendicular to the contour of a flexible backbone (Figure 3.31), or with cross-shaped mesogens inserted within a flexible or rigid polymer backbone (Figure 3.32). Interesting, all these polymers containing cross-shaped mesogens exhibit liquid crystalline mesomorphism. As the original authors have stated, it is not yet certain whether they should be considered as calamitic or discotic phases (Berg et al., 1986). These polymers present mesophases at lower temperatures than the unsubstituted homologous main chain thermotropic polymers.

3.10 Langmuir–Blodgett films of side chain LCPs

The conversion of preformed polymers into Langmuir–Blodgett films has received interest recently due to the ability to tailor-make polymer monolayers and multilayers (Bader et al., 1985; Ringsdorf, et al., 1987a; Tredgold, 1987). Preformed amphiphilic polymers containing hydrophilic groups in the backbone or in the side chains can be used to prepare monolayers, and X-, Y- or Z-type multilayers.

Y- and Z-type multilayers of side chain LCPs were prepared from preformed side chain LCPs containing hydrophilic groups in the backbone. Table 3.25 illustrates several examples of such LCPs obtained by the radical copolymerization of maleic anhydride with olefins containing mesogenic groups, followed by mono-esterification of the anhydride units with methanol (Vickers et al., 1985; Tredgold et al., 1987).

Preformed side chain LCPs containing hydrophilic groups attached at the end of the mesogenic group have been successfully used in the preparation of Langmuir–Blodgett monolayers which exhibit non-linear optical properties (Carr et al., 1987). One such series is illustrated in Figure 3.33. Langmuir–Blodgett multilayers were also obtained from thermotropic SC-LC polymalonates of oligooxyethylene glycols (Ringsdorf et al., 1987c). These polyesters contain mesogenic azobenzene units in the side chains, and

Table 3.25 Side chain LCPs used in the preparation of Langmuir–Blodgett multilayers (Vickers et al., 1985)

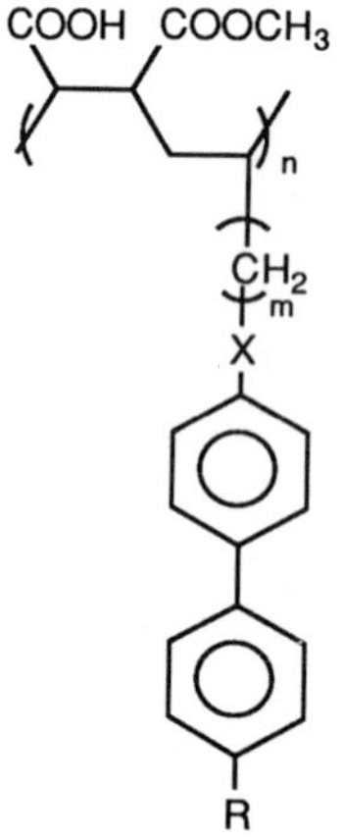

	R	X	m	$\bar{M}_n$
	—H	—COO—	9	12 000
	—H	—CO—	8	3 600
	—C_5H_{11}	—CO—	2	18 000
	—H	—CH_2—	8	14 000
	—C_5H_{11}	—CH_2—	2	18 000

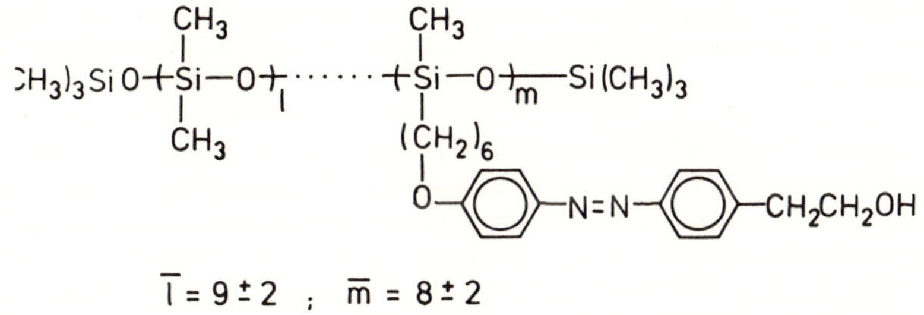

$$\bar{l} = 9 \pm 2 \; ; \; \bar{m} = 8 \pm 2$$

Figure 3.33 An example of a preformed liquid crystalline polysiloxane used in the preparation of Langmuir–Blodgett films (Carr *et al.*, 1987).

highly flexible and hydrophilic oligooxyethylene segments in the main chain. The photoinitiated *trans–cis* isomerization reaction of the azobenzene unit drastically influences the self-organization behavior of these materials in the bulk as well as at the gas–water interface.

Both monolayers and multilayers of side chain LCPs are of great interest in the general area of optical devices, and in particular in waveguides that take advantage of the films' nonlinear optical properties.

3.11 Side chain liquid crystalline copolymers (LCCs)

Although primarily chain copolymerization and polymer homologous reactions have been used to synthesize side chain liquid crystalline copolymers (LCCs), step copolymerization reactions may be used as well. In the case of chain copolymerization reactions, both the copolymer composition and its sequence distribution are dictated by the reactivity ratios of the monomers, and are conversion-dependent. Therefore, with the exception of azeotropic copolymerizations, the copolymer composition differs from that of the monomer feed in the initial reaction mixture. Knowledge of reactivity ratios allows prediction of the copolymer composition for a certain monomer feed at low or moderate conversions. Nevertheless, under extreme circumstances the copolymerization of a monomer pair can lead to either a mixture of two homopolymers even at low conversions, or to mixtures of heterogenous copolymers at high conversions. For example, the cationic copolymerization of a pair of vinyl ethyl ether monomers containing constitutionally isomeric side groups results in a mixture of two copolymers. Since one of the copolymers is nematic and the other is smectic, they are not miscible and the two copolymers were easily separated and characterized independently (Percec and Tomazos, 1987; Percec, 1988*a*). A recent review discussing the principles of copolymerization should be consulted by non-experts in this field (Tirrel, 1986).

Because the formation of high-molecular-mass polymers by step growth copolymerizations requires 100% conversion of the monomers, the copolymer composition is normally equal to the composition of the feed mixture. Nevertheless, the sequence distribution dictated by the polymerization kinetics is rapidly lost, particularly in polyesterifications and polyamidations performed under conditions where interchange and equilibration reactions complete with the formation of the polymer chain.

The same can be said about polymer homologous reactions. If complete conversion is attained, the copolymer composition corresponds to that of the reactants. Nevertheless, the sequence distribution is again dictated by the reactivity of the monomers.

To date, no LCC experiment available in the literature discusses the effect of the sequence distribution of a certain copolymer composition on phase transitions. In fact, most of the experiments do not even consider the real copolymer composition, but rather assume that the composition is that of the initial monomer feed. However, since copolymerization represents a very effective way of tailoring phase transitions in LCPs, we will qualitatively discuss LCCs by selecting a few representative systems.

3.11.1 *LCCs containing mesogenic and non-mesogenic structural units*

Statistical copolymerization of a mesogenic monomer with a non-mesogenic monomer dilutes the concentration of mesogenic units in the polymer. Subsequently, the copolymer loses its mesomorphic behavior below a minimum mesogen concentration, which depends on the nature of both monomers. Both the glass transition and isotropization temperatures are affected by the non-mesogenic comonomer content. When T_g for the non-mesomorphic homopolymer is lower than that of the mesomorphic homopolymer, an increase in the non-mesogenic monomer content can lead to an increase in the mesophase thermal stability. Further increases in the non-mesogenic monomer content causes the mesophase thermal stability to decrease and eventually vanish. An example of this behavior has already been discussed in section 3.8 (Figure 3.25) (Engel *et al.*, 1985). Generally, the nature of the mesophase is not influenced by the addition of a non-mesogenic comonomer; however, changes have been reported (Shibaev and Platé, 1984). As little as 20 mole% mesogenic units in the copolymer is usually sufficient to maintain its liquid crystallinity when the mesogenic monomer contains a single mesogenic group. Table 3.26 presents some selected examples of copolymers together with their phase transitions (Platé *et al.*, 1984b). The results from this table show that the minimum amount of mesogenic monomer required to preserve the liquid crystalline mesophase increases with increasing length of the alkyl group in the non-mesogenic monomer. For example, the methacrylate containing 22 carbons in the alkyl group requires 75 mole% mesogenic monomer to form a mesophase. Additional examples of LCCs based on polysiloxane backbones are discussed in chapter 4 of this book.

An additional class of copolymers from non-mesogenic monomers and monomers containing mesogenic groups are alternating copolymers. In this case, the polymeric unit derived from the non-mesogenic monomer contributes to the formation of a new structural unit containing a mesogenic side group. This structural unit is different from that resulting from the homopolymerization of the mesogenic monomer. Therefore, these copolymers behave like liquid crystalline homopolymers since they contain a mesogenic group in each structural unit. The first examples of alternating LCCs were liquid crystalline side chain polysulphones prepared by radical copolymerization of sulphur dioxide with 1-alkenes containing mesogenic groups (Braun *et al.*, 1987):

$$(SO_2 + CH_2 = CHX \xrightarrow{R^{\cdot}} [SO_2 - CH_2CHX]_n -, X = \text{mesogen})$$

A very important class of copolymers is obtained by copolymerizing mesogenic monomers with non-mesogenic or mesogenic monomers containing dyes. Liquid

Table 3.26 Transition temperatures of some liquid crystalline copolymers (Platé *et al.*, 1984*b*)

Comonomer M_1	Comonomer M_2 and its content in copolymer (mole %)	Phase transitions (°C)
$CH_2{=}C(CH_3)$ | CO NH | $(CH_2){-}COO{-}Chol^{**}$	$CH_2{=}CH$ | $COOC_4H_9$	
	0	g 120 lc* 180 i
	58	g 65 lc 160 i
	63	g 60 lc 140 i
	83	g 20 lc 100 i
$CH_2{=}C(CH_3)$ | CO NH | $(CH_2){-}COO{-}Chol^{**}$	$CH_2{=}C(CH_3)$ | $COOC_4H_9$	
	10	g 115 lc 180 i
	33	g 105 lc 170 i
	60	g 85 lc 160 i
$CH_2{=}C(CH_3)$ | CO NH | $(CH_2){-}COO{-}Chol^{**}$	$CH_2{=}CH$ | $COOC_{10}H_{21}$	
	25	g 90 lc 180 i
	42	g 70 lc 170 i

*lc = liquid crystalline
**Chol = cholesteryl rest

crystalline systems containing dyes are of interest for device applications. In low-molar-mass host–guest mixtures, a dye molecule is dissolved in a liquid crystalline phase. Because of solubility problems, such a system can accommodate only very low concentrations of dye. In addition, the solubility of the dye is temperature-dependent. These problems were solved by binding the liquid crystal and dye molecules to the same polymer backbone by copolymerization of the corresponding dye and mesogen-containing monomers, respectively. In this case, a host–guest system which is temperature-independent and which has a dye content as high as desired can be easily designed. An additional interesting characteristic of such a system is the formation of anisotropic glasses below the copolymer's glass transition temperature; the liquid crystalline order of low-molar-mass host–guest systems is destroyed by crystallization at lower temperatures. Ringsdorf *et al.* advanced this concept of liquid crystalline copolymers containing dyes (Ringsdorf and Schmidt, 1984; Ringsdorf *et al.*, 1985, 1986, 1987*b*; Engel *et al.*, 1985). Two representative examples are presented in Figure 3.34, together with the copolymers' mesomorphic behavior.

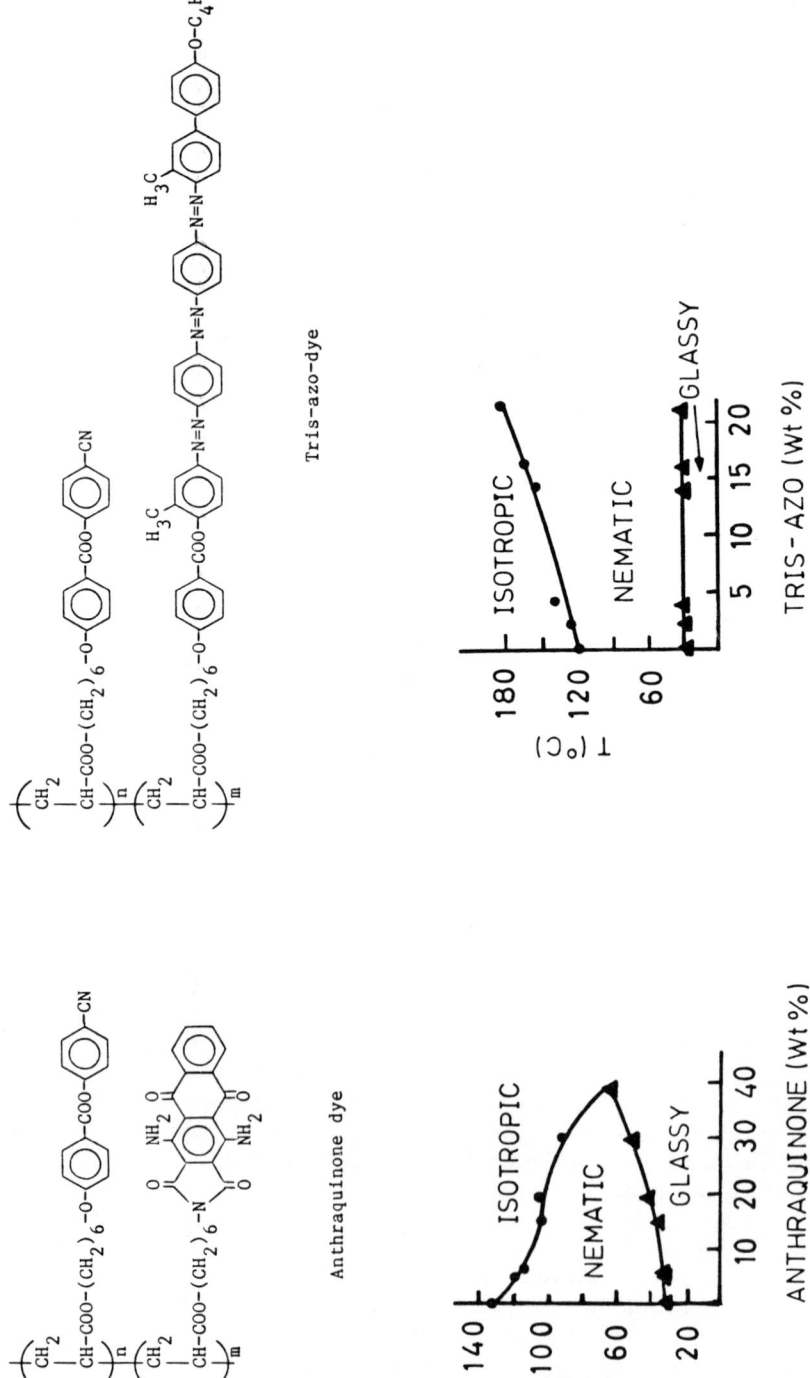

Figure 3.34 Phase behaviour of dye-containing liquid crystalline copolymers (Engel et al., 1985).

A final category of LCCs containing mesogenic and non-mesogenic comonomers is based on non-mesogenic monomers containing spiropyran groups (Cabrera and Krongauz 1987a, b). Spiropyran groups undergo photo- and thermal-induced reversible isomerization into merocyanine units (Figure 3.35). Atactic polymers containing spiropyran side groups give rise to a new type of 'zipper' crystallization upon isomerization due to the self-assembly of mesogenic groups into giant molecular stacks (Krongauz and Goldburt, 1981; Goldburt *et al.*, 1984; Wismontski and Krongauz, 1985).

Copolymers obtained from the thermotropic spiropyran acrylamide and cyanophenylbenzoate monomers presented in Figure 3.35 exhibit thermochromic properties (Cabrera and Krongauz, 1987a, b). Upon heating into the isotropic melt, the LCC displays strong dynamic birefringence if it is sheared or perturbed mechanically. This birefringence results from the formation of a transient liquid crystalline phase in which the mesogenic groups are arranged parallel to each other. The macromolecules are

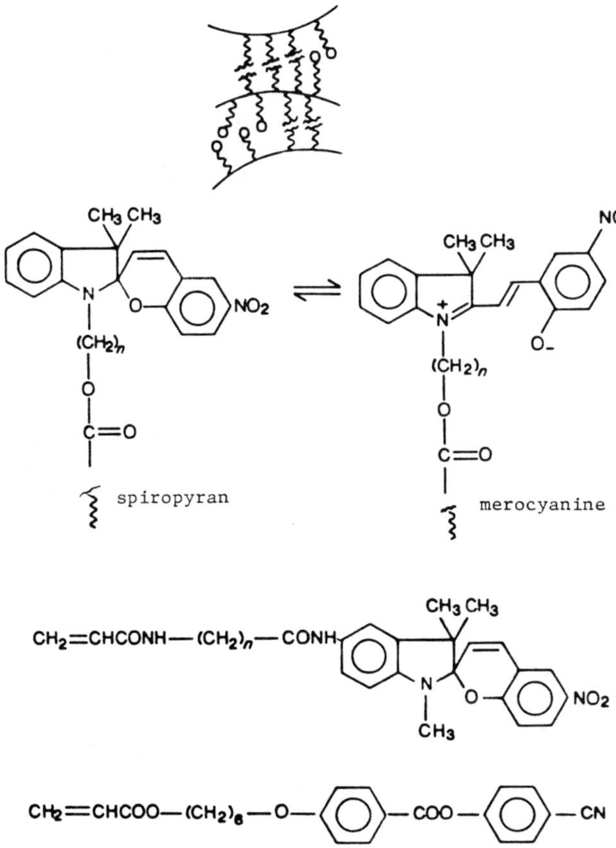

Figure 3.35 Photo and thermal induced reversible isomerization of copolymers containing non-mesogenic spiropyran groups (Cabrera and Krongauz, 1987b).

physically cross-linked into a reversible network by aggregation of the merocyanine isomer.

Novel photo- and thermochromic liquid crystalline copolysiloxane-containing mesogenic and spiropyran side groups have also been prepared (Cabrera *et al.*, 1987*a*; see also Chapter 4).

3.11.2 *LCCs containing mesogenic structural units*

There are at least four main LCC categories which can be considered:

(i) LCCs from monomer pairs containing identical mesogens and polymerizable groups, but different spacer lengths

(ii) LCCs from monomer pairs containing identical mesogens and spacer lengths, but different polymerizable groups

(iii) LCCs from monomer pairs containing dissimilar mesogens, but either similar or different spacer lengths and polymerizable groups

(iv) LCCs from monomer pairs containing constitutional isomeric mesogenic units, similar or dissimilar spacers and polymerizable groups.

Cases (i) and (ii) were previously discussed by Shibaev and Platé (1984) and will therefore be mentioned only briefly here. Figure 3.36 presents the isotropization temperatures and enthalpies of two sets of copolymers containing 4′-cyanobiphenyl mesogens; one with different spacers (PA-5-PA-11) (case i), and one with identical spacers but different polymerizable groups (PA-5-PM-5) (case ii). Copolymer PA-5-PA-11 exhibits iso-tropization temperatures and enthalpies that are essentially the compositional mass averages of those of the individual homopolymers. This copolymer becomes smectic when it contains approximately 30 mole% of the monomeric units with long spacers. Therefore, the type of mesophase is dictated by the spacer length in this case. In contrast, both the plots of the isotropization temperature and enthalpy of copolymer PA-5-PM-5 versus its composition pass through a minimum. At about 70 mole% MA-5 units, the mesophase becomes smectic. It seems that in this case the change in mesophase is controlled mostly by the flexibility of the polymer backbone.

A large number of monomer pairs containing dissimilar mesogens and either identical or different spacers (case iii) were studied by Hardy *et al.* (Nyitrai, 1977; Hardy *et al.*, 1979, 1982; Horvath *et al.*, 1985*b*). In most cases, the plot of the liquid crystalline transition temperatures versus copolymer composition has a minimum; in several systems it is discontinuous. Although few of these copolymerizations were performed in both solution and thermotropic liquid crystalline phases, both types of copolymeriz-ations apparently follow the same composition regularities. However, copolymeriz-ation in the liquid crystalline phase results in lower rates of copolymerization.

Recently, monomer pairs whose mesogenic units are constitutional isomers (case iv) have been copolymerized (Percec *et al.*, 1987*b*, 1988; Percec, 1988*b*; Percec and Tomazos, unpublished data). Copolymerization transforms monotropic mesophases into enantiotropic mesophases, depresses side chain crystallization tendencies, de-creases the phase transition enthalpies, and at the same time preserves the original mesophase of the parent polymers.

Although even the copolymerization of the simplest nonmesogenic comonomer pairs is mechanistically disputed, more thorough studies of the copolymerization of

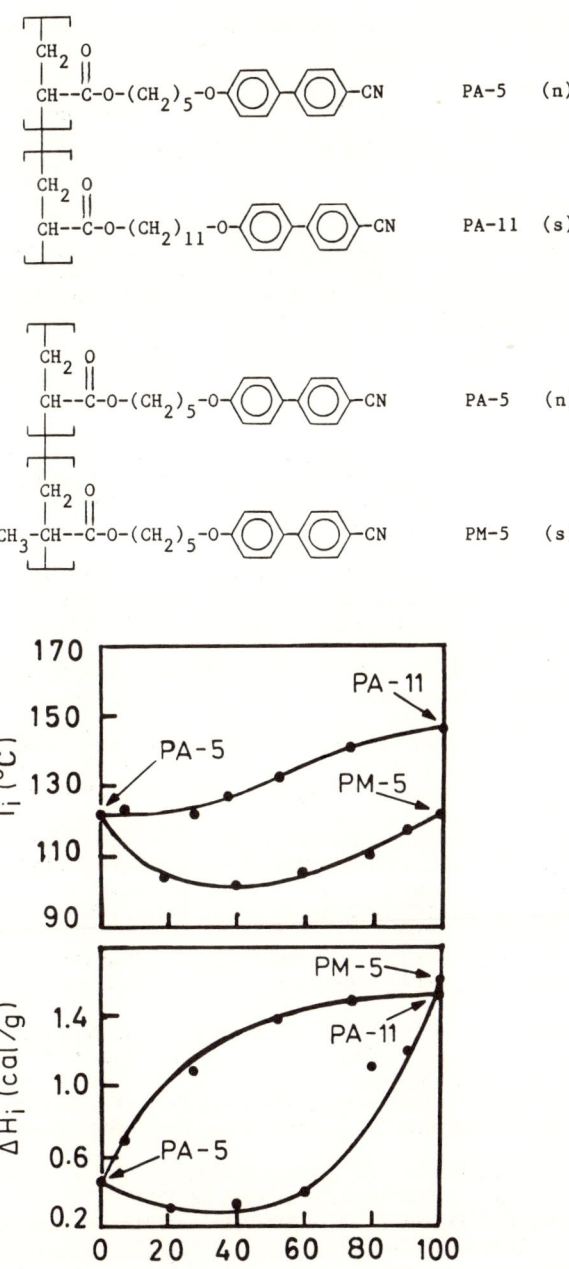

Figure 3.36 Clearing temperatures and the corresponding changes in enthalpy as a function of copolymer composition of 4-cyanobiphenyl-based mesogenic copolymers (Shibaev and Platé, 1984).

mesogenic monomers should be continued in order to solve problems of both fundamental and applied interests.

Another interesting topic in the field of LCCs is the synthesis of chloesteric copolymers by copolymerization of chiral with nonchiral monomers. Cholesteric side chain LCPs will be discussed in Chapter 9 of this book.

Acknowledgements

Financial support for work in this field from the National Science Foundation and the Office of Naval Research is gratefully acknowledged. The authors also express their gratitude to Professor H. Ringsdorf, Dr R. Zentel, Mr B. Reck and Mr V. Krone of the Institute of Organic Chemistry, University of Mainz, for their careful reading of this manuscript and for many helpful and critical discussions.

References

Abdullah, H.M., Gray, G.W. and Toyne, K.J. (1985) *Mol. Cryst. Liq. Cryst.* **124**, 105.

Alimoglu, A.K., Ledwith, A., Gemmell, P.A., Gray, G.W. and Lacy, D. (1984) *Polymer* **25**, 1342.

Apfel, M.A., Finkelmann, H., Janini, G.M., Laub, R.J., Lühmann, B.H., Price, A., Roberts, W.L., Shaw, T.J. and Smith, C.A. (1985) *Anal. Chem.* **57**, 651.

Attard, G.S. and Williams, G. (1986) *Chem. Brit.* 919.

Auman, B.C., Percec, V., Schneider, H.A. and Cantow, H.J. (1987a) *Polymer* **28**, 1407.

Auman, B.C., Percec, V., Schneider, H.A., Jishan, W. and Cantow, H.J. (1987b) *Polymer* **28**, 119.

Baccaredda, M., Magagnini, P.L., Pizzirani, G. and Giusti, P. (1971) *J. Polym. Sci. Polym. Lett.* **9**, 303.

Bader, H., Dorn, K., Hupfer, B. and Ringsdorf, H. (1985) *Adv. Polym. Sci.* **64**, 1.

Bata, L. (ed.) (1981) *Advances in Liquid Crystal Research and Applications.* vol. 2, Pergamon, New York.

Berg, S., Krone, V. and Ringsdorf, H. (1986) *Makromol. Chem., Rapid Commun.* **7**, 381.

Billard, J., Dubois, J.C., Tinh, N.H. and Zann, A. (1978) *Nuov. J. Chim.* **2**, 535.

Blumich, B., Boeffel, C., Harbison, G.S., Yang, Y. and Spiess, H.W. (1987) *Ber. Burgenges. Phys. Chem.* **91**, 1100.

Blumstein, A. (ed.) (1978) *Mesomorphic Order in Polymers and Polymerization in Liquid Crystalline Media*, ACS Symp. Ser. **74**, American Chemical Society, Washington DC.

Blumstein, A. (ed.) (1985) *Polymeric Liquid Crystals.* Plenum, New York.

Blumstein, A. and Hsu, E.C. (1978) in *Liquid Crystalline Order in Polymers*, ed. Blumstein, A., Academic Press, New York, 105.

Blumstein, A., Blumstein, R.B., Clough, S.B. and Hsu, E.C. (1975a) *Macromolecules* **8**, 73.

Blumstein, A., Clough, S.B., Patel, L., Kim, L.K., Hsu, E.C. and Blumstein, R.B. (1975b) *ACS Polym. Prepr.* **16**(2), 241.

Boeffel, C., Hisgen, B., Pschorn, U., Ringsdorf, H. and Spiess, H.W. (1983) *Israel J. Chem.* **23**, 388.

Boeffel, C., Spiess, H.W., Hisgen, B., Ringsdorf, H., Ohm, H. and Kirste, R.G. (1986) *Makromol. Chem., Rapid Commun.* **7**, 777.

Braun, D., Herr, R.P. and Arnold, N. (1987). *Makromol. Chem., Rapid Commun.* **8**, 359.

Bresci, V., Frosini, A., Lupinacci, D. and Magagnini, P.L. (1980) *Makromol. Chem., Rapid Commun.* **1**, 183.

Cabrera, I. and Krongauz, V. (1987a) *Nature (London)* **326**, 582.

Cabrera, I. and Krongauz, V. (1987b) *Macromolecules* **20**, 2713.

Cabrera, I., Krongauz, V. and Ringsdorf, H. (1987) *Angew. Chem. Int. Edn. Engl.* **26**, 1178.

Carr, N. and Gray, G.W. (1985) *Mol. Cryst. Liq. Cryst.* **124**, 27.

Carr, N., Goodwin, M.J., McRoberts, A.M., Gray, G.W., Marsden R. and Scrowston, R.M. (1987) *Makromol. Chem., Rapid Commun.* **8**, 487.

Carr, N., Gray, G.W. and Kelly, S.M. (1985) *Mol. Cryst. Liq. Cryst.* **129**, 301.

Chandrasekhar, S. (1982) in *Advances in Liquid Crystals*, vol. 5, ed. Brown, G.H., Academic Press, New York, 47.

Chandrasekhar, S. (1983) *Phil. Trans. R. Soc. London* **A309**, 93.

Chandrasekhar, S., Sadashiva, B.K. and Suresh, K.A. (1977) *Pramana* **9**, 471.

Chapoy, L.L. (ed.) (1985) *Recent Advances in Liquid Crystalline Polymers*. Elsevier Applied Science, London.

Chen, S. and Maa, Y.F. (1988) *Macromolecules* **21**, 904.

Clough, S.B., Blumstein, A. and DeVries, A. (1977) *ACS Polym. Prepr.* **18**(2), 1.

Cowie, J.M.G., Hag, Z. and McEwen I.J. (1979) *J. Polym. Sci. Polym. Lett. Ed.* **17**, 771.

Cowie, J.M.G., Hag, Z., McEwen, I.J. and Velickovic, J. (1981) *Polymer* **22**, 327.

Crivello, J., Deptolla, M. and Ringsdorf, H., (1988) *Liquid Crystals* **3**(2), 235.

Cser, F. (1984) in *Liquid Crystals and Ordered Fluids*, eds. Griffin, A.C. and Johnson, J.F. vol. 4, Plenum, New York, 945.

Cser, F., Nyitrai, K., Horvath, J. and Hardy, G. (1985) *Eur. Polym. J.* **21**, 259.

Davidson, P., Keller, P. and Levelut, A.M. (1985) *J. Phys.* **46**, 939.

Demus, D. and Zaschke, H. (1984) *Flussige Kristalle in Tabellen*, vol. 2, VEB Deutscher Verlag für Grundstoffindustrie, Leipzig.

Demus, D., Diele, S., Hauser, A., Latif, I., Selbmann, Ch. and Weissflog, W. (1985) *Cryst. Res. Technol.* **20**, 1547.

Destrade, C., Gasparoux, H., Foucher, P., Tinh, N.H., Malthete J. and Jacques, J. (1983) *J. Chim. Phys.* **80**, 137.

Destrade, C., Mouden M.C. and Malthete, J. (1979) *J. Phys. Paris* **40**, C3–17; C3–21.

Diele, S., Hisgen, B., Reck B. and Ringsdorf, H. (1986) *Makromol. Chem., Rapid Commun.* **7**, 267.

Diele, S., Oelsner, S., Kuschel, F., Hisgen, B., Ringsdorf H., and Zentel, R. (1987) *Makromol. Chem.* **188**, 1993.

Diele, S., Oelsner, S., Kuschel, F., Hisgen, B. and Ringsdorf, H. (1988) *Mol. Cryst. Liq. Cryst.* **155**, 399

Dietrich, H.J. and Steiger, E.L. (1972) *Mol. Cryst. Liq. Cryst.* **16**, 263.

Dubois, J.C. (1978) *Annls. Phys.* **3**, 131.

Dubois, J.C. and Billard, J. (1984) in *Liquid Crystals and Ordered Fluids*, vol. 4, eds. Griffin, A.C. and Johnson, J.F. Plenum, New York, 1043.

Dubois, J.C., Decobert, G., LeBarny, P., Esselin, S., Friedrich, C. and Nöel, C. (1986) *Mol. Cryst. Liq. Cryst.* **137**, 349.

Duran, R. and Gramain, P. (1987) *Makromol. Chem.* **188**, 2001.

Duran, R. and Strazielle, C. (1987) *Macromolecules* **20**, 2853.

Duran, R., Gramain, P., Guillon, D. and Skoulios, A. (1986) *Mol. Cryst. Liq. Cryst. Lett.* **3**, 23.

Duran, R., Guillon, D., Gramain, A. and Skoulios, A. (1987a) *Makromol. Chem., Rapid Commun.* **8**, 181.

Duran, R., Guillon, D., Gramain A. and Skoulios, A. (1987b) *Makromol. Chem., Rapid Commun.* **8**, 321.

Eich, M., Wendorff, J.H., Reck, B. and Ringsdorf, H. (1987) *Makromol. Chem., Rapid Commun.* **8**, 59.

Eidenschink, R. (1985) *Mol. Cryst. Liq. Cryst.* **123**, 57.

Elbert, R., Laschewsky, A. and Ringsdorf, H. (1985) *J. Am. Chem. Soc.* **107**, 4134.

Elias, H.G. (ed.) (1977) *Polymerization in Organized Systems*, Gordon and Breach, New York.

Endres, B.W., Wendorff, J.H., Reck, B. and Ringsdorf, H. (1987) *Makromol. Chem.* **188**, 1501.

Engel, M., Hisgen, B., Keller, R., Kreuder, W., Reck, B., Ringsdorf, H., Schmidt, H.W. and Tschirner, P. (1985) *Pure Appl. Chem.* **57**, 1009.

Everit, D.R.R., Care, C. and Wood, R. (1987) *Mol. Cryst. Liq. Cryst.* **153**, 55.

Ferry, J.D. (1980) *Viscoelastic Properties of Polymers*, 3rd edn., John Wiley, New York.

Finkelmann, H. (1982) in *Polymer Liquid Crystals*, eds. Ciferi, A., Krigbaum, W.R. and Meyer, R.B. Academic Press, New York, 35.

Finkelmann, H. (1983) *Phil. Trans. R. Soc. London* **A309**, 105.

Finkelmann, H. (1987a) *Angew. Chem. Int. Edn. Engl.* **26**, 816.

Finkelmann, H. (1987b) in *Thermotropic Liquid Crystals*, ed. Gray, G.W., John Wiley, New York, 145.

Finkelmann, H. and Rehage, G. (1984) *Adv. Polym. Sci.* **60/61**, 99.

Finkelmann, H., Happ, M., Portugall, M. and Ringsdorf, H. (1978a) *Makromol. Chem.* **179**, 2541.

Finkelmann, H., Ringsdorf, H., Siol, W. and Wendorff J.H. (1978b) in *Mesomorphic Order in Polymers and Polymerization in Liquid Crystalline Media*, ACS Symp. Ser. **74**, ed. Blumstein, A., American Chemical Society, Washington DC, 22.

Finkelmann, H., Ringsdorf, H. and Wendorff J.H. (1978c) *Makromol. Chem.* **179**, 273.
Finkelmann, H., Luhmann, B., Rehage, G. and Stevens H. (1984) in *Liquid Crystals and Ordered Fluids*, vol. 4, eds. Griffin, A.C. and Johnson, J.F., Plenum, New York, 715.
Flory, P.J. (1939) *J. Amer. Chem. Soc.* **61**, 3334.
Frosini, A., Levita, G., Lupinacci, D. and Magagnini, P.L. (1981) *Mol. Cryst. Liq. Cryst.* **66**, 21.
Geib, H., Hisgen, B., Pschorn, U., Ringsdorf, H. and Spiess, H.W. (1982) *J. Amer. Chem. Soc.* **104**, 917.
Gemmell, P.A., Gray, G.W. and Lacey, D. (1985a) *Mol. Cryst. Liq. Cryst.* **122**, 205.
Gemmell, P.A., Gray, G.W., Lacey, D., Alimoglu, A.K. and Ledwith, A. (1985b) *Polymer* **26**, 615.
Goldburt, E.S., Shvartsman, F. and Krongauz, V.A. (1984) *Macromolecules* **17**, 1876.
Gray, G.W. (1962) *Molecular Structure and the Properties of Liquid Crystals*, Academic Press, London.
Gray, G.W. (1979) in *The Molecular Physics of Liquid Crystals*, eds. Luckhurst, G.R. and Gray, G.W., Academic Press, London, chapters 1 and 12.
Gray, G.W. (1981) *Mol. Cryst. Liq. Cryst.* **63**, 3.
Gray, G.W. (1982) in *Polymer Liquid Crystals*, eds Ciferi, A., Krigbaum, W.R. and Meyers, R.B. Academic Press, London, 1.
Gray, G.W. (1983) *Phil. Trans. R. Soc. London* **A309**, 77.
Gray, G.W. (1985) *Proc. R. Soc. London* **A402**, 1.
Gray, G.W. and McDonnell, D.G. (1979) *Mol. Cryst. Liq. Cryst.* **53**, 147.
Gray, G.W. and Winsor, P.A. (1974) in *Liquid Crystals and Plastic Crystals*, vol. 1, eds Gray, G.W. and Winsor, P.A. Ellis Horwood, Chichester, vol. 1., 4.1.
Gray, G.W., Lacey, D. Nestor, G. and White, M.S. (1986) *Makromol. Chem., Rapid Commun.* **7**, 71.
Griffin, A.C. and Johnson, J.F. (eds.) (1982) *Liquid Crystals and Ordered Fluids*, vol. 4, Plenum, New York.
Hahn, B. and Percec, V. (1987) *Macromolecules* **20**, 2961.
Hahn, B. and Percec, V. (1988) *Mol. Cryst. Liq. Cryst.* **157**, 125.
Hahn, B., Wendorff, J.H., Portugall, M. and Ringsdorf, H. (1981) *Coll. Polym. Sci.* **259**, 875.
Haramoto, Y. and Kamogawa, H. (1983) *J. Chem. Soc., Chem. Commun.* 73.
Haramoto, Y. and Kamogawa, H. (1985a) *Mol. Cryst. Liq. Cryst.* **131**, 101.
Haramoto, Y. and Kamogawa, H. (1985b) *Mol. Cryst. Liq. Cryst.* **131**, 201.
Haramoto, Y., Kazawa, K.A. and Kamogawa, H. (1984a) *Bull Chem. Soc. Japan* **57**, 3173.
Haramoto, Y., Nobe, A. and Kamogawa, H. (1984b) *Bull. Chem. Soc. Japan* **57**, 1966.
Haramoto, Y., Sano M. and Kamogawa, H. (1986) *Bull. Chem. Soc. Japan* **59**, 1337.
Hardy, G., Cser, F. and Nyitrai, K. (1979) *Israel J. Chem.* **18**, 233.
Hardy, G., Cser, F., Nyitrai, K. and Bartha, E. (1982) *Ind. Eng. Chem. Prod. Res. Rev.* **21**, 321.
Helfrich, W. and Heppke, G. (eds.) (1980) *Liquid Crystals of One- and Two-Dimensional Order*, Springer Series in Chemical Physics **11**, Springer Verlag, Berlin.
Hessel, F. and Finkelmann, H. (1985a) *Polym. Bull.* **14**, 375.
Hessel, F. and Finkelmann, H. (1985b) *Polym. Bull.* **15**, 349.
Hessel, F., Herr, R.P. and Finkelmann, H. (1987) *Makromol. Chem.* **188**, 1597.
Hisgen, B. and Ringsdorf, H. (1983) *13th Freiburger Arbeitstagung Flüssigkristalle*, Freiburg.
Horvath, J., Cser, F. and Hardy, G. (1985a) *Progr. Coll. Polym. Sci.* **71**, 59.
Horvath J., Nyitrai, K., Cser, F. and Hardy, G. (1985b) *Eur. Polym. J.* **21**, 251.
Hsu, C.S. and Percec, V. (1987a) *Polym. Bull.* **17**, 49.
Hsu, C.S. and Percec, V. (1987b) *Polym. Bull.* **18**, 91.
Hsu, C.S. and Percec, V. (1987c) *Makromol. Chem., Rapid Commun.* **8**, 331.
Hsu, C.S. and Percec, V. (1987d) *J. Polym. Sci., Polym. Chem. Ed.* **25**, 2909.
Hsu, C.S. and Percec, V. (1988a) *J. Polym. Sci., Polym. Chem. Ed.* (in press).
Hsu, C.S. and Percec, V. (1988b) *Makromol. Chem.* **189**, 1141.
Hsu, C.S., Rodriguez-Parada, J.M. and Percec, V. (1987a) *Makromol. Chem.* **188**, 1017.
Hsu, C.S., Rodriguez-Parada, J.M. and Percec, V. (1987b) *J. Polym. Sci. Polym. Chem. Ed.* **25**, 2425.
Huu, T.N., Dubois, J.C., Malthete, J. and Destrade, C. (1978) *C.R. Acad. Sci. Paris* **286c**, 463.
Kapitza, H. and Zentel, R. (1988) *Makromol. Chem.* (in press).
Kelker, H. and Hatz, R. (1980) *Handbook of Liquid Crystals*, Verlag Chemie, Weinheim.
Keller, P. (1984) *Macromolecules* **17**, 2937.
Keller, P. (1985a) *Makromol. Chem., Rapid Commun.* **6**, 707.
Keller, P. (1985b) *Mol. Cryst. Liq. Cryst. Lett.* **2**, 101.

Keller, P. (1985c) *Macromolecules* **18**, 2337.

Keller, P. (1987) *Macromolecules* **20**, 462.

Keller, P., Carvalho, B., Cotton, J.P., Lambert, M., Moussa, F. and Pepy, G. (1985) *J. Phys. Lett.* **46**, L-1065.

Kelly, S.M. and Schad, H. (1985) *Helv. Chim. Acta.* **68**, 1444.

Kim, C. and Allcock, H.R. (1987) *Macromolecules* **20**, 1726.

Kirste, R.G. and Ohm H.G. (1985) *Makromol. Chem., Rapid Commun.* **6**, 179.

Koide, N. (1986) *Mol. Cryst. Liq. Cryst.* **139**, 47.

Konstantinov, I.I., Sitnov, A.A., Grebneva, V.S. and Amerik, Y.B. (1983) *Eur. Polym. J.*, **19**, 327.

Kostromin, S.G., Shibaev, V.P. and Platé, N.A. (1987) *Liquid Crystals* **2**, 195.

Kostromin, S.G., Simitsyn, V.V., Talroze, R.V. and Shibaev, V.P. (1984) *Polym. Sci. USSR* **26**, 370.

Kostromin, S.G., Simitsyn, V.V., Talroze, R.V., Shibaev, V.P. and Platé, N.A. (1982a) *Makromol. Chem., Rapid Commun.* **3**, 809.

Kostromin, S.G., Talroze, R.V., Shibaev, V.P. and Platé, N.A. (1982b) *Makromol. Chem., Rapid Commun.* **3**, 803.

Kreuder, W. and Ringsdorf, H. (1983) *Makromol. Chem., Rapid Commun.* **4**, 807.

Kreuder, W., Webster, O.W. and Ringsdorf, H. (1986) *Makromol. Chem., Rapid Commun.* **7**, 5.

Krigbaum, W.R. (1985) *J. Appl. Polym. Sci., Appl. Polym. Symp.* **41** 105.

Krongauz, V.A. and Goldburt, E.S. (1981) *Macromolecules* **114**, 1382.

Kunitake, T., Nayai, M., Yanagi, H., Takarabe, K. and Nakashima, N. (1984) *J. Macromol. Sci. -Chem.* **A21**, 1237.

Laschewsky, A., Ringsdorf, H., Schmidt, G. and Schneider, J. (1987) *J. Amer. Chem. Soc.* **109**, 788.

Le Barny, P. (1987) *Thin Solid Films* **152**, 99.

Le Barny, P., Dubois, J.C., Friedrich, C. and Nöel, C. (1986a) *Polym. Bull.* **15**, 341.

Le Barny, P., Ravanx, G., Dubois, J.C., Parneix, J.P., Njenmo, R., Legrand, C. and Levelut, A.M. (1986b) *SPIE* **682**, 56.

Lipatov, Yu. S., Tsukruk, V.V. and Shilov, V.V. (1983) *Polym. Commun.* **24**, 75

Lipatov, Yu. S., Tsukruk, V.V. and Shilov, V.V. (1984) *J. Macromol. Sci. Rev. Macromol, Chem. Phys.* **C24**, 173.

Luckhurst, G.R. and Gray, G.W. (1979) in *The Molecular Physics of Liquid Crystals*, eds. Luckhurst, G.R. and Gray, G.W. Academic Press, London, 1.

Lupinacci, D., Frosini, V. and Magagnini, P.L. (1980) *Makromol. Chem., Rapid Commun.* **1**, 671.

Magagnini, P.L. (1981) *Makromol. Chem., Suppl.* **4**, 223.

Magagnini, P.L., Marchetti, A., Matsa, F., Pizzirani, G. and Turchi, G. (1974) *Eur. Polym. J.* **10**, 585.

Mathias, L.J. and Carraher, C.E. Jr. (1984) *Crown Ethers and Phase Transfer Catalysis in Polymer Science*, Plenum, New York.

Mauzac, M., Hardouin, F., Richard, H., Achard, M.F., Sigaud, G. and Gasparoux, H. (1986) *Eur. Polym. J.* **22**, 137.

Monnerie, L. (1985) *Pure Appl. Chem.* **57**, 1563.

Morawetz, A. (1987) in *Chemical Reactions on Polymers*, eds. Benham, J.L. and Kinstle, J.F. ACS Symp. Ser. **364**, American Chemical Society, Washington DC, 317.

Nestor, G., White, M.S., Gray, G.W., Lacey, D. and Toyne, K.J. (1987) *Makromol. Chem.* **188**, 2759.

Nyitrai, K., Cser, F., Lengyel, M., Seyfried, E. and Hardy, G. (1977) *Eur. Polym. J.* **13**, 673.

Ober, C.K., Jin, J.I. and Lenz, R.W. (1984) *Adv. Polym. Sci.* **59**, 103.

Osman, M.A. (1982a) *Mol Cryst. Liq. Cryst. Lett.* **72**, 291.

Osman, M.A. (1982b) *Mol. Cryst. Liq. Cryst. Lett.* **82**, 47, 295.

Osman, M.A. (1983a) *Z. Naturforsch.* **38a**, 693.

Osman, M.A. (1983b) *Z. Naturforsch.* **38a**, 779.

Osman, M.A. and Huynh-Ba, T. (1983) *Helv. Chim. Acta.* **66**, 1786.

Paleos, C.M., Filippakis, S.E. and Leonidopoulou, G.M. (1981) *J. Polym. Sci., Polym. Chem. Ed.* **19**, 1427.

Paleos, C.M., Leonidopoulou, G.M., Filippakis, S.E., Malliaris, A. and Dais, P. (1982) *J. Polym. Sci., Polym. Chem. Ed.* **20**, 2267.

Percec, V. (1987) in *Phase Transfer Catalysis. New Chemistry, Catalysts and Applications*, ACS Symp. Ser. **326**, ed. Starks, C.M., American Chemical Society, Washington DC, 96.

Percec, V. (1988) *Makromol. Chem., Symp.* **13–14**, 397.

Percec, V. (1988b) *Mol. Cryst. Liq. Cryst.* **155**, 1.
Percec, V. and Pugh, C. (1987) in *Encyclopedia of Polymer Science and Engineering* vol. 10, eds. Mark, H.F., Bikales, N.M., Overberger, C.G., Menges, G. and Kroschwitz, J., John Wiley, New York, 432.
Percec, V. and Tomazos, D. (1987) *Polym. Bull.* **18**, 239.
Percec, V., Rodriguez-Parada, J.M. and Ericsson, C. (1987a) *Polym. Bull.* **17**, 347.
Percec, V., Rodriguez-Parada, J.M., Ericsson, C. and Nava, H. (1987b) *Polym. Bull.* **17**, 353.
Percec, V., Hsu, C.S. and Tomazos, D. (1988) *J. Polym. Sci., Polym. Chem. Ed.* **26**, 2047.
Pezzin, G., Ceccorulli, G. and Leonidopoulou, G.M. (1975) *Macromolecules* **8**, 762.
Platé N.A. and Shibaev, V.P. (1984) *Makromol. Chem., Suppl.* **6**, 3.
Platé, N.A. and Shibaev, V.P. (1987) *Comb-Shaped Polymers and Liquid Crystals*, Plenum, New York.
Platé, N.A., Friedzon, Y.S. and Shibaev, V.P. (1985) *Pure Appl. Chem.* **57**, 1715.
Platé, N.A., Talroze, R.V. and Shibaev, V.P. (1984a) *Pure Appl. Chem.* **56**, 403.
Platé, N.A., Talroze, R.V. and Shibaev, V.P. (1984b) *Makromol. Chem. Suppl.* **8**, 47.
Platé, N.A., Talroze, R.V., Freidzon, Y.S. and Shibaev, V.P. (1987) *Polym. J.* **19**, 135.
Poliks, M.D., Park, Y.W. and Samulski, E.T. (1987) *Mol. Cryst. Liq. Cryst.* **153**, 321.
Portugall, M., Ringsdorf, H. and Zentel, R. (1982) *Makromol. Chem.* **183**, 2311.
Pugh, C. and Percec, V. (1985) *ACS Polym. Prepr.* **26**(2), 303.
Pugh, C. and Percec, V. (1986a) *Polym. Bull.* **16**, 513.
Pugh, C and Percec, V. (1986b) *Polym. Bull* **16**, 521.
Pugh, C. and Percec, V. (1987) in *Chemical Reactions on Polymers*, eds. Benham, J.L. and Kirste, J.F. ACS Symp. Ser. **364**, American Chemical Society, Washington DC, 97.
Reck, B. and Ringsdorf, H. (1985). *Makromol. Chem., Rapid Commun.* **6**, 291.
Reck, B. and Ringsdorf, H. (1986) *Makromol. Chem., Rapid Commun.* **7**, 389.
Rim, P.B. (1985) *J. Macromol. Sci. -Phys.* **B23**, 549.
Rim, P.B. (1986) *Polym. Commun.* **27**, 199.
Ringsdorf, H. and Schmidt H.W. (1984) *Makromol. Chem.* **185**, 1327.
Ringsdorf, H. and Schneller, A. (1981) *Br. Polym. J.* **13**, 43.
Ringsdorf, H. and Schneller, A. (1982) *Makromol. Chem., Rapid Commun.* **3**, 557.
Ringsdorf, H., Schmidt, H.W., Baur, G. and Kiefer, R. (1985) in *Recent Advances in Liquid Crystalline Polymers*, ed. Chapoy, L.L., Elsevier Applied Science, London. 253.
Ringsdorf, H., Schmidt, H.W., Baur, G., Kiefer, R. and Windscheid, F. (1986) *Liq. Cryst.* **1**, 319.
Ringsdorf, H., Schmidt, G. and Schneider, J. (1987a) *Thin Solid Films* **152**, 207.
Ringsdorf, H., Schmidt, H.W., Ellingsfeld, H. and Etzbach, K.H. (1987b) *Makromol. Chem.* **188**, 1355.
Ringsdorf, H., Schneider, J. and Schuster, A. (1987c) *3rd Int. Conf. on Langmuir–Blodgett Films*, Gottingen, F.R.G., July 26–31 (1987).
Ringsdorf, H., Schlarb, B. and Venzmer, J., (1988) *Angew. Chem. Int. Edn. Engl.* **27**, 113.
Rodriguez-Parada, J. and Percec, V. (1986) *J. Polym. Sci. Polym. Chem. Ed.* **24**, 1363.
Rodriguez-Parada, J. and Percec, V. (1987) *J. Polym. Sci. Polym. Chem. Ed.* **25**, 2269.
Rodriguez-Parada, J., Percec, V. and Saeva, F. (1981) *Matromol. Chem. Suppl.* **5**, 58.
Seto, K., Takahashi, S. and Tahara, T. (1985a) *J. Chem. Soc., Chem. Commun.* **122**.
Seto, K., Takahashi, S. and Tahara, T. (1985b) *Mol. Cryst. Liq. Cryst., Lett.* **2**, 197.
Shibaev, V.P. and Platé, N.A. (1977) *Polym. Sci. USSR* **19**, 1065.
Shibaev, V.P. and Platé, N.A. (1984) *Adv. Polym. Sci.* **60/61**, 173.
Shibaev, V.P. and Platé, N.A. (1985) *Pure Appl. Chem.* **57**, 1589.
Shibaev, V.P., Kostromin, S.G. and Platé, N.A. (1982) *Eur. Polym. J.* **18**, 651.
Singler, R.E., Willingham, R.A., Lenz, R.W., Furukawa, A. and Finkelmann, H. (1987) *Macromolecules* **20**, 1727.
Sperling, L.H. (1986) *Introduction to Physical Polymer Science*, John Wiley, New York.
Spiess, H.W. (1985) *Pure Appl. Chem.* **57**, 1617.
Springer, J. and Weigelt, F.W. (1983a) *Makromol. Chem.* **184**, 1489.
Springer, J. and Weigelt, F.W. (1983b) *Makromol. Chem.* **184**, 2635.
Springer, J. and Weigelt F.W. (1985) in *Recent Advances in Liquid Crystalline Polymers*, ed. Chapoy, L.L., Elsevier Applied Science, London, 233.
Stevens, H., Rehage, G. and Finkelmann, H. (1984) *Macromolecules* **17**, 851.
Strzelecki, L. and Liebert, L. (1973) *Bull. Soc. Chim. France* 597.

Strzelecka, H., Gionis, V., Rivory, J. and Flandrois, S. (1983) *J. Phys., Paris* **44C3**, 1201.
Sucrow, W., Lüschen, R. and Risse, A. (1985) *Z. Naturforsch.* **B40**, 420.
Sugai, S., Kamashima, K., Makino, S. and Noguchi, J. (1966) *J. Polym. Sci. A-2* **4**, 183.
Takatsu, H., Takeuchi, K. and Sato, H. (1984) *Mol. Cryst. Liq. Cryst.* **111**, 311.
Takenaka, S., Masuda, Y. and Kusabayashi, S. (1986) *Chem. Lett.* 751.
Talroze, R.V., Shibaev, V.P. and Platé, N.A. (1983) *Polym. Sci. USSR* **25**, 2863.
Tinh, N.H., Gasparoux, H. and Destrade, C. (1985). *Mol. Cryst. Liq. Cryst.* **123**, 271.
Tirrel, D.A. (1986) in *Encyclopedia of Polymer Science and Engineering*, vol. 4, eds. Mark, H.F.,
    Bilales, N.M., Overberger, C.G., Menges, G. and Kroschwitz, J., John Wiley, New York, 192.
Toyne, K.J. (1987) in *Thermotropic Liquid Crystals*, ed. Gray, G.W., John Wiley, New York, 28.
Tredgold, R.H. (1987) *Thin Solid Films* **152**, 223.
Tredgold, R.H., Young, M.C.J., Hodge, P. and Khoshdel, E. (1987) *Thin Solid Films* **151**, 441.
Tsukruk, V.V., Shilov, V.V. and Lipatov, Yu.S. (1985) *Acta Polymerica* **36**, 403.
Tsukruk, V.V., Shilov, V.V. and Lipatov, Yu.S. (1986) *Macromolecules* **19**, 1308.
Tsutsumi, A., Hikichi, K., Takahashi, T., Yamashita, Y., Matsushima, N., Kanake, M. and
    Kaneko, M. (1973) *J. Macromol. Sci. -Phys.* **8**, 413.
Vickers, A.J., Tredgold, R.H., Hodge, P., Khoshdel, E. and Girling, I. (1984) *Thin Solid Films* **8**, 47.
Vickers, A.J., Tredgold, R.J., Hodge, P., Khoshdel, E. and Girling, I. (1985) *Thin Solid Films*, **134**,
    43.
Wassmer, K.H., Ohmes, E. Portugal, M., Ringsdorf, H. and Kothe, G. (1985) *J. Amer. Chem. Soc.*
    107, 1511.
Weissflog, W. and Demus, D. (1983) *Cryst. Res. Technol.* **18**, K21.
Weissflog, W. and Demus, D. (1984) *Cryst. Res. Technol.* **19**, 55.
Weissflog, W. and Demus, D. (1985) *Mol. Cryst. Liq. Cryst.* **129**, 235.
Wendorff, J.H. (1978) in *Liquid Crystalline Order in Polymers*, ed. Blumstein, A., Academic Press,
    New York, 1.
Westphal, S., Diele, S., Mädicke, A., Kuschel, F., Scheim, U., Rühlmann, K., Hisgen, B. and
    Ringsdorf, H. (1988) *Makromol. Chem., Rapid Commun.* **9**, 489.
Wismontski, T.K. and Krongauz, V.A. (1985) *Macromolecules* **18**, 2124.
Yamashita, Y., Tsutsumi, A., Hikichi, K. and Kaneko, M. (1975) *Polym. J.* **8**, 114.
Zaschke, H., Isenberg, A. and Vorbrodt H.M. (1984) in *Liquid Crystals and Ordered Fluids*, vol. 4,
    eds. Griffin, A.C. and Johnson, J.F. Plenum, New York, 75.
Zentel, R. and Ringsdorf, H. (1984) *Makromol. Chem., Rapid Commun.* **5**, 393.
Zentel, R., Reckert, G. and Reck, B. (1987a) *Liquid Crystals* **2**, 83.
Zentel, R., Schmidt, G.F., Meyer, J. and Benalia, M. (1987b) *Liquid Crystals* **2**, 651.
Zhou, Q.F., Li, H.M. and Feng, X.D. (1987) *Macromolecules* **20**, 233.

4 Synthesis and properties of side chain liquid crystal polysiloxanes

G.W. GRAY, School of Chemistry, The University of Hull, Hull HU6 7RX, UK.

4.1 Introduction

Recognition that polymers may exhibit liquid crystal properties is comparatively recent—some 20 years—but already such polymers command great theoretical and experimental interest, not only because of the deeper insight that their study lends to understanding of ordering processes in both polymers and liquid crystals, but also because of their potential in the field of applications. Indeed, possible technological uses for liquid crystal polymers have been pursued with unusual vigour, obviously a consequence of the great technological impact of their counterparts of low molar mass (LMM) in the field of electro-optical displays. Certainly, much of the research on liquid crystal polymers (LCP) in recent years has been intimately bound up with materials science and the quest for polymers with structures tailored for the best combination of properties deriving from the fluid but anisotropic behaviour of LCs and the polymer-specific characteristics of macromolecules.

LCPs belong to two distinct classes, in each of which the concept of the flexible spacer unit is vital. These are the main chain LCPs and side chain LCPs, abbreviated in this chapter to MCLCP and SCLCP respectively. As already made clear in preceding chapters, SCLCPs have a rather flexible polymer backbone to which are appended, usually via flexible spacer groups, the mesogenic side chains. The pendent groups in these comb-like polymers therefore have molecular structures compatible with LC or mesophase formation as in LMM systems (Toyne, 1987; Gray, 1982). As a consequence, if the mesogenic side chains are rod-like (calamitic) in nature, the SCLCP may, depending upon its detailed structure, exhibit any of the accepted calamitic phases: nematic (N), cholesteric (N*), or smectic (S), including the various smectic polymorphic types (S_A, S_B, S_C, etc.) (Gray and Goodby, 1984). Similarly, discotic side chains (SCs) give discotic phases and amphiphilic SCs give amphiphilic phases.

4.2 Structural types (homo- and copolymers)

SCLCPs may also be of different types as determined by the nature of the backbone. The backbones that have been most commonly employed are those of the acrylate, methacrylate, and siloxane types. This chapter narrows consideration to the polysiloxanes for which the diversity of structure is nevertheless still wide, not only because the pendent groups may be varied widely from one homopolymer to another, but also because copolymers are possible by virtue of using two or more different pendent groups on a homopolymer backbone or a backbone which is itself copolymeric in composition and has one or more than one type of pendent group. These possibilities for SCLC polysiloxanes are summarized in the structures below, where R_1, R_2 represent the mesogenic SCs. The structures for the copolymers are intended to represent statistical systems, although block copolymers can also be conceived.

Homopolymer

$$---(SiMe—O)_n---$$
$$|$$
$$R_1$$

SC copolymers

$$---(SiMe—O)_x---(SiMe—O)_y---$$
$$| \qquad |$$
$$R_1 \qquad R_2$$

Backbone copolymers

$$---(SiMe—O)_x---(SiMe_2—O)_y---$$
$$|$$
$$R_1$$

Backbone/side chain copolymers

$$---(SiMe—O)_x---(SiMe—O)_y---(SiMe_2—O)_z---$$
$$| \qquad |$$
$$R_1 \qquad R_2$$

Typical values for $n, x + y$, and $x + y + z$ would range from 30 to 120.

The occurrence of an LC phase in LCPs depends upon the ability of the mesogenic groups to arrange themselves anisotropically. This is opposed by the tendency for the backbone to adopt a statistical chain conformation. If the mesogenic groups are attached directly and rather rigidly to the backbone, the dynamics of the backbone dominate and no LC phases are formed. If the mesogenic groups are appended to the backbone by flexible spacer groups such as $(CH_2)_n$, the mesogenic groups may now adopt the anisotropic ordering of an LC phase. This does not mean that the nature of that order is uninfluenced by the backbone, or that the backbone is uninfluenced by the ordering of the side chains (Casagrande *et al.*, 1984; Strzelecki *et al.*, 1973). However, the decoupling effect makes anisotropic SC ordering possible, and the result is an SCLCP. (For a detailed discussion see Chapter 3.)

Achievement of ordering of the SCS is also affected by the flexibility of the backbone itself. In this respect, polysiloxanes are particularly good, as they have a backbone of alternating silicon and oxygen atoms:

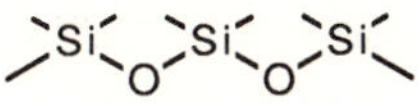

Although the Si—O bond is strong (bond dissociation energy $= 452\,kJ\,mol^{-1}$, as compared to $226\,kJ\,mol^{-1}$ for Si—Si and $333\,kJ\,mol^{-1}$ for Si—C), accounting for the high thermal stability of poly(organosilanes) (West, 1980), the rotational potential energy of the Si—O bond is low. As a result of this, and the low intermolecular forces consequent upon the shielding effect of the organic groups on the backbone, conformational changes in the backbone may occur easily, assisted by the possibility

Table 4.1 Comparison of various SCLCP transition temperatures.

Polymer	$T_g(K)$	$T_{N \rightarrow I}(K)$
—CH$_2$—CMe— $\quad\quad$ | $\quad\quad$ CO$_2$R	369	394
—CH$_2$—CH— $\quad\quad$ | $\quad\quad$ CO$_2$R	320	350
—O—SiMe— $\quad\quad$ | $\quad\quad$ CH$_2$R	276	334

R = —(CH$_2$)$_2$O—⬡—CO$_2$—⬡—OMe

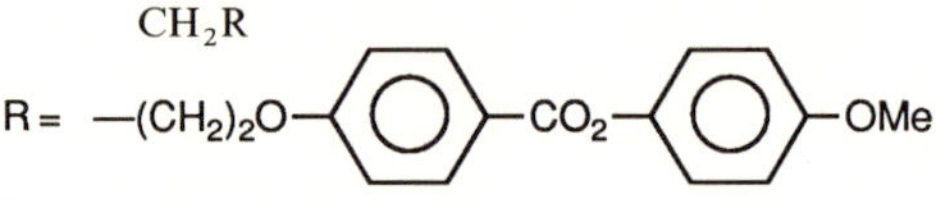

for wide variations in the Si—O—Si bond angle. These effects are much greater than those for the backbones of hydrocarbon polymers.

As far as comparisons can be made between polymers with different backbones and varying in $\overline{DP}$ and polydispersity, SCLC polysiloxanes do have lower T_g values and lower mesophase transition temperatures than their acrylate/methacrylate counterparts. This suggests that ordering of the SCs by the more flexible siloxane backbone is less good. Data from Finkelmann and Rehage (1984) illustrate this point well in relation to $T_{N \rightarrow I}$ values (Table 4.1).

Further evidence for the greater flexibility of the O—Si—O unit comes (Creed *et al.*, 1987) from LMM systems of structure

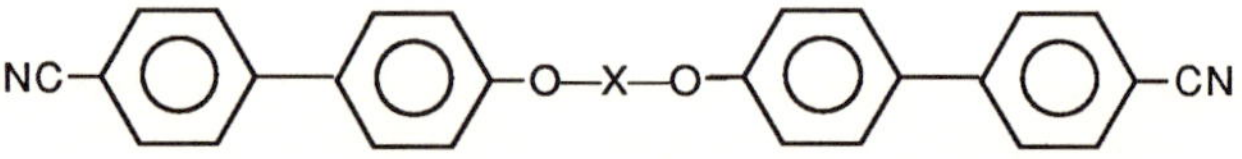

NC—⬡—⬡—O—X—O—⬡—⬡—CN

where   X = PM = —(CH$_2$)$_5$—,   EO = —(CH$_2$)$_2$O(CH$_2$)$_2$—,   DS = —CH$_2$SiMe$_2$OSiMe$_2$CH$_2$—.

$T_{N \rightarrow I}$ falls in the order PM > EO > DS, and this may be interpreted as a failure of the O—Si—O unit to propagate effectively a linear structure from one half of the molecule to the other.

The lower mesophase transition temperatures for polysiloxanes are not in fact disadvantageous, as these values can be corrected upwards by a suitable choice of mesogenic SCs.

The lower T_g values are, however, an advantage if we require the polymer to be an LC at room temperature, as for example with cholesteric polysiloxanes needed to monitor temperatures at or below ambient.

After it became clear that SCLC polymers were no competitors to LMM LC materials in applications in fast switching electro-optical displays, because of their very slow response times, attention switched to their applications as materials into which information could be written and then stored. This can be done effectively by writing data at a temperature above T_g and storing the data at temperatures below T_g (in the glassy state). This requires polymers with T_g above ambient temperatures. In this situation, the choice may be for an acrylate polymer. However, there are strong advocates for smectic polysiloxanes (Coles and Simon, 1985a, b, 1986), using these, the data may be stored in the viscous smectic state, with T_g below room temperature. This has been done very successfully (McArdle et al., 1987), but doubts do remain about the durability of the data storage, as already rather stringent criteria are indicated for archivability of storage devices, and doubts that these are met by smectic storage have been expressed (see also Chapter 13).

The arguments relating to higher v lower T_g values cannot be regarded as conclusive, however, and the prospects for SCLC polysiloxanes that are in the mesophase at ambient temperatures in the area of applications are very good. Moreover, from the standpoint of studies of the physical properties of polymer LC phases, there is definite merit in the ability to conduct measurements at ambient temperatures.

SCLC polysiloxanes therefore remain an area of strong research activity, and the discovery by Ringsdorf et al. (1982, 1987, 1988), that copolymers (4.1)

$$---(SiMe_2-O)_y---(SiMe-O)_x---$$

$$|$$
$$R$$

(4.1)

$$\text{where} \quad R = -(CH_2)_mO-\bigcirc-\bigcirc-OMe$$

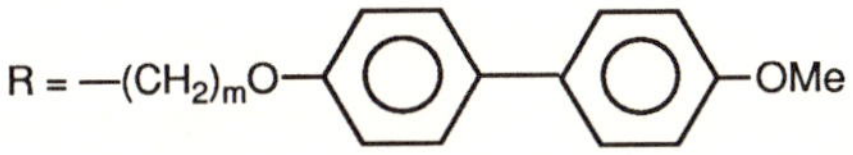

derived from a poly(hydrogenmethyldimethylsiloxane) had glass transitions well below those of their homopolymeric counterparts with the same mesogenic side chains, and indeed well below 0°C, was one of very great interest. Results in Table 4.2 show the extent of the effects and a comparison of the S → I values for a constant value of m, say, 11, shows the diminishing smectic thermal stability as the population of mesogenic groups becomes smaller and the backbone becomes correspondingly more flexible, i.e., less influenced by the mesogenic ordering processes.

SCLC polysiloxanes need not of course show glass transitions; often they exhibit crystalline properties. This fact, and the influence of longer spacers in increasing (i) the degree of anisotropic LC ordering from nematic to smectic and (ii) the tendency of the polymer to crystallize, are shown by the idealized DSC data (Finkelmann and Rehage,

Table 4.2 Transition temperatures for copolymers (4.1)

y/x	m	Phase transitions (K)
3/1	3	g 228 I
3/1	5	g 243 S 296 I
3/1	11	g 242 S 350 I
5/1	3	g 220 I
5/1	5	g 225 S 279 I
5/1	11	g 227 S 337 I
10/1	3	g 161 I
10/1	5	g 159 I
10/1	11	g 159 S 257 I

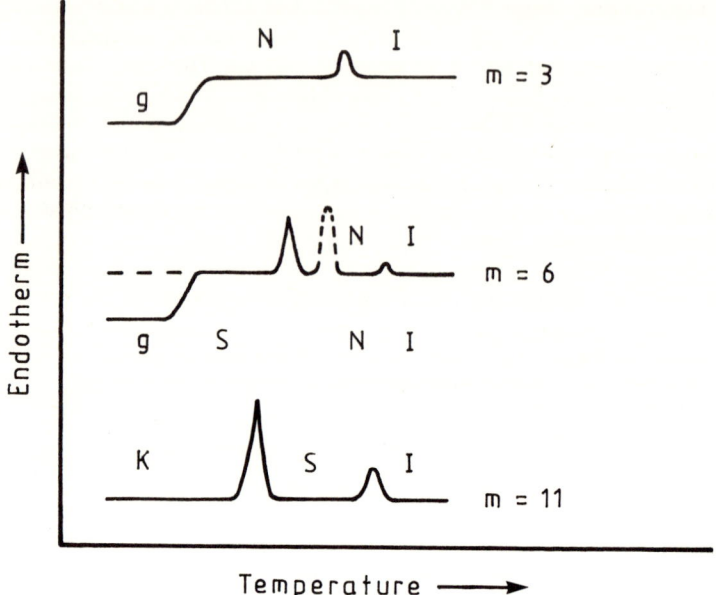

Figure 4.1 DSC curves (g = glass, S = smectic, N = nematic, K = crystal, I = isotropic) for

$$--- (O—Si(Me)—(CH_2)_mO-\text{⟨O⟩}-CO_2-\text{⟨O⟩}-OMe)_n ---$$

1980*a, b*, 1984; Hawthorne, 1986) in Figure 4.1. Maintaining the SC constant except for increasing the spacer length from 3 to 6, we pass from a polymer with a glass transition and an enantiotropic N phase, to one with glass, S, and N phases above T_g. On cooling and annealing, however, the N phase crystallizes, and on reheating, only a crystal–N transition, as shown by the dotted peak, and the N → I transition would be observed. At $m = 11$, the polymer is crystalline and has a melting point, at which temperature an enantiotropic S phase is formed.

4.3 Methods of synthesis

4.3.1 *General aspects*

SCLC polysiloxanes are usually prepared by a *polymer modification* process, commencing with a preformed polymer backbone which contains reactive functional groups (Si–H) to which the mesogenic groups are appended. During the reaction, monitoring is carried out to ensure complete reaction of the Si–H functions with the terminal alkene which is to constitute the mesogenic SC. The process, shown in Figure 4.2, is known as hydrosilylation.

The preformed backbone may be chosen to produce a homopolymer, i.e. every Si carries a reactive H and all the alkene molecules are the same, or a backbone copolymer, in which case the backbone has a statistical distribution of —SiH(Me)— —O— and SiMe$_2$—O— groups in a particular ratio.

The poly(hydrogenmethylsiloxane), i.e. homopolymer backbone, or the

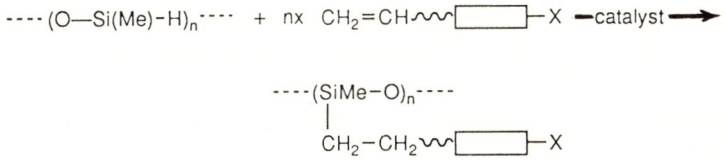

Figure 4.2 The hydrosilylation reaction to give a SCLCP

poly(hydrogenmethyldimethylsiloxane), i.e. copolymer backbone, is usually purchased, and the materials sold commercially are normally polydisperse. They are usually manufactured to meet some viscosity requirement, and not to have a specific value of $\overline{DP}$ or $\overline{M}_w/\overline{M}_n$. Such backbone material gives less than well-defined SCLCP, but there is an added problem that the $\overline{DP}$ values for commercial backbones are quite low, particularly for copolymer backbones. Even in the case of poly(hydrogenmethylsiloxanes) they are usually in the range 35–60 or at most 95–120. This puts the SCLCPs in the molar mass range in which their properties are dependent upon $\overline{DP}$; that is T_g, phase transition temperatures, etc. are influenced by $\overline{DP}$. Doubt therefore arises about the validity of comparisons of the physical properties of materials made by different researchers unless the characteristics of the backbone used and the SC polymer are specified, as already emphasized by Percec and Pugh (Chapter 3). An awareness of the problem is growing, and this is important because SC polysiloxanes required for speciality applications must have a guaranteed performance from one preparative batch to another.

It was originally suggested (Finkelmann and Rehage, 1984) that for a nematic polymer, the transition temperatures would rise only slightly above a $\overline{DP}$ of 10. Stevens *et al.* (1984) proposed a similar critical $\overline{DP}$ value for polymers with smectic transitions, and this view was further supported by Apfel *et al.* (1987). This has tended to condition workers to the view that, within limits, $\overline{DP}$ of the polysiloxane is not important.

However, recent work (Hawthorne, unpublished) on polysiloxanes prepared from poly(hydrogenmethylsiloxane) (PHMS) backbone carefully fractionated by GPC (Hawthorne and Semlyen, unpublished) suggests that in the cases studied (homopolymers involving three different side chains), T_g, $T_{N \to I}$, and $T_{S \to I}$ values are not constant at $\overline{DP} > 10$. T_g still rises quite steeply up to $\overline{DP}$ about 75, levelling thereafter, but increases of 1–3° are still observed as $\overline{DP}$ moves up to 100 and above. The $T_{N \to I} v. \overline{DP}$ curve rises less steeply, but levels less quickly, and 5° increases were observed from $\overline{DP}$ 99 to 134. The $T_{S \to I}$ curves were similar, but more level for $\overline{DP} > 100$.

Concurrent studies with these polymers on the influence of $\overline{M}_w/\overline{M}_n(\gamma)$ on transition temperatures showed that changes in γ from 1.9 to 1.25 may cause increase of 5° in T_g, 4° in $T_{N \to I}$ and 5° in $T_{S \to I}$.

These results should alert workers to the problems of $\overline{DP}$ and $\overline{M}_w/\overline{M}_n$ for SCLC polysiloxanes, and encourage definition of these values for all products. Other difficulties which can be overcome (see sections 4.3.2 and 4.4), but may have been less than fully appreciated in the past, are:

(i) Retention of excess of the alkene in the polymer.
(ii) The danger of cross-linking involving SC substituents such as CN
(iii) For SC copolymers, the need to demonstrate that the ratio of the side groups is in

fact the same as the stoichiometric ratio of reactant alkenes. With bulky SCs, this may not be so, and side chain ratios can be affected by reaction conditions.

Factors (i), (ii) and (iii) clearly affect product reproducibility, vital if the properties of these polymers are to be correlated with structure and they are to play a role in sophisticated applications.

4.3.2 *Catalyst types*

The hydrosilylation reaction, i.e. the addition of a silane to an alkene, is a reaction of importance in organo-silicon chemistry. It may proceed thermally (Barry *et al.*, 1947) by irradiation (Sommer *et al.*, 1947) or in the presence of a variety of catalysts (Benkeser, 1966; Speier *et al.*, 1956) of which many have been studied. Of these, the most extensively explored, and that used originally by Finkelmann and by Ringsdorf in their pioneering studies of SC polysiloxanes, is hexachloroplatinic acid ($H_2PtCl_6 \cdot xH_2O$) in propan-2-ol. From a range of catalysts including Pt on carbon and $RuCl_3$, this catalyst was chosen as highly effective by Speier *et al.* (1957) for the preparation of *n*-pentylsilanes from pent-1-ene.

This catalyst is not, however, without its problems, as it is capable of causing migration of double bonds; Speier *et al.* reported the production in high yield of *n*-pentylsilanes from pent-2-ene. Also, Finkelmann *et al.* (1983) found that chloroplatinic acid caused cross-linking through the reaction of any cyano-function in the SC with the PHMS backbone.

Indeed, until quite recently, successful reports of the preparation of cyano-substituted SCLC polysiloxanes had come from only two sources (Ringsdorf *et al.*, 1982, 1987, 1988; Gemmell *et al.*, 1985). As reported by Gray *et al.* (1986), the problem of cross-linking can, however, be overcome by the use of *freshly prepared* Speier's catalyst ($H_2PtCl_6 \cdot xH_2O$ in propan-2-ol), and the use of no more catalyst than is required to give effective hydrosilylation. In this way, white polymers, free from cross-linking and containing cyano-functions, may be obtained consistently. This is important if polysiloxanes having SCs of strong positive dielectric anisotropy are required.

Further comments on bond migration, α- or β-addition to the alkene, and cross-linking in relation to Speier's catalyst, are made in section 4.3.3.

Other catalysts for hydrosilylation have been sought and used, and some are sold commercially. Petrarch, Karlsruhe, FRG, have marketed some specialist catalysts for this purpose, and Johnson Matthey Chemicals, UK, list some ten catalysts (including H_2PtCl_6 hydrate) derived from Pd, Pt, or Rh, for example $[PtCl_2(PPh_3)_2]$, $[Pd(PPh_3)_4]$, $[Rh_4(CO)_{12}]$ and $[RhCl(PPh_3)_3]$.

A further catalyst of interest is dichloro(*endo*-dicyclopentadiene)Pt(II) and also the Pd(II) analogue. This catalyst (II) is used by Apfel *et al.* (1985) for the production of high-temperature mesomorphic polysiloxane solvents for glc stationary phases. (See Chapter 14).

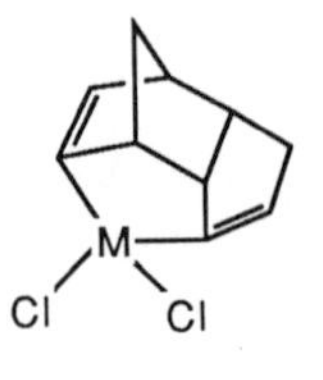

(II) M = Pt or Pd

This catalyst must be prepared, and Apfel *et al.* did not find that the procedure of Stille and Fox (1970) worked well. They developed a different route based upon methods of Kharasch and Ashford (1936) and Drew and Doyle (1971). However, in our hands (Gray and White, unpublished), this procedure too has caused problems. The catalyst does, however, appear to be a good one for the preparation of white polymers. A related catalyst is a divinyltetramethyldisiloxane platinum marketed by Petrarch (de Marignan *et al.*, 1987).

4.3.3 *Mechanistic aspects*

At the time of Speier's observations (1956), little was understood of the mechanism of hydrosilylation. It was simply noted that products were consistent with electrophilic attack of the Si fragment at the 1-carbon of the alkene, with H^- addition of the hydrogen. Later, more information emerged. A schematic mechanism (cf. the Oxo process) for hydrosilylation was proposed by Chalk and Harrod (1965). In this (Figure 4.3) they proposed that if the rates at which the Si fragment and the H^- attack the alkene are comparable, the process is a concerted addition and does not lead to isomerization. If, however, the rate of attack by the Si fragment is much less than that of the H^-, the hydride–alkene complex has a long enough lifetime for a reversible alkyl–alkene equilibrium to be established, giving (Chalk, 1970) isomerization as shown in Figure 4.4. Additionally, both k_2 and k_3 must be large compared to the rate at which reduction by the ligand occurs to give metallic Pt.

As noted by Chalk and Harrod, anomalous addition of H^- to the double bond has also to be considered; for terminal alkenes, this means α- or 1-addition of the Si fragment. Although this is unusual, the reduction of allyl chloride or acetate to prop-1-

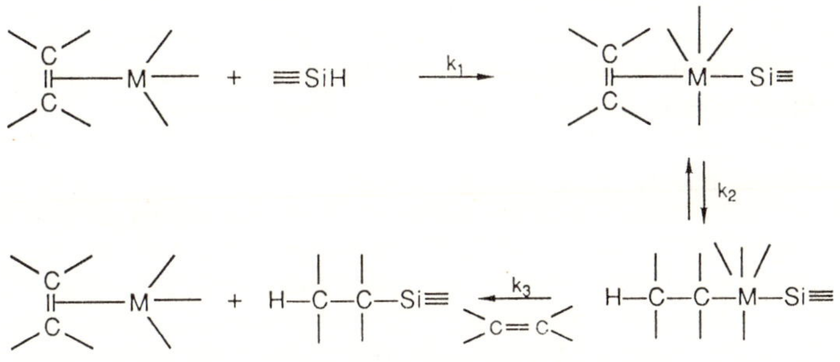

Figure 4.3 Schematic mechanism for hydrosilylation (after Chalk and Harrod, 1965)

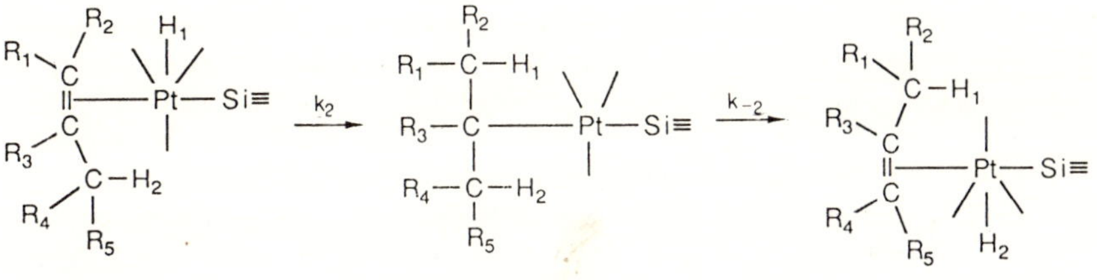

Figure 4.4 Alkene isomerization following alkyl–alkene equilibrium (see Chalk, 1970)

ene by silanes in the presence of such catalysts is most readily explained by the addition of H^- to the allyl chloride to give a secondary carbanion, i.e. 1-addition of H^-.

The significance of 1-addition of the Si fragment in hydrosilylation reactions of terminal alkenes with poly(hydrogenmethylsiloxanes) is not clear. It has been reported to occur to the extent of some 15% in the case of allyl esters (identified by ^1H NMR). The author and his group have no firm evidence of its occurrence with H_2PtCl_6 and a range of terminal alkenes, but its positive elimination at very low concentrations cannot always be certain. It is perhaps significant that the evidence for anomalous addition appears to arise (de Marignan et al., 1987) with allylic systems.

Benkeser and Kang (1970) produced evidence confirming the gradual oxidation of the propan-2-ol to propanone in Speier's catalyst solutions, and that after two weeks 99% of the Pt(IV) had been reduced to Pt(II). The proposed Pt(II) complex (III) so formed was examined spectroscopically, but a solution of Speier's catalyst had quite different spectral characteristics. By examining a solution of the authentic dimer in propan-2-ol with a small amount of added Et_4NCl, evidence was obtained for the solvolysis shown:

(III)   (IV)

It was suggested that the Cl^- from the reaction producing the dimer stabilised the anionic complex (IV). In summary, they proposed that Speier's catalyst contains $H[(C_3H_6)PtCl_3]$ which under hydrosilylation conditions undergoes the catalytic cycle of Chalk and Harrod (Figure 4.2).

These results should be considered in the light of new observations (Gray et al., 1986) relating to the effectiveness of aged and freshly prepared solutions of H_2PtCl_6 in propan-2-ol. If fresh solutions are used, it was suggested that $PtCl_6^{2-}$ may be the catalyst whereby white SCLCP with cyano-groups in the SCs may be made reproducibly and without cross-linking. However, with aged catalyst, most of the Pt is in the form Pt(II). Then, metallic Pt appears to form, and makes the polymer dark. Indeed, aged catalyst solution and PHMS in toluene evolve of H_2 and darken. Fresh catalyst solution does not do this. Yet Lewis and Lewis (1986) have evidence that silane reduction of catalysts such as (1, 5-cyclooctadiene)$_2$Pt(0), dichloro-(1, 5-cyclooctadiene)Pt(0), and a Dow Corning (Lo and Chandra, 1986) Pt(0) complex occurs with concomitant formation of highly active colloidal Pt; the latter is the key step in the hydrosilylation. It is suggested, moreoever, that this also occurs with H_2PtCl_6 in isopropyl alcohol (IPA) and that the Chalk and Harrod process is more likely to operate for authentic homogeneous catalysts such as Pt(PPh$_3$)$_4$. It seems clear that generalizations cannot yet be made about the mechanisms involved under diverse conditions and with different catalysts of different age, and that more investigative research is needed. In the author's group we always use *fresh* Speier's catalyst.

The nature of the cross-linking noted above is not clear; it could involve interaction between Si—H groups, hydrosilylation at the cyano-group (Finkelmann et al., 1983a), Pt catalysed cleavage of Si—C bonds (Akhrem et al., 1983), or a reaction (Lipowitz and Bowman, 1973), such as

$$2\,—Si—H + Pt(II) + [O] \rightarrow —Si—O—Si— + Pt(0) + H_2$$

4.4 Reproducibility of the synthetic method with Speier's catalyst

4.4.1 *Importance in applications*

The importance of reproducibility of SCLC polysiloxanes to be used in speciality applications has already been stressed. By reproducibility is meant the consistent production of a polymer which (i) is a linear macromolecule, (ii) is white, i.e. free from Pt contamination, and has no cross-linking, and (iii) possesses transition temperatures that are constant within experimental error. By speciality applications are meant those in which the polymers are used as data storage systems, non-linear optical materials, Langmuir–Blodgett films, or ferro- or pyro-electric systems.

It is first assumed that the hydrosilylation process will be carried out using a PHMS (**V**) or other backbone which is capped at its extremities and free from low-molar-mass

$$Me_3Si—O—[SiH(Me)]_n—SiMe_3 \qquad \text{(V)}$$

cyclics and oligomers. This ensures (i) above. Such backbones are made by cohydrolysis of methyldichlorosilane ($MeSiHCl_2$) and trimethylchlorosilane (Me_3SiCl) under neutral conditions (Zichy, 1965; Lee, 1966). This yields both linear and cyclic materials, according to:

$$MeSiHCl_2 + 2H_2O \rightarrow MeSiH(OH)_2 + 2HCl$$
$$nMeSiH(OH)_2 \rightarrow HO—(SiMeHO)_nH + (n-1)H_2O$$
$$\text{or} \rightarrow (SiMeHO)_n + nH_2O$$

Commercially available PHMS backbones are frequently polydisperse and often contain low-molar-mass cyclic and linear oligomers. In our recent work we have kept the number of batches of PHMS used to a minimum. One was Dow DC 1107 fluid (Dow Corning Chemical Co.), which was cleaned by a careful fractional precipitation procedure as described in detail by Hawthorne (1986). Careful thermostatting is used during the procedure, and agitation of the vessel is avoided while separation is occurring (24 h at 293 K). A sample of the polymer is then isolated and analysed by GPC. Some six precipitations are usually used to produce a backbone polymer free from the low-molar-mass materials detected by GPC in the original PHMS fluid. The precipitated product was also end-group analysed by ^{1}H NMR. The second source of PHMS was Wacker Chemie (München, FRG); analysis showed this material to be of good quality and reasonable polydispersity.

Other structurally different poly(hydrogensiloxane) prepolymers like PS 122.5 (**VI**)

$$Me_3Si—O—(SiH(Me)O)_a\text{---}(SiMe_2O)_b—SiMe_3 \qquad \text{(VI)}$$

were obtained from sources such as Petrarch Systems (Karlsruhe, FRG) and used after

Table 4.3 Poly(hydrogensiloxanes)

Source	Structure	By GPC*			By ^{1}H NMR**		
		$\bar{M}_w$	$\bar{M}_n$	$\bar{M}_w/\bar{M}_n$	$\bar{M}_n$	$\overline{DP}$	Ratio (a/b)
Dow DC 1107	V	9950	3100	3.2	2560	40 ± 3	—
Wacker	V	8050	3550	2.3	2920	46 ± 3	—
Petrarch PS 122.5	VI	2400	1300	1.9	1100	13 ± 2	1/1.17
*Carried out at RAPRA **Averages of 5 integrations							

E

confirming the quality of the material by GPC. Specifications for the two PHMS backbones and the copolymer backbone are given in Table 4.3.

Using such backbones, point (ii) above is then determined by the reproducibility of the synthetic method as discussed below.

4.4.2 General considerations

The conditions employed in the Hull group for hydrosilylation using Speier's catalyst in IPA have been described in the literature (Gray et al., 1986). Toluene is the solvent for the PHMS, and SC alkene is in 10% molar excess with respect to the Si—H content of the backbone. The Pt to C=C ratio may require variation from $1:10^{-3}$ to $1:10^{-6}$, depending on the SC used. The best ratio must be determined by experiment, i.e. the ratio giving smooth hydrosilylation with no or minimal Si—H detectable in the reaction mixture by IR spectroscopy* after an appropriate reflux time (oxygen-free atmosphere), no cross-linking, and no deposition of metallic Pt. More of the freshly prepared Speier's catalyst is needed for SC alkenes which carry a cyano-substituent, and for a given structure of the mesogenic core of the SC alkene, substantially more catalyst is needed for a spacer of $m = 3$ compared with that for $m = 6$.

Precipitation of the polymer may occur during the reaction, particularly with cyano-terminal SCs; this should not be interpreted as meaning cross-linking, for such products are soluble in dichloromethane. Good stirring is, however, needed in these cases.

When a small residual Si—H absorption remains in the infrared spectrum and does not diminish on further heating, it is possible to mop up the remaining Si—H by adding oct-1-ene (2 molar excess over the original content of Si—H groups). The resulting polymer is now strictly a copolymer. However, the population of Si—H groups is very small when this procedure is used, and test cases have revealed no differences in physical properties for polymers made with and without added oct-1-ene.

By these procedures, linear SCLC polysiloxanes which are white or pale cream in colour and free from cross-linking can be prepared with confidence using Speier's catalyst (H_2PtCl_6 in IPA). Active esters as an alternative route to SCLCPs are noted in section 4.6.

4.4.3 Side chain precursor

By attention to the points raised in sections 4.3.1 and 4.3.2, the wide variations in transition temperatures (S → I values varying between 61 and 90°) that are evident in the literature for SCLC polysiloxanes of identical structure are avoided. (For further examples of variations see Nestor et al. (1987). However, even when PHMS free from low-molar-mass cyclics and oligomers, and of given $\bar{M}_w/\bar{M}_n$ and $\overline{DP}$, are used for repeat preparations of a SCLC homopolymer with the same SCs, transition temperatures and other parameters for the products may vary quite appreciably. The problem lies in the work-up procedure and whether this removes all unreacted excess of the SC precursor. Doping studies show that very small amounts of low-molar-mass materials have a profound effect on the polymer properties. Total removal depends critically on the

---

*For a discussion of the relative merits of IR spectroscopy and ^{1}H NMR for detection of residual Si—H, see Nestor et al. (1987).

nature of the polymer and the number and efficiency of the precipitations used in the work-up. The procedure usually uses methanol to precipitate the polymer from solution in dichloromethane; in some cases diethyl ether is superior, requiring fewer precipitations. Soxhlet extraction of the polymer is also effective in dissolving out SC precursor. Many authors fail to provide details of number of precipitations, but it seems that two or three are accepted as standard. Some workers do, however, use a larger number. In some cases GPC is used to isolate and purify the polymer, and Ringsdorf *et al.* (1984) recognize the importance of monitoring products by TLC to follow the course of removal of alkene. By use of TLC, we (Nestor *et al.*, 1987) have established that three precipitations are usually not enough to rid the polymer of alkene.

The data in Table 4.4 show that after three precipitations (alkene still detectable by TLC), the T_g and $T_{S \to I}$ values vary appreciably. After 8–10 precipitations, however (no alkene detectable by TLC), the T_g values are more or less constant. The S → I values also represent a narrower distribution from 161 to 168°C. The higher glass and S → I transition temperatures and the narrower DSC peak widths show that purer products have been obtained by elimination of alkene within the detection limits of TLC (estimated at better than 0.06 wt%). The polymers used in these experiments were derived from PHMS of $\bar{M}_w/\bar{M}_n$ 2.3–3.2 (see Table 4.3). It could be argued that the precipitation procedures may also fractionate the polymer and change $\bar{M}_w$ and $\bar{M}_n$. It is emphasized, however, that the precipitations to remove alkene are carried out quite differently from those designed to fractionate a polymer. The former procedure is carried out quite rapidly with a volume/volume ratio of precipitating solvent of c. 5:1. It seemed unlikely that any fractionation was occurring, and a careful GPC check on one polymer at different stages in the precipitation procedure has confirmed this.

In summary, therefore, the use of linear polymer backbone free from low-molar-mass contaminants and of $\overline{DP}$ approximately 2 or 3, coupled with the total removal of

Table 4.4 Effect of number of precipitations on the thermal transitions (°C) for the polymer of structure

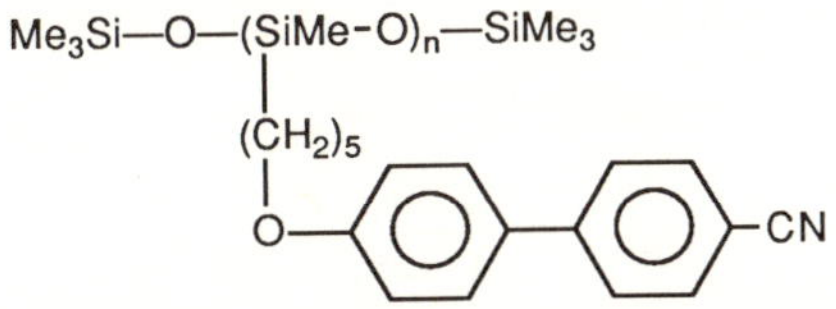

Me₃Si—O—(SiMe-O)ₙ—SiMe₃
         |
        (CH₂)₅
         |
        O—⟨benzene⟩—⟨benzene⟩—CN

3 precipitations			8–10 precipitations		
T_g	$T_{S \to I}$	$P_w{}^*$	T_g	$T_{S \to I}$	$P_w{}^*$
8	156	24	15	164	17
7	155	23	15	161	15
6	154	26	16	163	16
10	160	22	16	167	14
8	154	31	16	162	19
13	163	24	16	168	21

*Baseline peak width from DSC traces

alkene, yields consistent polymers with consistent glass transitions and more reproducible mesophase transition temperatures.

4.4.4 Role of $\overline{DP}$ and $\overline{M}_w/\overline{M}_n$

Hawthorne et al. (1986) used controlled precipitation procedures to fractionate PHMS backbone (DC 1107). The linear product was then fractionated by preparative GPC and the fractions characterized by analytical GPC, GLC and ^{29}Si NMR. Results for four of the fractions and derived SCLC polysiloxanes are given in Table 4.5, and clearly emphasize the effect of $\overline{DP}$ at values much greater than 10. Constancy is approached only at around $\overline{DP}$ 100. Very different results may be obtained for polymers with similar and very low values of $\overline{M}_w/\overline{M}_n$. Results such as these stress the importance of quoting data for $\overline{DP}$ and $\overline{M}_w$ and $\overline{M}_n$, without which comparisons of SCLC polysiloxanes cannot be made on a secure basis.

Table 4.5 Data for polymers of the structure shown below and derived from fractionated PHMS backbone

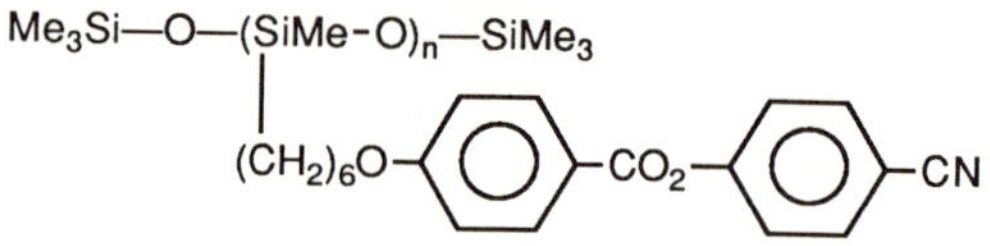

Me$_3$Si—O—(SiMe-O)$_n$—SiMe$_3$

(CH$_2$)$_6$O—⟨O⟩—CO$_2$—⟨O⟩—CN

Backbone		Polymer		
$\overline{M}_w/\overline{M}_n$	$\overline{M}_n$(GPC) (g mol^{-1})	$\overline{DP}$	T_g(°C)	T_{cl}(°C)
1.15	1760	40	2	128
1.15	3460	55	9	137
1.06	5200	84	13	142
1.12	6580	107	14	145

Table 4.6 SCLCPs derived from PHMS backbones of different polydispersity

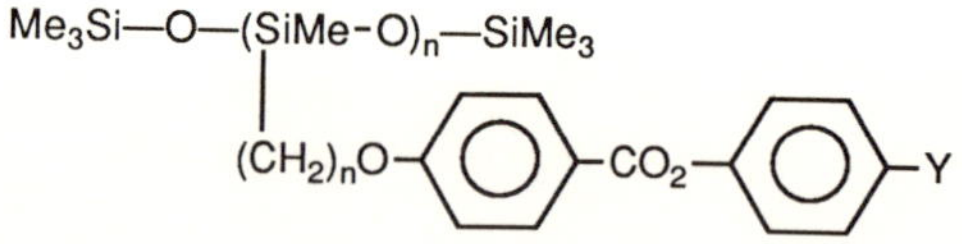

Me$_3$Si—O—(SiMe-O)$_n$—SiMe$_3$

(CH$_2$)$_n$O—⟨O⟩—CO$_2$—⟨O⟩—Y

Polymer	Backbone	$\overline{M}_w/\overline{M}_n$	T_g(°C)	T_{cl}(°C)	P_w(°C)
	A	15.5	−2	73	26
$n = 3; Y = OEt$	B	1.9	5	79	20
	C	1.35	10	81	18
	A	15.5	−12	117	25
$n = 6; Y = CN$	B	1.9	−3	123	20
	C	1.15	2	128	16

A: DC 1107 fluid as supplied; B: DC 1107 fluid cleaned by precipitation and all SC precursor removed from SCLCP; C: preparative GPC fraction of backbone B.

It is obviously expensive to use polymers derived from GPC-fractionated PHMS backbone. The use of uncleaned backbone cannot, however, be recommended, and a compromise using backbone of the kind specified in Table 4.3, coupled with careful alkene removal, seems to be an acceptable procedure to obtain reasonably reproducible polymers at an acceptable economic level. Results relevant to these procedural possibilities are in Table 4.6.

4.4.5 Homo- and copolymers

What has been said above also applies to SC copolymers prepared from homopolymeric PHMS. Two points must, however, be kept in mind. Firstly, the alkenes are used in a 10 mol% excess. When two alkenes are used, the SC ratio could therefore range from 1.1:0.9 to 0.9:1.1. When reactivity and steric factors differ, the deviation from 1:1 may be greater. Some estimate (spectroscopic) of the SC ratio should be obtained in order to characterize a copolymer appropriately. There may also be deviations from a statistical distribution of SC. Secondly, copolymers are effectively mixtures with different SC distributions, and the mesophase transitions are likely to be associated with broader biphasic regions.

Backbone copolymers. These are normally prepared from commerical backbones such as poly(hydrogenmethyldimethylsiloxane), e.g. Petrarch PS 122.5 (**VI**). For these, the DP is often low; batch characterization ($\overline{DP}$ and $\overline{M}_w/\overline{M}_n$) is therefore essential if product reproducibility is to be obtained. It is, of course, possible to prepare copolymer backbones by cohydrolysis of $MeSiHCl_2$ and Me_2SiCl_2 in the presence of Me_3SiCl; low-molar-mass linear and cyclic materials are removed by distillation. For these and commercial materials, ^{29}Si NMR, a relatively simple method (Hawthorne et al., 1986, 1988), gives a good approximation for the $SiHMe:SiMe_2$ ratio. By these methods, copolymer backbones with $MeH:Me_2$ ratios of 76:24 and 55:45 with $\overline{DP}$ about 105 and 60 respectively have been made (Hawthorne et al., 1986, 1988). Results in Table 4.7 stress the effect of $a:b$ ratio and DP upon thermal properties. Preparation of SCLCP from such copolymers follows earlier procedures.

Table 4.7 Data for the copolymer

$$Me_3Si-O-(SiMe-O)_a ---- (SiMe_2-O)_b-SiMe_3$$

$$(CH_2)_6O-\langle\bigcirc\rangle-CO_2-\langle\bigcirc\rangle-CN$$

Backbone		Polymer		
$a:b$	$\overline{DP}$	T_g	$T_{K \rightarrow S}$	$T_{S \rightarrow I}$
1.1:1	15	−10	7	98
1.2:1	60	+1	42	121
3.2:1	105	+10	57	145

4.4.6 *Alternative synthetic approach*

High-molar-mass poly(disubstituted siloxanes) may be made by ring opening and polymerization, for example of octamethylcyclotetrasiloxane, under acid- or base-catalysed conditions, giving a mixture of linear and macrocyclic materials of broad $\bar{M}_w/\bar{M}_n$. Linear product is increased by controlled addition of hexamethyldisiloxane. The products are equilibrium mixtures, and contain virtually no hexamethylcyclo-trisiloxane. This is attributed to ring strain in the trimer, leading to a greater rate of ring opening than for tetramer and higher cyclics. Lee *et al.* (1969) and Holle and Lehnen (1975) showed that low conversion anionic polymerization of hexamethylcyclo-trisiloxane yields linear polysiloxanes of narrow $\bar{M}_w/\bar{M}_n$, implying that all reactions were at trimer Si—O bonds. As conversion increases, however, reaction of the silanolates, from trimer ring opening, with chain Si—O bonds will increase and the product will be characterized by an equilibrium distribution containing macrocyclics (Rahalkar *et al.*, 1984).

Poly(dimethylsiloxanes), for example, can be made with molar masses from 25 000 to 300 000 (Hawthorne *et al.*, unpublished) and $\bar{M}_w/\bar{M}_n < 1.5$. Commercial (Petrarch) poly(hydrogenmethylcyclosiloxanes) can be obtained, and at York University, UK, pure cyclics with n up to 20 have been made. Ring opening and polymerization of mesogenically substituted cyclics of this kind may hold out prospects for making linear SCLC polysiloxanes with controlled and high molar mass and low $\bar{M}_w/\bar{M}_n$.

4.5 LC properties of side chain polysiloxanes

For SCLC polysiloxanes, structural correlations with LC and other physical properties are uncertain unless $\overline{DP}$, $\bar{M}_w/\bar{M}_n$, and purity are defined (e.g. Nestor *et al.*, 1987). For most known SCLC polysiloxanes, these data are not recorded; in this section therefore, structure/property relations are confined to broad generalizations.

4.5.1 *Homopolymers*

Enormous contributions to knowledge about homopolymers from PHMS have been made by Finkelmann and his colleagues (Finkelmann, 1982, 1983, 1987a, b; Finkelmann and Rehage, 1982; Finkelmann *et al.*, 1983b).

Owing to the linkage of the mesogenic moiety to the backbone, the polymers show a shift to more ordered LC phases compared to the SC precursor, or if both show the same phases, the LC polymer transition temperatures will be higher. Thus, a nematic monomer gives a smectic polymer, and a monomer with a low $T_{N \to I}$ gives a higher $T_{N \to I}$ polymer.

The role of the flexible spacer was shown by examples such as (**VII**). With $m = 1$, increase in n from 3 to 6 changes a purely nematic polymer to one with a smectic and a nematic phase with a higher $T_{N \to I}$. With $n = 3$, increasing m from 1 to 6 changes a

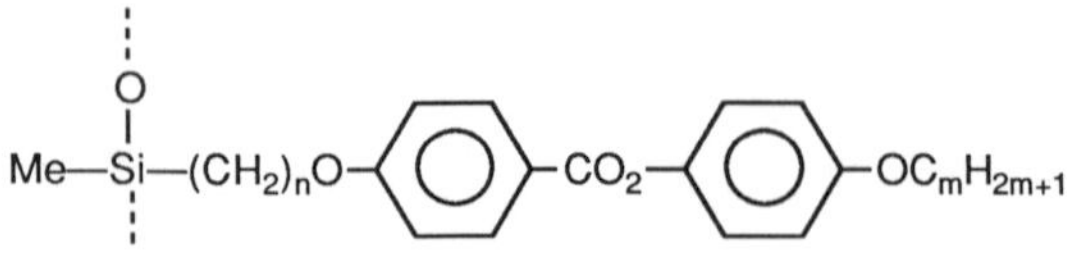

$$Me-\underset{\underset{\vdots}{O}}{\overset{\overset{\vdots}{O}}{Si}}-(CH_2)_nO-\bigcirc-CO_2-\bigcirc-OC_mH_{2m+1}$$

(VII)

nematic polymer to a smectic polymer with a much higher clearing temperature. Due to the enhanced LC order from monomer to polymer, smectic phases are the predominant phases shown by SC polysiloxanes; only short spacers and short terminal chains promote nematic behaviour. Very long spacers ($n = 11$) tend to give smectic polymers and to introduce semicrystalline behaviour (see, however, Chapter 3).

Production of cholesteric *homopolymers* has been difficult, as the chirality has usually been introduced through a branched terminal alkyl group more complex than the short chains needed for nematic (N*) phases.

SC siloxane *copolymers* can, however, be made using a nematic promoting SC and different amounts of a chiral SC. A much stronger induction of twist occurs in polymers compared with that in a mixture of the same composition of the SC monomers (Finkelmann and Rehage, 1980a, b), and with increasing spacer lengths, the twisting power decreases as decoupling increases and rotation of the SCs about their long axes becomes less hindered.

In many respects, SCLC polysiloxanes resemble their LMM counterparts (Toyne, 1987; Gray, 1982; Gemmell *et al.*, 1985), and not surprisingly therefore, increase in length of the core of the mesogenic SC strongly enhances LC properties. In polymer (VII) for example, with $n = 3$ and $m = 1$, replacing the right-hand phenyl ring with a biphenyl ring system ($—C_6H_4—C_6H_4—$) changes a glassy polymer with a modest $T_{N \to I}$ of 61°C to a crystalline polymer, $T_{N \to I}$ 319°C (Apfel *et al.*, 1985).

Also as in LMM mesogens, the change from a terminal alkyl to a terminal alkoxy group enhances clearing temperatures (as does the change from spacer $(CH_2)_{n-1}$ to $(CH_2)_nO$).

Finally, lateral substitution in the core of the SC depresses clearing temperatures, but more significantly may change a smectic polymer to a nematic polymer and extinguish crystalline tendencies (Table 4.8) (Nestor *et al.*, 1987).

Bearing in mind the fascinating variations in LC properties and physical properties that may be brought about in LMM systems and related directly to molecular structural changes (because pure systems with defined molecules are being compared) (Toyne, 1987; Gray 1982), it would be satisfying to make such comparisons for SCLC polysiloxanes and be able to tune their properties to specific needs. Generally speaking, however, data do not exist for polymers (with defined macromolecular parameters) that allow their physical properties to be compared across the board with any confidence with respect to often quite minor structural differences (but see section 4.5.3). It is hoped that this situation will change, to allow full advantage to be taken of results of

Table 4.8 Effects of lateral substitution in a SCLC polysiloxane (first example of a nematic cyano-terminal polysiloxane

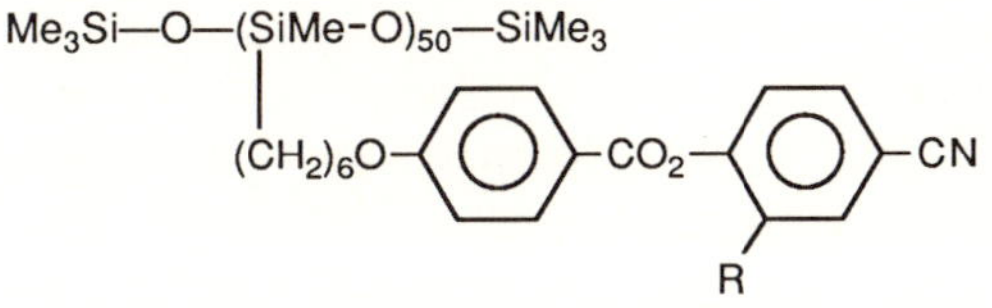

Me$_3$Si—O—(SiMe-O)$_{50}$—SiMe$_3$
$(CH_2)_6O$—⬡—CO_2—⬡—CN
with R substituent

R	T_g(°C)	K → S$_A$(°C)	S$_A$ → I(°C)	N → I(°C)
H	—	55	181.5	—
Me	7	—	—	48

interesting structural changes that are being made to SCs by several workers: cyclohexane, 1, 3-dioxan or 1, 3-dithian (Gemmell *et al.*, 1985; Murza *et al.*, 1985; Seto *et al.*, 1984; Zaschke *et al.*, 1984), or 1, 3, 2-dioxaborinane (Hahn and Percec, 1987) rings for benzene rings. The latter study, using PHMS backbone ($\bar{M}_n = 4500 - 5000$), stemmed from observations that SC polysiloxanes with $(CH_2)_{11}$ spacers and 2, 5-disubstituted *trans*-1, 3-dioxan rings gave no SC crystallization (Hsu *et al.*, 1987a, b) and produced interesting new polymers, containing 1, 3-dioxan or 1, 3, 2-dioxaborinane rings, which exhibit chiral smectic C phases (Percec and Pugh, Chapter 3).

As indicated above, SC polysiloxanes may give polymorphic smectic phases in addition to N(N*) phases. S_A phases are common, and examples of S_C and S_C^* phases are emerging as interest in ferroelectric phases grows. Polymers with ordered S_B and S_E phases are also documented (Noël, 1983).

The identification of N and different S phases by optical microscopy is less easy for polymers than for LMM materials. X-ray data are therefore very important in reaching unambiguous classifications. However, Shibaev and Platé (1984) note that because of the enhanced ordering in polymer LC phases, layered structures can persist in the N and isotropic states, and that as the molecular mass is increased in nematic acrylates, smectic characteristics appear. The inference that N polymers may become smectic at high $\overline{DP}$ values has not to the author's knowledge been drawn for polysiloxanes, for which very high $\overline{DP}$ values do not usually arise. It does seem, however, that there is a closer analogy between N and S phases in LC polymers than in LMM systems.

4.5.2 *Copolymers*

Structure/property correlations for backbone and/or SC copolymers are even more difficult. $\overline{DP}$, $\bar{M}_w/\bar{M}_n$, and purity still need to be defined, but in addition, good comparisons can only be made for polymers that have a statistical composition and for which the ratio of components is known. There are few copolymer systems for which this level of information is available.

Copolymers are nonetheless very interesting. The achievement of glass–mesophase transitions as low as $-120°C$ with siloxane backbone copolymers has already been noted (Ringsdorf and Schneller, 1982; Diele *et al.*, 1988). Such copolymers also have the advantage of lower viscosities and lower crystalline tendencies due to the interpolated $SiMe_2$ segments.

SC copolymers also afford the opportunity to fine-tune the polymer's properties by varying the ratio of the mesogenic SCs. An interesting example (Gemmell *et al.*, Nestor *et al.*, 1987) is a copolymer from PHMS in which one SC (*y*) carries a lateral Me group (anti-smectic) and one (*x*) does not. As the *x*:*y* ratio is changed from 35:15 to 16:34, we move from a smectic to a purely nematic polymer; the glass transition temperature is little affected. Also, in the case of chiral polymers, the N* pitch can be controlled by the population of chiral SCs employed.

4.5.3 *Results from controlled synthesis*

When *pure* SC polysiloxanes are made from a defined backbone source, intercomparisons can be made (e.g. Richard *et al.*, 1988; Keller, 1987). As examples made from

Wacker backbone (Table 4.3) the following polymers were made with SCs **(VIII)** (Nestor *et al.*, unpublished):

$$-(CH_2)_mO-\langle\bigcirc\rangle-CO_2-\langle\bigcirc\rangle-Y$$

where
$m = 5$ or 6
$Y = CN$, F or CF_3; $X = Z = H$
$Y = CN$; $X = F$; $Z = H$
$Y = CN$; $X = H$; $Z = F$ **(VIII)**

The order of decreasing $S_A \rightarrow I$ temperature as Y changes, with $X = Z = H$, is $CF_3 > CN > F$. In the last two cases, this is as for LMM systems, in which terminal CF_3 converts purely N esters into S materials with slightly lower $S \rightarrow I$ temperatures (Misaki *et al.*, 1981). The smectic promoting properties of CF_3 (Coates and Gray, 1976) are therefore more marked in the polymers, where with $Y = CF_3$, ordered B and A smectic phases occur.

The influence of lateral fluorine on smectic properties in LMM systems is not well documented. A 2-F decreases the $S_A \rightarrow I$ temperature, but a 3-F has a variable effect, sometimes even increasing the $S_A \rightarrow I$ temperature. With the polymers, $X = F$ strongly decreases the $S_A \rightarrow I$ temperature (by *c.* 30°), and $Y = F$ has a smaller effect (*c.* 20°). Note that, in the polymers, the smaller lateral F (unlike Me) does not suppress S phases, another example of enhanced order from LMM to polymer system.

Similar trends were obtained with SCs **(VIII)** attached to the copolymer backbone specified in Table 4.3; phase types shown did not change.

More studies of this kind should extend knowledge of structure effects on SCLCP, most important if we are to learn to design such polymers with properties tailored to specific needs, such as smectic C polymers, of which an example (Mauzac *et al.*, 1986) is:

$$Me_3Si-O-(SiMe-O)_n-SiMe_3$$
$$|$$
$$(CH_2)_{11}O-\langle\bigcirc\rangle-OOC-\langle\bigcirc\rangle-OMe$$

$$T_g 24°; T_m 56°; S_C \rightarrow S_A 60°; S_A \rightarrow I, 134°C \quad \text{(not chiral)}$$

Superior smectic C SC polymers (no crystallization) are certainly needed.

4.5.4 *Non-linear optical effects*

For second harmonic non-linear optical (NLO) processes, lithium niobate is a good material, but organics with extended π-electron systems and forming non-centrosymmetric crystals can have more pronounced NLO properties and switch faster. The need to have crystals is limiting, and a way around this would be to use the glassy state of a SCLC polymer into which has been locked a non-centrosymmetric arrangement of suitably structured SCs, for example:

$$-(CH_2)_nO-\langle\bigcirc\rangle-N=N-\langle\bigcirc\rangle-CN$$

Poling the polymer as it cools from the isotropic state, or possibly advantageously in a mesophase, then quenching into the glassy state, would preserve the polar order.

Evidence of positive results with acrylates has been obtained; LC polysiloxanes of this kind are also known (Gray *et al.*, 1987), but T_g values are very low.

An alternative approach is to use a polysiloxane with SCs which terminate in an amphiphilic group, so that a monolayer with SC polar order can be laid down and transferred to a substrate (Si, Al, etc). This has been done with siloxane (**IX**), where $m = 9 \pm 2$ and $n = 8 \pm 2$ (Carr *et al.*, 1987). The polymer forms a stable monolayer, and

$$\text{Me}_3\text{Si—O—(SiMe}_2\text{—O)}_m \text{---(SiMe-O)}_n\text{—SiMe}_3$$

for the first time in a polymer of this type, second harmonic generation has been demonstrated ($\beta = 35.10^{-49}\text{C}^3\text{m}^3\text{J}^{-2}$). Stable multilayers are readily produced, but these are Y-type, leading to cancellation of the polarity along the SC. To obviate this problem, another polysiloxane with axial polarity in the SC opposite to that in (**IX**) has been made. Deposition (Y-type) of alternating monolayers of the two polymers will give a superlattice with macroscopic polar order. Studies of this system are still continuing, but a greatly enhanced β value is expected.

4.5.5 *Chromophoric, photochromic and photosensitive polymers*

In the previous section, attention was drawn to polymers having SCs with conjugated structures. Such polymers (Gray *et al.*, 1987) are pale yellow or orange in colour, but much more strongly chromophoric groups—dye-based SC—could be used. For example, Ringsdorf *et al.* (1986, 1987) (see also Quotschalla *et al.*, 1987) studied acrylate SC copolymers with SCs derived from anthraquinone or trisazo dyes. For the copolymers, the dye order parameter is higher for the glassy S state than for the glassy N state, but order parameters for the monomeric dyes in the glassy homopolymer are still higher. The technical significance of having highly aligned, dichroic dyed polymers is obvious (Saeva, 1974), and additionally, the effects of molecular ordering on photochemical and photophysical processes can be probed (Whitten *et al.*, 1977).

Krongauz and Goldburt (1981) (see also Aldoshin and Atovmyan, 1987) have studied the effects of spiropyran to merocyanine photoinduced conversion in the SC of methacrylate homo- and copolymers upon crystallization and the accompanying colourless-to-violet photochromic effect. Domain stacking of merocyanine groups was proposed in explanation of the cooperative nature of the crystallization and conversion processes. Cabrera and Krongauz (1987) extended such studies to polyacrylates with a random distribution of mesogenic SC and spiropyran SCs (22%), and noted a rheo-optic effect above T_{cl} attributed to a network, again, involving an aggregation of merocyanine groups. (cf. Chapter 3).

Further developments in this area have involved polysiloxanes (Cabrera *et al.*, 1987, 1988). Spiropyran functional groups interfere with hydrosilylation, and a copolymer was prepared from PHMS backbone with x SCs of structure (**X**) and y SCs of the active ester type (see section 4.4.2). Interaction of the *N*-hydroxysuccinimide ester

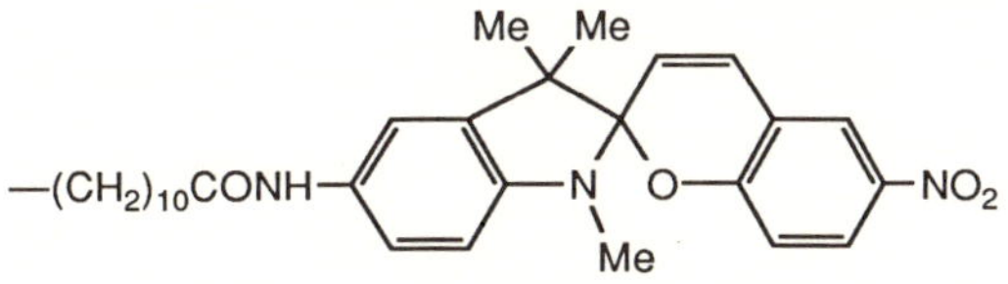

$$-(CH_2)_{10}CONH-\underset{Me}{\overset{Me\ Me}{N}}-O-NO_2 \qquad \textbf{(XI)}$$

with 5'-amino-1', 3', 3'-trimethyl-6-nitro-2', 3'-dihydrospiro[2H-chromen-2, 2'-indene] yielded the copolymer with the additional SCs (**XI**). T_g is about 10°C and T_{cl} is strongly dependent upon spiropyran content. A cast film is strongly birefringent (pale pink at room temperature). Light of $\lambda > 500$ nm changes the colour to pale yellow, and light of $\lambda = 365$ nm gives a deep red. These and other changes are associated with spiropyran–merocyanine interconversion and point the way to novel applications in imaging technologies.

The functionalization of siloxane copolymers with photosensitive cinnamoyl ester-containing SC has also been reported (Lemaitre *et al.*, 1987; Creed *et al.*, 1988). Considerable care had to be taken to control the hydrosilylation process to avoid reaction between the labile ester groups and Si—H groups. Here, the photoinduced reaction is concerned with cross-linking; this work is part of a continuing programme at Lille, France on UV-curable silicone polymers.

4.6 Polysiloxane elastomers

In 1976, Maher *et al.* (1976) noted that organic-soluble siloxane gums could be UV-cured to give materials varying from soft rubbers to hard glasses. Several years elapsed however, before Finkelmann *et al.* (1981) showed that cross-linking of SCLCP yielded form-retaining LC networks. Mechanical deformation of the rubbers oriented the mesogenic groups and made the opaque samples transparent. In these polymers the ratio SiMe$_2$:SiMeMesogen was 60:60 minus the number of cross-linking units (6 to 12). DSC traces were qualitatively similar to those for the analogous linear polymers, but transition temperatures were often reduced. These elastomers were quite different from those reported by Tsutsui *et al.* (1981) in the same year; their pure networks did not have thermotropic LC properties.

This work by Finkelmann has attracted much interest, generating the prospect of a whole range of new elastomers combining LC properties with shape stability and elasticity (Loth and Euschen, 1988). Activity has also extended to acrylates (Nishikubo *et al.*, 1987). Within Finkelmann's group, the properties of polysiloxane elastomers have been studied with special reference to their orientation by mechanical forces. A detailed coverage of this work is given in Chapter 10 of this book.

4.7 Discotic polysiloxanes

By analogy with LMM LC systems, SCLCP with disc-shaped pendent groups attached by flexible spacers to a polymer backbone exhibit discotic LC phases. These were first reported by Kreuder and Ringsdorf (1983). Work by the Bordeaux group on the synthesis of hexasubstituted triphenylenes with unequal lengths of alkyl chains (Tinh *et al.*, 1981; Destrade *et al.*, 1981) encouraged Kreuder and Ringsdorf to make monofunctionalized triphenylenes (**XII**). Hydrosilylation of PHMS and mixed polysiloxane backbones (SiMe2:SiMeH = 1:1), with 35 and 70 silicon atoms respectively, gave polymers which, after annealing, had the transition temperatures given below.

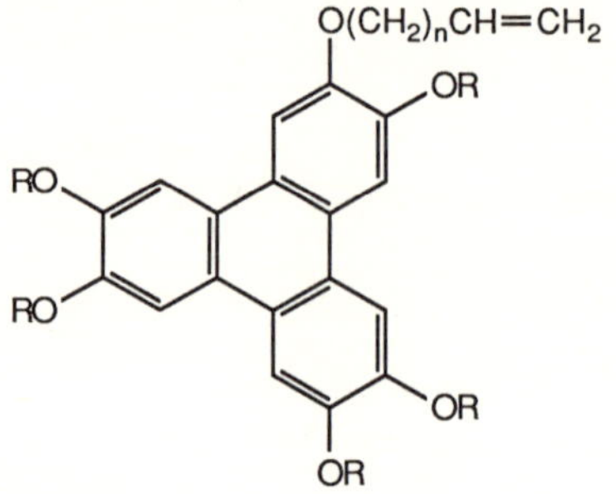

$$O(CH_2)_nCH=CH_2$$

(XII)

R	n	Backbone type	Transition temperatures (°C)
C_5H_{11}	9	homopolymer	g-24 $D_1$27 $D_2$35 $D_3$42 I
C_5H_{11}	9	copolymer	g-31 $D_1$26 $D_2$32 $D_3$41 I

Following this work, synthesis of other discotic SCLCPs has been encouraged. As yet, no applications for these materials are known, but speculations have been made (Kreuder and Ringsdorf, 1983) about their potential, especially if they contain fluorescent or metal-containing disc-shaped SCs (see also Sirlin *et al.*, 1988 and Ringsdorf and Wustefeld, 1987).

4.8 Polysiloxanes with forked or laterally attached side chains

(i) *LC polymers with forked side chains.* An example of a polysiloxane with this kind of structure (**XIII**) is shown below (Engel *et al.*, 1985). When $x = 2$ and $y = 6$, the transition temperatures are: g 18 S 100 N 113 I(°C). These materials are extremely interesting, particularly in view of the microphase model of highly decoupled systems discussed by Percec and Pugh in the preceding chapter (sections 3.7 and 3.2.3).

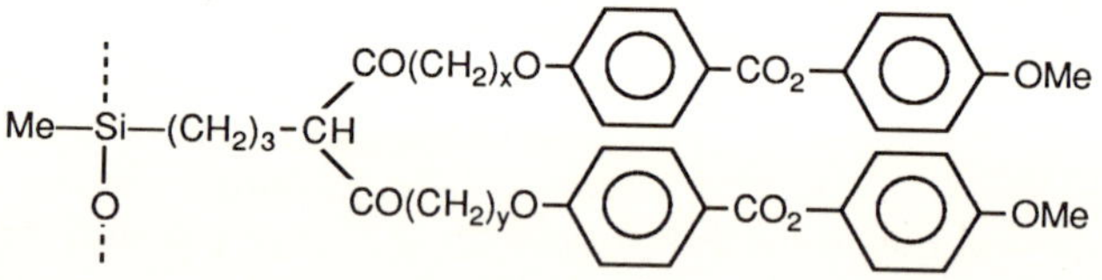

(XIII)

(ii) *LC polymers with laterally attached side chain.* Following observations by Demus and Weissflog (1984) that LMM calamitic mesogens carrying long laterally attached alkyl chains give uniaxial N phases, Finkelmann reasoned that flexible backbone polymers linked by spacers to a lateral point on each of the mesogenic groups would, through the restricted rotation motions of the SCs, give biaxial mesophases. The synthesis of methacrylate polymers of this kind was first realized by Hessel and Finkelmann (1985) (see also Hessel *et al.*, 1987) and the N phase were indeed biaxial.

Polysiloxane analogues are readily conceived and several have been made in the author's group (Hill *et al.*, unpublished, see also Keller, 1987). Examples made from Wacker backbone (see Table 4.3) are:

$Me_3SiO-(SiMe-O)_n-SiMe_3$

$(CH_2)_5$

O

$C_8H_{17}O-\langle\bigcirc\rangle-CO_2-\langle\bigcirc\rangle-\langle\bigcirc\rangle-X$

$X = C_3H_7$ g 11.2 N 31 i

OC_9H_{19} g   48 N 58 i

References

Akhrem, I., Christovalova, N. and Vol'pin, M. (1983) *Russ. Chem. Rev.* (Engl. Transl.) **52**, 542.

Aldoshin, S. and Atovmyan, L. (1987) *Mol. Cryst. Liq. Cryst.* **149**, 251.

Apfel, M., Finkelmann, H., Janini, G., Laub, R., Luhman, B-H, Price, A., Roberts, W., Shaw, T. and Smith, C. (1985) *Anal. Chem.* **57**, 651.

Barry, A., De Pree, L., Gilkey, J. and Hooke, D (1947) *J. Amer. Chem. Soc.* **69**, 2916.

Benkeser, R. (1966) *Pure Appl. Chem.* **13**, 133.

Benkeser, R. and Kang, J. (1970) *J. Organomet. Chem.* **185**, C9.

Cabrera, I. and Krongauz, V. (1987) *Nature* (London) **326**, 582.

Cabrera, I., Krongauz, V. and Ringsdorf, H. (1987) *Angew Chem. Int. Edn.* **26**, 1178.

Cabrera, I., Krongauz, V. and Ringsdorf, H. (1988) *Mol. Cryst. Liq. Cryst.* **155**, 221.

Carr, N., Goodwin, M., McRoberts, A., Gray, G., Marsden, R. and Scrowston, R. (1987) *Makromol. Chem. Rapid Commun.* **8**, 487.

Casagrande, C., Fabre, P., Veyssie, M., Weill, C. and Finkelmann, H. (1984) *Mol. Cryst. Liq. Cryst* **113**, 193.

Chalk, A. (1970) *Trans. N.Y. Acad. Sci.* **32**, 481.

Chalk, A. and Harrod, J. (1965) *J. Amer. Chem. Soc.* **87**, 16.

Coates, D. and Gray, G. (1976) *J. Chem. Soc. Perkin* 300.

Coles, H. (1985) *Faraday Disc. Chem. Soc.* **79**, 201.

Coles, H. and Simon, R. (1985a) in *Recent Advances in Liquid Crystalline Polymers*, ed. Chapoy, L., Elsevier Applied Science, London.

Coles, H. and Simon, R. (1985b) *Mol. Cryst. Liq. Cryst.* **1**, 75.

Coles, H. and Simon, R. (1986) *Mol. Cryst. Liq. Cryst.* **3**, 37.

Creed, D., Gross, J., Sullivan, S., Griffin, A. and Hoyle, C. (1985) *Mol. Cryst. Liq. Cryst.* **149**, 185.

Creed, D., Griffin, A., Gross, J., Hoyle, C. and Venkararam, K. (1988) *Mol. Cryst. Liq. Cryst. Bull.* **3**, 34.

de Marignan, G., Teyssie, D. and Boileau, S. (1987) *Int. Conf. on Liquid Crystalline Polymers, Bordeaux*, Abstr. 6PI.

Destrade, C., Tinh, N., Gasparouz, H., Malthete, J. and Levelut, A. (1981) *Mol. Cryst. Liq. Cryst.* **71**, 111.

Diele, S., Oelsner, S., Kuschel, F., Hisgen, B. and Ringsdorf, H. (1988) *Mol. Cryst. Liq. Cryst. Bull.* **3**, 63.

Drew, D. and Doyle, J. (1971) in *Inorganic Synthesis*, ed. Cotton, F., vol. XIII, McGraw-Hill, New York.

Engel, M., Hisgen, E., Keller, R., Kreuder, W., Reck, B., Ringsdorf, H., Schadt, H.-W. and Tschirner, P. (1985) *Pure Appl. Chem.* **57**, 109.

Finkelmann, H. (1982) in *Polymer Liquid Crystals*, eds. Ciferri, A., Krigbaum, W. and Meyer, R., Academic Press, New York.

Finkelmann, H. (1983) *Phil. Trans. Roy. Soc. London* **A309**, 105.

Finkelmann, H. (1987a) *Angew. Chem. Int. Edn.* **26**, 816.

Finkelmann, H. (1987b) in *Thermotropic Liquid Crystals*, ed. Gray. G., CRAC Series, vol. 22, John Wiley, Chichester.

Finkelmann, H. and Rehage, G. (1980a) *Makromol. Chem. Rapid Commun.* **1**, 733.

Finkelmann, H. and Rehage, G. (1980b) *Makromol. Chem. Rapid Commun.* **1**, 31.

Finkelmann, H. and Rehage, G., (1982) *Makromol. Chem. Rapid Commun.* **3**, 859.
Finkelmann, H. and Rehage, G. (1984) *Adv. Polym Sci* **60/61**, 99.
Finkelmann, H. Kock, H.-J. and Rehage, G. (1981) *Makromol. Chem. Rapid Commun.* **2**, 317.
Finkelmann, H. Kiechle, U. and Rehage, G. (1983a) *Mol. Cryst. Liq. Cryst.* **94**, 343.
Finkelmann, H., Benthack, H. and Rehage, G. (1983b) *J. Chem. Phys.* **80**, 16.
Gemmell, P., Gray, G. and Lacey, D. (1985) *Mol. Cryst. Liq. Cryst.* **122**, 205.
Gray, G. (1982) in *Polymer Liquid Crystals*, eds. Ciferri, A., Krigbaum, W. and Meyer, R., Academic Press, New York.
Gray, G. and Goodby, J. (1984) *Smectic Liquid Crystals: Textures and Structures*, Leonard Hill [Blackie], Glasgow and London.
Gray, G., Lacey, D., Nestor, G. and White M. (1986) *Makromol. Chem. Rapid Commun.* **7**, 71.
Gray, G., Lacey, D., Toyne, K. and White, M. (1987) *Int. Conf. on Liquid Crystalline Polymers, Bordeaux*, Abstr. 12P1.
Hahn, B. and Percec, V. (1987) *Macromolecules* **20**, 2961.
Hawthorne, W. (1986) unpubl. PhD. thesis, University of York, UK.
Hessel, V. and Finkelmann, H. (1985) *Polym. Bull.* **14**, 3751.
Hessel, V. and Finkelmann, H. (1986) *Polym. Bull.* **15**, 3491.
Hessel, V. Herr, R.-P. and Finkelmann, H. (1987) *Makromol Chem.* **188**, 1597.
Holle, H. and Lehnen, B. (1975) *Eur. Polym. J.* **11**, 663.
Hsu, C., Rodriguez-Parada, J. and Percec, V. (1987a) *Makromol. Chem.* **188**, 1017.
Hsu, C., Rodriguez-Parada, J. and Percec, V. (1987b) *J. Polym. Sci. Part A (Polym. Chem.)* **25**, 2425.
Keller, P. (1987) *Int. Conf. on Liquid Crystalline Polymers, Bordeaux*, Abstr. 20P1.
Keller, P., Hardouin, F., Mauzac, M. and Achard, M. (1988) *Mol. Cryst. Liq. Cryst.* **155**, 171.
Kharasch, M. and Ashford, T. (1936) *J. Amer. Chem. Soc.* **58**, 1733.
Kreuder, W. and Ringsdorf, H. (1983) *Makromol. Chem. Rapid Commun.* **4**, 807.
Krongauz, V. and Goldburt, E. (1981) *Macromolecules* **14**, 1382.
Lee, C. (1966) *J. Organomet. Chem.* **6**, 620.
Lee, C., Frye, C. and Johannson, O. (1969) *Polym. Preprints, ACS* **10(2)**, 1361.
Lemaitre, E., Coqueret, X., Mercier, R., Lablache-Combier, A. and Loucheux, C. (1987) *J. Appl. Polym. Sci.* **33**, 2189.
Lewis, L. and Lewis, N. (1986) *J. Amer. Chem. Soc.* **108**, 7228.
Lipowitz, J. and Bowman, S. (1973) *J. Org. Chem.* **38**, 162.
Lo, P. and Chandra, G. (1986) *Amer. Chem. Soc. Nat. Mtg., INOR*, Abstr. 36, New York.
Loth, H. and Euschen, A. (1988) *Makromol. Chem. Rapid Commun.* **9**, 35.
Maher, J., Schank, R. and Pfister, G. (1976) *Appl. Phys. Letts.* **29**, 293.
Mauzac, M., Hardouin, F., Richard, H., Achard, M., Sigaud, G. and Gasparoux, H. (1986) *Eur. Polym. J.* **22**, 131.
McArdle, C., Clark, M., Haws, C., Wiltshire, M., Parker, A., Nestor, G., Gray, G., Lacey, D. and Toyne, K. (1987) *Liquid Crystals* **2**, 573.
Misaki, S., Takamatsu, S., Suefuji, M., Mitote, T. and Matsamura, M. (1981) *Mol. Cryst. Liq. Cryst.* **66**, 123.
Murza, M., Safarov, M. and Miftakhova, G. (1980) *Zh. Org. Khim* **21**, 1550 Engl. Transl., *J. Org. Chem. USSR* 1412 (1985).
Nestor, G., White, M., Gray, G., Lacey, D. and Toyne, K. (1987) *Makromol. Chem.* **188**, 2759.
Nishikubo, T., Saita, S. and Fujii, T. (1987) *J. Polym. Sci. Part A (Polym. Chem.)* **25**, 1339.
Noël, C. (1985) in *Recent Advances in Liquid Crystalline Polymers*, ed. Chapoy, L.L., Elsevier Applied Science, London.
Quotschalla, U. and Haase, W. (1987) *Mol. Cryst. Liq. Cryst.* **153**, 83.
Rahalkar, R., Lamb, J., Harrison, G., Barlow, A., Hawthorne, W., Semlyen, J., North, A. and Pethrick, R. (1984) *Proc. Roy. Soc. London* **394**, 207.
Richard, H., Mauzac, M., Tinh, N., Sigaud, G., Achard, M., Hardouin, F. and Gasparoux, H. (1988) *Mol. Cryst. Liq. Cryst.* **155**, 141.
Ringsdorf, H. and Schneller, A. (1982) *Makromol. Chem. Rapid Commun.* **3**, 557.
Ringsdorf, H. and Schmidt, H.-W. (1984) *Makromol. Chem. Rapid Commun.* **185**, 1327.
Ringsdorf, H., Schmidt, H.-W., Baur, G., Kiefer, R. and Windscheid, F. (1986) *Liquid Crystals* **1**, 319.
Ringsdorf, H., Schmidt, H.-W., Eilingsfeld, H. and Etzbach, K.-H. (1987) *Makromol. Chem.* **188**, 1355.

Ringsdorf, H. and Wustefeld, R. (1987) *Int. Conf. on Liquid Crystalline Polymers, Bordeaux*, Abstr. 10P1.

Saeva, F. (1974) *Pure Appl. Chem.* **38**, 25.

Seto, K., Takahashi, S. and Tahara, T. (1984) *Bull. Chem. Soc. Jpn.* **57**, 1966, 3173.

Shibaev, V. and Platé, N. (1984) *Adv. Polym. Sci.* **60/61**, 173.

Sirlin, C., Bosio, L. and Simon, J. (1988) *Mol. Cryst. Liq. Cryst. Bull.* **3**, 49.

Sommer, L., Pietrusza, E. and Whitmore, F. (1947) *J. Amer. Chem. Soc.* **69**, 188.

Speier, J., Zimmerman, R. and Webster, J. (1956) *J. Amer. Chem. Soc.* **78**, 2278.

Speier, J., Webster, J. and Barnes, G. (1957) *J. Amer. Chem. Soc.* **79**, 974.

Stevens, H., Rehage, G. and Finkelmann, H. (1984) *Macromolecules* **17**, 851.

Stille, J. and Fox, D. (1970) *J. Amer. Chem. Soc.* **92**, 1274.

Strzelecki, L. and Liebert, L. (1973) *Bull. Soc. Chim. Fr.* 597.

Tinh, N., Bernaud, M., Sigaud, G. and Destrade C., (1981) *Mol. Cryst. Liq. Cryst.* **65**, 307.

Tinh, N., Mauzac, M., Richard, H. and Sigaud, G. (1987) *Int. Conf. on Liquid Crystalline Polymers, Bordeaux*, Abstr. 22P1.

Toyne, K. (1987) in *Thermotropic Liquid Crystals*, ed. Gray, G. CRAC Series, vol. 22, John Wiley, Chichester.

Tsutsui, T. and Tanaka, R. (1981) *Polymer* **22**, 117.

Weissflog, W. and Demus, D. (1984) *Cryst. Res. and Technol.* **19**, 55.

West, R. (1980) *J. Chem. Educ., Organosilicon Chemistry*, Parts I and II, 57, 165, 334.

Whitten, D., Hopf, H., Quina, F., Sprintschnik, G. and Sprintschnik, H. *Pure Appl. Chem.* **49**, 379.

Zaschke, H., Isenberg, A. and Vorbrodt, H. (1984) in *Liquid Crystals and Ordered Fluids*, vol. 4, eds. Griffin, A. and Johnson, J., Plenum, New York.

Zichy, E. (1965) *J. Organomet. Chem.* **4**, 411.

5 The chiral smectic C liquid crystal side chain polymers

P. LE BARNY and J.C. DUBOIS, Laboratoire Central de Recherches, Thomson-CSF, Domaine de Corbeville, 91400 Orsay Cedex, France

5.1 Introduction

The subject of liquid crystals, now almost a century old, does not appear to be a unique field of physics and chemistry, but rather a set of fields with their own means of investigation and applications. For example, lyotropic liquid crystals which form ordered solutions above a critical concentration value, in appropriate solvents (and below a critical temperature value), are quite different from thermotropic liquid crystals which arise by heating an anisotropic fluid, before leading to a liquid. Thermotropic systems have experienced two important developments in the recent past: first, realization of thermotropic liquid crystal polymers; secondly, the discovery of smectic phases exhibiting a non-vanishing spontaneous polarization P_S, such as the chiral smectic C phase (S_C^*).

A natural evolution was, therefore, synthesis of an S_C^* liquid crystal polymer with the chiral centre in its side chains. This was done for the first time by Shibaev *et al.* (1984). Interest in S_C^* liquid crystal polymers (LCPs) lies in the possibility of having a new type of ferroelectric polymer whose alignment can be frozen into the glassy state. This new field of the liquid crystal science is in its infancy; nevertheless, some interesting results have already been recorded.

In this chapter, some families of S_C^* LCPs are described. The ferroelectric properties and the potential applications of these new materials are discussed. Because of the strong relationships between the low-molecular-mass liquid crystals (LMWLCs) and the side chain LCPs, the most relevant information pertaining to structure/property relations of S_C^* LMWLCs are reported firstly after some general considerations. This discussion is then used as a foundation to build an understanding of what is needed in S_C^* LCPs.

5.2 General considerations

Mesophases obtained from elongated molecular structures can be roughly divided into nematic and smectic phases. The nematic mesophase, where the long axes of the molecules are only approximately parallel to a director $\vec{n}$, is the less ordered one.

Smectic mesophases are more ordered and have molecular packing in layers which exhibit a complex polymorphism with respect to the positional ordering of the molecules within the layers and the molecular tilt orientational ordering. Now, 12 different smectic phases have been identified (Gray, 1985; Patel and Goodby, 1987a). In four of them, the average direction of the long molecular axes is perpendicular to the planes of the layers (S_A, $S_{B_{hex}}$, $S_{B_{cryst}}$ and S_E).

In seven others, the long molecular axes are tilted with respect to the layers normal (S_C, S_I, S_F, S_G, S_H, S_J and S_K).

Tilted smectic structures built up with optically active molecules give rise to chiral smectic phases. For the less ordered smectics or smectic liquid crystals (S_C^*, S_F^* and S_I^*), a twisted superstructure arises where the tilt director is turned through an azimuthal angle in passing from one layer to the next. In this way, the tilt direction forms a helix whose axis is parallel to the layer normal. Finally for the more ordered smectics or smectic crystals (S_G^*, S_H^*, S_J^* and S_K^*) only a long-range positional ordering of the molecules remains, the helical ordering being suppressed (Brand and Cladis, 1984a, b).

In addition, due to their symmetry properties, chiral tilted smectic phases possess a *non-vanishing component of the polarization* $\overline{P}_S$ (Meyer *et al.*, 1975). Interest in ferroelectric liquid crystals and more precisely in S_C^*, increased substantially when Clark and Lagerwall put forward the possibility of utilizing a ferroelectric liquid crystal in an electro-optical device (Clark and Lagerwall, 1980a, b). A bistable fast switching display device operating in the S_C^* phase was demonstrated. Since then, the number of papers dealing with the chemical, applied and theoretical aspects of S_C^* has increased exponentially.

5.3 Low-molecular-mass chiral smectic C compounds

5.3.1 *Structure of the S_C^* phase*

The molecular arrangement for the S_C^* phase is shown in Figure 5.1. The molecules are packed in layers, and there is no long-range regularity of packing of the centres of gravity of the molecules in the planes of the smectic layers. The molecular long axes of the molecules are tilted with respect to the normal to the layer planes. Because of the chirality of the molecules, the tilt directions form a helical distribution on moving from layer to layer.

The tilt angle depends on the chemical nature of the mesogens and decreases as the temperature increases.

When the $S_C^* \rightarrow S_A$ phase sequence is observed on heating, tilt angles are given by the

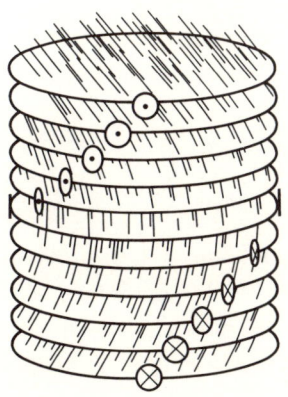

Figure 5.1 Structure of the S_C^* phase.

following power law expression (Martinot-Lagarde *et al.*, 1981):

$$\theta(T) = \theta_0(T_c - T)^\beta$$

where $\theta(T)$ is the optical tilt angle at temperature $T(°C)$
        θ_0 is a constant
        T_C is the $S_C^* \rightarrow S_A$ transition temperature (°C)
and T is the actual temperature (°C).

Materials with the $S_C^* \rightarrow N^*$ phase sequence exhibit a higher saturated tilt angle of approximately 45°. This value is relatively temperature-independent, except near the $S_C^* \rightarrow N^*$ phase transition temperature, where it decreases abruptly to zero.

Let us consider one isolated layer of the S_C^* phase; the only symmetry element existing is a two-fold axis of rotation C_2. The C_2-axis is parrallel to the layer planes and perpendicular to the plane. Thus, a non-vanishing component of the polarization vector $\vec{P}_s$ exists in the direction of the C_2-axis (Figure 5.2*a*). Two directions are possible for the spontaneous polarization to act along. According to the convention of Clark and Lagerwall (1984), the polarization is positive $(P_s(+))$ when $\vec{P}_s$, $\vec{z}$ and $\vec{n}$ form a right-handed system; the polarization is negative $(P_s(-))$ when $\vec{P}$, $\vec{z}$ and $\vec{n}$ form a left-handed system (Figure 5.2*b*).

What we have just described is a local property of one layer. Due to the helical structure of the mesophase, a bulk S_C^* phase does not exhibit any net spontaneous polarization. For this reason, the term helielectric, instead of ferroelectric, has been suggested to describe this type of behaviour (Brand and Cladis, 1984*b*). A macroscopic non-vanishing polarization can be obtained only by unwinding the helix.

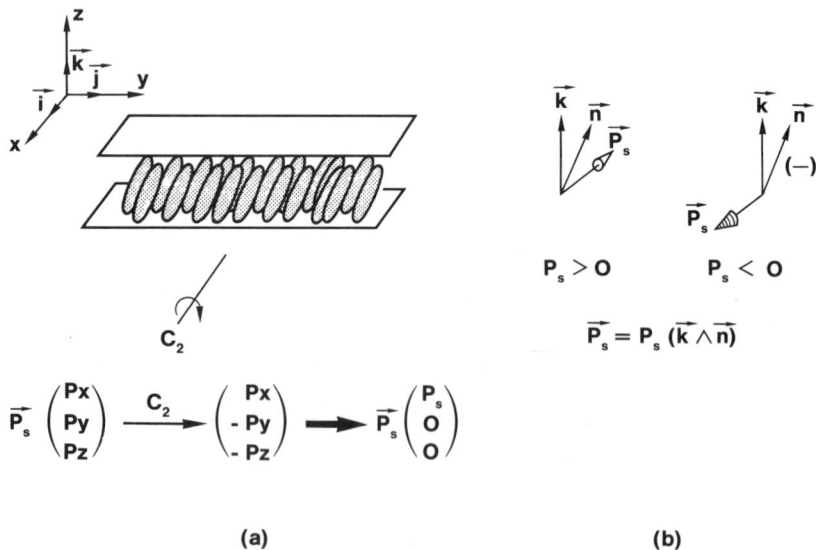

(a)   **(b)**

Figure 5.2(*a*) C_2-symmetry operation applied to a single S_C^* layer. (*b*) Sign convention for P_s.

5.3.2 *Molecular structure of* S_C^* *LMWLC*

Since 1976, several hundred LMWLCs exhibiting a monotropic or an enantiotropic S_C^* phase have been synthesized. The general structure of these compounds can be described by the following formula:

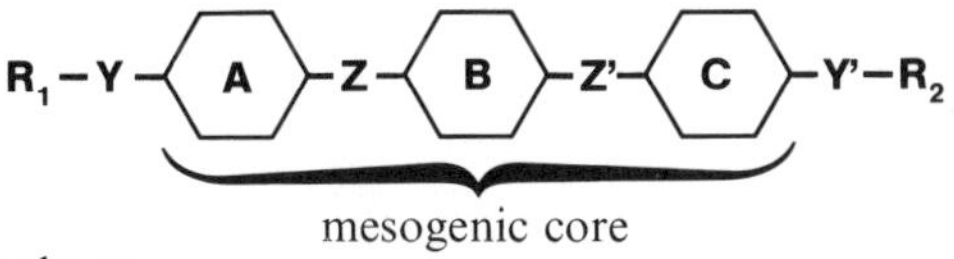

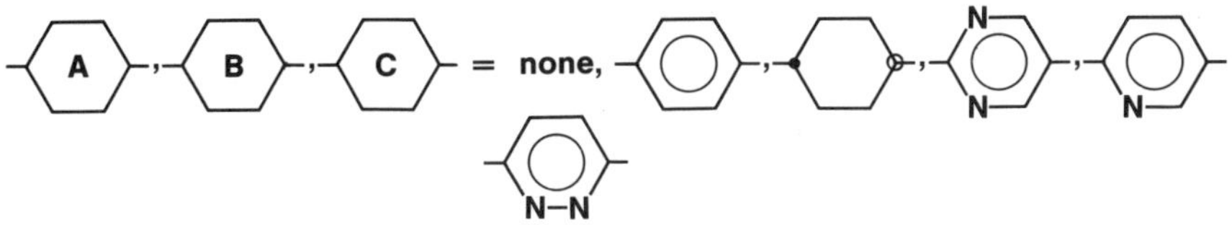

Z, Z' = none, $-COO-$, $-O-CH_2-$, $-CH_2-CH_2-$, $-CH=N-$, $-CH=CH-$

Y, Y' = none, $-O-$, $-COO-$, $-COS-$, $-CH=CH-COO-$, $-CH=C-COO-$,
$\qquad\qquad\qquad\qquad\qquad\qquad\qquad\qquad\qquad\qquad\qquad$ $|$
$\qquad\qquad\qquad\qquad\qquad\qquad\qquad\qquad\qquad\qquad\quad$ Cl

$-CH=C-COO-$
$\quad\quad\ |$
$\quad\quad$ CN

$R_1, R_2 = C_nH_{2n+1}-$; $C_nH_{2n+1}-\overset{*}{C}H-C_mH_{2m}-$;
$\qquad\qquad\qquad\qquad\qquad\qquad\quad |$
$\qquad\qquad\qquad\qquad\qquad\qquad\ X$

$\qquad\qquad C_nH_{2n+1}-\overset{*}{C}H-\overset{*}{C}H-C_mH_{2m}-$
$\qquad\qquad\qquad\qquad\qquad\underset{O}{\diagdown\ \diagup}$

with X = CH_3 , Cl , Br , CN , CF_3.

The mesogenic core contains at least two aromatic groups. Sometimes one of the benzene rings is laterally substituted by a bromine, chlorine or fluorine atom. Until now, the chiral centre has been included solely in the terminal groups and in most cases the molecular structure involves only one asymmetric carbon. Chiral terminal groups are currently synthesized from a limited number of commercially available optically active precursors, such as:

(i) Alcohols (S-(−)-2-methyl-1-butanol, 1-methyl-1-propanol, 2-octanol)
(ii) 2-methylbutanoic acid
(iii) S-ethyl lactate
(iv) Terpenols ((−)menthol,(+)menthol,(−)borneol,(−)-neo-menthol,(+)-fenchol, (−)-isolongifolol)
(v) Amino acids.

Chemical modifications of the chiral precursors are possible without appreciable racemization. Some examples are given in Figures 5.3 to 5.5.

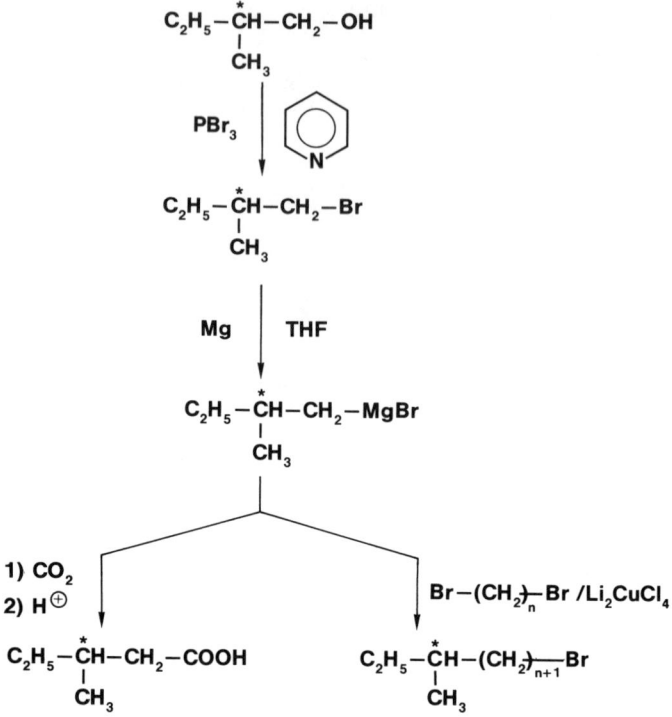

Figure 5.3 The synthesis of optically active S-3-methylpentanoic acid and S-methyl alkyl bromides from S-(−)-2-methyl-1-butanol.

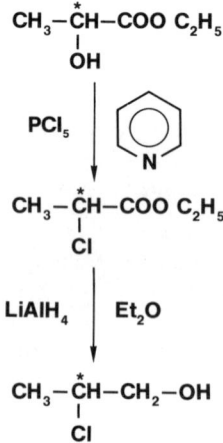

Figure 5.4 The synthesis of R-2-chloro-1-propanol from S-ethyl lactate.

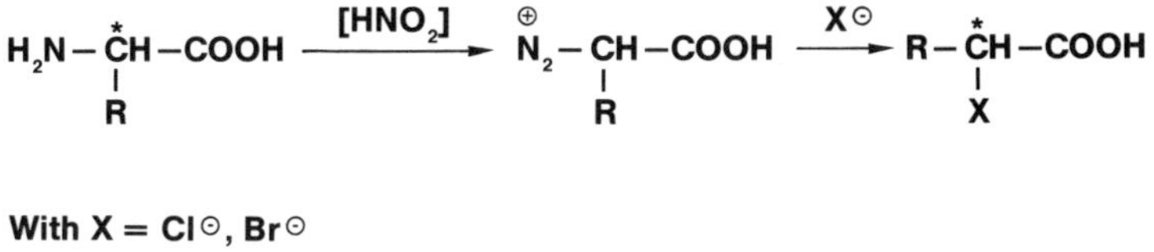

$$H_2N - \overset{*}{C}H - COOH \xrightarrow{\text{[HNO}_2\text{]}} \overset{\oplus}{N_2} - CH - COOH \xrightarrow{X^{\ominus}} R - \overset{*}{C}H - COOH$$
$$\qquad\qquad R \qquad\qquad\qquad\qquad R \qquad\qquad\qquad\qquad X$$

With X = Cl$^{\ominus}$, Br$^{\ominus}$

Figure 5.5 The synthesis of optically active α-substituted acids, from amino acids.

Finally, few examples of S_C^* compounds having a strong dipolar end group have been published (Destrade et al., 1985):

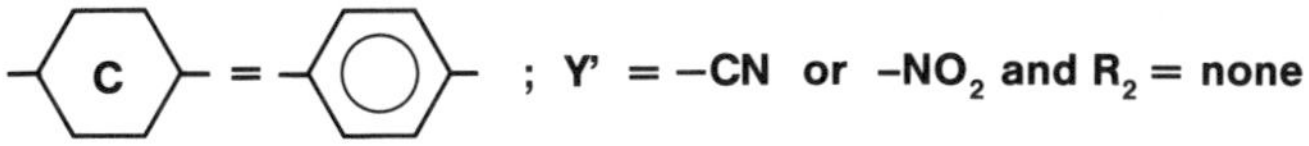

$$-\hexagon{C} - = -\hexagon{\bigcirc}- \quad ; \; Y' = -CN \text{ or } -NO_2 \text{ and } R_2 = \text{none}$$

5.3.3 Main structural factors influencing S_C^* LMWLC properties

The increased number of S_C^* LMWLC compounds has allowed the possibility of studying the relationship between the molecular structure and the following S_C^* properties: phase sequencing, magnitude of the spontaneous polarization, and helical twist sense. These properties are described in more detail in what follows.

5.3.3.1 *Phase sequencing.* When the chiral centre is located in the terminal group attached to the electron-deficient end of the mesogenic core a $N^* - S_C^*$ sequence is commonly observed, but if it is attached to the more electron-rich moiety, the $S_A - S_C^*$ sequence is usually obtained (Goodby and Leslie, 1984).

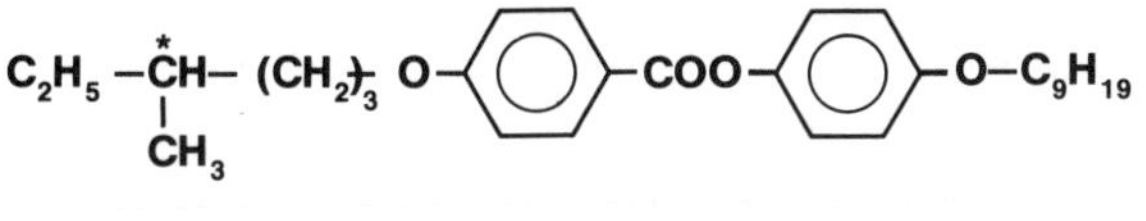

$$C_2H_5 - \overset{*}{C}H - (CH_2)_3 \; O - \bigcirc - COO - \bigcirc - O - C_9H_{19} \qquad \text{(I)}$$
$$\qquad\qquad CH_3$$

C 32 S$_c$* 45.4 N* 57.2 L

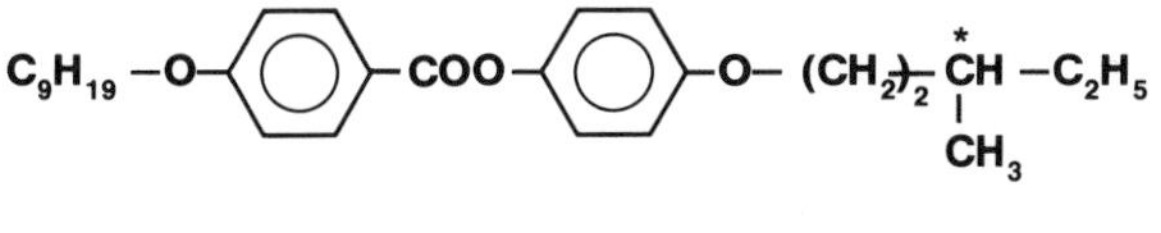

$$C_9H_{19} - O - \bigcirc - COO - \bigcirc - O - (CH_2)_2 \; \overset{*}{C}H - C_2H_5 \qquad \text{(II)}$$
$$\qquad\qquad\qquad\qquad\qquad\qquad\qquad CH_3$$

C 41 (S$_3$) 6 S$_c$* 49 S$_A$ 59 N* 62 L

5.3.3.2 *Magnitude of the spontaneous polarization.* The molecular processes responsible for the spontaneous polarization of the S_C^* mesophase are not well understood. A plausible assumption is that an asymmetry in the molecular rotation around the direction exists and therefore some rotational states are preferred, leading to a dipolar ordering (Lagerwall and Dahl, 1984). The model presented by Walba et al. (1986) is based on a novel kind of molecular recognition related to a diastereomeric guest–host complexation. The origin of the spontaneous polarization would be due to a barrier to rotation (determined by the molecular structure) in the molecular rotation around $\bar{n}$.

Table 5.1 P_S and θ values for series (**III**) (Otterholm *et al.*, 1987*a*)

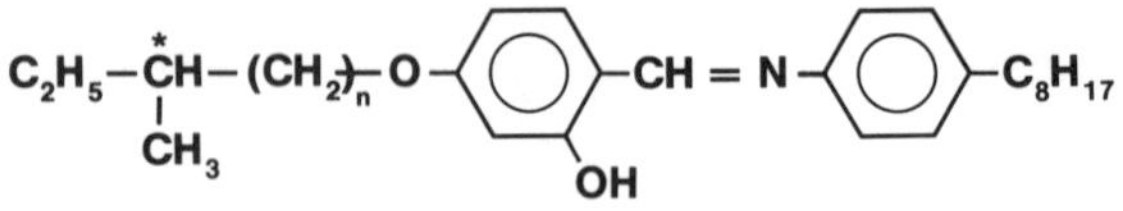

n	P_S (nC/cm^2)	θ (°)	P_S/θ (nC/cm^2 rad)
1	3	8	22
3	2	40	2.9
5	0.1	35	0.16

The following relationships between the spontaneous polarization and the molecular structure have been established.

(i) The stronger the dipole at a given chiral centre, the larger the absolute value of the spontaneous polarization.

(ii) Depending upon the strength of the polar group attached to the asymmetric carbon, two situations have to be considered. When the chiral centre is located in a non-polar terminal group, it appears that the closer the chiral centre is to the mesogenic core, the higher the absolute value of the spontaneous polarization (Table 5.1). In this case, the dipoles adding up to P_S originates from the mesogenic core.

If a polar group is directly attached to the asymmetric carbon, it seems that the absolute value of the spontaneous polarization is fairly independent of the position of the chiral centre in the terminal group, whereas a quite important stereopolar coupling takes place. This has been demonstrated by studying the family of compounds (**IV**) (Otterholm *et al.*, 1987*a*).

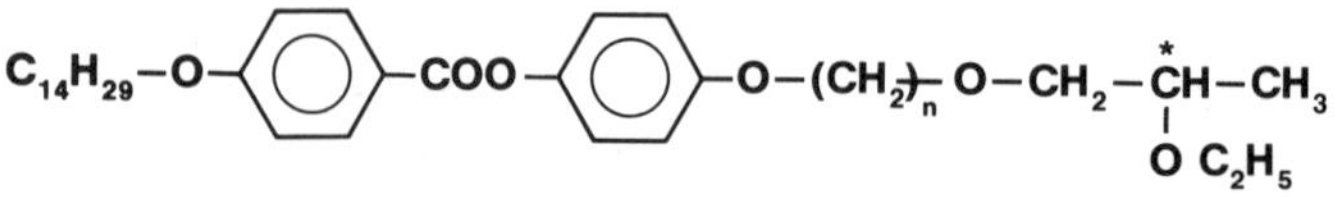

(iii) The closer the asymmetric carbon is to the centre of the molecule, the higher the absolute value of the spontaneous polarization, as exemplified in Table 5.2 (Bradshaw *et al.*, 1986). This is due to an increase in steric hindrance which leads to less freedom of movement of the chiral centre.

Table 5.2 Spontaneous polarization as a function of terminal group length for compounds (**V**)

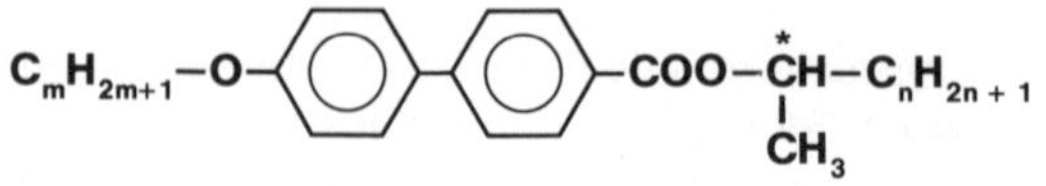

m	n	P_S (nC/cm^2)
8	2	17
8	5	48
8	6	92
9	6	100

(iv) As the rotation of the mesogenic core in relation with the terminal groups is unbiased, the core structure does not greatly contribute to the magnitude of P_S. Thus, P_S is mainly produced by the chiral moiety of the molecule which is generally decoupled from the core.

(v) The behaviour of the absolute value of the spontaneous polarization as a function of temperature is quite similar to that of the tilt angle. If the $S_C^* \to S_A$ phase sequence is exhibited by the materials, P_S is given by the following power law expression:

$$P_S(T) = P_S^0(T - T_C)^\alpha$$

where $P_S(T)$ is the polarization at T (°C)
P_S^0 is a constant
T_C is the $S_C^* \to S_A$ phase transition temperature (°C)
T is the actual temperature (°C).

By studying three different homologous series, Otterholm et al. (1987b) have shown

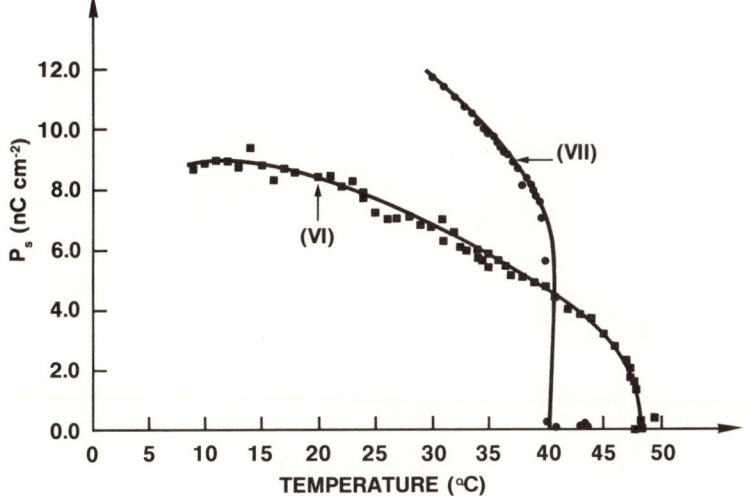

Figure 5.6 Spontaneous polarization for two isomeric compounds v. temperature. Compound (VI) exhibits a $S_C^* \to S_A$ phase sequence, whereas compound (VII) exhibits a $S_C^* \to N^*$ phase sequence.

that the quotient polarization/tilt angle is practically independent of T except near the $S_C^* \to S_A$ phase transition.

Materials having the $S_C^* \to N^*$ phase sequence possess a much larger absolute value of their spontaneous polarization than the previous compounds.

In addition, P_S generally increases rapidly and sometimes linearly as a function of $(T_C - T)$.

The behaviour of the two different types of compounds is depicted in Figure 5.6 (Patel and Goodby, 1987a).

Until now, the highest spontaneous polarization ever found ($P_S = 280\,\mathrm{nC\,cm^{-2}}$ at $T = 43.5^\circ\mathrm{C}$) has been obtained with compound (VIII) (Sakurai et al., 1986; Bahr and Heppke, 1986).

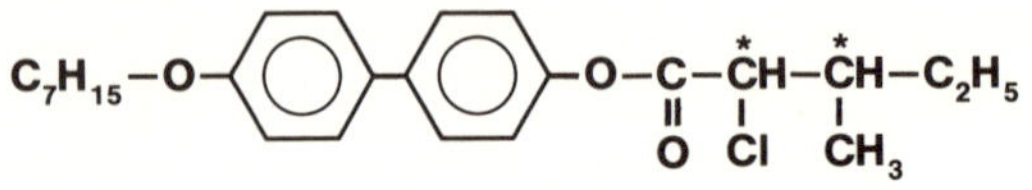

$$C_7H_{15}-O-\text{⬡—⬡}-O-\underset{\substack{\| \\ O}}{C}-\overset{*}{C}H-\overset{*}{C}H-C_2H_5 \qquad \text{(VIII)}$$

5.3.3.3 *Helical twist sense.* From studies of more than $100\,S_C^*$ compounds, and assuming that the molecules are in the all *trans* conformation with their mesogenic core more tilted than the overall molecular structure, Patel and Goodby (1987) have displayed some interesting relationships between the sign of the spontaneous polarization ($P_S(+)$ or $P_S(-)$), the helical twist sense (lefthand, LH, or righthand, RH, helix), the configuration of the chiral centre (R or S), the parity displacement of the chiral centre from the core (odd, o, or even, e) and the bond polarization direction at the chiral centre ($+I$ or $-I$). These results are summarized in Table 5.3.

For compounds in which the core is less tilted than the overall structure, an opposite polarization sign is obtained. These rules are semi-empirical in origin, and recently Otterholm et al. (1987b) have synthesized S_C^* molecules which do not exhibit the twist sense and the sign of P_S alternations with the parity displacement of the chiral centre from the core, as expected from Table 5.3.

5.3.4 *Applications*

5.3.4.1 *Display devices.* The twisted nematic effect (Schadt and Helfrich, 1971) is still the most widely used of all liquid crystal electro-optic effects. But its rather low

Table 5.3 Relationship between spontaneous polarization and molecular architecture

Absolute configuration	Parity	Inductive effect	Twist sense	Rotation of plane polarized light	Polarization sign
S	e	$+I$	LH	d	$P_S(-)$
S	o	$+I$	RH	l	$P_S(+)$
R	e	$+I$	RH	l	$P_S(+)$
R	o	$+I$	LH	d	$P_S(-)$
S	e	$-I$	RH	l	$P_S(+)$
S	o	$-I$	LH	d	$P_S(-)$
R	e	$-I$	LH	d	$P_S(-)$
R	o	$-I$	RH	l	$P_S(+)$

response time (some tens of milliseconds), its poor viewing angle and its limited multiplexing capability do not make this effect suitable for large area displays. For this reason a great deal of study has been devoted to ferroelectric display devices since the discovery of the fast switching bistable electro-optic device operating in the S_C^* phase.

Ferroelectric properties which appear when the helix of the S_C^* phase is unwound can be obtained in different ways (Lagerwall and Dahl, 1984), but the most closely studied approach is the helix-free surface-stabilized structure involving the effects of boundary conditions. The acronym SSFLC (surface stabilized ferroelectric liquid crystal) is now universally used to describe this type of device. The SSFLC display is characterized by:

(i) A homogeneous alignment of molecules of negative $\Delta\varepsilon$ at each electrode, with the smectic layers standing perpendicular to the confining glass plates ('bookshelf geometry')

(ii) A cell thickness lower than the helical pitch of the S_C^* phase, avoiding the helix formation and giving rise to two molecular orientations of opposite polarization directions as shown in Figure 5.7. Since these two states are surface stabilized, bistability occurs.

The cell is sandwiched between crossed polarizers (one tilt direction is parallel to the direction of one polarizer). The electro-optical effect is achieved by applying an external electric field across the cell that switches the molecules from one tilt direction to the other as the field is reversed. The state with the director parallel to either polarizer appears dark in transmission (extinction). The other state is bright (transmission).

Since the electro-optic effect is due to the linear coupling between the polarization and the applied field, the switching time τ of the director, given roughly by the following expression, is very fast (some microseconds):

$$\tau \sim \frac{\gamma \sin \theta}{P_S \cdot E}$$

where γ is the rotational viscosity and E is the electric field strength.

SSFLC displays therefore exhibit very attractive properties, but they are not commercially available at the present time, because some difficulties concerning the display technology have to be overcome, namely:

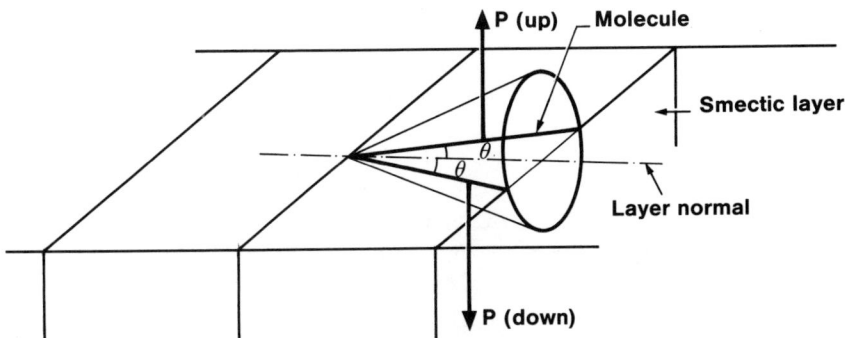

Figure 5.7 The two possible orientations of the spontaneous polarization in the SSFLC device.

(i) Obtaining a cell thickness less than $2\,\mu m \pm 0.2\,\mu m$ over a large area

(ii) Surface treatment leading to homogeneous alignment of the molecules

(iii) Understanding of the polar interaction between a S_C^* liquid crystal and its confining surfaces

(iv) Stability of the display under mechanical stress

(v) Obtaining grey levels

(vi) Existence of several intermediate states (splay states, twist states), zigzag defects and 'chevron' local layer structure (Rieker *et al.*, 1987).

Broad-range S_C^* mixtures exhibiting the convenient physicochemical properties have begun to appear (Geelhaar, 1988; Sage *et al.*, 1988) and we can expect the advent of the first commercial SSFLC displays in the next few years.

5.3.4.2 *Other applications.* The SSFLC devices possess the following properties attractive for electro-optical applications:

(i) Large permanent birefringence (allowing individual electro-optic elements to be small in size)

(ii) Low switching energy (leading to low power dissipation)

(iii) Low-threshold switching voltage

(iv) High switching speed (some microseconds)

(v) Intrinsic two-state memory

(vi) Low cost (in comparison with inorganic materials).

SSFLC devices can be used in single-cell operation as shutters and modulators (Armitage *et al.*, 1986), in linear arrays as printed heads (Umeda *et al.*, 1985) in matrix array as real-time masks and in optical processing systems (Pagano-Stauffer *et al.*, 1986; Handschy *et al.*, 1987).

These applications will not be extensively discussed here.

5.4 Chiral smectic C liquid crystal side chain polymers

The study of S_C^* side chain LCPs is at an early stage, and at present only few families of side chain LCPs exhibiting a S_C^* phase have been published. The physical properties of these are still practically unknown at this time.

5.4.1 *Chemical structure of known S_C^* side chain LCPs*

The first S_C^* side chain LCP synthesized was the polymethacrylate P5*N (Shibaev *et al.*, 1984), the exact formula of which was published later (Shibaev and Platé, 1985).

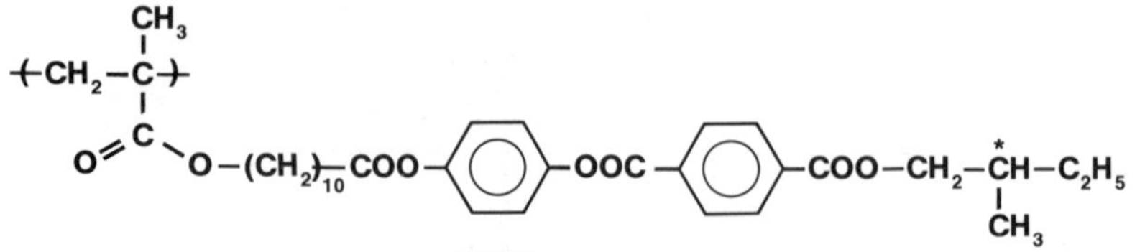

P5*M

G 20 - 30 S$_c$* 73 - 75 S$_A$ 83 - 85 L

Since then, several structural modifications have been attempted to obtain a S_C^* side chain LCP; these mainly concerned the polymer backbone (polyacrylate, polysiloxanes) and the mesogenic moiety. In all cases, the chiral centre has been located in the terminal group. (S(−)-2-methyl-1-butanol has principally been used as starting material to insert the chiral group in the side chain. Recently, some S_C^* side-chain LCPs have been synthesized from R(−)-2-octanol, 2-chloro-1-propanol and 4-methyl-2-chloro-1-pentanol. Although combined liquid crystal polymers and cross-linked combined liquid crystal polymers are quite different in nature from side chain LCPs, they will be discussed in this section.

5.4.1.1 *S_C^* polyacrylates and polymethacrylates.* In order to study the influence of the polymer backbone and the spacer length on the possibility of obtaining ferroelectric smectic LCPs, we first synthesized a number of polymers having the structure AX$_n$ (Decobert *et al.*, 1985). We then focused our attention on the influence of the mesogenic

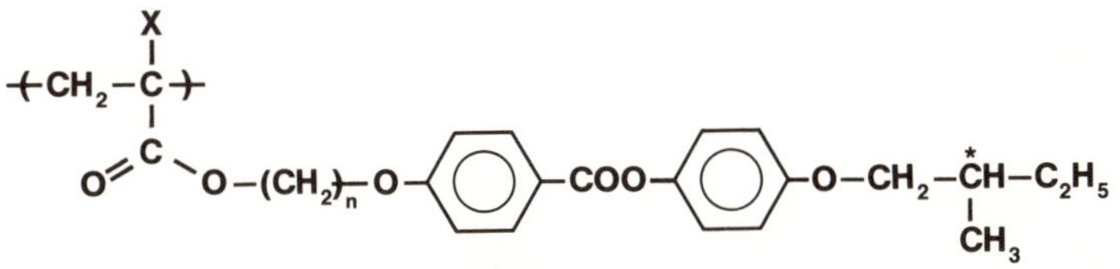

AXn

X = H, CH$_3$, Cl

n = 2, 6, 11

core on the phase behaviour of chiral side chain LCPs, and prepared new families of polymers BX$_n$ (Guglielminetti *et al.*, 1986). Corresponding monomers were synthesized

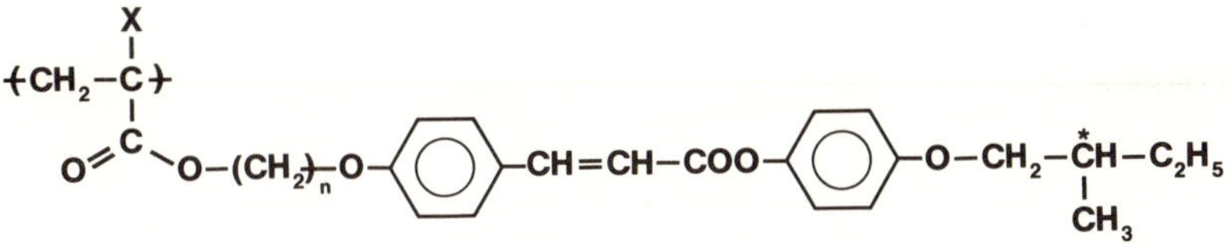

BXn

X = H, CH$_3$, Cl

n = 2, 6, 11

by standard methods avoiding racemization of the chiral centre. The synthetic route used is described in Figure 5.8. The lack of solubility of cinnamate derivatives has somewhat complicated their synthesis and purification.

Monomers derived from 4-hydroxycinnamic acid exhibit a complex smectic polymorphism when n = 6 and 11 (Guglielminetti *et al.*, 1986), whereas monomers derived from 4-hydroxybenzoic acid are very often non-mesogenic or display mainly a S$_A$ mesophase (Decobert *et al.*, 1985).

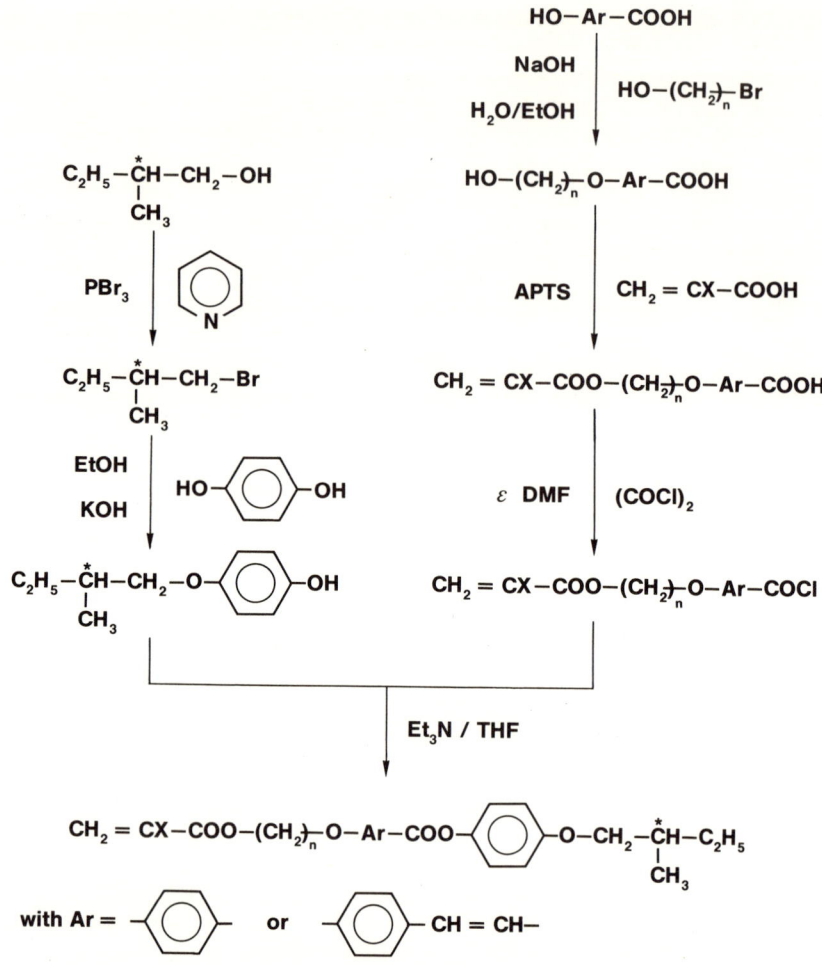

Figure 5.8 Synthetic route for the preparation of monomers.

Polymerizations were carried out under vacuum, using azo-bis-isobutyronitrile as free radical initiator.

Polymers AX_n and BX_n have been characterized by polarized optical microscopy, DSC and X-ray diffraction. The results are summarized in Table 5.4 (Decobert *et al.*, 1986; Esselin *et al.*, 1987; Esselin *et al.*, 1988). It appears that:

(i) Most of these polymers are semi-crystalline in character.
(ii) The glass transition temperature decreases as the length of the spacer increases; this is due to the effect of internal plasticizing and to a higher mobility of the main chain, since the bulky mesogenic core is moved away from the polymer backbone. Polymer AH 11 exhibits an abnormal behaviour in this respect, since its T_g (32°C) is higher than that of polymer AH 6 (16°C). A possible explanation lies in a partial crystallization of the side chain (Esselin *et al.*, 1987).

Table 5.4 Thermal behaviour of polymers AX_n and BX_n

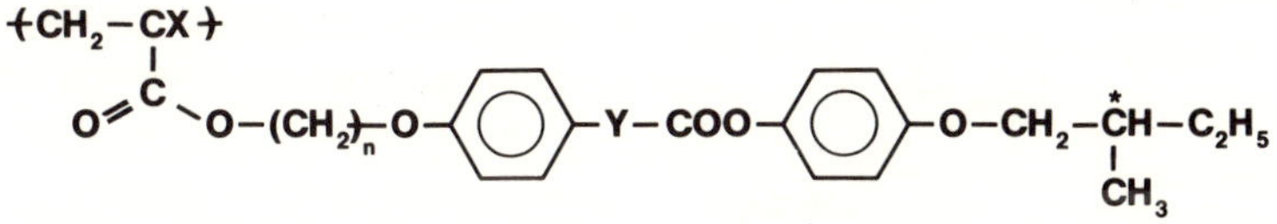

$$+CH_2-CX+$$
$$\overset{\displaystyle |}{\underset{\displaystyle O}{C}}\diagdown O-(CH_2)_n-O-\bigcirc\!\!\!-Y-COO-\bigcirc\!\!\!-O-CH_2-\overset{*}{C}H-C_2H_5$$
$$\qquad\qquad\qquad\qquad\qquad\qquad\qquad\qquad\qquad\qquad\qquad\;\; |$$
$$\qquad\qquad\qquad\qquad\qquad\qquad\qquad\qquad\qquad\qquad\qquad CH_3$$

						Thermal behaviour				
Y	X	n	T_g	T_m		$T_{S_1S_a}$		T_c	ΔH_0 kJ/m.u.[c]	ΔS J/K m.u.[c]
—	H	2	65		S_C^*	110	S_{A_2}	146 L	1.8	4.29
—	H	6	15	46[a]			S_{A_1}	87 L	4	11.1
—	H	11	32	59			S_{A_1}	108 L	6.6	17.2
—	CH$_3$	2	S_C^* 110				S_{A_2}	155 L	3.5	8.2
—	CH$_3$	6	40				S_{A_1}	90 L	5.0	13.8
—	CH$_3$	11	10–40[b]	58	S_C^*	90	S_{A_1}	106 L	6.9	18.3
—	Cl	2	97	172				L	4.8	
—	Cl	6	37	94				L	5.0	
—	Cl	11	10–40[b]	62			S_A-like	112 L	7.5	19.4
—CH=CH—	CH$_3$	2	125				S_{A_2}	172 L	2.8	6.2
—CH=CH—	CH$_3$	6	50	92			S_{A_1}	125 L	3.5	8.8
—CH=CH—	CH$_3$	11	35	65	S_C^*	115	S_{A_1}	142 L	5.4	12.9
—CH=CH—	Cl	11	35	76			S_A-like	140 L	4.6	11.1

(a) Small endotherm which appears only after sample annealing at 30°C for 6 h.
(b) These glass transitions were hard to detect
(c) m.u.: monomer unit

(iii) The stiffness of the polymer backbone increases when passing from the polyacrylates to the polymethacrylates or polychloroacrylates, in agreement with the observed increase in T_g. The effects of $X = CH_3$ and $X = Cl$ are nearly the same.

(iv) The thermal stability of mesophases of polymers BX_n is higher than that of polymers AX_n having the same X substituent and the same value of n.

(v) For a given series, the replacement of hydrogen in the polymer backbone by a methyl group or a chlorine atom seems to increase the orientational order in the side chains as suggested by the observed gain in entropy change at the clearing point.

(vi) 'Bilayer' structures are observed for polymers where $n = 2$, whereas 'monolayer' structures are found for polymers having longer flexible spacers.

(vii) Poly-α-chloroacrylates do not exhibit S_C^* phase, but the only liquid crystal properties they show are an S_A-like structure where some organization seems to occur in the smectic layers. This organization could be due to the chlorine atoms in the polymer backbone, which would impose a local ordering of the side chains.

(viii) Finally, one polyacrylate (AH-2) and three polymethacrylates (A-CH$_3$,-2, A-CH$_3$-11 and B-CH$_3$-11) exhibit a S_C^* mesophase.

Recently, the polymethacrylate P6XM having the 4-methyl-2-chloro-pentyl moiety as terminal group has been published (Shibaev et al., 1987).

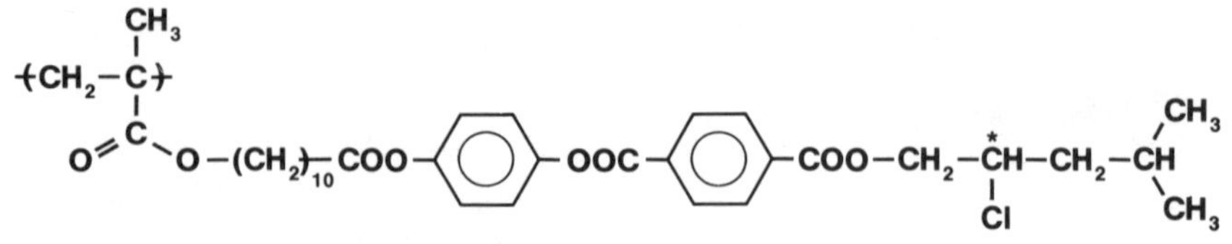

P 6XM

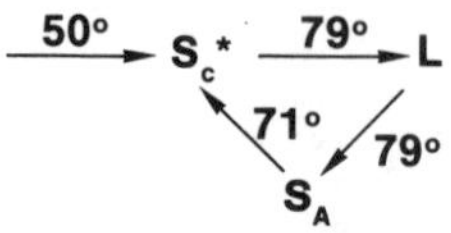

This polymer seems to possess an enantiotropic S_C^* phase between 50°C and 79°C and a monotropic S_A phase between 79°C and 71°C.

5.4.1.2 S_C^* *polysiloxanes.* Liquid crystal polysiloxanes are known to display lower viscosity and lower glass transition temperature than polyacrylates and polymethacrylates having the same side chain. Two laboratories have recently succeeded in obtaining S_C^* smectic polysiloxanes. Hahn and Percec (1987) have reported the synthesis and preliminary data on the characterization of the first examples of chiral smectic liquid crystal polysiloxanes containing 2, 5-disubstituted *trans* 1, 3-dioxane and 2, 5-disubstituted-1, 3, 2-dioxaborinane in their side chains. The quite complicated

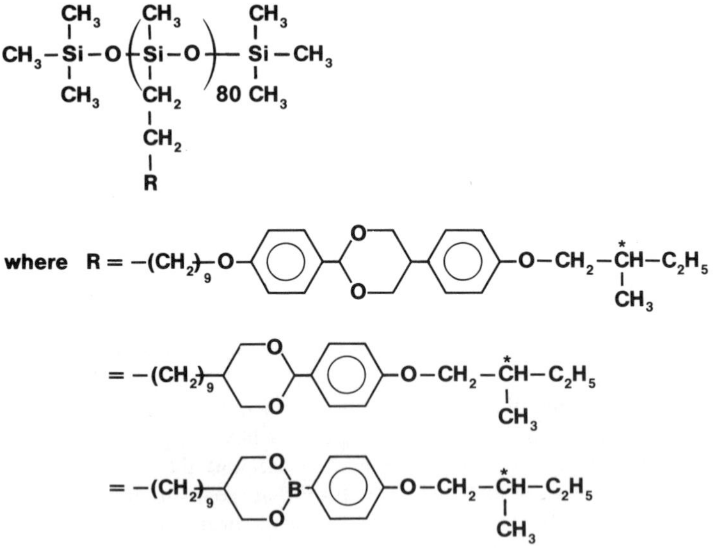

Polysiloxanes I

thermal behaviour of these polymers ($A - C$) is summarized in Table 5.5. Identification of the mesophases was carried out only by optical microscopy but the results have to be checked by X-ray investigations.

Table 5.5 Thermal behaviour of polysiloxanes I

Side chain	Scan	T_g	Heating			Cooling		
			T_1 $\Delta H_1/\Delta S_1$	T_2 $\Delta H_2/\Delta S_2$	T_c $\Delta H_c/\Delta S_c$	T_c $\Delta H_c/\Delta S_c$	T_2 $\Delta H_2/\Delta S_2$	T_1 $\Delta H_1/\Delta S_1$
A	1	30	73^a 1.8/5.2^a	78^c 0.005/0.014^b	140^a 1.6/3.9^a			
	2	30	69^c 0.56/1.6	S_C^* ?	152 1.2/2.8	149 1.0/2.4	65^d 0.05/0.15	38 0.1/0.4
B	1	4	50^a 0.4/1.2^a		91^a 0.2/0.5^a			
	2	4	50 0.03/0.10	S_C^*	97 0.2/0.6	97 0.2/0.6		48 0.05/0.2
C	2 (S_2)	-7	19 0.1/0.4	S_C^*	85 1.5/4.3	S_A ? 79 1.5/4.2		17 0.1/0.4

ΔH in kcal/m.u; ΔS in cal/m.u. K
(a) After annealing above T_g; (b) overlapped transitions, part of ΔH_2 is overlapped by ΔH_1;
(c) overlapped transitions; $\Delta H_c = \Delta H_1 + \Delta H_2$;
(d) overlapping transitions.

Polysiloxane C is the first side chain LCP ever synthesized to have a boron atom in its structure. 1, 3, 2-dioxaborinane rings possess the advantage over 1, 3-dioxane rings that the boron atom is trivalent, therefore *cis-trans* configuration isomerism does not exist and the synthesis does not require the separation of the two isomers.

A second S_C^* side chain polysiloxane has been prepared by Keller (1988a). It belongs to the chiral polysiloxane II family. The thermal behaviour of polysiloxanes II is

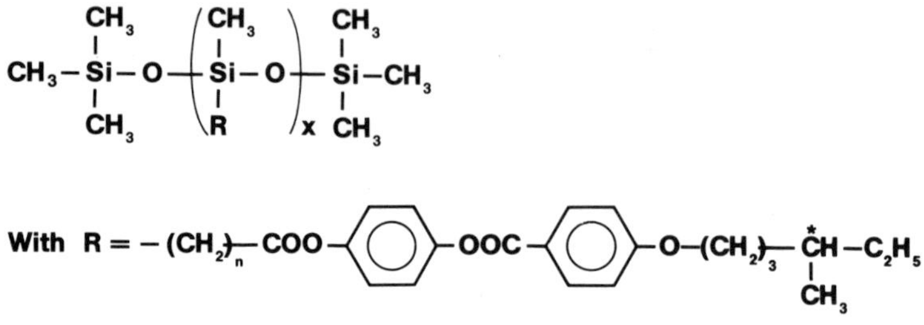

With $R = -(CH_2)_n-COO-\langle\bigcirc\rangle-OOC-\langle\bigcirc\rangle-O-(CH_2)_3-\overset{*}{C}H-C_2H_5$ with CH_3

Polysiloxanes II

reported in Table 5.6.

Table 5.6 Thermal behaviour of polysiloxanes II

n	x	Phase transitions
4	80	K 50 S_A 143 L
6	80	K 88 S_A 146 L
10	80	K 78 S_3 90 S_C^* 106 S_A 138 L
10	25	K 75 S_3 90 S_C^* 105 S_A 133 L
10	36	K 75 S_3 90 S_C^* 105 S_A 134 L

It appears that

(i) Short spacers are not favourable for the realization of the S_C^* phase.
(ii) Polysiloxanes II are crystalline in character; this quite unusual behaviour has already been observed in other polysiloxane series (Keller, 1988b; Rötz et al., 1987).
(iii) An increase of the average degree of polymerization x affects neither the transition temperature nor the mesomorphism of polysiloxanes II. This bahaviour confirms previous published results (Finkelmann and Rehage, 1984a).

5.3.1.3 *Combined S_C^* LCPs.* Combined LCPs are known to form different smectic phases (Reck and Ringsdorf, 1985) and their chemical structure can easily be modified, so they appeared to be good candidates for obtaining S_C^* LCPs.

Zentel et al. (1978a) synthesized the first series of S_C^* combined polymers, CLCP-I. The thermal behaviour of the polymers is summarized in Table 5.7.

Table 5.7 Thermal behaviour of CLCP-I

R_1	n	R_2	Molecular weight	Thermal behaviour
—	6	—	56 000	K_1 151 K_2 154 L
—N=N—	6	—	3200	K 108 S_C^* 115 N* 128 L
—N=N—	2	—	21 000	K 150 L
—	6	—N=N—	30 000	K 130 S_C^* 136 L
—N=N—	6	—N=N—	36 000	K 107 S_C^* 111 N* 137 L

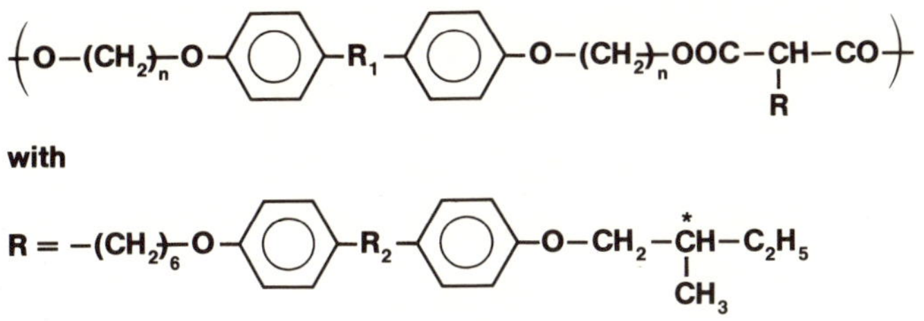

CLCP - I

Since then, two other series of combined LCPs (CLCP-II and CLCP-III) derived from substituted malonates have been prepared (Bualek *et al.*, 1988; Kapitza *et al.*, 1988). The aim of polymers CLCP-II is to obtain S_C^*-combined LCPs with a potential higher

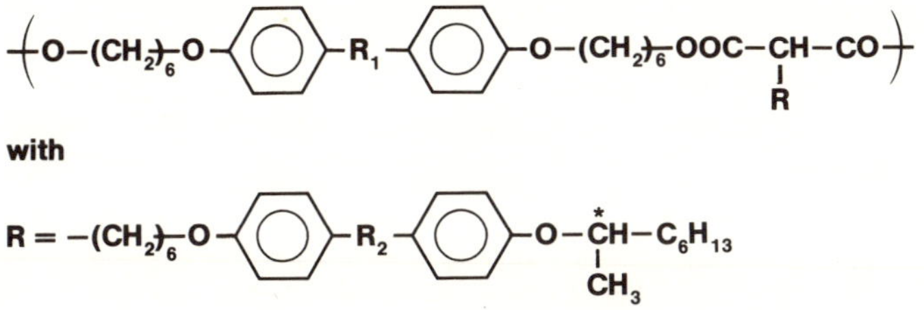

CLCP - II

spontaneous polarization than that supposed to be exhibited by CLCP-I series. Indeed, 2-octanol is used as chiral terminal group instead of 2-methyl-1-butanol (see section 5.3.3.2). As for polymers CLCP-III, they were synthesized to obtain S_C^* combined LCPs with a broad mesophase temperature range and low melting temperatures, owing to lateral substitutions of the benzene rings of the mesogenic cores.

Several polymers belonging to CLCP-II and CLCP-III series exhibit a S_C^* phase. They are listed in Tables 5.8 and 5.9. The oxygen atom at the azoxy group which acts as a lateral substituent, suppresses the crystallization of polymers CLCP-II. Only substitution in the main chain leads to the S_C^* phase. CLCP-III polymers with lateral substituents have low melting points as expected. Note that the authors have reservations regarding the identification of smectic phases of CLCP-II and CLCP-III.

F

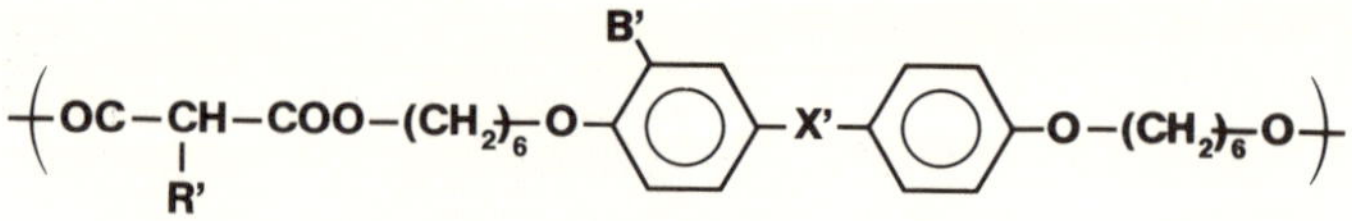

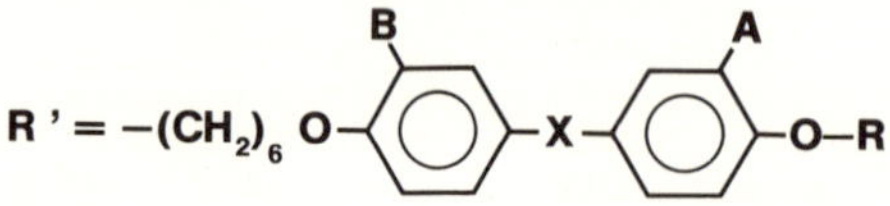

$$R' = -(CH_2)_6 \: O - \bigcirc\!\!\stackrel{B}{} - X - \bigcirc\!\!\stackrel{A}{} - O - R$$

with :

A, B, B' = Br, H

X, X' = none, $-N = N-$, $-N = N-$
                              $\downarrow$
                              O

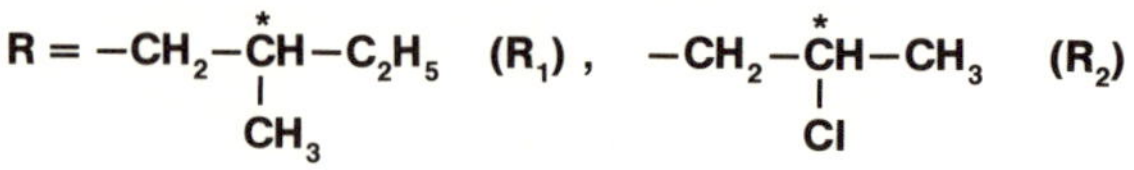

$R = -CH_2 - \overset{*}{C}H - C_2H_5$ (R_1) , $-CH_2 - \overset{*}{C}H - CH_3$ (R_2)
            $|$ $|$
            CH_3 Cl

CLCP - III

Table 5.8 Thermal behaviour of CLCP-II

R_1	R_2	Molecular weight	Thermal behaviour
$-N=N-$	—	62 000	K 79 S_{A_2} 88 N* 95 L
$-N=N-$ $\downarrow$ O	—	28 000	G 9 $S_{C_1}^*$ 67 S_{A_2} 92 N* 98 L
$-N=N-$	$-N=N-$	58 000	K 72 (S_C^* 57) S_{A_2} 93[a] N* 104 L
$-N=N-$ $\downarrow$ O	$-N=N-$	56 000	G 20 S_{C_2} 70[a] S_{A_2} 98 N* 108 L

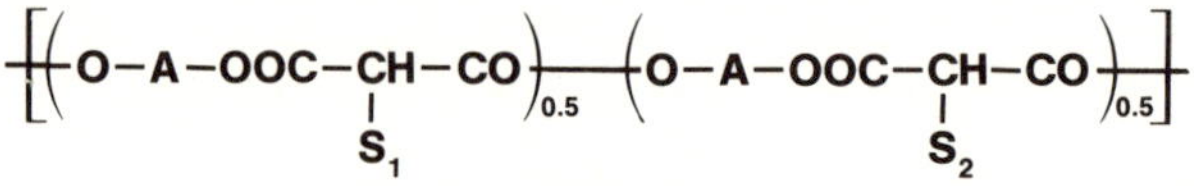

where

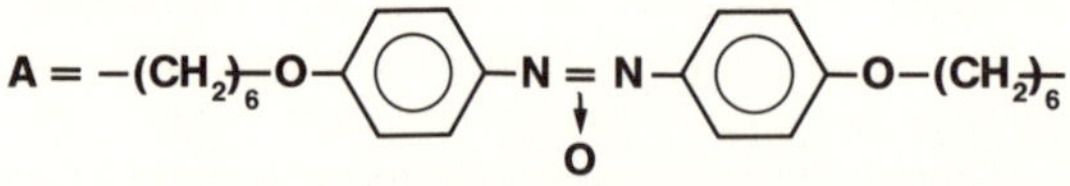

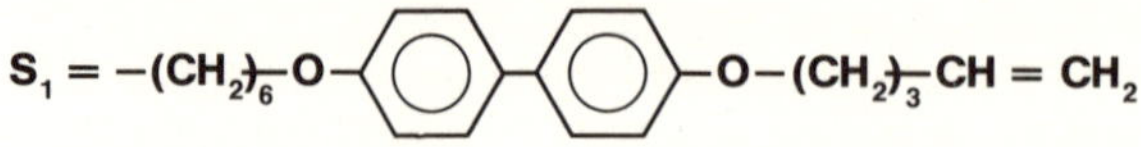

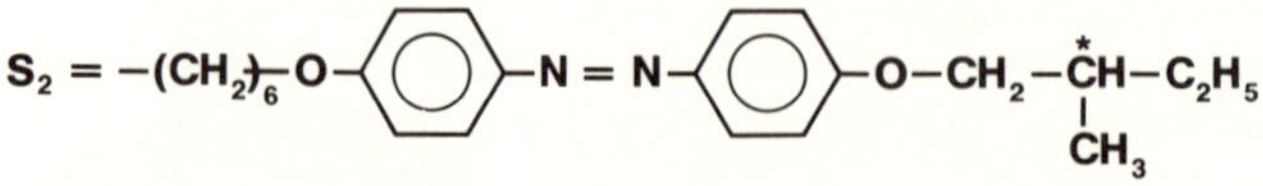

$$A = -(CH_2)_6 - O - \bigcirc - N = N - \bigcirc - O - (CH_2)_6 -$$
$$\qquad\qquad\qquad\qquad \downarrow$$
$$\qquad\qquad\qquad\qquad O$$

$$S_1 = -(CH_2)_6 - O - \bigcirc - \bigcirc - O - (CH_2)_3 - CH = CH_2$$

$$S_2 = -(CH_2)_6 - O - \bigcirc - N = N - \bigcirc - O - CH_2 - \overset{*}{C}H - C_2H_5$$
$$\qquad\qquad\qquad\qquad\qquad\qquad\qquad\qquad\qquad | $$
$$\qquad\qquad\qquad\qquad\qquad\qquad\qquad\qquad\qquad CH_3$$

CLCP - IV

Table 5.9 Thermal behaviour of CLCP-III

A	B	B'	X	X'	R	Molecular weight	Thermal behaviour
H	H	H	—	—N=N—	R_2	56 000	K 109 S_C^* 149 L 2.4^a 2.8
H	H	H	—	—N=N— ↓O	R_2	49 000	K 57 S_C^* 140 S_A 149 L 0.9 0.1 2.1
H	H	H	—N=N— ↓O	—	R_1	23 000	K 77 S_C^* 126 S_A 131 L 3.3 0.8 0.6
H	H	Br	—	—	R_1	40 000	K 59 S_C^* 105 L 1.6 2.9
H	H	H	—	—N=N—	R_1	43 000	K 112 S_C^* 129 N* 132 L
H	H	H	—	—N=N— ↓O	R_1	40 000	K 63 S_C^* 124 N* 133 L 1.7 1.1 0.7
H	H	Br	—N=N—	—	R_1	31 000	K 56 S_C^* 91 L 1.5 2.2
H	H	H	—N=N—	—N=N— ↓O	R_1	43 000	K 53 S_C^* 107 N* 138 L 1.5 1.4 1.1
H	H	H	—N=N—	—	R_1	30 000	K 130 S_C^* 136 L 2.7 1.6

K: crystalline or highly ordered smectic phase
(a): transition enthalpies (kJ/mol)

5.4.1.4 *Cross-linked S_C^* LCPs.* Several slightly cross-linked LCPs derived from polysiloxane (Finkelmann *et al.*, 1984*b*), polyacrylate and polymethacrylate (Zentel and Reckert, 1986; Canessa *et al.*, 1986) have been recently studied. Interest in these elastometric materials lies in the possibility of reversibly orienting liquid crystalline phases by small mechanical strains (Zentel and Reckert, 1986; Mitchell *et al.*, 1987; Schätzle and Finkelmann, 1986; Gleim and Finkelmann, 1987; Zentel *et al.*, 1987*b*; Zentel and Benalia, 1987). This property may become of considerable importance in the case of cross-linked LCPs having a S_C^* phase if the stretching of such a network were to reversibly unwind the helical structure, giving rise to a macroscopic polarization (Zentel, 1987). The first and, until now, the only cross-linked LCP having a S_C^* phase has been very recently prepared by Zentel (1988). This elastomer derives from a combined copolymer, which has been cross-linked by reaction of some of the ethylenic double bonds of the copolymer, with the Si—H groups of an oligo(dimethyl siloxane).

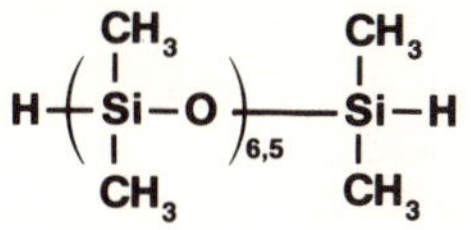

$$\text{H}\!-\!\!\left(\!\begin{array}{c}\text{CH}_3\\|\\\text{Si}-\text{O}\\|\\\text{CH}_3\end{array}\!\right)_{\!\!6,5}\!\!\!-\!\!\begin{array}{c}\text{CH}_3\\|\\\text{Si}-\text{H}\\|\\\text{CH}_3\end{array}$$ **(10 mol. %)**

The resulting material has the following thermal behaviour: G 15 S_C^* 100 N* 128 L. Preliminary X-ray investigations have shown that it was possible to freeze in, at room temperature, an orientation induced by a stretching made in the S_C^* phase. It appeared that the mesogenic groups were oriented parallel to the fibre axis and the smectic layers were tilted with respect to the fibre axis.

5.4.2 Identification of the chiral smectic C mesophase

As for LMWLCs, the structure of LCPs are deduced from thermal analysis, optical microscopy observations and X-ray measurements. However, due to differences in viscosity, relaxation response and thermal stability, identification of LCP mesophases is more difficult than in the case of a LMWLC. All aspects of the structure determination of the LCPs are discussed in detail in Chapter 6. In this section we wish just to emphasize the obstacles encountered in studying S_C^* phases.

When a well-defined texture is obtained by observing a supposed S_C^* phase of a LCP, it is then possible to distinguish the tilted phase from the S_A or N* phases. But it is sometimes impossible to recognize the S_C^* phase from the S_F^* or S_I^* phases, because these two more ordered phases exhibit some of the S_C^* textures such as the focal-conic fan texture and the pseudo-homeotropic texture (only for the S_F^* phase).

Until now, X-ray investigations on smectic side chain LCPs have been performed only on powder samples or, in some cases, on oriented fibres. Attempts to obtain a monodomain by using a strong magnetic field have failed.

X-ray diffraction patterns of powder samples obtained from a S_C^* LCP present a broad diffuse outer ring reflecting the absence of ordering within the layer planes and one or several well-defined inner rings which are related to the lamellar thickness d (Figure 5.9a).

These patterns are very similar to that obtained from a S_A mesophase.

If $d < L$ (L is the length of the side chain in its most extended conformation) one can conclude that the phase under study is a tilted one.

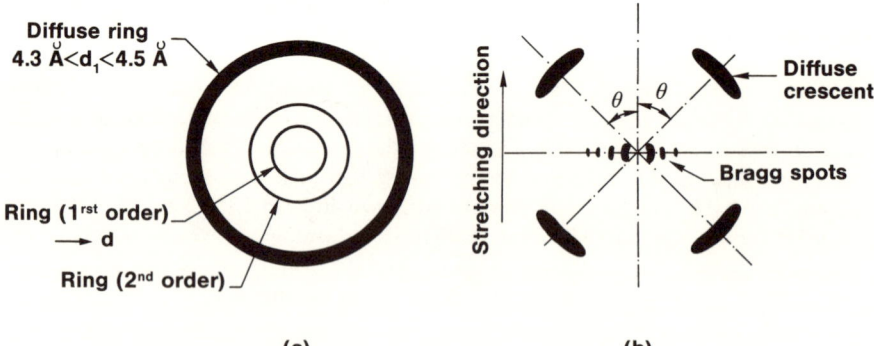

(a)   **(b)**

Figure 5.9(a) Schematic X-ray diffraction pattern of a S_C^* LCP powder sample. (b) Schematic X-ray diffraction pattern of polymer AH 2 obtained from a stretched oriented fibre drawn out of the S_C^* phase.

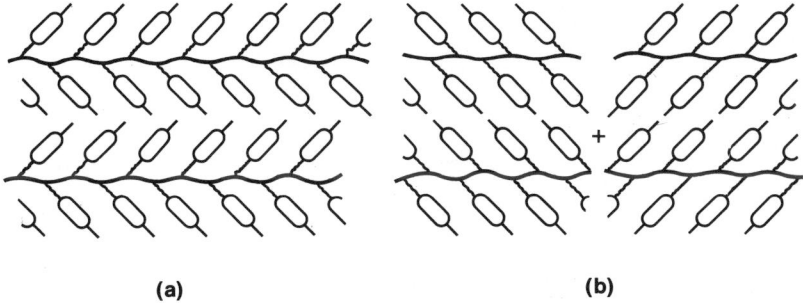

(a)   (b)

Figure 5.10 Possible models for side chain ordering in the S_C^* phase of polymer AH 2 (Decobert *et al.*, 1986).

But if $L < d < 2L$, it is impossible to distinguish between a tilted phase and the S_{Ad} phase. In this case, further X-ray investigations on stretched oriented fibres are needed. Figure 5.9*b* shows an example of X-ray diffraction pattern obtained from an oriented fibre drawn out of the S_C^* phase of polymer AH-2.

This pattern allows one to conclude that the phase under investigation is tilted. The six small-angle Bragg spots correspond to the three first orders of reflection on the layer plane. They are located on the equator, which indicates that the smectic layers and, as a consequence, the main chains are parallel to the fibre axis. The four diffuse crescents at large angles indicates a liquid-like order within the layers. The two distinct orientations for the mesogenic side chain observed could be explained by the models depicted in Figure 5.10.

Distinction between S_C^* and the more ordered structures (S_F^* and S_I^*) necessitates X-ray diffraction from a monodomain, the X-rays being perpendicular to the layers, in order to highlight the hexagonal local order which exists in S_F^* and S_I^*.

5.4.3 *Ferroelectic properties of chiral smectic C polymers*

Determination of the magnitude of the spontaneous polarization is of prime importance for the subsequent uses of ferroelectric materials. Several methods exist for measuring the magnitude of the spontaneous polarization, and have been recently reviewed by Martinot-Lagarde (1988). At present, the methods involving the electric reversal of the spontaneous polarization are widely used to determine P_S for S_C^* LMWLCs. Shibaev *et al.* (1984, 1987) have used the pyroelectric method to perform the same type of measure on S_C^* LCPs. The principle of pyroelectric measurement is as follows.

The typical dependence of polarization of ferroelectric materials on temperature is depicted in Figure 5.11. At some temperature known as the Curie temperature, this polarization vanishes to zero. The derivative of P versus T at a particular temperature is the pyroelectric coefficient $\gamma (\gamma = dP/dT)$; which can be measured using the dynamic Chynoweth technique (1956). Integrating γ with respect to temperature leads to the polarization. The ferroelectric material inserted between two conductive parallel electrodes is first oriented such that its polar axis is normal to the electrodes. It is then

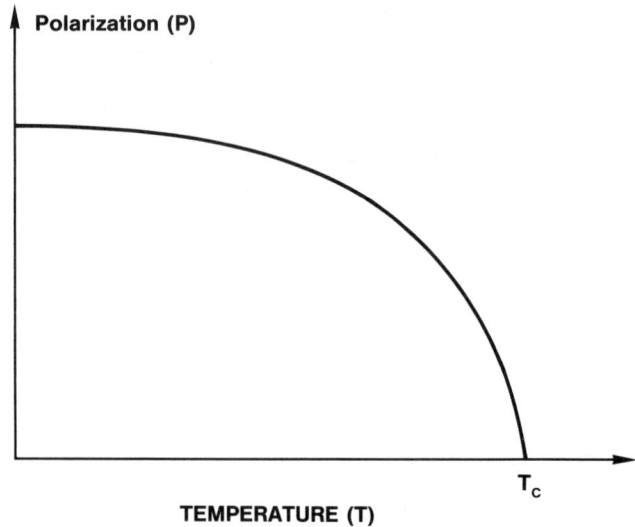

Figure 5.11 Temperature dependence of polarization of ferroelectric materials.

subjected to a small periodic variation of temperature, δT, produced by a chopped light source which causes a weak change δP in polarization. δP appears as a change on the capacitor formed by the ferroelectric material and the two electrodes. One obtains:

$$\gamma = \frac{-CV}{A_h \delta T}$$

with C: capacitance of the sample cell

V: voltage between the two electrodes

A_h: heated area.

δT depends upon the light source and the thermal properties of the material.

For S_C^* LMWLC compounds, orientation of the sample is achieved either by way of a surface treatment (the cell thickness being lower than the pitch of the S_C^* mesophase) or a DC electric field (which can be maintained during the entire heating and cooling rates) (Glass *et al.*, 1986).

We have tried to measure the spontaneous polarization of polymers AH-2 and A-CH$_3$-11, using the electric reversal of P_S (Martinot-Lagarde, unpublished). Unfortunately we failed in trying to align our polymers. In other respects, the S_C^* polymethacrylates P5*M and P6XM allowed the determination of their spontaneous polarization using the pyroelectric method (Shibaev *et al.*, 1984; Shibaev *et al.*, 1987). The temperature dependence of γ and P_S is reported in Figures 5.12 and 5.13. P5*M has been aligned with a DC electric field as low as 0.3 V μm^{-1} (Blinov *et al.*, 1987).

The maximum value of P_S in the S_C^* phase of P5*M is equal to about 0.8 nC cm^{-2}. Such a value is characteristic for typical LMWLC compounds bearing a small dipole at the chiral centre. For example, the P_S value of DOBAMBC is 3 nC cm^{-2} at $T_c - T = 10°C$ (Terashima, 1986).

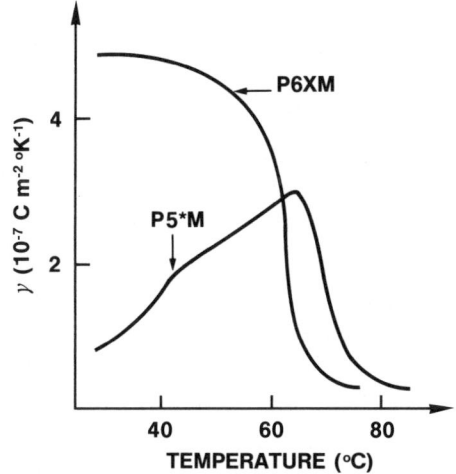

Figure 5.12 Temperature dependence of the pyroelectric coefficient for polymers P5*M and P6XM.

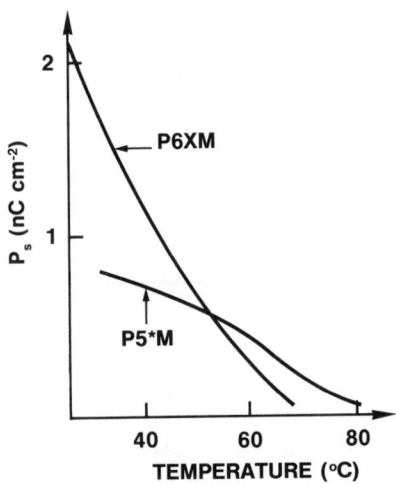

Figure 5.13 Temperature dependence of the spontaneous polarization for polymers P5*M and P6XM.

$$C_{10}H_{21}-O-\langle\bigcirc\rangle-CH=N-\langle\bigcirc\rangle-CH=CH-COO-CH_2-\overset{*}{C}H-C_2H_5$$
$$CH_3$$

DOBAMBC

On the other hand, the value of P_S in the S_C^* phase of P6XM seems lower than expected by comparison with LCWLC, since a strong polar atom is located at the chiral centre and the terminal group is longer than that of P5*M. For example, HOBACPC has a polarization that is approximately an order of magnitude higher

$$C_{13}H_{27}-O-\!\!\!\bigcirc\!\!\!-CH=N-\!\!\!\bigcirc\!\!\!-CH=CH-COO-CH_2-\overset{*}{C}H-CH_3$$
$$\underset{Cl}{|}$$

HOBACPC

than that for DOBAMBC (Patel *et al.*, 1987). This surprising result may be due to an insufficient poling of P6XM sample.

5.4.4 *Potential applications of S_C^* liquid crystal polymers*

Applications of S_C^* side chain LCPs are likely to exploit the possibility of unwinding the helix to obtain a monodomain having a non-vanishing spontaneous polarization. This point has not yet been clearly demonstrated. On the other hand, S_C^* LCPs will have to compete with S_C^* LMWLCs and true ferroelectric polymers such as polyvinylidene fluoride (PVF_2) and vinylidene fluoride-trifuloroethylene copolymers $P(VF_2/VF_3)$ which exhibit a very high spontaneous polarization. Nevertheless, four applications can be considered for S_C^* LCPs and are considered in what follows: display devices, transducers, pyroelectric detectors, and non-linear optics.

5.4.4.1 *Display devices.* It is premature to suppose that S_C^* LCPs will find an application in display devices, because their switching times are not fully known (see, however, Uchida *et al.*, 1987). The storage effect argument does not seem really decisive, since S_C^* LMWLCs also show bistability. Perhaps, with S_C^* LCPs, we can expect the integrity of stored information to be superior than with S_C^* LMWLCs. In principle the surface treatment would also be less critical than in the case of S_C^* LMWLC.

5.4.4.2 *Transducers.* Potential possibilities of reversibly unwinding the helix of a S_C^* elastomer by stretching give rise to a macroscopic polarization which would allow the materials to be used as piezoelectrics. But these materials will have to compete with PVF_2 which has a piezoelectric coefficient (the derivative of P_S versus the stress σ), d_{33}, of about $30\,pC\,N^{-1}$ (Micheron, 1979). This coefficient is roughly proportional to P_S. It is unlikely that S_C^*-cross-linked LCPs would exhibit a P_S of the order of magnitude of that of PVF_2 ($P_S = 6.10^3\,nC\,cm^{-2}$) (Micheron, 1979), but they could generate some interest owing to their higher Young's modulus.

5.4.4.3 *Pyroelectric detectors.* A pyroelectric detector consists essentially of a slice of pyroelectric material sandwiched between two electrodes and connected to a suitable high-impedance amplifier. Pyroelectric detectors are commonly used for thermal detection of infrared radiation. They are particularly able to convert infrared images into a spatially replicated pattern which can be then read with an electron beam (pyroelectric vidicon) or with a charge-coupled device (CCD). The nature of pyroelectric material used strongly affects the performance of the detector. The five

Table 5.10 Comparison of the pyroelectric and dielectric properties of S_C^* LMWLCs at their Curie points T_c (with a 9 V applied bias), and at 5°C below T_c, with those of conventional pyroelectric materials. ε', real part of the dielectric constant; ε'', imaginary part of the dielectric constant.

Material	γ (nC cm^{-2} K^{-1})	γ/ε'	$t_g\delta = \dfrac{\varepsilon''}{\varepsilon'}$	γ/ε''
LT	19	0.41	0.003	136
TGS	30	0.6	0.008	75
SBN	85	0.14	0.02	7
PVF$_2$	3	0.3	0.02	15
(IX)				
T_c (9 V bias)	3.5	0.8	0.005	160
$T_c - 5°C$	0.4	0.11	0.015	7.3
(X)				
T_c (9 V bias)	0.8	1.7	0.004	425
$T_c - 5°C$	0.7	0.18	0.003	60

most widely used pyroelectric materials are triglycine sulphate (TGS), lithium tantalate (LT), strontium barium niobate (SBN), ceramic materials and PVF$_2$.

Recently, Glass *et al.* (1986) have shown that S_C^* LMWLCs could compare very favourably under biased conditions with the previous well-established materials, because of their low thermal capacity, their processability, their ability to form intimate electrical contact with CCD and their good voltage responsivity. Their main disadvantages are a small pyroelectric coefficient and a rapid thermal relaxation with the cell walls. As an indication, Table 5.10 gives the most relevant parameters, such as the pyroelectric coefficient γ (determining the high current responsivity), the ratio γ/ε' (connected with the high voltage responsivity) and the ratio γ/ε'' (connected with a high signal/noise) for two S_C^* LMWLCs **(IX)** and **(X)** at the $S_C^* \rightarrow S_A$ transition temperature and at 5°C below the transition. Also given are the same parameters for some single-crystal conventional pyroelectric materials at room temperature.

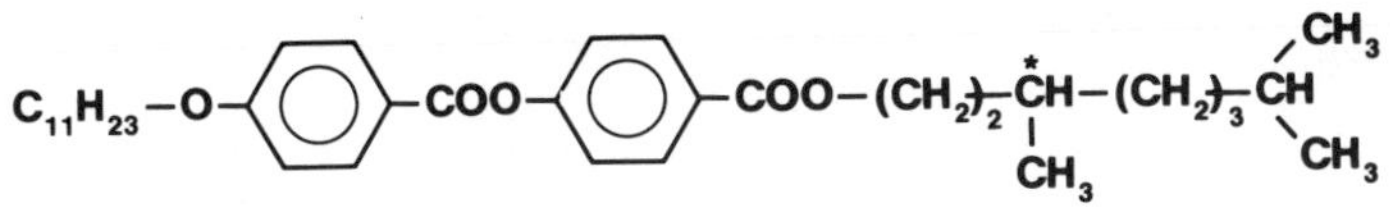

$$C_{11}H_{23}-O-\langle\bigcirc\rangle-COO-\langle\bigcirc\rangle-COO-(CH_2)_2\overset{*}{C}H-(CH_2)_3CH\overset{CH_3}{\underset{CH_3}{}} \qquad \textbf{(IX)}$$

$$C_{10}H_{21}-O-\langle\bigcirc\bigcirc\rangle-COO-CH_2-\overset{*}{C}H-C_2H_5 \qquad \textbf{(X)}$$
$$\underset{CH_3}{|}$$

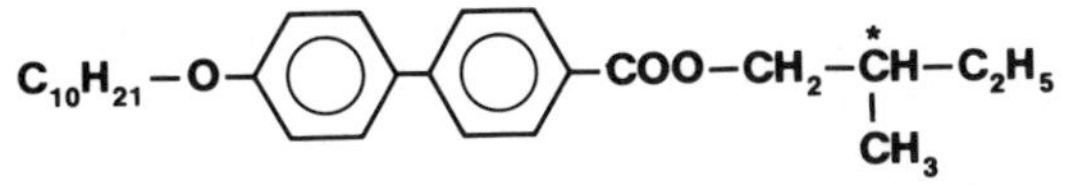

We can expect that figures of merit of S_C^*-LCPs would of the same order of magnitude as those of S_C^*-LMWLCs. But the processability of LCPs may be better than that of LMWLCs since they can be easily deposited on to various surfaces by spin coating. On the other hand, the stability of the alignment in the polymeric film could be increased.

5.4.4.4 *Non-linear optics (NLO)*. It is well-known that $\chi^{(2)}$, the second-order optical susceptibility, has a non-vanishing value only in an aligned non-centrosymmetric medium (cf. Chapter 12). $\chi^{(2)}$ is responsible for frequency conversion (second harmonic generation) electro-optic effects and frequency mixing. Organic materials which offer unique properties are very much in focus today. Ordered structures of molecules with large non-linear responses, suitable for second harmonic generation, can be obtained from single crystals, Langmuir–Blodgett films, and molecularly doped and aligned polymers.

The attraction of polymeric films lies in their relative ease of formation of non-linear guided wave devices. Until now, the polymeric materials studied have been a solid solution of a non-linear optical dye in a host nematic LCP (Meredith *et al.*, 1982) or amorphous polymer (Singer *et al.*, 1987). The centre of inversion symmetry of these solid solutions is destroyed by applying a DC electric field at a temperature above the glass transition temperature, followed by freezing the structure into the glassy state of the material.

Use of S_C^*-LCPs for second-order non-linear effects can be considered in two different ways:

(i) The internal field due to the polar alignment of the host S_C^*-LCPs would orient the guest non-linear optical dye. Unfortunately, the expected dipolar order in a S_C^*-LCPs would be low. This fact is reflected for example by the measured value of P_S of HOBACPC which is 100 times lower than the calculated maximum attainable value of P_S in the absence of rotational averaging (Otterholm *et al.*, 1987a).

(ii) The NLO moiety is directly linked to the backbone, and the S_C^* side chain copolymer thus obtained would then be aligned. In addition, relaxation of the orientational order due to the diffusion of the dye in the solid solution would be avoided. NLO copolymers exhibiting N or S_A phases have already been synthesized (Le Barny *et al.*, 1986; Griffin *et al.*, 1988).

It is worth noting that an encouraging result has already been obtained in the S_C^* LMWLC field by Shtykok *et al.* (1985). These authors have demonstrated the feasibility of phase-matched second harmonic generation in the S_C^* phase of DOBAMBC.

5.5 Conclusion

The study of ferroelectric side chain liquid crystal polymers is in its infancy. Nevertheless, in less than four years, it has been demonstrated that polyacrylate, polymethacrylate polysiloxane and polymalonate polymer backbones could lead to S_C^*. LCPs. Of course, many other chemical structure variations have to be tested in order to have a better understanding of the relationship between the chemical architecture and the properties of the LCPs, namely:

(i) Variation of the nature of the terminal group (length, position of the chiral centre, strength of the polar group attached to the asymmetric carbon)
(ii) Variation of the nature of the mesogenic core
(iii) Introduction of the chiral centre in the spacer, the terminal group being achiral.

The results already obtained by studying S_C^*-LMWLCs will be very useful in this respect.

On the other hand, a tremendous amount of work has to be done to gain a better knowledge of the physical properties of the S_C^*-LCPs, such as helical pitch, helical twist sense, and the sign and the magnitude of the spontaneous polarization. Do S_C^*-LCPs follow the same rules as S_C^*-LMWLCs? Some exceptions have already been encountered; for example, in the polysiloxanes series II published by Keller, the chiral centre is located in the terminal group attached to the electron-deficient end of the mesogenic core, but no sequence S_C^*-N* is observed.

Finally, development of S_C^*-LCPs is strongly connected to the possibility of obtaining a monodomain having a non-vanishing spontaneous polarization. It is to be hoped that, owing to the flexibility of their backbone, polysiloxanes will allow this achievement in the near future.

References

Armitage, D., Thackara, J., Clark, N. and Handschy, M. (1986) *SPIE Liquid Crystals and Spatial Light Modulator Materials*, **684**, 67.

Bahr, C. and Heppke, G. (1986) *Mol. Cryst. Liq. Cryst. Lett.* **4(2)**, 31.

Billard, J., Dubois, J.C., Nguyen, Yuu Tinh and Zann, A. (1978) *Nouv. J. Chim.* **2(5)**, 535.

Blinov, L., Bailkalov, V., Barnik, M., Beresnev, L., Pozhidayev, E. and Yablonsky, S. (1987) *Liquid Crystals* **2(2)**, 121.

Bradshaw, M., Constant, J. and Raynes, E. (1986) *Ann. Conf. British Liquid Crystal Society*, Manchester.

Brand, H. and Cladis, P. (1984a) *Mol. Cryst. Liq. Cryst.* **114**, 207.

Brand, H. and Cladis, P. (1984b) *J. Phys. (Paris) Lett.* **45**, L217.

Bualek, S. and Zentel, R. (1988) *Makromol. Chem.* (in press).

Canessa, G., Reck, B., Reckert, G. and Zentel, R. (1986) *Makromol. Chem. Macromol. Symp.* **4**, 91.

Chandrasekhar, S., Sadashiva, B. and Suresh, K. (1977) *Pramana* **9**, 471.

Chynoweth, A. (1956) *J. Appl. Phys.* **27**, 78.

Clark, N. and Lagerwall, S. (1980a) US Patent 4,367,924.

Clark, N. and Lagerwall, S. (1980b) *Appl. Phys. Lett.* **36**, 899.

Clark, N. and Lagerwall, S. (1984) *Ferroelectrics* **59**, 25.

Decobert, G., Soyer, F. and Dubois, J.C. (1985) *Polym. Bull.* **14**, 179.

Decobert, G., Dubois, J.C., Esselin, S. and Noël, C. (1986) *Liquid Crystals* **1**, 307.

Demus, D. and Zaschke, H. (1984) *Flüssige Kristalle in Tabellen II*, Deutscher Verlag für Grundstoffindustrie, Leipzig.

Destrade, C., Malthete, J. and Nguyen, Huu Tinh (1985) *Mol. Cryst. Liq. Cryst.* **127**, 273.

Esselin, S., Bosio, L., Noël, C., Decobert, G. and Dubois, J.C. (1987) *Liquid Crystals* **2**, 505.

Esselin, S., Noël, C., Decobert, G. and Dubois, J. C. (1988) *Mol. Cryst. Liq. Cryst.* (in press).

Finkelmann, H. and Rehage, G. (1984a) in *Advances in Polymer Sciences* **60/61**, ed. Gordon, M., Springer Verlag, Berlin etc.

Finkelmann, H., Kock, H., Gleim, W. and Rehage, G. (1984b) *Makromol. Chem. Rapid Commun.* **5**, 287.

Geelhaar, T. (1988) *Ferroelectrics* (in press).

Glass, A., Patel, J., Goodby, J., Olson, D. and Geary, J. (1986) *J. Appl. Phys.* **60(8)**, 2778.

Gleim, W. and Finkelmann, H. (1987) *Makromol. Chem.* **188**, 1489.

Goodby, J. and Leslie, T. (1984) *Mol. Cryst. Liq. Cryst.* **110**, 175.

Gray, G. (1985) *Proc. R. Soc. London.* **A402**, 1.

Griffin, A., Bhatti, A. and Hung, K. (1988) in *Non-Linear Optical Electro-Active Polymers*, ed. Prasad, P., Plenum, New York.

Guglielminetti, J. M., Decobert, G. and Dubois, J.C. (1986) *Polym. Bull.* **16**, 411.

Hahn, B. and Percec, V. (1987) *Macromolecules* **20(12)**, 2961.

Handschy, M., Johnson, K., Cathey, W. and Pagano-Stauffer, L. (1987) *Optics Lett.* **12(8)**, 611.

Hessel, F. and Finkelmann, H. (1985) *Polym. Bull (Berlin)* **14**, 375.

Kapitza, H. and Zentel, R. (1988) *Makromol. Chem.* (in press).

Keller, P. (1988a) *Ferroelectrics* (in press).

Keller, P. (1988b) *Mol. Cryst. Liq. Cryst.* (in press).

Lagerwall, S. and Dahl, I. (1984) *Mol. Cryst. Liq. Cryst.* **114**, 151.

Le Barny, P., Ravaux, G., Dubois, J.C., Parneix, J.P., Njeumo, R., Legrand, C. and Levelut, A.M. (1986) Molecular and Polymeric Opto-electronic Materials: Fundamentals and Applications, *SPIE Proc.* 682, San Diego, 56.

Martinot-Lagarde, P., Duke, R. and Durand, G. (1981) *Mol. Cryst. Liq. Cryst.* 75, 249.

Martinot-Lagarde, P. (1988) *Ferroelectrics* (in press).

Meredith, G., Van Dusen, J. and Williams, D. (1982) *Macromolecules* 15, 1385.

Meyer, R., Liebert, L., Strezlecki, L. and Keller, P. (1975) *J. Phys. (Paris) Lett.* 36, L69.

Micheron, F. (1979) *Revue Technique Thomson-CSF* 11(3).

Mitchell, G., Davis, F. and Ashman, A. (1987) *Polymer* 28, 639.

Otterholm, B., Nilsson, M., Lagerwall, S. and Skarp, K. (1987a) *Liquid Crystals* 2(6), 757.

Otterholm, B., Alstermark, C., Flatischler, K., Dahlgren, A., Lagerwall, S. and Skark, K. (1987b) *Mol. Cryst. Liq. Cryst.* 146, 189.

Pagano-Stauffer, L., Johnson, K., Clark, N. and Handshy, M. (1986) *SPIE Liquid Crystals and Spatial Light Modulator Materials*, 684, 88.

Patel, J. and Goodby, J. (1987) *Optical Eng* 26(5), 373.

Reck, B. and Ringsdorf, H. (1985) *Makromol. Chem. Rapid Commun.* 6, 291.

Rieker, T., Clark, N., Smith, G., Parmar, D., Sirota, E. and Safinya, C. (1987) *Phys. Rev. Lett.* 59(23), 2658.

Rötz, U., Lindau, J., Reinhold, G. and Kuschel, F. (1987) *Z. Chem.* 27, 293.

Sage, I., Jenner, J., Chambers, M., Bradshaw, M., Brimmell, V., Constant, J., Hughes, J., Raynes, E., Samra, A., Gray, G., Lacey, D., Toyne, K., Chan, L., Shenouda, I. and Jackson, A. (1988) *Ferroelectrics* (in press).

Sakurai, T., Mikami, N., Higuchi, R., Honma, M., Dzazi, M. and Yoshino, K. (1986) *J. Chem. Soc., Chem. Commun.* 978.

Schadt, M. and Helfrich, W. (1971) *Appl. Phys. Lett.* 18, 127.

Schätzle, J. and Finkelmann, H. (1987) *Mol. Cryst. Liq. Cryst.*, 142, 85.

Shibaev, V., Kostromin, S., Platé, N., Ivanov, S., Vetrov, V. and Yakovlev, I. (1983) *Polymer* 24, 364.

Shibaev, V., Kozlovsky, M., Beresnev, L., Blinov, L. and Platé, A. (1984) *Polymer Bull.* 12, 299.

Shibaev, V. and Platé, N. (1985) *Pure and Appl, Chem.* 57(11), 1589.

Shibaev, V., Kozlovski, M., Platé, N., Beresnev, L. and Blinov, L. (1987) *Vissokomol Soedin* 29(7), 1470.

Shtykok, N., Barnik, M., Beresnev, L. and Blinov, L. (1985) *Mol. Cryst. Liq. Cryst.* 124, 379.

Singer, K., Kuzyk, M. and Sohn, J. (1987) *J. Opt. Soc. Amer. B* 4(6), 968.

Terashima, K., Ichihashi, M., Kikuchi, M., Furukawa, K. and Inukai, T. (1986) *Mol. Cryst. Liq. Cryst.* 141, 237.

Uchida, S., Morita, K. and Hashimoto, K., (1987) *Eur. Pat. Appl.*, EP228703.

Umeda, T., Hori, Y. and Mukoh, A. (1985) *SID Digest* 16, 373.

Walba, D., Slater, S., Thurmes, W., Clark, N., Handschy, M. and Supon, F. (1986) *J. Amer. Chem. Soc.*, 108, 5210.

Zentel, R. (1987) *Progr. Colloid and Polymer Sci.* 75, 239.

Zentel, R. (1988) *Liquid Crystals* (in press).

Zentel, R. and Reckert, G. (1986) *Makromol. Chem.* 187, 1915.

Zentel, R. and Benalia, M. (1987) *Makromol. Chem.* 188, 665.

Zentel, R., Reckert, G. and Reck, B. (1987a) *Liquid Crystals* 2(1), 83.

Zentel, R., Schmidt, G., Meyer, J. and Benalia, M. (1987b) *Liquid Crystals* 2, 651.

6 Macroscopic structural characterization of side chain liquid crystal polymers

C. NOËL, Laboratoire de Physicochimie Structurale et Macromoléculaire, Paris, France

6.1 Introduction

The discovery of the side chain liquid crystalline polymers (SCLCPs) was a stimulating event in macromolecular science, not only because of their potential as new materials for electronic displays and non-linear optics, but also because of their academic interest in the theoretical scheme of structural order in fluid phases. In considering the effects of change in molecular structure on the properties of these polymers we have a more difficult subject to tackle, because of the different types of mesophase. Low-molar-mass liquid crystals (LMMLCs) are conventionally classified into nematics, cholesterics and smectics. As already mentioned by Le Barny and Dubois (Chapter 5) there are more than ten recognized smectic modifications and these are denoted by S_A, S_B, $S_C \ldots S_L$ (Gray and Goodby, 1984). Positive identification of the type of mesophase is thus an important step in the characterization of SCLCPs. Some of the smectic modifications are much less common than others, and, as a result, knowledge about the molecular arrangement in the different types of mesophase is variable. We will therefore concentrate attention on the relatively more disordered phases: nematic, cholesteric, smectic A and smectic C.

The primary aim of this chapter is to present to the reader how the different types of mesophase are identified and distinguished from one another. An attempt is made to integrate the information available from the thermal data, textural phenomena, miscibility tests and X-ray diffraction patterns in a discussion of the specific features of the nematic, cholesteric and smectic phases exhibited by SCLCPs. Results obtained by these methods on SCLCPs are compared with those obtained for LMMLCs. The properties specific to polymers such as polydispersity, chain flexibility, etc. are discussed in terms of their influence on the macroscopic behaviour of the polymer, for example mesophase formation, transition temperatures including the glass transition, and the alignment of the materials. Recent work on the effect of the pendent mesogenic groups on the polymer backbone conformation is described.

6.2 Characterization of SCLCPs by differential scanning calorimetry

The essential features of the phase behaviour of SCLCPs are relatively well established. Most of these polymers are non-crystalline in character: the DSC trace exhibits a glass transition characteristic of the polymer backbone and a first-order transformation from the mesophase to the isotropic phase due to the mesogenic side chains. It is to be noted that with sufficiently long flexible spacers, partial crystallization of the side chains may occur (Esselin et al., 1987; Wunderlich et al., 1988). The macromolecular

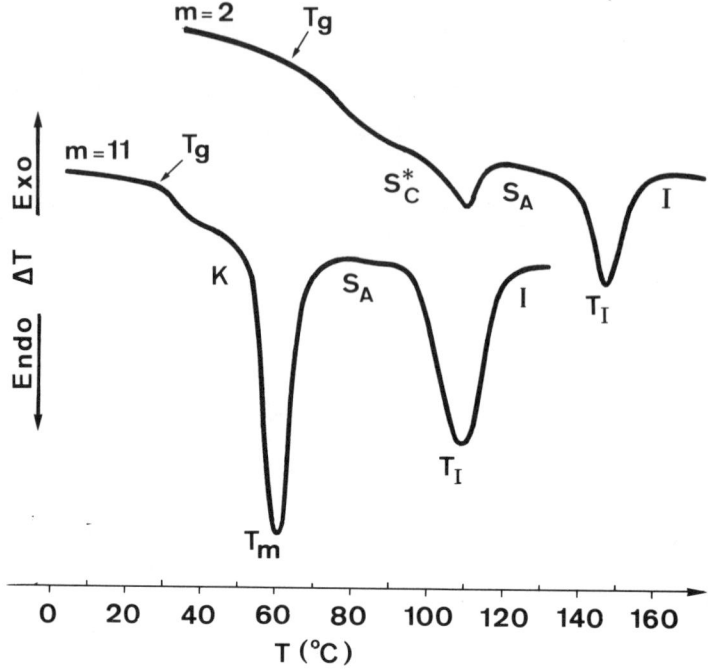

Figure 6.1 DSC heating traces of polymers:

(Decobert *et al.*, 1986)

chain becomes only a connecting backbone with little consequence for the thermodynamics and kinetics of crystallization. In this case the DSC trace also shows a melting
endotherm (Figure 6.1). Hsu and Percec (1987) have reported biphasic SCLC
polysiloxanes containing trans-11-[2-(4-cyanophenyl)-1,3-dioxan-5-yl] undecyl side
groups which exhibit two glass transitions (cf. Chapter 3 and references therein, e.g.
Cowie *et al.*). In these systems the side chains relax in a cooperative way and
independently of the main chain. Polymers containing alkyl groups shorter than
heptyl do not undergo phase separation, whereas those containing alkyl groups longer
than undecyl give rise to side chain crystallization. A given SCLCP may have several
mesomorphous phases and, so far as can be gathered from observations, these are
in general separated from one another by first-order transition points (Figure 6.1).

The transition temperatures are dependent on the sample history and molecular
mass, so polymer samples must be heat-treated in the same manner and be of
sufficiently high molecular mass for comparison to be possible. For the former, it has
been shown that a fraction of amorphous material may be present together with the
mesophase, the relative amounts of the two phases varying with the thermal history
(Pracella *et al.*, 1984). In some cases, the isotropic melt can even be quenched in ice-

water to give a fully amorphous glass (Frosini et al., 1974). For the latter, it has been found that transition temperatures increase rapidly with molecular mass before levelling off, as has been previously discussed in Chapter 3 and by various authors (Finkelmann and Rehage, 1984; Stevens et al., 1984; Shibaev and Platé, 1985). The shorter the flexible spacer, the stronger is the dependence at low degrees of polymerization.

The structure of SCLCPs can be modified in numerous ways, as a large degree of variation is possible for the chemical structure of the main chain, the nature and the length of the flexible spacer, the molecular structure of the mesogenic group and the size and the polarity of the terminal group. Systematic studies of a number of types of SCLCPs have established that certain systematic trends in liquid crystal behaviour accompany particular changes in molecular structure, as has been highlighted by Percec and Pugh in this text.

A statistical average of the clearing entropies of over 100 LCs (Wunderlich et al., 1984, 1988) yielded values of approximately 2 ± 1, 7 ± 5 and $15 \pm 7 \, \mathrm{J \, K^{-1} \, mol^{-1}}$ for LMMLCs, SCLCPs and main chain liquid crystalline polymers (MCLCPs), respectively. In MCLCPs flexible spacers do not play the rôle of solvents but, rather, participate actively in the ordering process. In the liquid crystalline state, the mesogenic groups of neighbouring molecules tend to align on average parallel to one another. This has significant conformational consequences for the spacers that are in a rather extended state (Samulski et al., 1984; Müller and Kothe, 1985; Ghanem, 1987). In this regard, however, it is important to note that the distribution of chain sequence extension strongly depends on odd–even character of the number of atoms in the spacer (Blumstein et al., 1983; Abe, 1984; Müller and Kothe, 1985). Less effect is observed for SCLCPs. Indeed, even though the spacers experience on one end the ordering tendency of the mesogen, they are linked at the other end to a more or less flexible macromolecular chain that tries to achieve a random-coil conformation. From a detailed analysis of deuteron-NMR spectra, Spiess et al. established that the order parameter decreases from mesogen to spacer to polymer backbone (Boeffel et al., 1983; Spiess, 1985; Pschorn et al., 1986 and Chapter 8).

Thermodynamic data for an interesting series of polymers (1) reported recently by Esselin et al. (1988) is summarized in Table 6.1:

$$\begin{array}{c} {+}CH_2{-}CX{+} \\ | \\ COO{+}CH_2{+}_m O{-}R{-}O{-}CH_2{-}\overset{*}{C}H{-}C_2H_5 \end{array} \qquad \begin{array}{c} CH_3 \\ | \end{array} \qquad \text{(1)}$$

X = H, CH₃, Cl

m = 2, 6, 11

R = ⟨◯⟩—COO—⟨◯⟩— (A),

—⟨◯⟩—CH=CH—COO—⟨◯⟩— (B)

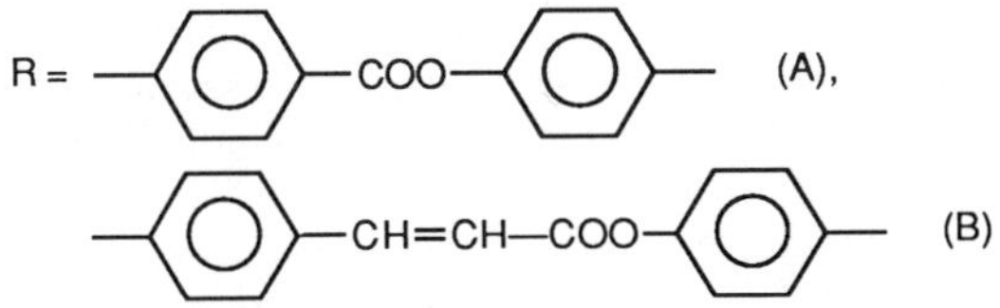

The ordered to isotropic phase transition has a smaller gain in entropy than expected for the fully oriented mesogen alone (20–50 J K⁻¹ mol⁻¹). Thus, the mesophases must

Table 6.1 Heats and entropies of clearing for selected smectic SCLCPs

Polymer R	X	m	T_1 (°C)	ΔH_1 kJ mol^{-1}	ΔS_1 J K^{-1} mol^{-1}
A	H	2	146	1.8	4.29
A	H	6	87	4	11.1
A	H	11	108	6.6	17.2
A	CH$_3$	2	155	3.5	8.2
A	CH$_3$	6	90	5.0	13.8
A	CH$_3$	11	106	6.9	18.3
A	Cl	11	112	7.5	19.4
B	CH$_3$	2	172	2.8	6.2
B	CH$_3$	6	125	3.5	8.8
B	CH$_3$	11	142	5.4	12.9
B	Cl	11	140	4.6	11.1

contain considerable disorder, even for the mesogens. For a given series, the replacement of hydrogen in the polymer backbone by a methyl group or a chlorine atom seems to result in higher orientational order in liquid crystals. As in LMMLCs, there is some indication that longer flexible spacers increase the transition entropy, but less than expected for fully extended aliphatic chains (9–10 J K^{-1} mol^{-1} for each methylene unit, Wunderlich *et al.*, 1984). Additive values to the entropy change of 0.75–1.4 J K^{-1} can be calculated for each methylene unit which implies that the alkyl chains have some conformational freedom in the smectic phase.

6.3 Optical microscopy of SCLCPs

6.3.1 *Textural assignments*

The term texture, as used by Friedel (1931*a*, *b*) designates the picture of a thin layer (< 0.3 mm) of liquid crystal observed with a polarizing microscope. The features of the various textures are caused by the existence of different kinds of defects. For a practical and relatively fast classification of LCs, the microscopic observations of the textures are most useful. This optical method has been used since the discovery of liquid crystals and led to their classification as nematics, cholesterics and smectics.

   In a book of this kind it is not possible to consider in detail the various optical features characteristic of each mesophase structure. Two relevant books with photographic illustrations have been published dealing deeply with this subject (Demus and Richter, 1978; Gray and Goodby, 1984). The primary aim of this section is therefore to present to the reader textures which are typical of nematic, cholesteric, smectic A and smectic C phases and to show how these types of mesophase can be identified and distinguished from one another. In this regard, however, it is important to note that microscopic observations are sometimes misleading owing to the difficulty with which SCLCPs give specific textures in the liquid crystalline state. This might be due to their multiphase nature (existence of polycrystalline and amorphous material), polydispersity and/or the high viscosities of the liquid crystalline melts. In most cases samples must be annealed for hours or days at a suitable temperature if textures reminiscent of those of LMMLCs are to be seen (Coles and Simon, 1985; Hsu *et al.*, 1987). Lack of patience can lead to a serious misclassification of the polymer texture!

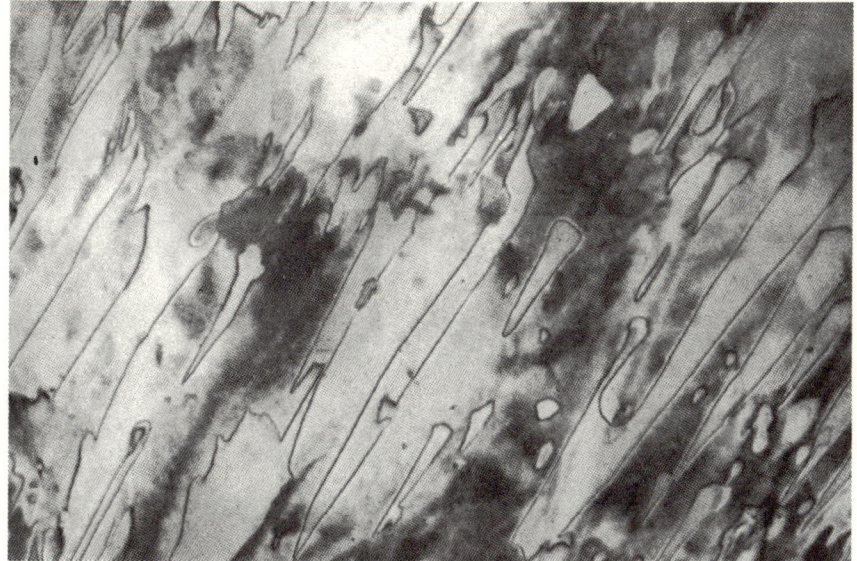

Figure 6.2 Threaded texture of copolymer 2 where $x = 7.1 \, \text{mol}\%$. Crossed polarizers. Magnification $\times 200$, $T = 107°C$.

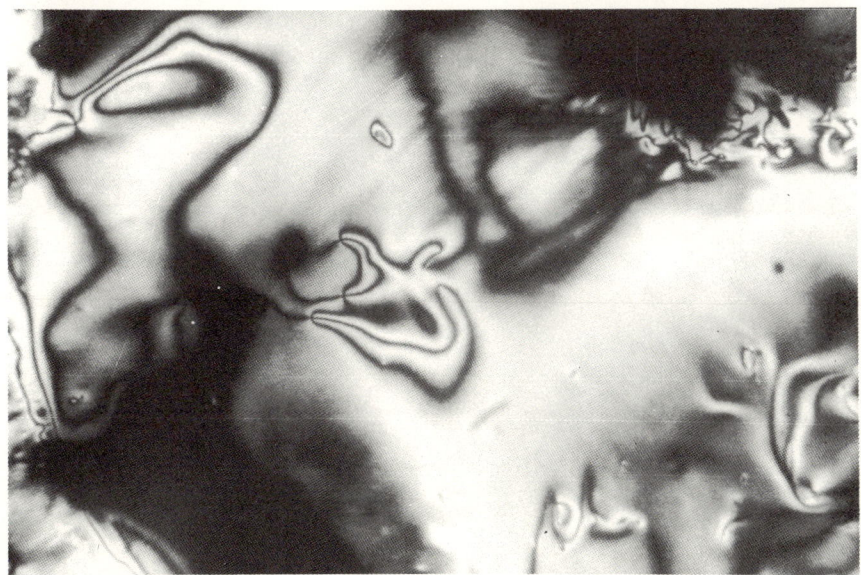

Figure 6.3 Schlieren texture of polymer

$$\begin{array}{c}
-\!\!\!-\!\!(\text{CH}_2\!\!-\!\!\text{CH})\!\!-\!\!\!- \\
| \\
\text{COO}\!\!-\!\!(\text{CH}_2)_6\!\!-\!\!\text{O}\!\!-\!\!\bigcirc\!\!-\!\!\bigcirc\!\!-\!\!\text{CN}
\end{array}$$

($S = \pm 1$, four dark brushes; $S = \pm 1/2$, two dark brushes). Crossed polarizers. Magnification $\times 200$, $T = 130°C$.

(i) *Nematic modifications.* Several SCLCPs when examined under a polarizing microscope exhibit threaded and/or Schlieren textures indicative of nematic phases (Figures 6.2 and 6.3) (Le Barny *et al.*, 1986; Noël *et al.*, 1988).

Usually, on cooling from the isotropic liquid, typical droplets appear at first (Figure 6.4) (Noël *et al.*, 1988) which after further cooling grow and coalesce to form large domains. Nematic droplets characterize a type-texture of the nematic phase, since they occur nowhere else.

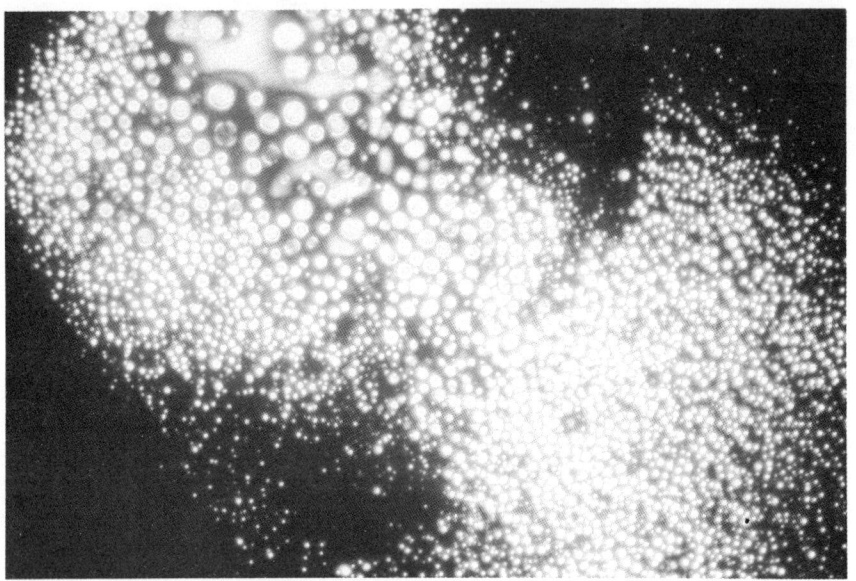

Figure 6.4 Nematic droplets from copolymer 2 where $x = 50.3$ mol%. Crossed polarizers. Magnification $\times 200$. $T = 107°C$.

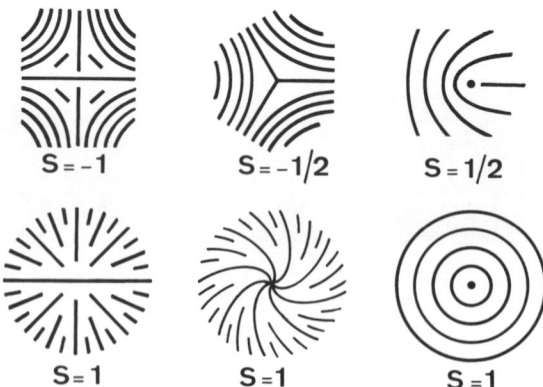

Figure 6.5 Schematic diagram of molecular trajectories associated with disclinations of strength $\pm 1/2$ or ± 1.

The appearance of the Schlieren texture between crossed polarizers is unmistakably characterized by dark brushes which start from points, in which the direction of extinction is not defined. Usually points with two or four brushes can be observed. These points indicate singularities, i.e. disclinations in the structure (Demus and Richter, 1978). The disclination strength S is connected with the number of dark brushes meeting at one point:

$$|S| = \text{number of brushes}/4$$

The sign of S is positive when the brushes turn in the same direction as the rotated polarizers, and negative when they turn in the opposite direction. Figure 6.5 shows some typical configurations (Frank, 1958). From the observation of singularities with $S = \pm 1/2$, the mesophase can be identified unambiguously as a nematic phase since these singularities occur nowhere else. Indeed, for the smectic C phase which can exhibit the Schlieren texture, all the point singularities are of the $S = \pm 1$ type.

Homeotropic texture (Figure 6.6) caused by a spontaneous orientation of the sample is also found in experiments using treated surfaces (Noël, 1985). This texture occurs when the long axes of the mesogenic groups are at right angles to the glass surfaces. In the homeotropic regions, the field of view using crossed polarizers and orthoscopic illumination remains uniformly dark as the preparation is turned. However, if the cover glass is touched, the originally dark field of view brightens instantly, thus distinguishing between homeotropic and isotropic texture.

These mesophases exhibit other nematic characteristics such as intense movement

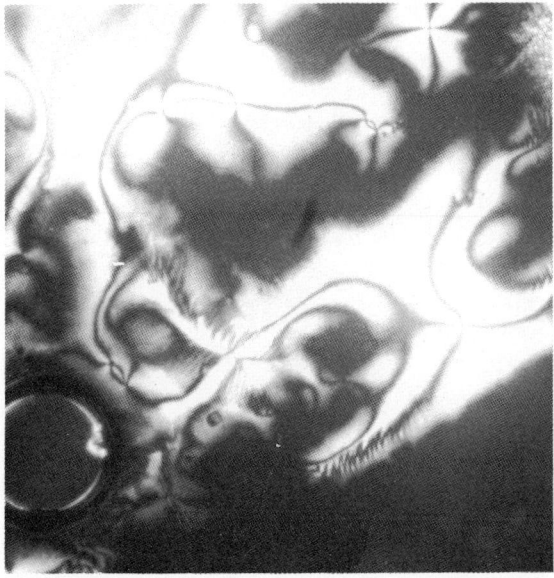

Figure 6.6 The Schlieren and homeotropic textures obtained by simple treatment of glass slides with boiling chromic-sulphuric acid, acetone and methanol (sequentially interspersed with water rinses) and rinsing with hot distilled water. Copolymer 2 where $x = 50.3 \, \text{mol}\%$. Crossed polarizers. Magnification $\times 200$, $T = 93.5°\text{C}$.

within the melt and scintillation effects due to a directly observable Brownian motion.

Quenching the polymers from their nematic phases to room temperature succeeds in 'freezing in' these textures (Talroze *et al.*, 1981; Finkelmann, 1982). In spite of the thermal shock, both the homeotropic alignment and the planar one can be retained in the glassy state.

(ii) *Cholesteric modifications.* Cholesteric polymers can be obtained in the form of mixtures of nematic polymers with an optically active substance. Another possibility is offered by introducing chirality into the molecular structure. This is commonly obtained by copolymerization (i) of a binary mixture of chiral monomers (Finkelmann *et al.*, 1978*a*), or (ii) of a nematogenic monomer with a chiral comonomer (Finkelmann *et al.*, 1978*b*; Platé and Shibaev, 1980; Shibaev *et al.*, 1981). These cholesteric liquid crystals can occur in the typical planar textures with oily streaks, moiré fringes and/or Grandjean lines. In the planar texture the helices are more or less aligned with their axes perpendicular to the glass surfaces. These cholesterics can show reflection colours. The wavelength (λ_R) of the light at the centre of the reflection band is, for perpendicular incidence, equal to the length of the helical pitch P multiplied by mean refractive index $\bar{n}$. The pitch of the helix is highly dependent on (i) the temperature of the preparation, (ii) the nature of the organic materials comprising the mesophase, and (iii) the content of the chiral component (Finkelmann and Rehage, 1984). The helical pitch P decreases as

Figure 6.7 The focal-conic texture of the smectic A phase of polymer 1 where $m = 11$ and

$$R = -\langle O \rangle - COO - \langle O \rangle -$$

Crossed polarizers. Magnification $\times 200$. $T = 79°C$. (From Decobert *et al.* 1986.)

the mole fraction of the chiral substance increases. Hence, λ_R can be adjusted between $300\,\text{nm} < \lambda_R < \infty$. The optical activity of a planar cholesteric layer is directly observable by the asymmetrical colour changes that appear upon rotating the analyser. In polarization microscopy, for a dextro-mesophase, a small clockwise rotation of the analyser shifts the colour to longer wavelengths.

More recently, it has been proved that it is possible to prepare cholesteric homopolymers (Finkelmann and Rehage, 1982). This possibility can lead to strongly twisted cholesterics which may exhibit non-planar textures. Indeed, while the defects and the textures of cholesterics with low twist and nematics show many similarities, strongly twisted cholesteric can be considered to have a quasi-layered structure. As a consequence, their textures often resemble those of smectics, especially smectic A (Demus and Richter, 1978). These cholesterics may appear in fan-shaped, focal conic or polygonal textures. This makes it difficult to interpret the observed textures in terms of cholesterics or smectics. In this regard, it is important to note that by adding a small amount of a nematic solute, the helical pitch can be increased and planar textures can be observed.

(iii) *Smectic modifications.* The textures of smectic polymers have not been inves-

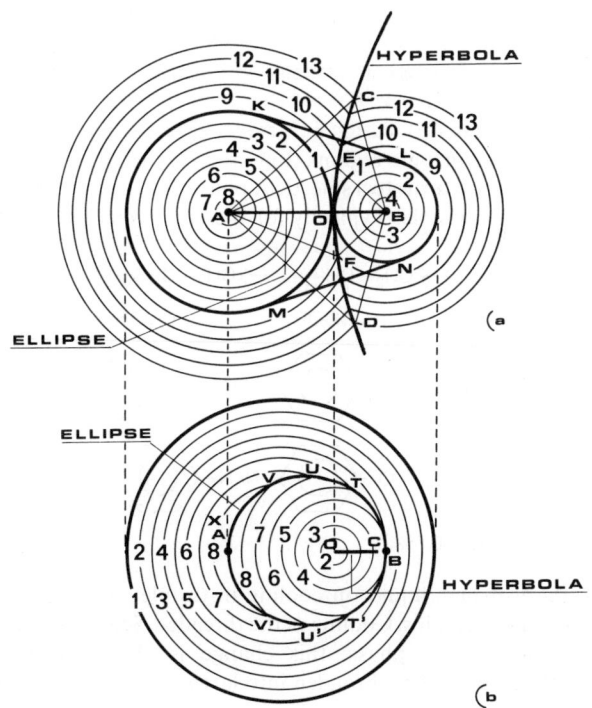

a) Cyclides in principal section
b) Cyclides in plan

Figure 6.8 Focal conic structure (Redrawn from Hartshorne, 1974).

tigated in detail. Often, textures occur whose characteristics are somewhat obscure and observable only with difficulty even after annealing the polymer films for hours or days (Coles and Simon, 1985). However, some SCLCPs exhibit smectic A and/or smectic C phases whose textures are reminiscent of those of LMMLCs (Hahn *et al.*, 1981; Shibaev *et al.*, 1982; Kostromin *et al.*, 1982; Decobert *et al.*, 1986; Esselin *et al.*, 1987, 1988; Noël *et al.*, 1988).

Smectic A phases are easily recognized by the typical simple focal conic texture (Figure 6.7). This term refers to the ellipses and hyperbolae possessing a common focus and which thus appear as lines of discontinuity in a molecular arrangement following Dupin cyclide surfaces. Figure 6.8 shows the geometrical basis of a focal conic texture, the parallel and equally spaced cyclides being numbered serially.

Smectic A phases often display textures which are probably based on the focal conic arrangement but in which the ellipses and hyperbolae cannot always be distinguished in the microscopic image. The chief examples are, firstly, the so-called *bâtonnets* (Figure 6.9); separate little objects having a characteristic elongation along an axis of rotational symmetry but otherwise a bewildering variety of shapes. Bâtonnets correspond to nematic droplets with regard to formation conditions. They are the form in which a smectic A phase may begin to separate at the clearing point on cooling the isotropic liquid. Upon further cooling, the bâtonnets grow and coalesce to form larger structures from which the focal conic texture finally forms. Secondly, there is the fan-like texture (Friedel's *plages en éventail*) which consists of irregularly disposed areas in which a fan-like pattern is revealed by using crossed polarizers. Thirdly, there are the oily streaks (Friedel's *stries huileuses*) which are long transversely striated bands consisting of chains of focal conic groups. They appear as bright bands or ribbons in the dark homeotropic regions (Figure 6.10) (Spassky *et al.*, 1988).

If a nematic modification turns to a smectic A phase, transient stripes in the form of a striated texture are often visible. Typically for the copolymer (**2**), the nematic phase

$$\text{---}\!\!\{\text{CH}_2\text{---}\text{CH}\}_{1-x}\quad \text{---}\!\!\{\text{CH}_2\text{---}\text{CH}\}_x$$
$$\qquad\quad\ \ \text{COOR}_1\qquad\qquad\ \text{COOR}_2$$

$$R_1 = \ -(CH_2)_6\text{---}O\text{---}\langle\bigcirc\rangle\text{---}COO\text{---}\langle\bigcirc\rangle\text{---}CN \tag{2}$$

$$R_2 = \ -(CH_2)_6\text{---}O\text{---}\langle\bigcirc\rangle\text{---}N{=}N\text{---}\langle\bigcirc\rangle\text{---}NO_2$$

$$x = 50.3\%$$

separates from the isotropic liquid on cooling in droplets (Figure 6.4) which grow, coalesce and form large domains. Cooling of the threaded-Schlieren texture produces a transition to the smectic A phase; this change is characterized by transition phenomena, mostly stripes, which broaden into larger areas (Figure 6.11) (Noël *et al.*, 1988).

Smectic A phases resemble nematic phases in the tendency to form homeotropic textures. The homeotropy may persist through a $N \rightarrow S_A$ transition, thus concealing the smectic A phase from microscopic observations. However, if the preparation is disturbed by touching the cover slip, oily streaks appear as bright bands or ribbons in

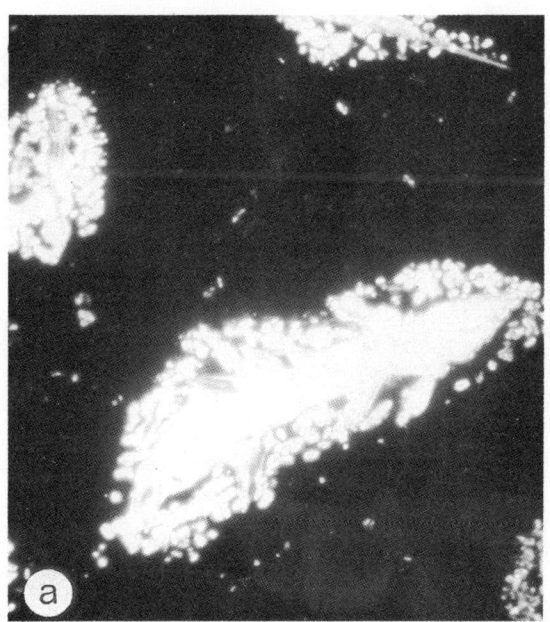

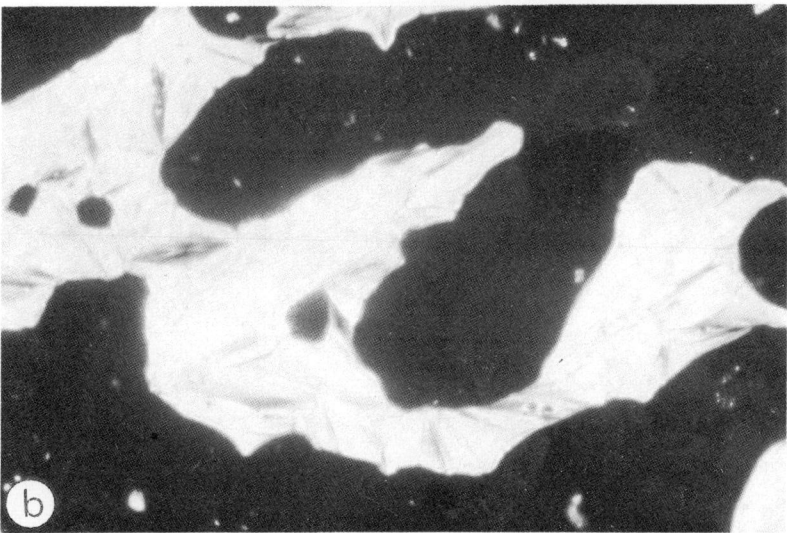

Figure 6.9 The separation of the smectic A phase in the form of bâtonnets from the isotropic liquid. Polymer 1 where $m = 2$ and

$$R = \underset{}{\bigcirc} - COO - \underset{}{\bigcirc} -$$

(a) $T = 143.3°C$; (b) $T = 141.8°C$.

Figure 6.10 Oily streaks, smectic A phase of polymer 3 where $X = (CH_2—CH_2—O)_2$.

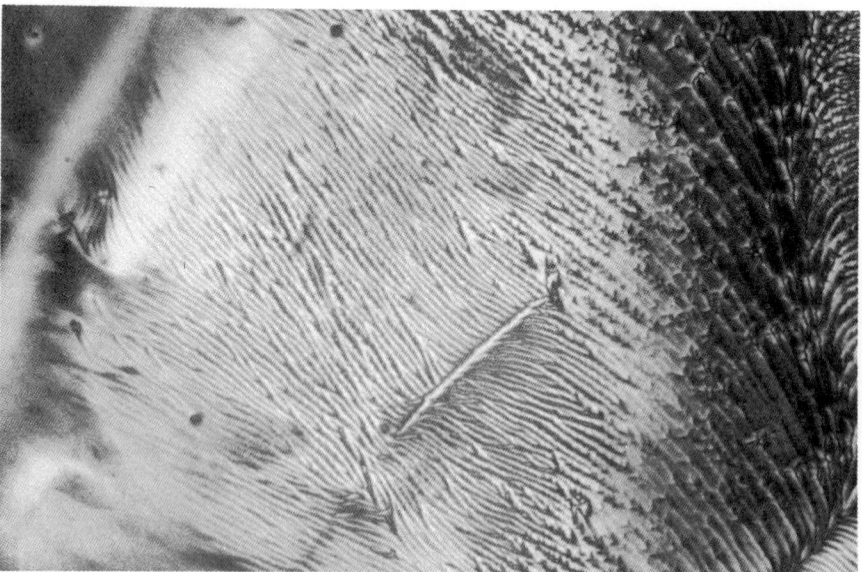

Figure 6.11 Nematic/smectic A transition. Copolymer 2 where $x = 50.3\,mol\%$. Crossed polarizers. Magnification $\times 200$, $T = 90°C$.

the dark homeotropic regions, thus distinguishing between a nematic and a smectic A phase.

Smectic C phases also form focal conic textures, though with characteristic breaks or interruptions in the pattern. As previously discussed, smectic C phases can exhibit Schlieren textures which display only points that have four dark brushes originating

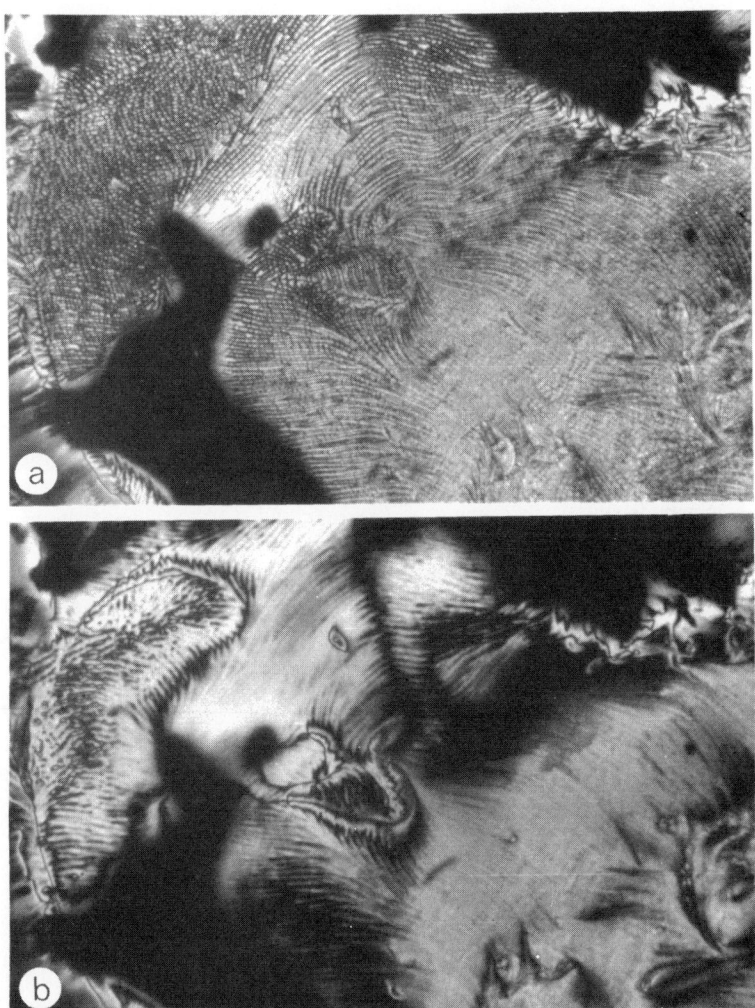

Figure 6.12(*a*) Smectic A phase, $T = 121°C$; (*b*), re-entrant nematic phase, $T = 75°C$. Polymer:

from them. If a smectic A phase is first formed then, the Schlieren texture will be obtained from the homeotropic regions of the A phase while the broken focal-conic texture will be obtained from the clear, simple focal-conic texture of the A phase.

Like nematic phases, smectic C phases can be twisted by dissolving an optically active compound in them or by introducing chirality into the molecular structure (Shibaev et al., 1984; Decobert et al., 1986; Esselin et al., 1987, 1988; Zentel et al., 1987). For polymers (1), the existence of chiral smectic C phases was partly established from the broken focal-conic fan textures. (Decobert et al., 1986). However, compared to the corresponding textures in LMMLCs (Gray and Goodby, 1984) and MCLCPs (Fayolle et al., 1979), these textures are poorly defined.

In LMMLCs with a terminal cyano- or nitro-group, it is evident from simple energy considerations that interactions between neighbouring dipoles can by no means be neglected. This effect favours an antiparallel arrangement of the permanent dipoles (Chandrasekhar, 1984). An important consequence of this type of correlation is that the mesophases of these materials often consist of 'bilayers', the molecules arranged in an antiparallel, overlapping interdigitated structure with a layer spacing of about 1.4 times the molecular length (Hardouin et al., 1983). As the temperature is varied, the molecular packing is slightly altered and the resultant subtle changes in the bilayer structure appear responsible for the occurrence of more than half-a-dozen distinct S_A and S_C phases (S_{A1}, S_{A2}, S_{Ad}, $S_{\tilde{A}}$, S_C, S_{C2}, $S_{\tilde{C}}$) and reentrant phases. Three terminally cyano-substituted SCLCPs are presently known to exhibit the unusual I—N—S_{Ad}—N_{re} sequence (Gubina et al., 1986; Le Barny et al., 1986a; Spassky et al., 1988). Typically, upon cooling from the isotropic melt, the high-temperature nematic phase begins to separate at the clearing point in the form of droplets which after further cooling grow, coalesce and reorganize their shape until a final Schlieren texture is established (Figure 6.3). Further reduction in temperature produces a chevron texture with characteristic transition bars. This texture changes on standing for some time into the stable fan-shaped and focal conic textures (Figure 6.12a). Both these forms of texture are typical of the smectic A phase. As the temperature falls, a Schlieren texture is progressively restored or a paramorphic fan-shaped texture is formed (Figure 6.12b).

$$X = -(CH_2)_{11}O; -(CH_2)-CH_2-O)_3$$

Polymers of structure (3) exhibit complex polymorphism involving different S_A and S_C phases (Spassky et al., 1988). Typically for the polymer where $X = -(CH_2—CH_2—O)_3$, the high-temperature S_A phase separates from the isotropic liquid in the form of bâtonnets which coalesce to produce a simple focal-conic fan texture (Figure 6.13a). As the temperature falls below 80°C a textural change is detected as a multiplication of the focal conics (Figure 6.13b). Note that such subtle textural modifications have been previously evidenced at the $S_{Ad} \rightarrow S_{A2}$ transition in compounds of the 4-n alkoxy phenyl-4' cyanobenzoyloxybenzoate series (Hardouin et al., 1985).

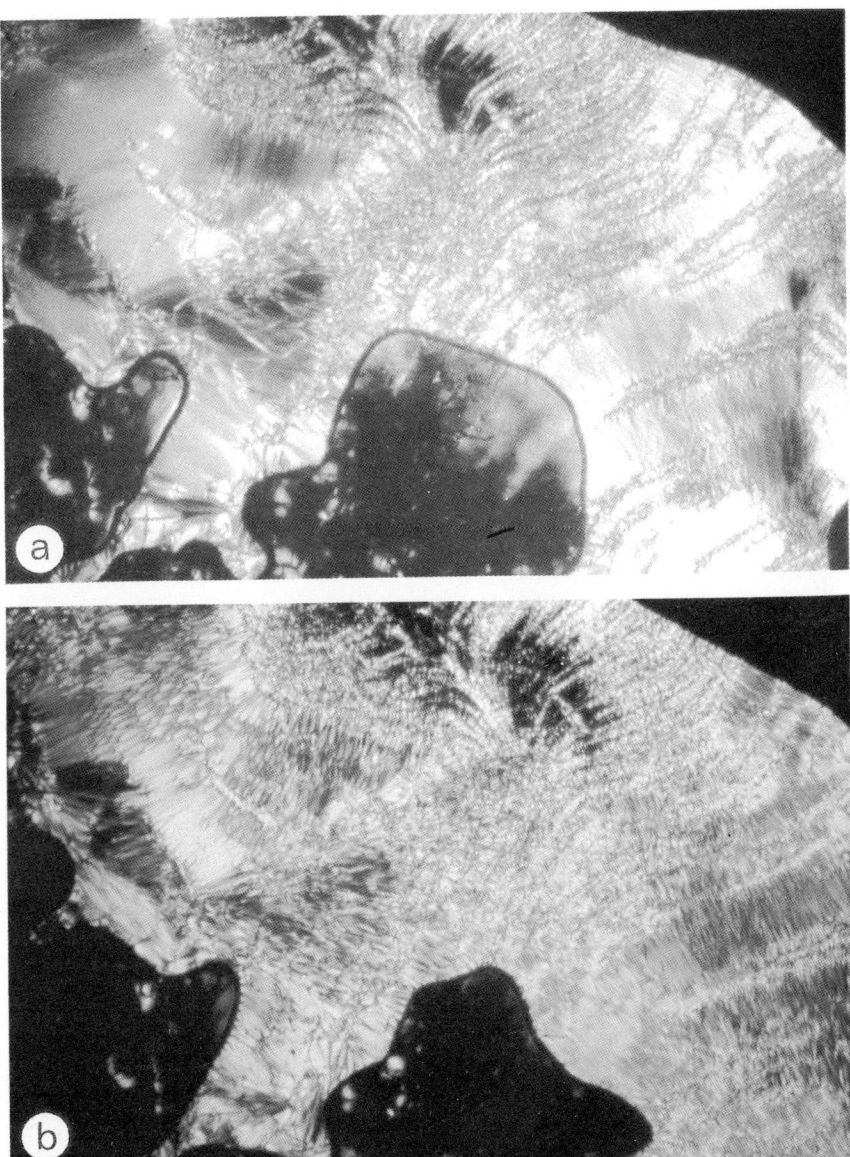

Figure 6.13(*a*) High-temperature and (*b*) low-temperature smectic A phases of polymer 3 where X = $+CH_2-CH_2-O+_3$. Crossed polarizers. Magnification × 200.

6.3.2 *Molecular alignment and conoscopic observations*

This section deals with the molecular alignment and consequent optical properties of nematic phases. Since the preparation of a uniformly aligned sample (monodomain) is prerequisite for most liquid crystal applications and physical investigations, a brief introduction is presented which reviews the alignment techniques.

In what follows, it is assumed that the reader is familiar with the principles of polarization microscopy and its application to the study of birefringent materials. Those wishing for further information on these subjects should consult one of the standard works, references to which are given in Hallimond (1970), Hartshorne and Stuart (1970) and Viney (1986).

(i) *Alignment of SCLCPs on treated surfaces.* A feature of most nematic phases when prepared between treated glass surfaces is that the molecules in contact with the glass have a striking tendency to attach themselves to it parallel to a preferred direction, and govern the orientation of those in the bulk of the phase. Two terms which will be mentioned often in this section are homeotropic and homogeneous (or planar). In homeotropic alignment, the mesogenic side groups are oriented with their long axes perpendicular to the surface. Homogeneous alignment results when liquid crystal molecules lie parallel to a surface; it becomes uniform when all the molecules at the surface point in the same direction. In cholesterics, the homeotropic alignment causes the helical axes to be randomly distributed in planes parallel to the substrate while, with a homogeneous alignment the helices are more or less aligned with their axes perpendicular to the surface. In smectic A mesophases, where the director is perpendicular to the layers, the homeotropic alignment has layers parallel to the planes and is usually quite transparent, as the fluctuations of orientation are limited by the rigidity of the planes. A homogeneous molecular alignment with planes perpendicular to the surfaces can also be obtained. It is possible to prepare 'monodomain' of the more complex smectic phases. However, in most cases, the ordering depends on the possibility of cooling from a uniformly oriented higher-temperature mesophase. The alignment techniques used for LMMLCs are reviewed in the papers of Kahn *et al.* (1973), Goodman (1974), Haller (1975), Guyon and Urbach (1976) and Cognard (1982). The first attempts to orient thin samples of polymers by these techniques were unsuccessful (Shibaev and Platé, 1984). In this regard, it is to be noted that the exact mechanisms of these techniques are not always understood and the conditions for obtaining uniform alignment depend crucially on a *tour de main*.

For mixtures of SCLCPs in LMMLCs homeotropic alignment was achieved by treating the surfaces with a lecithin solution (BDH-Egg Grade II at a concentration of 0.5% in chloroform) (Hopwood and Coles, 1985a, b; Sefton and Coles, 1985; Sefton *et al.*, 1985). The wetted surface was then allowed to dry naturally. For polymer concentrations up to 40% w/w good homeotropic alignment was observed.

Homeotropic alignment was also produced in experiments using surface active molecules covalently bound to the substrate (Mattoussi, 1987). The best results were obtained by reaction of hexadecanol with the superficial Si—OH groups formed by UV radiation under oxygen.

A number of techniques have been used to create reproducible planar alignment. It is convenient to separate them into two categories: (a) organic alignment layer materials and (b) inorganic films and substrates.

Polymer coatings on glass substrates can be used to align SCLCPs homogeneously. However, since a non-degenerate planar alignment is generally required, rubbing of the substrate after deposition of the polymer is used. Disclination-free planar alignment was achieved over areas of few mm² with a rubbed poly(vinyl alcohol) layer on the surface for mixtures of SCLCPs in LMMLCs (Hopwood and Coles, 1985a, b). Poly(vinyl alcohol) was applied from a 3% w/w solution (in distilled water), dried on a warm surface ($\sim 60°C$) and then rubbed unidirectionally using a rayon velvet-coated block. The alignment method was found to be suitable for all of the mixtures studied. Interestingly, it was observed that the higher-concentration mixtures (40% w/w polymer concentration) exhibited fewer non-uniformities in alignment when observed microscopically through crossed polars than the low-concentration mixtures. It should be noted that the rubbed poly(vinyl alcohol) introduces a very small surface tilt angle.

Finkelmann et al. (1983) reported that good planar alignment was achieved using oriented polyimide surface coatings. It should be noted, however, that the preparation of completely uniform homogeneous samples required the nematic polymers to be annealed for several hours at a temperature just below the clearing point. The efficiency of the aligning surface seemed to depend on the chemical structure of the polymer, i.e. the length of the flexible spacer and the type of polymer backbone itself. With increasing length of the spacer, the polymers behave more like LMMLCs and can be oriented more easily. It should be noted that upon cooling from a nematic phase uniformly oriented by a similar procedure (Haase and Pranoto, 1984), optically uniform preparations ($\sim 0.5\,cm^2$) of smectic A phases were also obtained (Haase and Pranoto, 1985).

The first known report of surface alignment of LCs was provided by Mauguin (1911) when he rubbed a raw glass plate with a piece of paper and obtained uniform homogeneous alignment of p-azoxyanisole. The first attempts to orient polysiloxanes by the rubbing technique were unsuccessful (Finkelmann et al., 1983). However, this method was successfully used by Casagrande et al. (1983) and Mattoussi et al. (1986a) who prepared samples, with thickness of about $100\,\mu m$, suitable for determination of the viscoelastic coefficients K_1 and γ_1 of polysiloxanes. On the other hand, SiO_x layers evaporated obliquely (at 60° incidence) led to uniformly aligned samples (Finkelmann et al., 1983). Microscopic observations indicated, however, that the director is not parallel to the glass plates, but tilted.

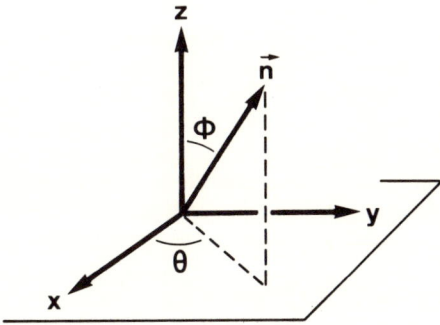

Figure 6.14 Coordinate geometry: xy is the plane of the substrate.

(ii) *Uniaxiality and biaxiality in nematic SCLCPs.* Most nematic phases are uniaxial. The ellipsoid of indices has a rotational symmetry about its long (extraordinary) axis which is parallel to the director $\vec{n}$. The geometry is defined in Figure 6.14 where $\vec{n}$ is characterized by the angles ϕ and θ. Homeotropic preparations ($\phi = 0$) are characterized by appearing dark between crossed polarizers. On the other hand, viewing planar samples ($\phi = \pi/2$, $\theta = 0$) from the top between crossed polarizers results in the observation of four positions of extinction. Classification of LC alignment into homeotropic or planar is an oversimplification. As mentioned above, the nematic director may make an angle with the substrate surface. The observations between crossed polarizers of such samples ($0 < \phi < \pi/2$; $\theta = $ constant) are the same as in the planar case. Such tests are insufficient to check the uniformity of alignment and observations of the conoscopic images formed in a monochromatic beam converging in the sample are needed (Figures 6.15, and 6.16) (Hartshorne and Stuart, 1970).

With the aid of conoscopic observations, Finkelmann (1980) has proved the positive uniaxial character (Figure 6.17) of SCLCPs:

$$-\!\!\left[CH_2\!-\!\underset{\underset{COO\{CH_2\}_m\!-O-\!\!\bigcirc\!\!-COO-\!\!\bigcirc\!\!-R}{|}}{C(CH_3)} \right]\!\!- \qquad m=6, R=CH_3 \quad (4)$$

and

$$-\!\!\left[Si(CH_3)\!-\!O \right]\!\!- \\ (CH_2)_m\!-O-\!\!\bigcirc\!\!-COO-\!\!\bigcirc\!\!-OCH_3 \qquad\qquad m=3 \qquad\qquad (5)$$

which exhibit textures reminiscent of those of LMMLCs. Similar to LMMLCs, the ordinary refractive index n_o is smaller than the extraordinary refractive index n_e thus causing a positive birefringence $\Delta n(\Delta n = n_e - n_o)$.

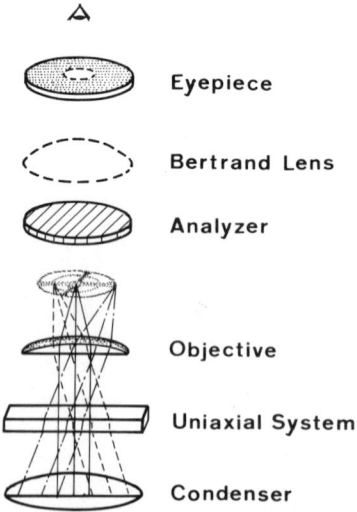

Eyepiece

Bertrand Lens

Analyzer

Objective

Uniaxial System

Condenser

Figure 6.15 Equipment for obtaining conoscopic figures.

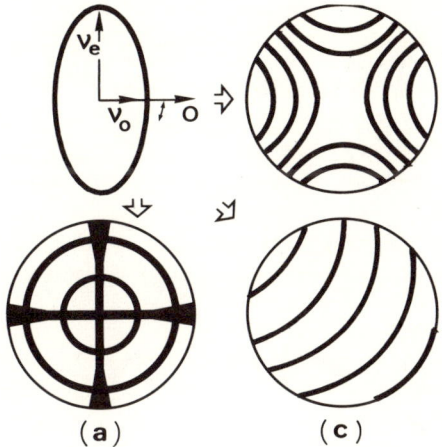

Figure 6.16 Ellipsoid of indices for a positive ($v_e > v_o$) uniaxial LC. The typical conoscopic images with average beam direction perpendicular to the plates are shown for (a) homeotropic; (b) planar and (c) oblique orientation.

Conoscopic figures were also obtained by Hopwood and Coles (1985b) for mixtures of SCLCPs and LMMLCs exhibiting only nematic phases. The cross and the clarity of the fringes indicated good homeotropic alignment of the mesogenic groups even at high polymer concentration. Insertion of a quarter wave plate showed that these nematic systems were optically positive and uniaxial.

It is interesting to note that contradictory results were reported by Finkelmann (1980) for polymethacrylate (**4**) where $m = 2$, and polyacrylate (**6**).

$$\text{---}\{CH_2\text{---}CH\}\text{---}$$
$$\underset{COO\text{---}\bigcirc\text{---}COO\text{---}\bigcirc\text{---}}{\mid} \qquad\qquad (6)$$

Although these polymers exhibit X-ray diffraction patterns consistent with nematic phases, no typical textures can be obtained and conoscopic observations show that these uniaxial media have negative birefringence ($n_e < n_o$). A possible explanation for this specific behaviour might be a disturbing effect of the polymer backbone owing to the short flexible spacers. This might correspond to the N_{II} phase predicted by Wang and Warner (1987) (cf. Chapter 2) when the drive toward parallel order of the mesogenic side chain moieties under the usual influence of steric and soft forces (described by the v_A term in the model) acting between these elements is comparable to the same nematic influence (v_B) acting between sections of the main chain. In such a phase, it is perhaps the main chain that explores directions about the ordering direction, Z and the side chains that are confined to the perpendicular plane ($S_A < 0$; $S_B > 0$).

The existence of biaxial nematic phases was predicted in 1970 by Freiser. The first-order transformation from the isotropic state to a uniaxial nematic phase is expected to be followed at lower temperature by a second-order transformation to a biaxial nematic phase in which the rotation of the molecules around their long axes becomes

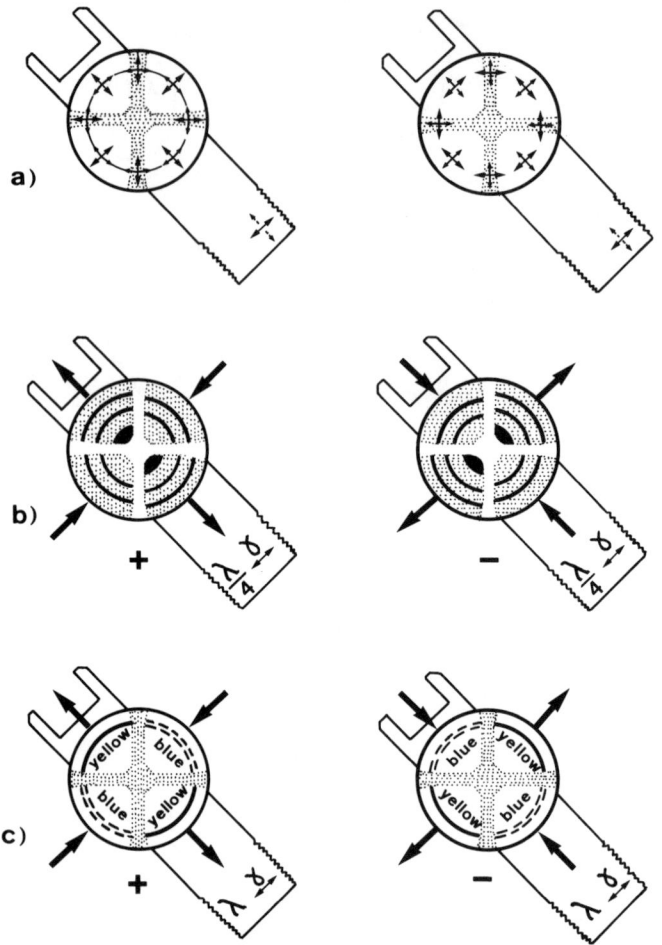

Figure 6.17(a) Basal interference figures of positive and negative uniaxial media, showing vibration directions in different quadrants; (b), (c), the same after insertion of a quarter wave plate and unit wave plate, respectively.

hindered and the mesogens are aligned as lath-like particles corresponding to the flat structure of the mesogenic groups. Such a biaxial nematic phase has never been observed for thermotropic LMMLCs because usually the onset of crystallization or a transition to a smectic phase intervenes at low temperature, thus precluding the formation of a biaxial nematic phase.

Recently, Zhou *et al.* (1987) and Hessel *et al.* (1987) have prepared SCLCPs with mesogenic groups laterally attached to the polymer backbone (**7**). DSC measurements, miscibility studies and X-ray investigations are consistent with nematic phases. However, microscopic observations under a polarizing microscope show that the textures of these polymers differ from those of conventional LMMLCs and SCLCPs. The samples annealed for several hours at a temperature just below the clearing point

$$-\{CH_2-CX\}-$$

$$\underset{O}{\overset{\parallel}{C}}$$

$$(CH_2)_m$$

$$R_1-COO-\langle\bigcirc\rangle-OOC-R_2 \qquad (7)$$

exhibit a texture with differently coloured domains separated by inversion walls. Uncovered thin samples form Schlieren textures which display only points that have four dark brushes originating from them ($|S| = 1$). This suggests that the observed phases are biaxial, i.e. that orientational correlations between molecules exist not only between their long axes but also between their short axes; hence benzene rings in adjacent molecules tend to stack parallel to each other. The question of defects is very different in uniaxial and biaxial nematics (Toulouse, 1977; Kléman, 1985); each of the three axes of the molecule plays the role of a director and has associated defects, but only defects of even integral strength can 'escape in the third dimension'; defects of odd integral strength are all topologically equivalent, which means that any defect of this type associated with a given 'axis' can turn continuously towards the more favourable 'axis' configuration. Such a possibility does not exist for the half-integral disclinations, which are of three different types. This might be related to the fact that half-integral disclinations have not been observed.

For polymers (7) where $X = CH_3$, $R_1 = R_2 = OCH_3$, OC_6H_{13} and $R_1 = OC_4H_9$, $R_2 = CN$ the conoscopic interference patterns obtained from homeotropic samples correspond to those of biaxial systems (Figure 6.18) (Hessel and Finkelmann, 1986). In the parallel position a split cross of zero birefringence is observed. After a 45° rotation of the microscope stage, the interference pattern of the 45° position appears which is

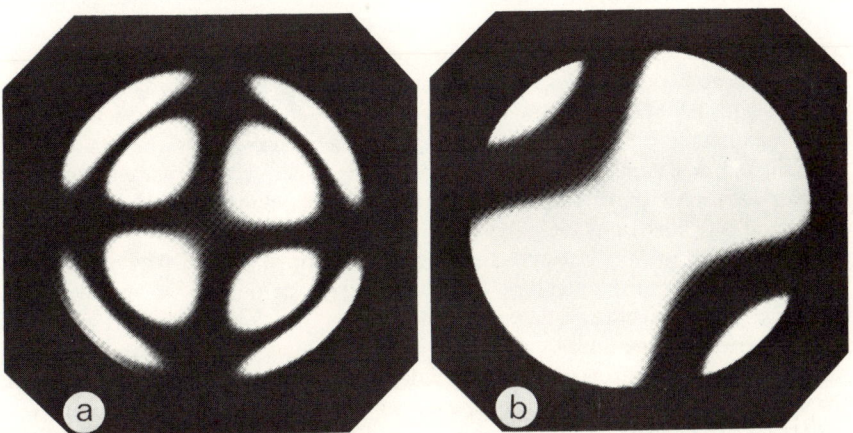

Figure 6.18 Basal interference figures of biaxial media. Sections perpendicular to the bisectrix of the acute angle. The cases depicted are those for thin sections of material and axial angle of about 10° (*a*) and 45° (*b*), respectively.

G

characterized by the maximum splitting of the two isogyres. A further rotation of 45° yields again a split cross but with the line of zero birefringence turned 90° from the original position. After a 180° rotation from the original position, the original interference figure is restored. The optical axial angle $2V$ is found to be about 10° for all the samples.

6.4 Miscibility studies

A complete classification of smectic phases by texture is not always possible. It can happen that similar textures are observed with two liquid crystalline states separated by a phase transition (Gray and Goodby, 1984). In case of LMMLCs an extremely useful and powerful tool for assessing the type of mesophase was found to be the determination of the isobaric temperature–concentration diagrams of binary mixtures. According to the rule of selective miscibility, all liquid crystalline modifications which exhibit an uninterrupted series of mixed crystals in binary systems can be marked with the same symbol (Sackmann and Demus, 1973; Gray and Winsor, 1974; Kelker and Hatz, 1980). It means that those liquid crystalline modifications that have the same designation in no case exhibit an uninterrupted series of mixed liquid crystals with the liquid crystalline states of another designation. While uninterrupted miscibility establishes isomorphism, the converse is not necessarily true. In addition, two mesophases, formed by the same compound but separated by a well-defined phase transition, are considered as different in their structures. In case of LMMLCs these two assumptions have been found to be consistent and have not led to contradictory results, although a large number of binary mixtures has been studied. Sackmann and Demus (1973) have observed in only one case a heterogeneous region between two modifications of the same type, but in over 100 cases they have found heterogeneous regions between modifications of different types. Temperature–composition phase diagrams for liquid crystalline mixtures can be generated from thermal data or, because of the various optical features characteristic of each mesophase structure, from observations of microscopic textures of the mesophases between crossed polarizers. The latter method, called the contact method (Kofler and Kofler, 1954) allows great rapidity in the assessment of the phase diagrams.

The applicability of the rule of selective miscibility has been examined for mixtures of SCLCPs with LMMLCs (Nyitrai et al., 1977; Cser et al., 1981; Ringsdorf et al., 1982; Finkelmann et al., 1982; Casagrande et al., 1982; Benthack-Thoms and Finkelmann, 1985; Sigaud et al., 1987). Unlimited miscibility is illustrated in Figure 6.19a. If the samples are freshly prepared, no eutectic is seen: the melting point of the reference compound is depressed as the polymer concentration is increased, whereas the glass transition of the polymer is lowered as the mole fraction of the LMMLC is increased, this being the so-called softening effect. If, however, the polymer is annealed for 3 months at room temperature, partial crystallization occurs and a eutectic is seen in the phase diagram (Figure 6.19b).

In considering the effects of change in molecular structure on the SCLCP/LMMLC phase diagrams we have a more difficult subject to tackle, because, contrary to LMMLCs and main chain LCPs, a gap of miscibility often exists in the nematic state. The first observation of a heterogeneous region between two nematics was made by Casagrande et al. (1982).

In 1980, a detailed lattice model for mixtures of thermotropic nematics was proposed

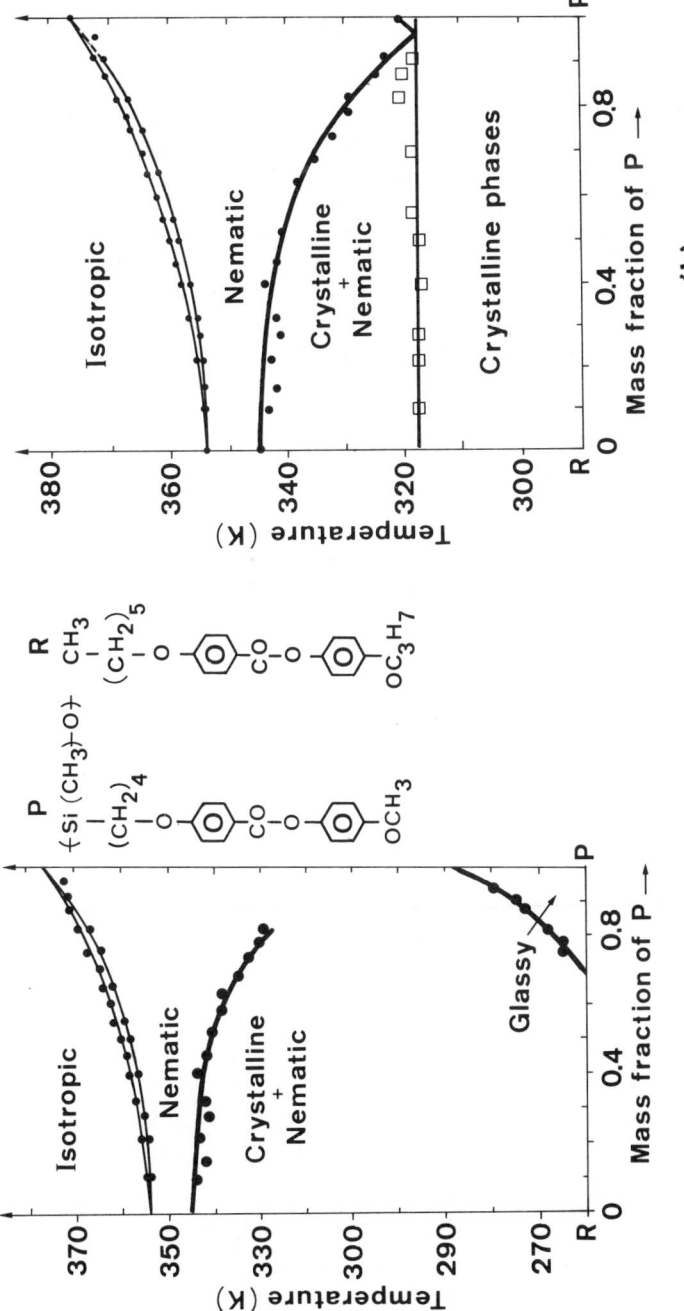

Figure 6.19 Isobaric phase diagram of a SCLCP and a nematic reference compound (from Benthack-Thoms and Finkelmann, 1985). (*a*) Freshly prepared samples; (*b*) after annealing the samples for 3 months at room temperature.

by Sivardière. In this model, each molecule occupies one lattice site and can be aligned into three separate directions. This choice of three possibilities ensures that the nematic–isotropic transition for a pure species is first order, as it should, and predicts an order parameter at transition $S_C = 0.5$, slightly too large but not implausible. The model requires four interaction parameters, which appears to be the minimal number for a realistic description. More recently, Brochard *et al.* (1984) enlarged the Sivardière description. They combined the Maier–Saupe theory (1959) of the nematic–isotropic transition and the Flory–Huggins theory of mixtures to describe the phase diagrams of two nematogens A + B. This improved slightly the predictions on the order parameters and allowed for two independent values S_A and S_B. This distinction between the order parameters S_A and S_B may be important when the components are chemically very different. For certain sets of parameters, rather complex diagrams were predicted when one component is a SCLCP. The miscibility in the nematic state is very small as soon as the nematic interaction parameter U_{AB} is smaller than U_{AA}. This is related to the most familiar case of a conventional polymer (polystyrene, polyethylene oxide) in a nematic solvent (Brochard, 1979). Figure 6.20 shows a phase diagram calculated for LMMLC-SCLCP mixture ($N_A = 1$, $N_B = 100$; $U_{AA} = U_{BB} = U$). On the other hand, if $U_{AB} > U_{AA}$, the miscibility of a SCLCP and a LMMLC is largely increased by the nematic alignment. In this connection systematic studies have been undertaken (Sigaud *et al.*, 1987; Benthack-Thomas *et al.*, 1985) on certain SCLCP/LMMLC mixtures. The work is still incomplete but available results confirm the conclusions reached about the rôle of solute–solute, solvent–solvent and solute–solvent interactions.

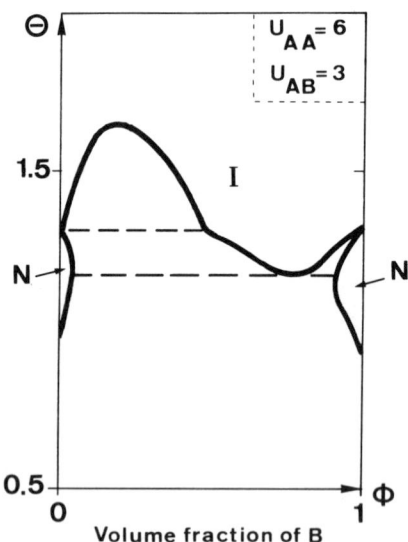

Figure 6.20 Phase diagram calculated for LMMLC–SCLCP mixture (degrees of polymerization $N_A = 1$, $N_B = 100$; nematic interaction parameters $U_{AA} = U_{BB}$).

6.5 Organization in SCLCPs

6.5.1 *Large and small angle X ray investigation*

X-ray diffraction provides information concerning the arrangement and mode of packing of molecules and the types of order present in a mesophase. For powder samples, the well-known Debye–Scherrer technique is used. This method gives all the reticular spacings but no information about the spatial orientation of these planes. Thus, it is often but not always possible to distinguish between the nematic and smectic phases of liquid crystals (smectic mesophases S_A and S_C may be confused). As will be stressed below, reliable characterization of molecular arrays by X-ray diffraction is possible only in aligned samples (Doucet, 1979; Leadbetter, 1979). Oriented specimens can be prepared by cooling in a strong magnetic field from the isotropic liquid phase into the nematic phase. Aligned S_A and S_C may then be prepared by careful cooling from the aligned nematic phase (Davidson *et al.*, 1985; Decobert *et al.*, 1985). This may in fact be very difficult and the result depends on the specimen and the conditions. An alternative procedure is by drawing fibre out of the mesophase and quenching the LC array in the glassy state so that it can be examined at room temperature.

Detailed reviews have appeared laying stress upon the liquid crystalline order in SCLCPs (Wendorff, 1978; Platé and Shibaev, 1980; Hoppner and Wendorff, 1984; Kostromin *et al.*, 1984; Lipatov *et al.*, 1984; Shibaev and Platé, 1984, 1985; Platé *et al.*, 1985; Tsukruk *et al.*, 1985; Zugenmaier, 1986; Azaroff, 1987; Mitchell *et al.*, 1987). We will therefore concentrate attention on the relatively more disordered phases: nematic, smectic A and smectic C.

(i) *Powder samples.* The diffraction pattern of a powder sample can be divided into inner rings at small angles, and outer rings at large angles. Small diffraction angles are taken to be of the order of $3°$ and correspond to distances of about $30 \, \text{Å}$. By large diffraction angles we mean angles of the order of $20°$ which correspond to distances of approximately $4.5 \, \text{Å}$ for the usual wavelengths ($CuK_\alpha = 1.54 \, \text{Å}$ and $CoK_\alpha = 1.79 \, \text{Å}$). The inner rings are indicative of longer layer spacings while the outer rings correspond to shorter preferred spacings occurring in the lateral packing arrangement of the molecules. The appearance of a broad halo or a sharp ring furnishes a qualitative indication of the degree of order.

Studies on thermotropic LMMLCs (Doucet, 1979; Leadbetter, 1979; Benattar *et al.*, 1983; De Vries, 1985) have shown that the most common mesophases can be divided into three groups according to the characteristics of their X-ray patterns at large diffraction angles. The first group is composed of the least-ordered phases (N, S_A and S_C). They give only one diffuse halo, which indicates that the lateral arrangement of the molecules is disordered. It should be noted, however, that in the smectic A and smectic C phases, the outer ring is somewhat less broad than in the nematic phase. In the second group we find the S_B, S_E, S_G, S_H, S_J and S_K (Gray and Goodby, 1984) whose diffraction patterns show one or a few Bragg reflections instead of a diffuse ring. Such sharp outer rings are related to the high degree of order within the layers. The S_F and S_I phases are intermediate between these two groups.

The X-ray patterns of nematics and smectics differ mainly in their characteristics at small angles. Nematic patterns present a diffuse ring corresponding to distances equal to the molecular length, which indicates that there is no order in the direction of the molecular long axes. It should be noted that in the nematic phase, the inner ring is somewhat stronger and less diffuse than in the isotropic phase. In contrast, X-ray patterns obtained from smectics present one or several sharp rings (usually two orders are observed) which is indicative of a periodic lamellar structure, corresponding to the smectic layers.

For SCLCPs, X-ray patterns were reported which are consistent with nematic phases (Shibaev *et al.*, 1982; Finkelmann and Rehage, 1984; Kostromin *et al.*, 1984; Shibaev and Platé, 1984, 1985; Noël *et al.*, 1988) (Figure 6.21). Usually they show only one diffuse halo at large angles which corresponds to average intermolecular spacings of approximately 4–6 Å. They may present at small angles a diffuse ring related to intramolecular interferences (Le Barny, 1986*b*; Noël *et al.*, 1988; Mitchell *et al.*, 1987).

In 1985, Freidzon *et al.* reported for a LC polyacrylate a X-ray diffraction pattern which did not correspond to any known liquid crystalline structure. It was characterized by a sharp ring in the wide-angle region. No reflection was detected in the small-angle region. From their X-ray diffraction studies, these authors concluded that this phase in which the mesogenic groups were packed in a hexagonal array but without translational order in the direction of their long axes, was of a novel kind and suggested it be given a new code letter N_B.

At a similar time to this report, Duran *et al.* (1986) made disclosures of another type of LC polymethacrylate which exhibited a phase that had Schlieren texture and X-ray pattern different from those reported for conventional nematics. The X-ray powder pattern consisted of two diffuse rings located at about 30 and 4.7 Å, respectively and a sharp reflection at 8.4 Å. The diffuse ring at 30 Å corresponded exactly to twice the length of an extended monomeric unit; this small-angle diffuse band was interpreted as an indication of cybotactic ordering. The diffuse ring at 4.7 Å was associated with the average lateral approach distance of the molecules. This phase was designated as a 'novel nematic' mesophase, though its detailed structure was left unresolved. Indeed, the sharp reflection at 8.4 Å could not be associated with any typical length in the polymer molecule. Quite recently Duran *et al.* (1987) proposed the concept of ribbon-like conformation of polymer chains and it became clear that the sharp reflection at 8.4 Å represents a stacking periodicity related to the thickness of the ribbons. This is consistent with the ribbons packed as shown in Figure 6.22. It is clear that this

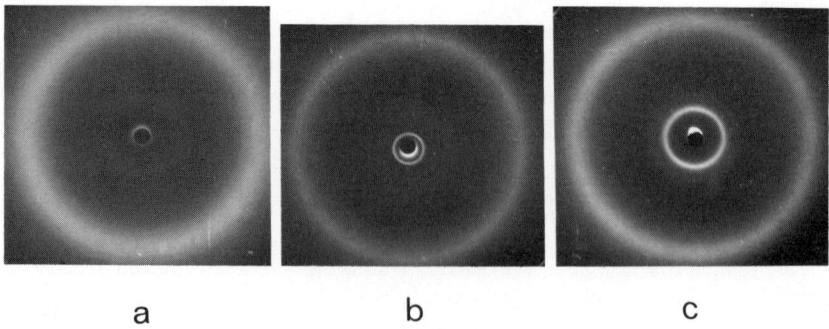

a   b   c

Figure 6.21 Typical X-ray diffraction patterns for unoriented nematic (*a*), smectic A (*b*), and smectic C (*c*) phases.

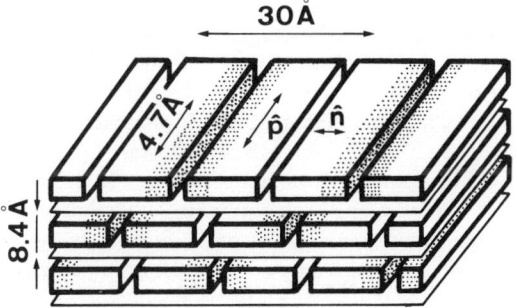

Figure 6.22 Structure of the 'novel nematic'. The darkened part of the ribbons correspond to the polymer backbone. (Redrawn from Duran *et al.*, 1987.)

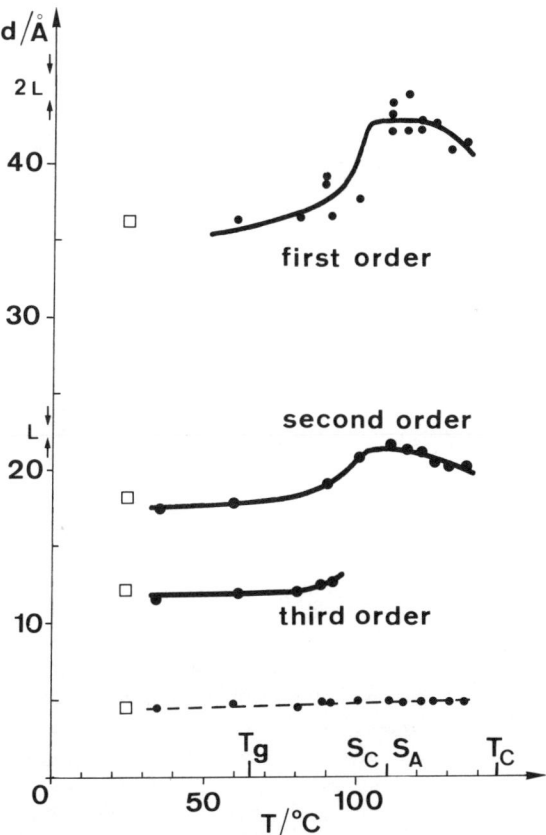

Figure 6.23 Intermolecular distance (------) and layer spacing (——) in polymer 1 where X = H and $m = 2$, as a function of temperature. Stretched oriented fibre. (From Decobert *et al.*, 1986.)

structure is intermediate between smectic and nematic or columnar, as it corresponds to the periodical stacking of two-dimensional palisades of ribbons.

Other X-ray patterns obtained from unoriented SCLCPs in the smectic state are characteristic of a disordered lamellar structure (Platé and Shibaev, 1980; Paleos *et al.*, 1982; Shibaev *et al.*, 1982; Alimoglu *et al.*, 1984; Kostromin *et al.*, 1984; Shibaev and Platé, 1984; Gemmell *et al.*, 1985; Richardson and Herring, 1985; Basu *et al.*, 1986; Decobert *et al.*, 1986; Mauzac *et al.*, 1986; Esselin *et al.*, 1987; Zentel *et al.*, 1987). They present one or two sharp inner rings corresponding to the lamellar thickness d, and a diffuse outer ring reflecting the absence of ordering within the layer planes. For S_C phases, diffraction patterns of powder samples are essentially the same as those for S_A. However, because of the tilt angle the layer spacings d are smaller. Hence, there is an abrupt decrease of d at the transition between a smectic A phase and a smectic C phase (Kostromin *et al.*, 1982; Decobert *et al.*, 1986; Mauzac *et al.*, 1986; Esselin *et al.*, 1987) (Figure 6.23). Simultaneously an intense 003 Bragg reflection usually appears. Most of the smectic polymers with terminally cyano-substituted side chains show the partially overlapping structure common in their low-molar-mass analogues (Shibaev *et al.*, 1982; Alimoglu *et al.*, 1984; Kostromin *et al.*, 1984; Richardson and Herring, 1985; Le Barny *et al.*, 1986*a*; Mauzac *et al.*, 1986; Lacoudre *et al.*, 1988; Spassky *et al.*, 1988). However, for SCLCPs this structure is not confined to cyano-derivatives (Gemmell *et al.*, 1985; Mauzac *et al.*, 1986). SCLCPs may exhibit A and C phases which have a monolayer structure, a perfect bilayer structure or an interdigitated structure depending on the nature of the polymer backbone, the number of atoms in the flexible spacer and the length of the terminal group (Lipatov *et al.*, 1984; Shibaev and Platé, 1985).

Finally, it is to be noted that SCLCPs may also form more ordered smectic phases such as S_F, S_I, S_B, S_E and S_G (Kostromin *et al.*, 1982; Talroze *et al.*, 1982; Shibaev and Platé, 1984, 1985; Mauzac *et al.*, 1986; Freidzon *et al.*, 1986; Duran *et al.*, 1987*b*; Frère *et al.*, 1988). For instance, various types of evidence lend support to the conclusion that polyacrylate having the repeating unit (**8**)

$$\text{—}[\text{CH}_2\text{—CH}]\text{—}$$
$$\text{COO—}(\text{CH}_2)_5\text{—COO—}\langle\bigcirc\rangle\text{—COO—}\langle\bigcirc\rangle\text{—OC}_3\text{H}_7 \qquad (8)$$

exhibits the following polymorphism:

$$\text{(annealing at} \qquad \text{—}S_G \xrightarrow{60°C} S_C \xrightarrow{95°C} N \xrightarrow{116°C} i$$
$$35°C \text{ for}$$
$$\text{several days)} \qquad\qquad S_F \nearrow 40°C$$

The X-ray scattering curves recorded at different temperatures are given in Figure 6.24. As previously discussed, the width of the wide-angle maximum decreases as the intralayer ordering increases. The S_C phase with a liquid-like arrangement of the mesogenic groups in the layers gives a diffuse maximum. The intralayer correlation length, $\xi_\perp$, calculated from the halfwidth of the maximum by assuming a Lorentzian line shape is of the order 5–10 Å. For the S_F phase, a sharp maximum is observed, which results from the hexagonal packing of the molecules within the layers; its width suggests a correlation length within the layers of 40 Å. The S_G phase gives a sharper maximum and additional weaker reflections which are indicative of a two-dimensional lattice in

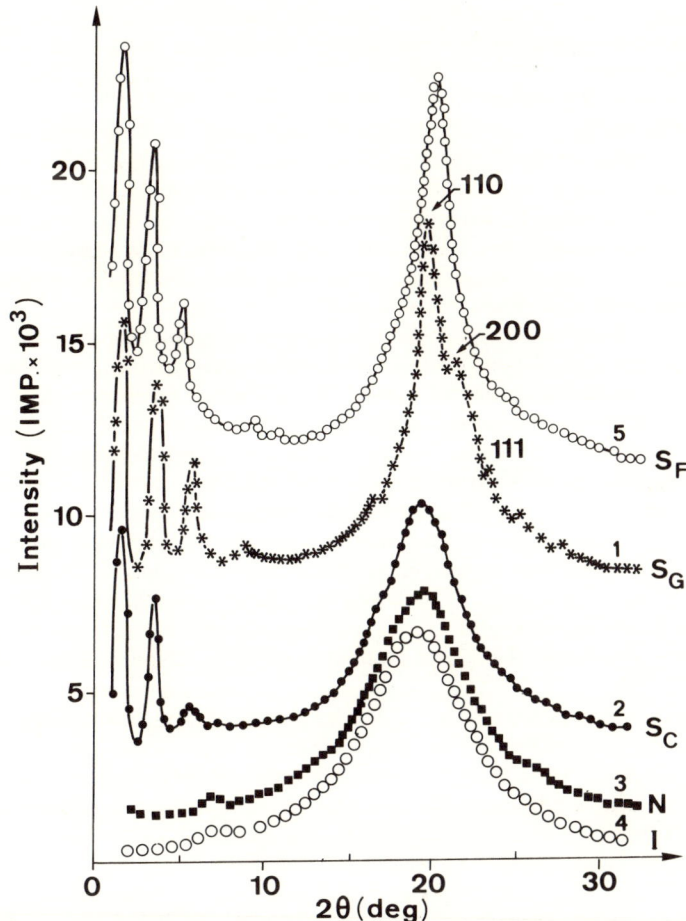

Figure 6.24 X-ray scattering curves recorded in the isotropic (I), nematic (N), smectic C (S$_C$), smectic F(S$_F$) and smectic G(S$_G$) phases. (Redrawn from Freidzon *et al.*, 1986.)

smectic layers with a symmetry differing from the hexagonal one. The S$_G$ phase has a monoclinic cell. The intralayer correlation length is approximately 70 Å which is smaller than that observed for conventional S$_G$ phases. This reflects noticeable disorder of the molecules. As can be seen from the position and intensity of the small-angle reflections, the tilted arrangement of the molecules is preserved.

Quite recently (Duran *et al.*, 1987*b*; Frère *et al.*, 1988) smectic E phases were observed for a polymethacrylate and two alternating copolymers of maleic anhydride with vinyl ethers, bearing a 4′-methoxy-4-biphenylyl group via an oligo(ethylene oxide) spacer. Smectic layers are single layers of ribbon-like polymer chains with all pendent groups arranged in a single row and pending on the same side of the backbone. Mesogenic groups are oriented up and down at random and polymer chains are aligned along the [110] direction of the rectangular lattice describing the packing of the pendent groups.

(ii) *Oriented samples*. X-ray diffraction patterns characteristic of oriented nematics were reported for polysiloxanes (Finkelmann and Rehage, 1984; Zugenmaier, 1986) polyacrylates (Shibaev and Platé, 1984, 1985; Zentel and Strobl, 1984; Le Barny, 1986*a*, *b*; Mitchell *et al.*, 1987; Freidzon *et al.*, 1988) polymethacrylates (Shibaev and Platé, 1984, 1985; Konstantinov *et al.*, 1984; Zentel and Strobl, 1984; Davidson *et al.*, 1985) and poly-α chloroacrylates (Decobert *et al.*, 1985). The marked anisotropy of the molecular organization in these samples is clearly shown by two symmetrical wide-angle crescents which arise from the spectral correlations between mesogenic units. Their radial and angular extensions are similar to those of conventional nematics. Additional smaller-angle-diffuse lines, associated with intramolecular interferences are sometimes seen in a perpendicular direction.

If the samples are aligned by cooling in a strong magnetic field from the isotropic liquid phase into the nematic phase, then the wide-angle crescents are located in a direction perpendicular to the field, which indicates that the mesogenic groups are oriented along the magnetic field, the flexible main chains being arranged arbitrarily in the bulk. In most cases, similar X-ray patterns are observed for oriented samples produced by drawing fibres out of the nematic phase. However, few X-ray patterns were reported (Zentel and Strobl, 1984) in which the crescents are seen in a direction parallel to the fibre axis. This implies that the mesogenic groups are perpendicular to the fibre axis, the macromolecular chains being preferentially aligned in the direction of extension.

Secondary nematic structures were reported for a few SCLCPs (Konstantinov *et al.*, 1984; Shibaev and Platé, 1984; Zentel and Strobl, 1984; Davidson *et al.*, 1985; Decobert *et al.*, 1985; Le Barny *et al.*, 1986*b*). These are polymers in which pronounced smectic order parameter fluctuations occur. This is especially true of certain polymers which transform to S_A or S_C phases at lower temperatures. In addition to the common features previously discussed, the diffraction pattern of the nematic then shows the development of enhanced order characteristic of smectic phase: symmetrical diffuse spots are observed at small angles. They are located either in a direction perpendicular to the imaginary line connecting the centres of the wide-angle crescents ($N \rightarrow S_A$) or they form pairs aligned on straight lines making an angle with respect to this imaginary line ($N \rightarrow S_C$) (Figure 6.25). These short-range order effects were called cybotactic groups by de Vries (1970*a*, *b*). The first X-ray scattering measurements of pretransitional S_C fluctuations in the nematic phase of a LC polysiloxane were reported by Nachaliel *et al.* (1987). Their results are consistent with other differential scanning calorimetry and PVT measurements (Frenzel and Rehage, 1983) in identifying the $N \rightarrow S_C$ transition as weakly first-order. The data were well described by the mean-field theory of Chen and Lubensky (1976) suggesting that the mesogenic side groups are only weakly coupled to the polymer backbone. Because of disorder in the polymer backbone, the correlation lengths $\xi_\parallel$ and $\xi_\perp$, in the vicinity of the nematic–smectic C phase transition, were weak functions of temperature and their magnitudes are smaller than those of conventional LCs.

Most SCLCPs are characterized by layered structures but, compared with LMMLCs, smectic polymers exhibit mainly modifications with liquid-like intralayer order, i.e. smectic A and/or smectic C (Kostromin *et al.*, 1982; Konstantinov *et al.*, 1984; Shibaev and Platé, 1984; Zentel and Strobl, 1984; Zugenmaier and Mügge, 1984; Davidson *et al.*, 1985; Decobert *et al.*, 1985; Diele *et al.*, 1986; Zugenmaier, 1986; Le Barny *et al.*, 1986*b*; Diele *et al.*, 1987; Sutherland *et al.*, 1987; Zentel *et al.*, 1987*a*, *b*;

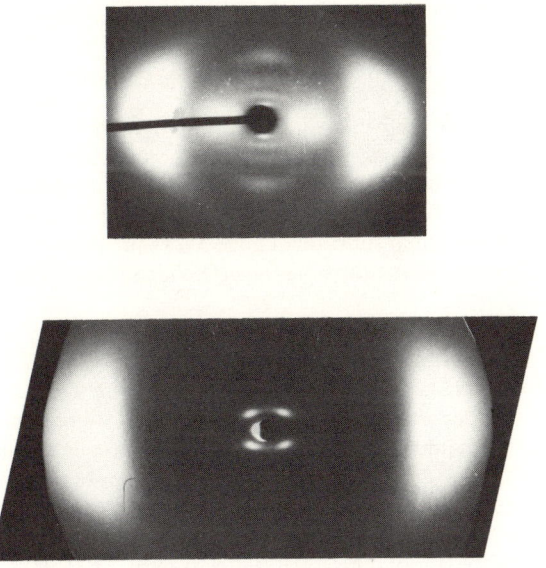

Figure 6.25 Typical X-ray diffraction patterns for nematics with cybotactic groups.

Davidson *et al.*, 1988; Freidzon *et al.*, 1988). Typical diffraction patterns for oriented samples of a S_A and a S_C phase are shown in Figure 6.26. The wide-angle crescents are qualitatively like those for a nematic but rather more intense and with a less pronounced arc. The order parameter increases from the nematic to the smectic phase. The very strong sharp reflections at small angles show the existence of extensive layer-like correlations. Different types of ordering may be obtained depending on the materials and the conditions. If the samples are oriented by careful cooling in a

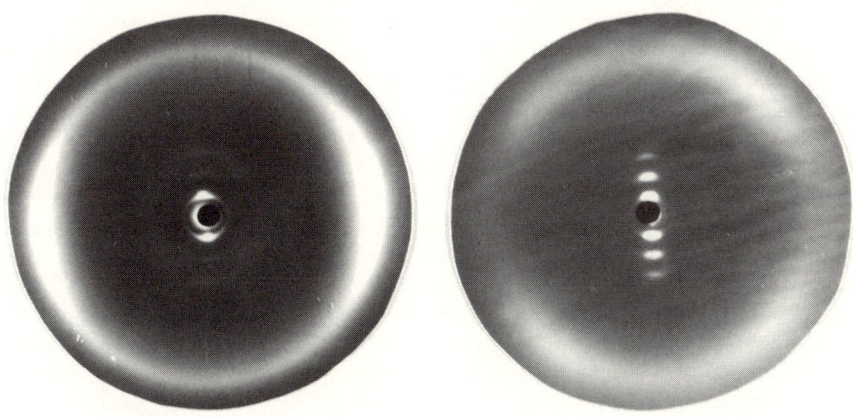

Figure 6.26 Typical X-ray diffraction patterns for a S_A and a S_C phase.

magnetic field, then the relative position of the wide-angle crescents and the small angle Bragg spots with respect to the direction of the field is indicative of side chains parallel to the field with perpendicular (S_A) or tilted (S_C) layers (Davidson *et al.*, 1985; Decobert *et al.*, 1985; Diele *et al.*, 1986, 1987). In contrast, when the samples are mechanically aligned by stretching the polymer, different X-ray patterns are usually obtained, indicating that the layers are parallel to the direction of extension (Zentel and Strobl, 1984; Decobert *et al.*, 1986; Zugenmaier, 1986; Esselin *et al.*, 1987; Sutherland *et al.*, 1987; Freidzon *et al.*, 1988). However, stretched samples of polymers having mesogenic groups with ring-CN terminal substituent gave X-ray patterns which were consistent with the side chains parallel to the fibre axis (Davidson *et al.*, 1985; Le Barny *et al.*, 1986b; Noël *et al.*, 1988).

In addition to these common features, the X-ray patterns for oriented samples of an S_A and an S_C phase can show parallel diffuse lines equally spaced versus q (Davidson *et al.*, 1985; Decobert *et al.*, 1985; Esselin *et al.*, 1987; Sutherland *et al.*, 1987). This is characteristic of disorder along the director. The extra scattered intensity arises from uncorrelated periodic columns which are out of the mean position in the layer plane (Doucet *et al.*, 1979). Such diffuse features do not usually appear in S_A and S_C diffraction patterns of LMMLCs. They often occur in smectic phases which exhibit three-dimensional order (for example smectics B), and so the most obvious explanation of their appearance in the X-ray diffraction patterns of SCLCPs is in terms of rigidity effect of the smectic layers due to the macromolecular nature of the compound.

A second diffuse zone can be seen (Konstantinov *et al.*, 1984; Davidson *et al.*, 1985; Diele *et al.*, 1986): four off-meridian spots are visible. Konstantinov *et al.* (1984) assigned these reflections to the formation of 'blocks' of mesogenic units. By comparison, this extra scattering was interpreted by Davidson *et al.* (1985) to be caused by a periodic modulation of adjacent layers, with a wave vector a parallel to the layer plane.

Different models of the S_A and the S_C phase stemmed from the numerous X-ray investigations on smectic SCLCPs (Strzelecki and Liebert, 1973; Shibaev and Platé, 1984, 1985; Davidson *et al.*, 1985; Azaroff, 1987; Diele *et al.*, 1987; Freidzon *et al.*, 1988). In general, the same arrangements as known from LMMLCs were proposed for the packing of the mesogenic side chains without making any definite proposals about the conformation of the polymer backbone. In most cases, the unstated assumption was that the polymer backbone forms a two-dimensional coil sandwiched between the smectic layers formed by the side chains. However, values determined by small-angle scattering methods (see 6.5.2) for the radii of gyration of the main chain in directions parallel and perpendicular to the director show that the main chain cannot stay strictly confined between two layers; each macromolecular chain must hop from a layer to an adjacent one, thus creating defects in the smectic planes. Such a behaviour has been predicted by Renz and Warner (1986) (see also Chapter 2). Quite recently, detailed X-ray diffraction studies by Davidson and Levelut (1988) conclusively proved the existence of such defects.

It should be noted that Zugenmaier and Mügge (1984) deduced a different model from X-ray investigations of polysiloxanes with mesogenic side groups in the crystalline and smectic state. According to these authors, the polymer backbone forms an ordered structure from which the mesogenic groups stick out at an angle of approximately 90°, depending on packing effects and chemical constitution of the side chains. The macromolecules (backbone and side chains) form elliptical bodies which can be shifted with respect to their long axes and which are not or only weakly correlated. Although this model which assumes a well-ordered conformation for the polymer backbone does not seem to agree with recent SANS studies (see 6.5.2), it may be considered valid for some polymers.

6.5.2 Conformation of SCLCPs as revealed by small-angle scattering methods

In terms of competition between rod order and internal entropy of the chains, a question naturally arises as to the influence of the pendent mesogenic groups on the polymer backbone conformation. Recently, Wang and Warner (1987) and Kunchenko and Svetogorsky (1986) modelled this competition taking main chains of various stiffness and side groups of various length. The theory predicts molecular conformational changes. The antagonistic influences of the nematic field and the internal entropy of the chain are resolved by a distorsion of chain statistics away from spherical. Depending on temperature, nematic coupling and stiffness, the chain becomes prolate (rod-like) or oblate (disc-like).

Experimental studies of the magnetic field effect on SCLCPs (5) where $m = 3–6$ either in the melt (Fabre et $al.$, 1984) or dissolved in a conventional nematic (Casagrande et $al.$, 1984; Weill et $al.$, 1986; Mattoussi et $al.$, 1986a) were recently reported. While the static properties (the elastic constants K_1 and K_3) were close to those of LMMLCs, the hydrodynamic behaviour resembled that of conventional polymers in the melt. The response time τ and the twist viscosity coefficient γ_1 which characterizes the coupling between the fluid and the nematic director, were several orders of magnitude longer than those obtained for LMMLCs. The analysis of the data, within the framework of a hydrodynamic model due to Brochard (1979), indicated that the polymer backbone would have a non-spherical conformation, the anisotropy defined by the ratio $R_\perp/R_\parallel$ of gyration radii being large enough to be measured by a small angle scattering method. The validity of this interpretation was checked by Mattoussi et $al.$ (1986b) who determined by small-angle X-ray scattering (SAXS) the radius of gyration of the main chain in directions parallel $(R_\parallel)$ and perpendicular $(R_\perp)$ to the director in dilute solutions of polymers (5) when dissolved in a conventional nematic. They found a significant difference between the two values, with $R_\parallel > R_\perp$. For a polysiloxane of similar structure (the —COO— groups are reversed to —OOC—) but in the undiluted melt, Moussa et $al.$ (1987) determined by small-angle neutron scattering (SANS) a ratio $R_\perp/R_\parallel$ of gyration radii of 0.75 in close agreement with the value of 0.73 reported by Mattoussi et $al.$ (1986). These results might serve as an experimental corroboration of the prolate conformation predicted by Wang and Warner (1987) for the $N_{II}(S_A < 0, S_B > 0)$ and $N_{III}(S_A > 0, S_B > 0)$ phases where the main chain is on average along the director (see Chapter 2).

It is of interest to note that the opposite effect was observed for SCLCPs (4) where $m = 6$ and $R = CN, OCH_3, OC_4H_9$ or OC_6H_{13}. For these polymers, with polymethacrylate backbone, $R_\parallel$ is smaller than $R_\perp$ (Kirste and Ohm, 1985; Keller et $al.$, 1985; Moussa et $al.$, 1987). This corresponds to the behaviour predicted by Wang and Warner (1987) for the N_1 phase when the side chains order the most strongly (v_A is large). Then, if v_f, which determines the extent to which side and main chains wish to be perpendicular, dominates over the nematic coupling, v_c, tending to make for alignment between the directions of the side and main chains, the main chain will on average be reduced to exploring the plane perpendicular to the director. From these SANS experiments it was also inferred that the anisotropy increases in the smectic state where the backbone appears more or less confined in one or two layers.

Quite recently, Kunchenko and Svetogorsky (1987) using a simple model (a comb-like freely jointed chain) have made an estimate for the second moment of the distribution of the mesogenic units relative to the backbone segments. For the experimentally-measured values $a = R_\perp/R_\parallel = 1.1$ (Keller *et al.*, 1985), or 1.25 (Kirste and Ohm, 1985) and an estimate for the nematic order parameter S'' of 0.5, the parameters S^f, reduced to the minimum of $-1/2$, are $1/3$ and $1/2$, respectively.

From studies performed on polymethacrylates (Keller *et al.*, 1985; Moussa *et al.*, 1987) it can be inferred that the anisotropy increases in the smectic state. A large variation of $R_\parallel$ is observed. Good agreement is obtained with the 'layer hopping' model of Renz and Warner (1986) based on de Gennes calculation (1982)

$$R_\parallel^2 \alpha \exp(-E/K_B T)$$

of the 'hairpin energy'. The activation energy deduced from this model is $\sim 80 \text{kJ mol}^{-1}$ which is ten times greater than that estimated for a smectic defect by Kunchenko and Svetogorsky (1986).

Further systematic studies of this type on a few SCLCPs are of interest and may suffice to establish certain laws and to detail the arrangement of the polymer backbone, flexible spacers and mesogenic cores in the nematic and smectic phases.

References

Abe, A. (1984) *Macromolecules* **17**, 2280.
Alimoglu, A.K., Ledwith, A., Gemmell, P.A., Gray, G.W. and Lacey D. (1984) **25**, 1342.
Azaroff, L.V. (1987) *Mol. Cryst. Liq. Cryst.* **145**, 31.
Basu, S., Rawas, A. and Sutherland, H.H. (1986) *Mol. Cryst. Liq. Cryst.* **132**, 23.
Benattar, J.J., Moussa, F. and Lambert, M. (1983) *J. Chim. Phys.* **80**, 99.
Benthack-Thoms, H. and Finkelmann, H. (1985) *Makromol. Chem.* **186**, 1895.
Blumstein, A., Blumstein, R.B., Gauthier, M.M., Thomas, O. and Asrar, J. (1983) *Mol. Cryst. Liq. Cryst. Lett.* **92**, 87.
Boeffel, C., Hisgen, B., Pschorn, U., Ringsdorf, H. and Spiess, H.W. (1983) *Israel J. Chem.* **23**, 388.
Brochard, F. (1979) *C.R. Hebd. Sci. Paris* **289B**, 229.
Brochard, F., Jouffroy, J. and Levinson, P. (1984) *J. de Physique* **45**, 1125.
Casagrande, C., Veyssié, M., Weill, C. and Finkelmann, H. (1982) *J. Phys. Lett.* **43**, L-671.
Casagrande, C., Veyssié, M., Weill, C. and Finkelmann, H. (1983) *Mol. Cryst. Liq. Cryst. Lett.* **92**, 49.
Casagrande, C., Fabre, P., Veyssié, M., Weill, C. and Finkelmann, H. (1984) *Mol. Cryst. Liq. Cryst.* **113**, 193.
Chandrasekhar, S. (1984) in *Polymers, Liquid Crystals and Low-Dimensional Solids*, eds. March, N. and Tosi, M., Plenum, New York, Chapters. 8 and 10.
Chen, J. and Lubensky, T.C. (1976) *Phys. Rev.* **A14**, 1202.
Coates, D. and Gray, G.W. (1975) *J. Phys. (Paris)* **36**, 365.
Cognard, J. (1982) *Mol. Cryst. Liq. Cryst., Suppl.* **1**, 1.
Coles, H.J. and Simon, R. (1985) in *Polymeric Liquid Crystals*, ed, Blumstein, A., Plenum, New York, 351.
Cser, F., Nyitrai, G., Hardy, Gy., Menczel, J. and Varga, J. (1981) *J. Polym. Sci. Polym. Symp.* **69**, 91.
Davidson, P. and Levelut, A.M. (1988) *J. de Phys.* **49**, (689), 193.
Davidson, P., Keller, P. and Levelut, A.M. (1985) *J. de Phys.* **46**, 939.
Decobert, G., Soyer, F., Dubois, J.C. and Davidson, P. (1985) *Polym. Bull.* **14**, 549.

Decobert, G., Dubois, J.C., Esselin, S. and Noël, C. (1986) *Liquid Crystals* **1**, 307.
De Gennes, P.G. (1982) in *Polymer Liquid Crystals*, eds. Cifferi, A., Krigbaum, W.R. and Meyer, R., Academic Press, New York.
Demus, D. and Richter, L. (1978) *Textures of Liquid Crystals*. Verlag Chemie, Weinheim.
De Vries, A. (1970) *Mol. Cryst. Liq. Cryst.* **10**, 31(a), 219(b).
De Vries, A. (1985) *Mol. Cryst. Liq. Cryst.* **131**, 125.
Diele, S., Hisgen, B., Reck, B. and Ringsdorf, H. (1986) *Makromol. Chem., Rapid Commun.*, **7**, 267.
Diele, S., Oelsner, S., Kuschel, F., Hisgen, B., Ringsdorf, H. and Zentel, R. (1987) *Makromol. Chem.* **188**, 1993.
Doucet, J. (1979) in *The Molecular Physics of Liquid Crystals*, eds. Luckhurst, G.R. and Gray, G.W., Academic Press, New York, Chapter 14.
Duran, R., Gramain, P., Guillon, D. and Skoulios, A. (1986) *Mol. Cryst. Liq. Cryst.* **3**, 23.
Duran, R., Guillon, D., Gramain, P. and Skoulios, A. (1987a) *Makromol. Chem., Rapid Commun.* **8**, 321.
Duran, R., Guillon, D., Gramain, P. and Skoulios, A. (1987b) *Makromol. Chem., Rapid Commun.* **8**, 181.
Esselin, S., Bosio, L., Noël, C., Decobert, G. and Dubois, J.C. (1987) *Liquid Crystals* **2**, 505.
Esselin, S., Noël, C., Decobert, G. and Dubois, J.C. (1988) *Mol. Cryst. Liq. Cryst.* **155**, 371.
Fabre, P., Casagrande, C., Veyssié, M. and Finkelmann, H. (1984) *Phys. Rev. Lett.* **53**, 993.
Fayolle, B., Noël, C. and Billard, J. (1979) *J. de Phys., Colloque* **C3**, 40, C3–485.
Finkelmann, H. (1980) in *Liquid Crystals of One- and Two-Dimensional Order*, eds. Helfrich, W., and Heppke, G., Springer Verlag, Berlin.
Finkelmann, H. (1982) in *Polymer Liquid Crystals*, eds. Ciferri, A., Krigbaum, W.R., and Meyer, R., Academic Press, New York.
Finkelmann, H. and Rehage, G. (1982) *Makromol. Chem. Rapid Commun.* **3**, 859.
Finkelmann, H. and Rehage, G. (1984) *Adv. Polym. Sci.*, **60/61**, 99.
Finkelmann, H., Ringsdorf, H., Siol, W. and Wendorff, H. (1978a) *Makromol. Chem.* **179**, 829.
Finkelmann, H., Koldehoff, J. and Ringsdorf, H. (1978b) *Angew. Chem.* **90**, 992.
Finkelmann, H., Kock, H.J. and Rehage, G. (1982) *Mol. Cryst. Liq. Cryst.* **89**, 23.
Finkelmann, H., Kiechle, U. and Rehage, G. (1983) *Mol. Cryst. Liq. Cryst.* **94**, 343.
Frank, F.C. (1958) *Disc. Faraday Soc.* **25**, 19.
Freidzon, Y.S., Boiko, N.I., Shibaev, V.P. and Platé, N.A., (1985) *Dokl. AN SSSR* **282**, 922.
Freidzon, Y.S., Boiko, N.I., Shibaev, V.P., Tsukruk, V.V., Shilov, V.V. and Lipatov, Y.S. (1986) *Polymer Commun.* **27**, 190.
Freidzon, Y.S., Talroze, R.V., Boiko, N.I., Kostromin, S.G., Shibaev, V.P. and Platé, N.A. (1988) *Liquid Crystals* **3**, 127.
Freiser, M.J. (1970) *Phys. Rev. Lett.* **24**, 1041.
Frenzel, J. and Rehage, G. (1983) *Makromol. Chem.* **184**, 1685.
Frère, Y., Yang, F., Gramain, P., Guillon, D. and Skoulios, A. (1988) *Makromol. Chem.* **189**, 419.
Friedel, G. (1931a) *Z. Kristallogr.* **79**, 26.
Friedel, G. and Friedel, E. (1931b) *J. Physique Radium (VII)* **2**, 133.
Frosini, V., Magagnini, P.L. and Newman, B.A. (1974) *J. Polym. Sci., Polym. Phys. Ed.* **12**, 23.
Gemmell, P.A., Gray, G.W. and Lacey, D. (1983) *Polymer Preprints* **24**, 253.
Gemmell, P.A., Gray, G.W., Lacey, D., Alimoglu, A.K. and Ledwith, A. (1985) *Polymer* **26**, 615.
Ghanem, A. Thèse de Docteur-ès-Sciences, Université Pierre et Marie Curie, Paris, 29 septembre 1987.
Goodman, L.A. (1974) *RCA Review* **35**, 477.
Gray, G.W. (1975) *J. Phys. (Paris)* **36**, 337.
Gray, G.W. and Winsor, P.A. (1974) *Liquid Crystals and Plastic Crystals*. Ellis Horwood, Chichester.
Gray, G.W. and Goodby, J.W. (1984) *Smectic Liquid Crystals* Leonard Hill [Blackie], Glasgow and London.
Gubina, T.I., Kostromin, S.G., Talroze, R.V., Shibaev, V.P. and Platé, N.A. (1986) *Vysokomol. Soed. Ser. B* **28**, 394.
Guyon, E. and Urbach, W. (1976) *4BBC Symp. Nomen. El Disp.*, eds. Kmetz and Von Willisen, Plenum, New York.
Haase, W. and Pranoto, H. (1984) *Progr. Colloid Polym. Sci.* **69**, 139.

Haase, W. and Pranoto, H. (1985) in *Polymeric Liquid Crystals*, ed. Blumstein, A., Plenum, New York, 313.

Hahn, B., Wendorff, J.H., Portugall, M. and Ringsdorf, H. (1981) *Colloid Polym. Sci.* **259**, 875.

Haller, I. (1975) *Thermodynamic and Static Properties of LCs* (*Progr. Solid State Chem.*) **10**, 103.

Hallimond, A.F. (1970) in *The Polarizing Microscope*, 3rd edn., Vickers Instruments, York.

Hardouin, F., Levelut, A.M., Achard, M.F. and Sigaud, G. (1983) *J. Chimie Phys.* **80**, 53.

Hardouin, F., Achard, M.F., Tinh, N.H. and Sigaud, G. (1985) *J. de Physique Lett.* **46**, L-123.

Hartshorne, N.H. (1974) in *Liquid Crystal and Plastic Crystals*, eds. Gray, G.W. and Winsor, P.A., Ellis Horwood, Chichester, 2, 24.

Hartshorne, N.H. and Stuart, A. (1970) in *Crystals and the Polarising Microscope*, 4th edn., Edward Arnold, London.

Hessel, F. and Finkelmann, H. (1986) *Polym. Bull.* **15**, 349.

Hessel, F., Herr, R.P. and Finkelmann, H. (1987) *Makromol. Chem.* **188**, 1597.

Hoppner, D. and Wendorff, J.H. (1984) *Angew. Makromol. Chem.* **125**, 37.

Hopwood, A.I. and Coles, H.J. (1985a) *Mol. Cryst. Liq. Cryst.* **130**, 281.

Hopwood, A.I. and Coles, H.J. (1985b) *Polymer* **26**, 1312.

Hsu, C.S. and Percec, V. (1987) *Makromol. Chem., Rapid Commun.* **8**, 331.

Hsu, C.S., Rodriguez-Parada, J.M. and Percec, V. (1987) *Makromol. Chem.* **188**, 1017.

Kahn, F.J., Taylor, G.N. and Schonborn, H. (1973) *Surface-Produced Alignment of Liquid Crystals*, (*Proc. IEE* **61**) 823.

Kelker, H. and Hatz, R. (1980) *Handbook of Liquid Crystals*. Verlag Chemie, Weinheim.

Keller, P., Carvalho, B., Cotton, J.P., Lambert, M., Moussa, F. and Pépy, G. (1985) *J. Physique Lett.* **46**, L-1065.

Kirste, R.G. and Ohm, H.G. (1985) *Makromol. Chem., Rapid Commun.* **6**, 179.

Kléman, M. (1985) *Faraday Disc. Chem. Soc.* **79**, 215.

Kofler, L. and Kofler, A. (1954) *Thermomikromethoden*. Verlag Chemie, Weinheim.

Konstantinov, I.I., Amerik, Y.B., Alexandrov, A.I. and Pashkova, T.V. (1984) *Mol. Cryst. Liq. Cryst.* **110**, 121.

Kostromin, S.G., Sinitzyn, V.V., Talroze, R.V., Shibaev, V.P. and Platé, N.A. (1982) *Makromol. Chem. Rapid Commun.* **3**, 809.

Kostromin, S.G., Sinitzyn, V.V., Talroze, R.V. and Shibaev, V.P. (1984) *Polym. Sci. USSR* **26**, 370.

Kunchenko, A.B. and Svetogorsky, D.A. (1986) *J. de Physique* **47**, 137.

Kunchenko, A.B. and Svetogorsky, D.A. (1987) *Liquid Crystals* **2**, 617.

Leadbetter, A.J. (1979) in *The Molecular Physics of Liquid Crystals*, eds. Luckhurst, G.R. and Gray, G.W., Academic Press, New York, Chapter 13.

Le Barny, P., Dubois, J.C., Friedrich, C. and Noël, C. (1986a) *Polym. Bull.* **15**, 341.

Le Barny, P., Ravaux, G., Dubois, J.C., Parneix, J.P., Njeumo, R., Legrand, C. and Levelut, A.M. (1986b) *Molecular and Polymeric Optoelectronic Materials: Fundamentals and Applications*, (Proc. SPIE 682) San Diego.

Lipatov, Y.S., Tsukruk, V.V. and Shibaev, V.P. (1984) *Rev. Macromol. Chem. Phys.* **C24**, 173.

Magagnini, P.L. (1981) *Makromol. Chem. Suppl.* **4**, 223.

Maier, W. and Saupe, A. (1959) *Z. Naturforsch.* **14a**, 882; **15a**, 287.

Mattoussi, H., Veyssié, M., Casagrande, C., Guedeau, M.A. and Finkelmann, H. (1986a) *Mol. Cryst. Liq. Cryst. Bull.* **1**, 254.

Mattoussi, H., Ober, R., Veyssié, M. and Finkelmann, H. (1986b) *Europhys. Lett.* **2**, 233.

Mattoussi, H. Thèse de Doctorat de l'Université Paris VI, Paris, 12 octobre 1987.

Mauguin, C. (1911) *Bull. Soc. Fr. Min* **34**, 71.

Mauzac, M., Hardouin, F., Richard, H., Achard, M.F., Sigaud, G. and Gasparoux, H. (1986) *Eur. Polym. J.* **22**, 137.

Mitchell, G.R., Davis, F.J. and Ashman, A. (1987) *Polymer* **28**, 639.

Moussa, F., Cotton, J.P., Hardouin, F., Keller, P., Lambert, M., Pépy, G., Mauzac, M. and Richard, H. (1987) *J. de Phys.* **48**, 1079.

Müller, K. and Kothe, G. (1985) *Ber. Bunsenges Phys. Chem.* **89**, 1214.

Nachaliel, E., Keller, E., Davidov, D., Zimmermann, H. and Deutsch, M. (1987) *Phys. Rev. Lett.* **58**, 896.

Noël, C., Friedrich, C., Léonard, V., Le Barny, P., Ravaux, G. and Dubois, J.C. (1988) *Makromol. Chem., Macromol. Symp.* (in press).

Nyitrai, K., Cser, F., Lengyel, M., Seyfried, E. and Hardy, Gy. (1977) *Eur. Polym. J.* **13**, 673.
Paleos, C.M., Margomenou-Leonidopoulou, G., Filippakis, S.E. and Malliaris, A. (1982) *J. Polym. Sci., Polym. Chem. Ed.* **20**, 2267.
Platé, N.A. and Shibaev, V.P. (1980) *J. Polym. Sci. Polym. Symp.* **67**, 1.
Platé, N.A., Freidzon, Y.S. and Shibaev, V.P. (1985) *Pure and Appl. Chem.* **57**, 1715.
Portugall, M., Ringsdorf, H. and Zentel, R. (1982) *Makromol. Chem.* **183**, 2311.
Pracella, M., De Petris, S., Frosini, V. and Magagnini, P.L. (1984) *Mol. Cryst. Liq. Cryst.* **113**, 225.
Pschorn, U., Spiess, H.W., Hisgen, B. and Ringsdorf, H. (1986) *Makromol. Chem.* **187**, 2711.
Renz, W. and Warner, M. (1986) *Phys. Rev. Lett.* **56**, 1268.
Richardson, R.M. and Herring, N.J. (1985) *Mol. Cryst. Liq. Cryst.* **123**, 143.
Ringsdorf, H., Schmidt, H.W. and Schneller, A. (1982) *Makromol. Chem. Rapid Commun.* **3**, 745.
Sackmann, H. and Demus, D. (1973) *Mol. Cryst. Liq. Cryst.* **21**, 239.
Samulski, E.T., Gauthier, M.M., Blumstein, R.B. and Blumstein, A. (1984) *Macromolecules* **17**, 479.
Sefton, M.S. and Coles, H.S. (1985) *Polymer* **26**, 1319.
Sefton, M.S., Bowdler, A.R. and Coles, H.S. (1985) *Mol. Cryst. Liq. Cryst.* **129**, 1.
Shibaev, V.P. and Platé, N.A. (1984) *Adv. Polym. Sci.* **60/61**, 173.
Shibaev, V.P. and Platé, N.A., (1985) *Pure Appl. Chem.* **57**, 1589.
Shibaev, V.P., Finkelmann, H., Kharitonov, A.V., Portugall, M., Platé, N. and Ringsdorf, H. (1981) *Polym. Sci. USSR* **23**, 1029.
Shibaev, V.P., Kostromin, S.G. and Platé, N.A. (1982) *Eur. Polym. J.* **18**, 651.
Shibaev, V.P., Kozlovsky, M.V., Beresnev, L.A., Blinov, L.M. and Platé, N.A. (1984) *Polym. Bull.* **12**, 299.
Sigaud, G., Achard, M.F., Hardouin, F., Mauzac, M., Richard, H. and Gasparoux, H. (1987) *Macromolecules* **20**, 578.
Sivardière, J. (1980) *J. de Phys.* **41**, 1081.
Spassky, N., Lacoudre, N., Le Borgne, A., Vairon, J.P., Jun, C.P., Friedrich, C. and Noël, C. (1988) *Makromol. Chem. Macromol. Symp.* (in press).
Spiess, H.W. (1985) *Adv. Polym. Sci.* **66**, 23.
Stevens, H., Rehage, G. and Finkelmann, H. (1984) *Macromolecules* **17**, 851.
Strzelecki, L. and Liebert, L. (1973) *Bull. Soc. Chim.* **2**, 597.
Sutherland, H.H., Basu, S. and Rawas, A (1987) *Mol. Cryst. Liq. Cryst.* **145**, 73.
Talroze, R.V., Kostromin, S.G., Shibaev, V.P. and Platé, N.A. (1981) *Makromol. Chem. Rapid Commun.* **2**, 305.
Talroze, R.V., Sinitzyn, V.V., Shibaev, V.P. and Platé, N.A. (1982) *Mol. Cryst. Liq. Cryst.* **80**, 211.
Toulouse, G. (1977) *J. de Phys. Lett.* **38**, L-67.
Tsukruk, V.V., Shilov, V.V. and Lipatov, Y.S. (1985) *Acta Polym.* **36**, 403.
Viney, C. (1986) *Polym. Eng. Sci.* **26**, 1021.
Wang, X.J. and Warner, M. (1987) *J. Phys. A: Math. Gen.* **20**, 713.
Weill, C., Casagrande, C., Veyssié, M. and Finkelmann, H. (1986) *J. de Phys.* **47**, 887.
Wendorff, J.H. (1978) in *Liquid Crystalline Order in Polymers*, ed. Blumstein, A., Academic Press, New York.
Wunderlich, B., Grebowicz, J. (1984) *Adv. Polym. Sci.* **60/61**, 1.
Wunderlich, B., Möller, M., Grebowicz, J. and Baur, H. (1988) *Adv. Polym. Sci.* (in press).
Zentel, R. and Ringsdorf, H. (1984) *Makromol. Chem. Rapid Commun.* **5**, 393.
Zentel, R. and Strobl, G.R. (1984) *Makromol. Chem.* **185**, 2669.
Zentel, R., Reckert, G. and Reck, B. (1987a) *Liquid Crystals*, **2**, 83.
Zentel, R., Schmidt, G.F., Meyer, J. and Benalia, M. (1987b) *Liquid Crystals* **2**, 651.
Zhou, Q.F., Li, H.M. and Feng, X.D. (1987) *Macromolecules* **20**, 233.
Zugenmaier, P. (1986) *Macromol. Chem., Macromol. Symp.* **2**, 33.
Zugenmaier, P. and Mügge, J. (1984) *Makromol. Chem. Rapid Commun.* **5**, 11.

7 Dielectric relaxation spectroscopy of liquid crystalline side chain polymers

C.M. HAWS and M.G. CLARK, GEC Hirst Research Centre, East Lane, Wembley, Middlesex HA9 7PP, UK
and
G.S. ATTARD, Department of Chemistry, University of Southampton, Southampton S09 5NH, UK

7.1 Introduction and motivation

The purpose of this chapter is to discuss the dielectric and conductivity properties of thermotropic liquid crystal side chain polymers as a function of frequency. Our interest ranges from DC up to microwave frequencies. Our motivation is both scientific and technological.

From a scientific point of view these properties are of interest because they give information on the structure and dynamics of the polymers and their packing in the fluid. We have as an implicit purpose to compare and contrast the polymer scientist's view of dielectric phenomena with that of the liquid crystal researcher. This is because these substances are often seen as embodying the (desirable and undesirable) characteristics of both classes of material. We favour the view that side chain LCPs may be thought of as assemblies of liquid crystal molecules strung together by polymer chains. The fundamental pointer to this view is that a flexible spacer between the mesogenic moiety and the polymer backbone is required for liquid crystalline phases to appear. In agreement with this view, LCP devices show all the phenomena familiar from low-molar-mass LC devices, but, since the polymer chains present a considerable impediment to molecular motions, with radically changed dynamics. We shall see that the dielectric spectra of LCPs may be understood by the same principles.

The technological reasons for interest in the dielectric spectra of LCPs are several. The DC conductivity due to extrinsic carriers is an index of the purity of the material. Purity is always an important factor in device lifetime and reliability. Furthermore, if it is desired to align a liquid crystal polymer by dielectric reorientation in an electric field, the conductivity must not be too great or it will not be possible to maintain the field across the film. Knowledge of the principal values of the permittivity tensor, $\varepsilon_{\parallel}$ and $\varepsilon_{\perp}$ in a nematic phase, is essential for quantifying the aligning action of an electric field. Since, as one might expect, the stringing together of the mesogenic groups by the polymer chain has a slowing effect on molecular motions, the characteristic relaxation frequencies at which the permittivities show decreases are lower in an LCP than in a low-molar-mass (LMM) LC. These relaxations are, of course, accompanied by increases in the dielectric loss, or, equivalently, by AC conductivity. Knowledge of the frequency variation of the dielectric properties is therefore important both for selecting suitable frequencies for electric field realignment and for identifying the scope for dielectric heating or situations in which the dielectric anisotropy $(\varepsilon_{\parallel} - \varepsilon_{\perp})$ changes from positive to negative sign with increasing frequency. This last effect can be exploited by two-frequency switching techniques in exact analogy with LMMLCs (Attard *et al.*, 1987*a*).

There are several well-known texts which adequately explain the principles and the practice of dielectric spectroscopy of isotropic fluids (Daniel, 1967; Böttcher *et al.*, 1973; Böttcher and Bordewijk, 1978). The salient points to be noted are that permittivity and loss phenomena may be described in a unified way by generalizing the permittivity to become a complex function of frequency $\omega/2\pi$:

$$\varepsilon(\omega) = \varepsilon'(\omega) - i\varepsilon''(\omega) \qquad (7.1)$$

where the real part ε' is the 'dielectric constant' and the imaginary part ε'' describes the dielectric loss corresponding to a conductivity

$$\sigma(\omega) = \omega\varepsilon''(\omega) \qquad (7.2)$$

The prototypic form for the frequency dependence due to a relaxation with frequency f_R is the Debye form

$$\varepsilon' = \varepsilon_\infty + (\varepsilon_s - \varepsilon_\infty)/(1 + \omega^2\tau^2) \qquad (7.3)$$

$$\varepsilon'' = (\varepsilon_s - \varepsilon_\infty)\omega\tau/(1 + \omega^2\tau^2) \qquad (7.4)$$

where

$$f_R = (2\pi\tau)^{-1} = \omega_R/2\pi \qquad (7.5)$$

The significance of ε_s, ε_∞, and ω_R is clear from Figure 7.1 which shows both the real part ε' and the loss ε'' plotted against $\ln \omega$, together with a plot of the AC conductivity $\sigma = \omega\varepsilon''$. An alternative representation is the Cole–Cole plot in which ε'' is plotted against ε' with ω as a parameter. In this plot, eqns (7.3) and (7.4) give a semicircle centred on the x-axis. Experimental loss data are often fitted to the Fuoss–Kirkwood equation

$$\cosh^{-1}(\varepsilon''_{max}/\varepsilon'') = \beta \ln(\omega\tau) \qquad (7.6)$$

where the empirical parameter β normally lies between zero and unity, $\beta = 1$ corresponding to the Debye form, eqns (7.3) and (7.4). Note that the decrement in the real part $\varepsilon'_s - \varepsilon'_\infty$ is given by

$$\varepsilon'_s - \varepsilon'_\infty = 2\varepsilon''_{max}/\beta \qquad (7.7)$$

In a liquid crystal the quantities ε', ε'' discussed above are tensors characterized by two principal values $\varepsilon_\parallel$ and $\varepsilon_\perp$ in nematic and smectic A phases, and by three distinct principal values in lower symmetry phases such as tilted smectics. The subscript identifying the principal value refers to the orientation of the measuring electric field relative to the director. Each of the equations (7.1)–(7.7) may be written down independently for each principal component.

Since the relaxation frequencies f_R relate to molecular motions, it is to be expected that they should increase strongly with increasing temperature. Thus in the study of solid polymers it has often been the practice, instead of scanning frequency, to measure at a fixed frequency and varied temperature. The resulting plots of ε'' versus temperature typically contain two or more loss peaks labelled α, β, γ with *decreasing* temperature. The sequence corresponds to *increasing* frequency at fixed temperature. Unfortunately, the most striking characteristic of both low-molar-mass and side chain polymeric liquid crystals is a further relaxation at lower frequency than any in the

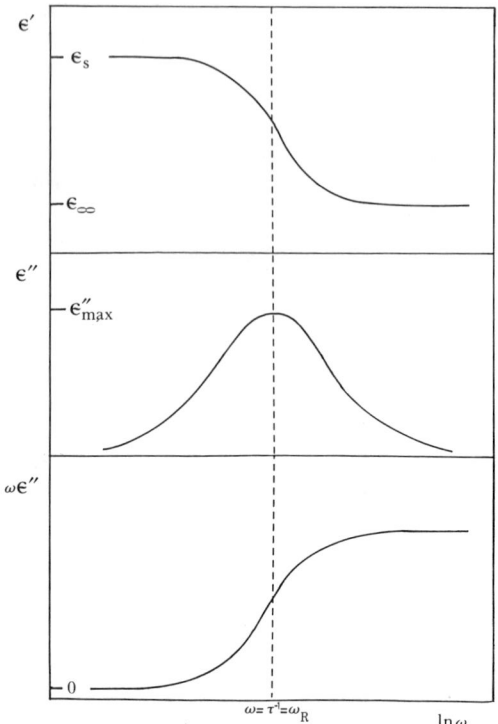

Figure 7.1 Schematic plots of real (ε') and imaginary (ε'') permittivity and AC conductivity, $\sigma = \omega\varepsilon''$, against frequency.

isotropic phase This, perforce, must be labelled δ, effectively destroying such logic as the notation has (see also Boyer, 1978).

Dielectric relaxation spectroscopy is a particular useful technique for the study of side chain LCPs. This is due to the dipolar nature of the usual mesogenic moieties and, in general, to the absence of coexistent amorphous and crystalline phases, a common complication in many polymer systems. The anisotropic permittivity and permeability facilitate alignment by electric and magnetic fields which, together with the ease of substitution of functional groups at various molecular sites, allows ready interpretation of the relaxation mechanisms. Furthermore, the information obtained gives insight into molecular processes which are also responsible for other properties, such as viscoelastic relaxations (Lamb, 1978), which are more difficult or tedious to measure.

7.2 Molecular models of dielectric relaxation

The molecular factors that determine the dielectric properties of LC materials may be understood by use of statistical mechanical theories which model the molecular physics of relaxation phenomena in anisotropic fluids as a basis for the interpretation of experimental data in terms of fundamental molecular properties. In order to set dielectric relaxation into context, we first outline a general approach to relaxation phenomena and to the information they provide on molecular dynamics and statics (Nordio and Segre, 1979a). At the heart of all spectroscopic measurements is an interaction between an external field and the molecules of the sample under investigation. In laboratory-fixed axes this interaction energy H' can be written

$$H' = \sum_{L=0}^{\infty} F^{(L)} \cdot \overline{X^{(L)}}^{*} \tag{7.8}$$

where $F^{(L)}$ is a field tensor of rank L (e.g. $L = 1$ for an electric field) and $X^{(L)}$ is the corresponding material susceptibility. The asterisk denotes complex conjugate and the overbar averaging over the molecular orientational distribution. If the orientational distribution is isotropic all $\overline{X^{(L)}}$ with $L > 0$ vanish. The orientationally averaged material susceptibility can be related to the molecular susceptibility $X^{(L)}_{mol}$ by

$$\overline{X^{(L,n)}} = \sum_{m=-L}^{L} X^{(L,m)}_{mol} \overline{D^{L}_{m,n}} \tag{7.9}$$

where n, m label the $2L + 1$ components of the tensors in a spherical basis, and $D^{L}_{m,n}$ denotes the rotation matrix describing the orientation of the molecule-fixed frame with respect to laboratory axes.

Following removal of the external field the system will return to its unperturbed equilibrium state. The dynamics of this dissipative process are contained in the macroscopic time correlation function $\overline{X^{(L)}(0)^{*} X^{(L)}(t)}$. Within the limits of linear response theory this can be related to molecular properties through the angular autocorrelation functions $g^{L}_{m,n}(t)$:

$$\overline{X^{(L,n)}(0)^{*} X^{(L,n)}(t)} = \sum_{m} X^{(L,m)*}_{mol} X^{(L,m)}_{mol} g^{L}_{m,n}(t) \tag{7.10}$$

where, for simplicity, we have assumed a nematic fluid of rod-like molecules. Fourier transformation of $g^{L}_{m,n}(t)$ yields spectral densities

$$j^{L}_{m,n}(\omega) = \frac{1}{2} \int_{-\infty}^{\infty} g^{L}_{m,n}(t) \exp(-i\omega t) \, dt \tag{7.11}$$

which determine the shapes, amplitudes, and frequency locations of the spectral features observed in experiments such as dielectric spectroscopy or NMR (Nordio and Segre, 1979a, b; see also Chapter 8).

The exact evaluation of the angular correlation function

$$g^{L}_{m,n}(t) = \overline{D^{L*}_{m,n}(\Omega_0) D^{L}_{m,n}(\Omega_t)} - \overline{D^{L*}_{m,n}} \, \overline{D^{L}_{m,n}} \tag{7.12}$$

where Ω_0 and Ω_t denote the Euler angles at time zero and time t, is a formidable task. Progress can be made by approximating $g^{L}_{m,n}(t)$ with a sum of amplitude and temporal factors

$$g^{L}_{m,n}(t) = \sum_{p} A^{L,p}_{m,n} \psi^{L,p}_{m,n}(t) \tag{7.13}$$

This form follows naturally if the molecular reorientation is assumed to be a Markov process obeying a master equation (Nordio and Segre, 1979a).

If the molecular reorientation in a nematic fluid is described by a diffusion model, then for dielectric relaxation ($L = 1$) eqns (7.10) and (7.13) become

$$\overline{\mu_\parallel(0)\mu_\parallel(t)} = \mu_l^2 g_{0,0}^1(t) + \mu_t^2 g_{0,1}^1(t) \tag{7.14}$$

$$\overline{\mu_\perp(0)\mu_\perp(t)} = \mu_l^2 g_{1,0}^1(t) + \mu_t^2 g_{1,1}^1(t) \tag{7.15}$$

$$g_{m,n}^1 = \sum_{p \neq 0} |M_{m,n}^P|^2 \exp(-\alpha_{m,n}^P D_\perp t) \tag{7.16}$$

In these equations $\mu_\parallel$ and $\mu_\perp$ denote electric dipole components parallel and perpendicular to the nematic director, while μ_l and μ_t are the components of the molecular electric dipole moment longitudinal and transverse to the 'long axis' (see Clark, 1985) of the molecule, $D_\perp$ is the transverse component of the molecular rotational diffusion tensor, and the $M_{m,n}^P$ are matrix elements of the rotational diffusion eigenfunctions. The modes (m, n) correspond to processes illustrated in Figure 7.2 as follows:

$$(0, 0) \equiv (\parallel, l) \qquad (0, 1) \equiv (\parallel, t)$$
$$(1, 0) \equiv (\perp, l) \qquad (1, 1) \equiv (\perp, t)$$

Each of these modes is a sum over many exponential processes with characteristic times τ_P where

$$1/\tau_P = \alpha_{m,n}^P D_\perp \tag{7.17}$$

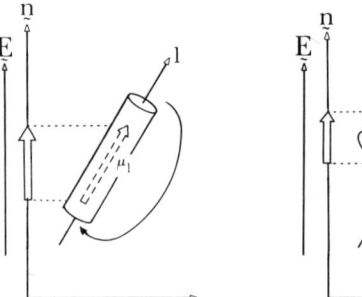
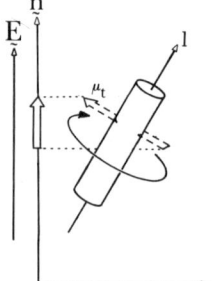
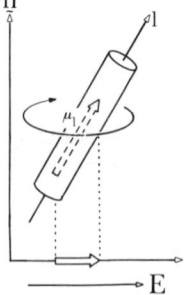
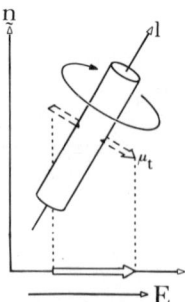

Figure 7.2 Rotational diffusion modes leading to dielectric relaxation in nematic liquids. The molecular 'long axis' is denoted by l.

Figure 7.3 shows a plot of the leading coefficient $\alpha_{m,n}^1$ against the molecular order parameter $\bar{P}_2$ for two values of the anisotropy of the rotational diffusion tensor κ where

$$\kappa = D_\parallel / D_\perp \qquad (7.18)$$

It is seen that for an anisotropic tensor ($\kappa = 11$) there are four distinct values of $\alpha_{m,n}^1$ for $\bar{P}_2 > 0$.

Of particular experimental significance is the observation that the curve for $\alpha_{0,0}^1$ ($\parallel, l$) drops rapidly with increasing $\bar{P}_2$, becoming isolated from the other $\alpha_{m,n}^1$ curves and the curves for $p > 1$. Thus whereas in general terms with $p > 1$ may be significant in eqn (7.16), the leading term alone is a reasonable approximation for $g_{0,0}^1$ leading to a Debye-like form, eqns (7.3) to (7.5), for this relaxation. The low frequency of the ($\parallel, l$) mode for large $\bar{P}_2$ is in accordance with rotational diffusion against the nematic potential (see Figure 7.2). Thus the theory successfully rationalizes the characteristic dielectric spectrum of nematics, namely an exceptionally low-frequency relaxation in $\varepsilon_\parallel$ (associated with $\mu_l \neq 0$) which has an accurately Debye-like form in a highly anisotropic fluid, together with a broader relaxation (being a mixture of a number of terms) which appears in both $\varepsilon_\parallel$ and $\varepsilon_\perp$ provided both μ_l and μ_t are non-zero.

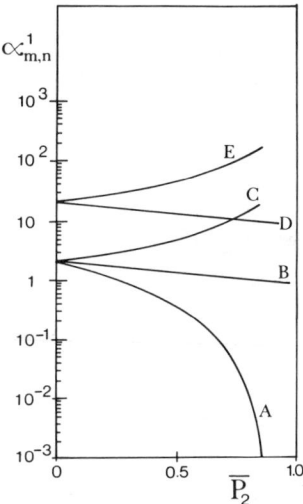

Figure 7.3 Plots of the coefficients $\alpha_{m,n}^1$, eqn (7.17), against the molecular order parameter $\bar{P}_2$ for two values of the anisotropy of the rotational diffusion tensor $D_\parallel/D_\perp = \kappa$.

A	$m, n = 0, 0$	$\kappa = 1, 11$
B	$m, n = 1, 1$	$\kappa = 1$
C	$\begin{cases} m, n = 0, 1 \\ m, n = 1, 0 \end{cases}$	$\begin{matrix} \kappa = 1 \\ \kappa = 1, 11 \end{matrix}$
D	$m, n = 1, 1$	$\kappa = 11$
E	$m, n = 0, 1$	$\kappa = 11$

An alternative, empirical, approach which has been widely used in dielectric theory is to take the right-hand side of eqn (7.13) as a single term with $\psi_{m,n}^{L}(t)$ given by the Kohlrausch–Williams–Watts function.

$$\psi_{m,n}^{L}(t) = \exp\left[-(t/\tau_{m,n}^{L})^{\beta}\right] \qquad (7.19)$$

where $0 < \beta \leqslant 1$. The relationship of this phenomenological function to molecular processes remains a subject for debate in the literature (e.g. Palmer *et al.*, 1984; Shlesinger, 1984; Klafter and Shlesinger, 1986).

It is difficult to establish quantitative relationship between molecular processes and the macroscopic permittivity $\varepsilon(\omega)$ because of the need to allow for the difference between the vacuum field and the actual field experienced by a molecule. This 'internal field effect' arises from the response of the remainder of the material to the applied field. Unfortunately, since $\varepsilon \gg 1$ for condensed matter, these 'corrections' are a substantial fraction of the whole effect and should desirably be integrated into the theory of permittivity in a more satisfactory way than is normally the case. One approach to this has been suggested by Madden and co-workers (Edwards and Madden, 1983; Madden and Kivelson, 1984).

To illustrate qualitatively the molecular factors determining $\varepsilon(\omega)$ we quote the expressions for $\varepsilon_{\parallel}$ and $\varepsilon_{\perp}$, at low frequencies, as given by the simple theory of Maier and Meier (1961):

$$\varepsilon_{\parallel} = 1 + (\rho h F/\varepsilon_0)\{\bar{\alpha} + \tfrac{2}{3}\overline{P_2}\Delta\alpha + (F/3k_BT)[\mu_l^2(1 + 2\overline{P_2}) + \mu_t^2(1 - \overline{P_2})]\} \qquad (7.20)$$

$$\varepsilon_{\perp} = 1 + (\rho h F/\varepsilon_0)\{\bar{\alpha} - \tfrac{1}{3}\overline{P_2}\Delta\alpha + (F/3k_BT)[\mu_l^2(1 - \overline{P_2}) + \mu_t^2(1 + \tfrac{1}{2}\overline{P_2})]\} \qquad (7.21)$$

where $\bar{\alpha}$ and $\Delta\alpha$ are the spherical mean and anisotropy of the molecular polarizability, ρ is the number density of molecules, and h and F are internal field correction factors. It is seen that both molecular polarizability and the molecular dipole contribute to the permittivity of the fluid. However, over the frequency range of interest the dielectric losses are due to electric dipole terms. The terms $\mu_l^2(1 + 2\overline{P_2})$, $\mu_l^2(1 - \overline{P_2})$, $\mu_l^2(1 - \overline{P_2})$, and $\mu_t^2(1 + \tfrac{1}{2}\overline{P_2})$ correspond, respectively, to the modes $(\parallel, l)$, $(\parallel, t)$, $(\perp, l)$, and $(\perp, t)$ identified above. In favourable cases estimates of the dipole components μ_l and μ_t and the molecular order parameter $\overline{P_2}$ can be deduced from the corresponding dielectric decrements (Bone *et al.*, 1984).

The behaviour of mixtures of mesogenic compounds may be broadly generalized as follows (Clark, 1985). Mixtures of chemically similar compounds show a single dielectric spectrum and can apparently be regarded as a single-pseudocomponent fluid. Pranoto *et al.* (1986) have reported a siloxane liquid crystalline copolymer consisting of a mixture of CN-terminated and methoxy-terminated phenyl benzoate side groups which shows a single unsplit δ peak. Mixtures of chemically different components show distinct low-frequency relaxations associated with each type of component. The relaxations are, of course, modified by the presence of the other type of component, but clear instances of this type of behaviour in LMM materials have been documented in, for example, mixtures of diesters with monoesters (Bata and Molnar, 1975), diesters with azoxy compounds (Barnick *et al.*, 1979) and cyanoterphenyls with cyanobiphenyls (Zeller, 1981). The theory has been examined for isotropic liquids by Kivelson and Madden (1983).

Although the preceding molecular field models of relaxation were developed for nematics, they are to a good approximation applicable to dielectric relaxation in smectic A phases. This is because from a molecular point of view the long range spatial periodicity characteristic of smectic A phases plays no part in relaxing the dipole moment, although it will influence the internal field effect. Thus, on passing from the nematic to the S_A phase, broad continuity would be expected with only slight changes in permittivities and relaxation rates, in agreement with experiment. In contradistinction, the lowering of symmetry in tilted smectic phases, particularly the chiral ones, introduces important new features (Durand and Martinot-Lagarde, 1980) which, unfortunately, space does not permit us to describe.

Even the change in dielectric parameters on passing from the mesophase into the isotropic phase can yield information on the underlying anisotropic molecular environment. The use of the temperature dependence of the spherical mean $\bar{\varepsilon}'$ of the low-frequency permittivity ε', and its continuity or discontinuity at the clearing point, as a guide to local molecular correlation, is well documented in the context of LMMLCs (Madhusudana and Chandrasekhar, 1975; Dunmur and Miller, 1980; Bradshaw and Raynes, 1983). Studies of the same kind for side chain LCPs would undoubtedly be of value. Further, as is evident from Figure 7.3, if the anisotropy of the diffusion tensor is sufficiently great, two relaxation frequencies may be observed in the isotropic $(\overline{P_2} = 0)$ phase.

The theories above have been expounded for rigid rod-like particles. Real mesogens, and particularly liquid crystalline side chain polymers, are neither rigid nor rod-like. Fortunately in most materials the major permanent dipoles are coupled to mesogenic moieties which are at least semi-rigid. Allowance for the low symmetry of these moieties results in additional order parameter components which refine the model but do not alter the basic physics of the relaxation processes. Thus we can use the models described in this section in the interpretation of the dielectric spectra of polymeric side chain mesogens.

Theories of dielectric relaxation in classical polymers (Böttcher and Bordewijk, 1978, section 7.4) focus on the cooperative motions of permanent dipoles associated with sites on the chain. Such sites may be of three kinds: dipoles perpendicular to the chain, dipoles parallel to the chain, and 'side chain' dipoles, which in this context means, for example, the dipole associated with the MeOOC moiety in poly(methylmethacrylate). For a chain of n units the theory seeks to calculate the quantity

$$\sum_{i=1}^{n} \sum_{j=1}^{n} \overline{\boldsymbol{\mu}_i(0) \cdot \boldsymbol{\mu}_j(t)}$$

This is a difficult task which can presently only be pushed to a conclusion at the price of simplifying assumptions. Such model calculations do rationalize the experimental observation of broad relaxations explicable by distributions of relaxation times (Böttcher and Bordewijk, 1978; Blythe, 1979; Block, 1979, and references therein).

7.3 Experimental aspects of dielectric relaxation spectroscopy

The dielectric properties of a material depend on both frequency and temperature, particularly in the neighbourhood of phase transitions. Accordingly, in order to characterize completely the dielectric properties of materials such as liquid crystalline polymers, we need to map the dielectric response at thermal equilibrium throughout the range of frequency and temperature over which the system is dielectrically active. The experimental data can be described by a surface in either $(\varepsilon', \log_{10} v, T)$ space or

$(\varepsilon'', \log_{10} v, T)$ space, as illustrated in Figure 7.4 (adapted from Araki *et al.*, 1988). We refer to this as the dielectric response surface of a material. It follows from the Kramers–Kronig relations that the complete shape of the dielectric response surface encodes the entire dielectric behaviour of a given sample.

In order to obtain a fully detailed dielectric response surface of liquid crystalline polymers, measurements must be made over the frequency range 10^{-4}–10^{8} Hz, and over temperatures ranging from the glass transition region up to, and including, the isotropic fluid.

Dielectric relaxation spectra over the frequency range 10–10^{8} Hz can be obtained by using one or a combination of several commercially available impedance analysers. Many of these can be interfaced with desktop computers, making it possible to acquire large numbers of data points rapidly, reliably and reproducibly. The sample is treated as a capacitor C_p with a conductance G_p in parallel. Although many impedance analysers claim to correct for lead effects if standard-length cables are used, it is essential above 10^{5} Hz, and often convenient below, to model the leads by introducing into the equivalent circuit an inductor L_s and a resistor R_s in series with the sample. The capacitance C and conductance G measured by the impedance analyser in parallel mode are related to the equivalent circuit parameters by

$$G/(G^2 + \omega^2 C^2) = G_p/(G_p^2 + \omega^2 C_p^2) + R_s \tag{7.22}$$

$$C/(G^2 + \omega^2 C^2) = C_p/(G_p^2 + \omega^2 C_p^2) - L_s \tag{7.23}$$

Plots of the left-hand sides of eqns (7.22) and (7.23) for an empty cell ($G_p = 0$) against ω and $1/\omega^2$, respectively, yield straight lines whose intercepts on the y-axis give R_s and L_s,

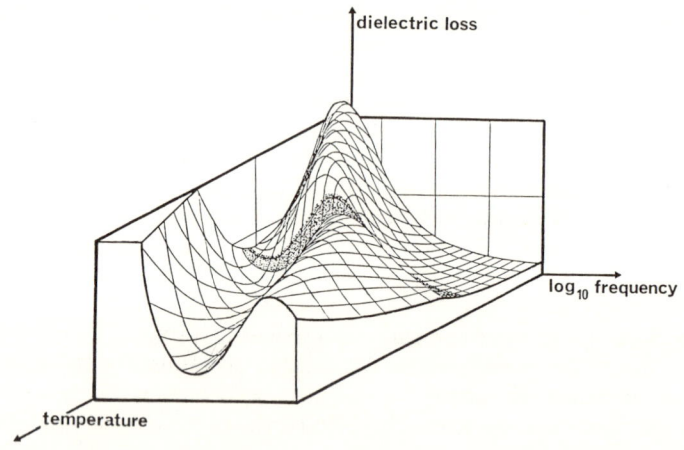

Figure 7.4 Dielectric response surface $(\varepsilon'', \log_{10} v, T)$ for the homopolymer —[SiO(CH$_3$)-((CH$_2$)$_6$OX)]$_n$— where $n \approx 35$ and X is

$$\text{—}\langle\bigcirc\rangle\text{—CO}_2\text{—}\langle\bigcirc\rangle\text{—CN}$$
$$\underset{\text{Me}}{}$$

(Shaded region indicates biphasic range at the transition).

respectively, with the slope of eqn (7.23) giving $1/C_p$ for the empty cell. Data points at low frequency should be avoided when fitting the straight lines, since for an empty cell the cell impedance will be too high. The values obtained for R_s and L_s are cross-checked by measurements on standard liquids. Another way of investigating dielectric response is by using direct-current step-response measurements (Daniel, 1967). Although in principle DC response measurements can provide information on dielectric properties over a very broad range of frequencies (10^{-4}–10^6 Hz), the experiments are often difficult to automate and so tend to be employed to study the very low-frequency regime ($< 10^{-1}$ Hz) which is inaccessible to other measurement techniques. One should, of course, also bear in mind the undersirability of repeatedly applying DC voltage to any liquid crystalline sample, since it may lead to electrochemical degradation.

The cell geometry most commonly employed in the dielectric studies of liquid crystalline materials is that of the parallel plate capacitor with the sample sandwiched between two electrodes. The capacitance for such a system is proportional to the electrode area and inversely proportional to the inter-electrode separation. The signal-to-noise ratio of the dielectric spectra may therefore be improved by decreasing the inter-electrode separation and/or increasing the area of the electrodes. Since it is more difficult to ensure that large-area electrodes are defect-free and parallel throughout, most dielectric cells consist of electrodes with areas $\leqslant 1 \, \text{cm}^2$. Constant sample dimensions can be maintained by employing a variety of spacer materials of known thickness. For example, PTFE films, Mylar films, chopped glass fibres and even aluminium foil, have been used to achieve sample thickness ranging from under $10 \, \mu\text{m}$ to $120 \, \mu\text{m}$. The measuring voltage applied should be less than about 0.2 V to avoid any possibility of dielectric reorientation of the director. Fortunately, since the Freedericksz threshold voltages (Blinov, 1983) are independent of cell thickness, the likelihood of the measuring voltage inducing reorientation is at least approximately independent of the inter-electrode separation.

Although useful results can be obtained using simple two-terminal glass cells with (low resistivity) ITO electrodes, guard electrodes may be used to give greater accuracy. These electrode configurations can be realized either as etched ITO surfaces on glass or as spring-loaded metal electrodes mounted in an earthed cell. The empirical correction of data from simple two-electrode cells for fringing-field effects has been discussed (in the context of LMM materials) by Clark et al. (1980). A difficulty with guard-electrodes cells is the precise alignment of the top and bottom electrodes relative to each other. Generally two-electrode cells are acceptable for routine studies. Glass cells have the advantage that it is possible to inspect the samples visually and under the polarizing microscope, the importance of which cannot be overemphasized.

Cells in which the top and bottom electrodes are permanently bonded in position are ideally suited for high-precision measurements since they can be calibrated by use of materials of known permittivity. However, changes in the electrode separation on filling can be a problem with glass cells, and with all demountable cells. Permanently bonded cells have to be filled by capillary action (possibly vacuum-assisted) at a high temperature with the LCP in its isotropic phase. This should be performed in a vacuum oven to reduce the possibility of chemical degradation or cross-linking of the polymer while it is hot. For smaller inter-electrode separations, filling can be remarkably time-consuming compared with the seconds or minutes required for LMMLCs, and the filling times of hours or days reported in the literature must raise concerns of sample degradation, particularly if a vacuum oven is not used. Polysiloxanes are better than polyacrylates for ease of filling, Haws *et al.* (1987) reporting a filling time of only 20 minutes for siloxane copolymers in 25 μm cells with total area > 1 cm². An alternative method of sample containment particularly suited to viscous liquids is to squeeze the sample between two separable electrodes, while taking care to avoid trapping air bubbles. The need for accurate and reproducible temperature control must be taken into account when designing cells and the leads to them, although LCPs have the advantage relative to LMMLCs that imposed alignments are retained, with the result that there is less often a need to make measurements in the presence of an aligning electric or magnetic field.

Samples intended for dielectric studies should be dry and thoroughly degassed. This treatment is particularly important if high voltages are to be applied, for instance for alignment, since it minimizes the chances of dielectric breakdown.

It will be evident from the discussion of section 7.2 that the dielectric spectra of macroscopically aligned mesophases have a simpler structure than the spectra of non-aligned mesophases. Director alignment in monomeric mesogens is commonly induced either by surface alignment treatments applied to the cell walls or by applying electric or magnetic fields to the mesophase. Although similar, albeit much slower, responses to electric fields have been observed for some side chain polymers, these materials cannot conveniently be aligned when they are in their liquid crystalline phases. In order to induce director alignment in side chain LCPs, the aligning field must be applied to the sample in its isotropic phase followed by cooling, often rather slowly, with the field applied, to a temperature at which the material is liquid crystalline. The cooling rate used in these thermal cycles can be crucial to the formation of well-aligned mesophases. As is described in greater detail in Chapter 11, an electric field can be used to two-frequency address certain materials. This leads to the formation of either homeotropically aligned mesophases or of planarly aligned mesophases depending on the frequency of the aligning field. Although there have been several reports of surface-induced alignment of LCPs after prolonged annealing (cf. the preceding chapter), surface treatment does not always lead to alignment and hence this method is not used in the routine preparation of well-aligned polymers.

7.4 Analysis and interpretation of dielectric spectra

One of the major difficulties encountered in the analysis and interpretation of the dielectric spectra of LCPs is that the various relaxation modes shown in Figure 7.2 often occur closely spaced in the frequency domain. Furthermore, as noted in section 7.2, the higher-frequency modes appear always to coalesce into broad bands. As a consequence the loss spectra observed experimentally are often broad, highly asymmetric, and without clearly resolvable features. If such spectra are to yield information they must be resolved into their various components, which through the models developed in section 7.2 can be linked directly to molecular processes.

Although the choice of line-shape function for fitting dielectric spectra is to some extent arbitrary, and a variety of functional forms have been proposed (Daniel, 1967; Böttcher and Bordewijk, 1978), the most common procedure is to fit the loss curves with the Fuoss–Kirkwood function, eqn (7.6), which we rewrite as

$$L(v) = \varepsilon° \operatorname{sech} \left[\beta \ln \left(v / v° \right) \right] \qquad (7.24)$$

where ε° is the amplitude of maximum loss, β ($0 < \beta \leqslant 1$) defines the width of the absorption (on a logarithmic frequency scale, width at half-height is $\log(7 + 4\sqrt{3})/\beta$ which equals $1.144/\beta$ for $\log_{10}$) and v° is the frequency of maximum loss. Since the spectrum due to n loss curves is simply

$$L_{\text{tot}}(v) = \sum_{i=1}^{n} \varepsilon_i^\circ \, \text{sech} \, [\beta_i \ln(v/v_i^\circ)] \tag{7.25}$$

experimental spectra can be fitted with eqn (7.25) by minimizing the sum of squares

$$S = \sum_j |L_{\text{exp}}(v_j) - L_{\text{tot}}(v_j)|^2 \tag{7.26}$$

where the subscript refers to the jth digitized frequency point and $L_{\text{exp}}(v_j)$ is the experimental loss at frequency v_j. Although the minimization of S can be achieved through a non-linear least-squares fitting algorithm, it is found that the most efficient method of data fitting involves the use of an interactive programme enabling the operator to direct the fitting of particular portions of the loss spectrum. This is a consequence of the particular form of unresolved loss spectra leading to a shallow global minimum for the convergence surface.

Since in principle we do not know *a priori* how many constituent loss curves are contained in a given experimental spectrum, it must be fitted with the minimum number of component curves. From examples such as those discussed in the following sections it is apparent that the experimental spectra of many side chain LCPs can be adequately fitted by the superposition of only two loss peaks plus a conductivity curve (section 7.6). Figure 7.5 shows an example. The lower-frequency loss curve typically has β close to unity (> 0.85), which corresponds to a process having a single relaxation time. By contrast, the higher-frequency peak is found to be quite broad with β much less (0.45–0.25). As noted in section 7.1, these component curves are often labelled δ and α

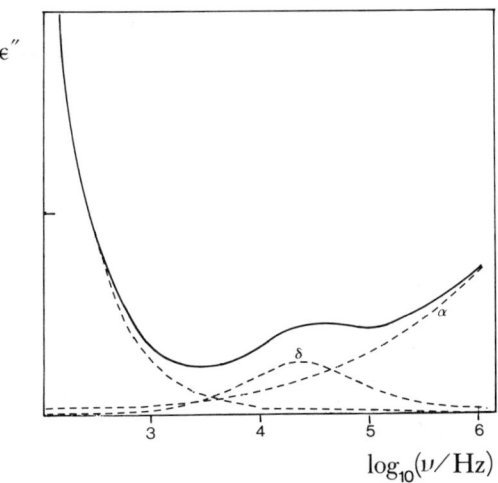

Figure 7.5 Dielectric loss data for polymer GN3/14 (Table 7.1) at 56°C fitted with a conductivity curve and two Fuoss–Kirkwood curves as described in the text.

respectively. Based on our discussion in section 7.2, the δ peak is identified with the $(\parallel, l)$ process. The α peak is likewise identified as the appropriate combination of the other processes $[(\parallel, t), (\perp, l)$ and $(\perp, t)]$ depending on the degree and nature of the director alignment. These assignments can be verified by studies of aligned samples. Since the δ peak is associated solely with $\varepsilon_{\parallel}$, an increase in the degree of homeotropic alignment will enhance this peak, whereas an increase in the degree of homogeneous alignment (a magnetic field can be used) will suppress it. Conversely, the α peak will be enhanced by homogeneous and decreased by homeotropic alignment, although it will be present in both provided that the side chain dipole has a transverse component. A sensitive fitting procedure greatly enhances the value of dielectric spectroscopy as a probe of effects such as phase separation. For example, in the polymers GN4/11 and GN4/16 (Table 7.1) Fuoss–Kirkwood fitting revealed a splitting of the δ loss into two peaks which we correlated with phase separation observed under the polarizing microscope.

Table 7.1 Structures and phase transitions of some siloxane LCPs cited as examples in other tables and the text; $a:b = 21:19$, $\overline{DP} \approx 35$

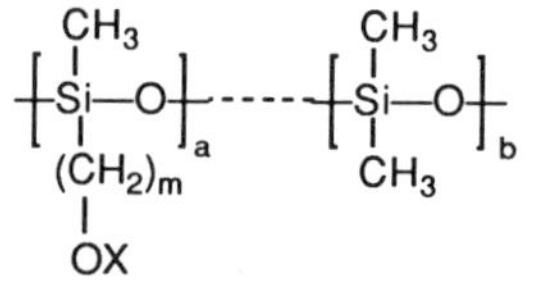

Code	m	X	Phase transitions (°C)[†]
GN3/3	5	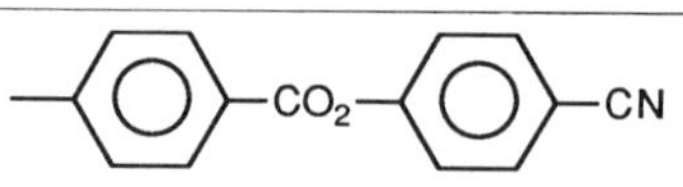	g-9K5S$_A$97[37]i
GN3/14	6	—⟨O⟩-CO$_2$-⟨O⟩-CN	g-125S$_A$85[22]i
GN3/15[‡]	8		g5S$_A$95[8]i
GN3/16	8	—⟨O⟩-CO$_2$-⟨O⟩-CN, Me	g-15S$_A$57[14]i
GN3/17[‡]	6		g-9S$_C^*$76[19]i
GN3/18	6	—⟨O⟩-CO$_2$-⟨O⟩-CO$_2$CH$_2$CH(Me)(Et)	g-25S$_A$30[17]i
GN3/19	5	—⟨O⟩-⟨O⟩-CN	g-14S$_A$95[20]i
GN3/22	6	—⟨O⟩-CO$_2$-⟨O⟩-OCH$_3$	g-15N52[10]i
GN4/11	6	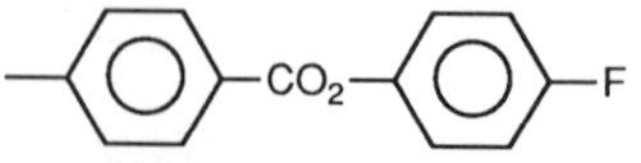 —⟨O⟩-CO$_2$-⟨O⟩-F	K39S$_A$72[33]i
GN4/16	6	—⟨O⟩-CO$_2$-⟨O⟩-CF$_3$	g3K75S$_A$110[46]i

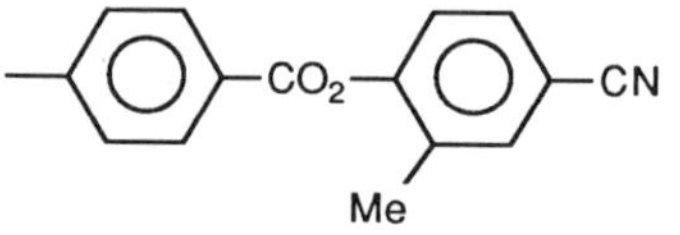

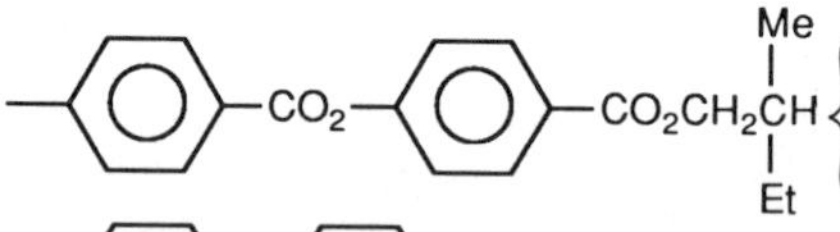

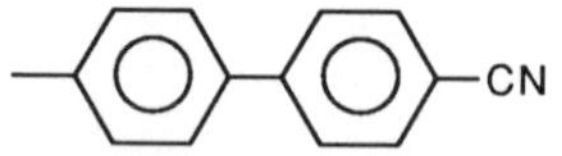

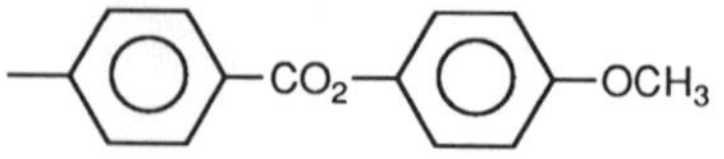

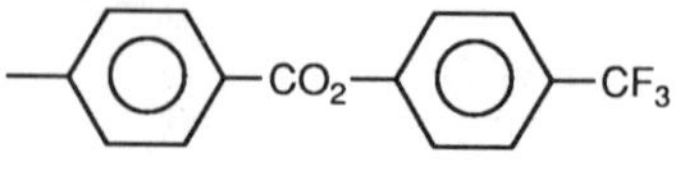

Table 7.1 (*Contd.*)

Code	m	X	Phase transitions (°C)[†]
GN4/17	5	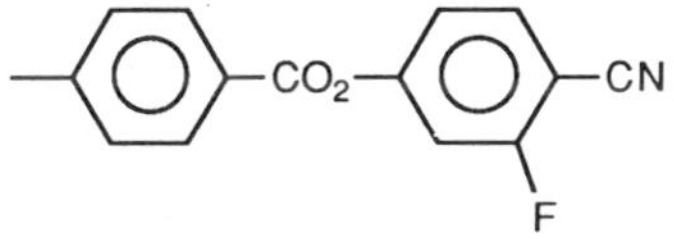	$g\text{-}14S_A80[19]i$
GN4/18	5	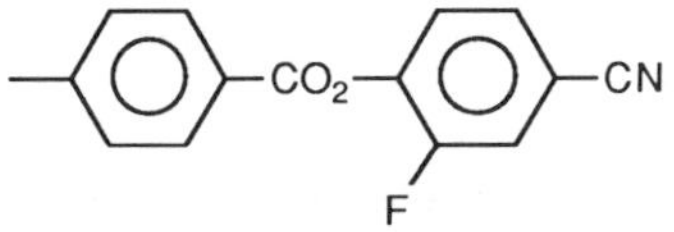	$g\text{-}4K13S_A65[17]i$
GN4/19	6	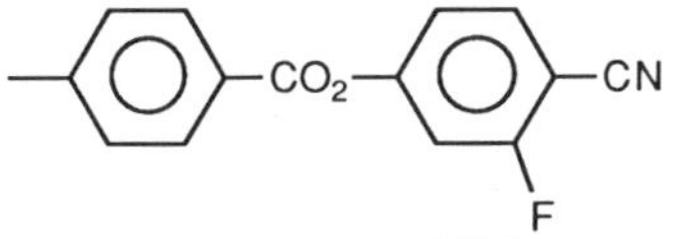	$g\text{-}19S_A78[15]i$
GN3/39*	8	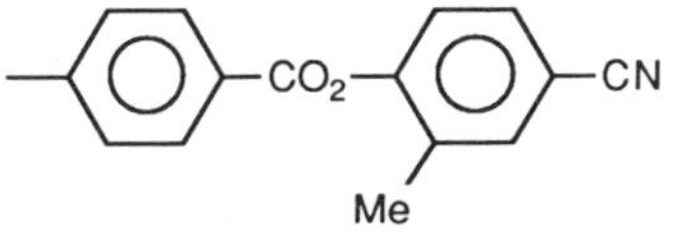	$g\text{-}23i$
GN3/40[‡]	8		$g1S_A80[13]i$

[†]All transitions determined by DSC at $10°C\,min^{-1}$. Mesophase to isotropic transition shown as $T_c[T_2 - T_1]$ where T_c is the maximum in the DSC trace and $T_2 - T_1$ the width of the biphasic region.
[‡]Homopolymer with formula $-[SiO(Y)((CH_2)_mOX)]_n-$ where $y = Me$ (GN3/15) or Et (GN3/40) with $n \approx 35$ or Y = Me (GN3/17) with $n \approx 40$.
*Copolymer with formula $-[SiO(CH_3)((CH_2)_mOX)]_a[SiO(CH_3)((CH_2)_3CN)]_b-$

The majority of dielectric measurements of side chain LCPs are made on non-aligned or partly aligned samples. The resulting spectra are then a superposition of the principal spectra $\varepsilon_\parallel(\omega)$ and $\varepsilon''_\perp(\omega)$ which evidently contain information on the degree of director misalignment. It is thus possible to use dielectric relaxation spectroscopy as a non-optical method of monitoring director alignment (Attard and Williams, 1986a) and to assess the likely thermal stability of information thermo-optically stored on LCP films (Haws et al., 1987). Unfortunately, its quantitative interpretation is difficult. The spatial variation of the director $n(r)$ gives a dielectrically non-uniform sample because of the anisotropy of ε. However, the problems of an inhomogeneous dielectric and its magnetic analogue are of long standing (e.g. Maxwell, 1904; Böttcher and Bordewijk, 1978, section 98; Hashin and Shtrikman, 1962; Boyd, 1983). All that can be done in general is to give bounds for the sample permittivity. More complete a priori knowledge of the director distribution $n(r)$ leads to tighter bounds. The simplest approach, giving widest bounds, is to assume either uniform electric field E or uniform displacement D both taken to be normal to the sample film, to give, respectively

$$\varepsilon_U = (1/d) \int \varepsilon(r)\, d^3r \qquad (7.27)$$

and

$$(1/\varepsilon_L) = (1/d) \int [1/\varepsilon(r)] \, d^3r \tag{7.28}$$

where the integrals are taken over a unit area of sample film of thickness d, U = upper, L = lower, and

$$\varepsilon(r) = \varepsilon_\| \cos^2 \theta + \varepsilon_\perp \sin^2 \theta \tag{7.29}$$

for a nematic, where θ is the angle between the director and E or D. In the case that $n(r)$ is a function only of the coordinate z perpendicular to the sample film, then the problem is identical to that of a laminar LCD (Deuling, 1972) and the lower bound, eqn (7.28), applies exactly.

Misalignment of the director may be described by the ordering tensor (de Gennes, 1974):

$$q_{\alpha\beta} = \tfrac{1}{2}(3n_\alpha n_\beta - \delta_{\alpha\beta}) \tag{7.30}$$

Equation (7.29) then becomes

$$\varepsilon(r) = \tfrac{1}{3}\varepsilon_\|(1 + 2q_{zz}) + \tfrac{2}{3}\varepsilon_\perp(1 - q_{zz}) \tag{7.31}$$

and eqn (7.27) may be written

$$\varepsilon_U = \tfrac{1}{3}\varepsilon_\|(1 + 2\overline{q_{zz}}) + \tfrac{2}{3}\varepsilon_\perp(1 - \overline{q_{zz}}) \tag{7.32}$$

where $\overline{q_{zz}}$ is the *director* order parameter, a macroscopic quantity which must be carefully distinguished from the (microscopic) *molecular* order parameter $\overline{P_2}$.

In the upper bound limit the sum of the dielectric decrements, eqn (7.7), associated with the δ and α peaks takes a simple form. By use of eqns (7.20) and (7.21), and ignoring any change in the internal field factors between the δ and α peaks, we obtain

$$\Delta\varepsilon_\delta + \Delta\varepsilon_\alpha = (\rho h F^2/3\varepsilon_0 k_B T) \left[\mu_l^2(1 + 2\overline{q_{zz}\,P_2}) + \mu_t^2(1 - \overline{q_{zz}\,P_2}) \right] \tag{7.33}$$

This is mathematically equivalent to homeotropic alignment with an *effective* order parameter $\overline{q_{zz}\,P_2}$. Since both $\overline{q_{zz}}$ and $\overline{P_2}$ lie in the range $(-\tfrac{1}{2}, 1)$, the effective order parameter will be closer to zero than either.

Unfortunately the lower bound limit, eqn (7.28) is not susceptible to such a simple closed-form treatment unless we assume that θ takes only the values 0, π and $\pm\tfrac{1}{2}\pi$. Then eqn (7.28) becomes

$$(1/\varepsilon_L) = (1 + 2\overline{q_{zz}})/3\varepsilon_\| + 2(1 - \overline{q_{zz}})/3\varepsilon_\perp \tag{7.34}$$

which rearranges to

$$\varepsilon_L = 3\varepsilon_\|\varepsilon_\perp/[2\varepsilon_\|(1 - \overline{q_{zz}}) + \varepsilon_\perp(1 + 2\overline{q_{zz}})] \tag{7.35}$$

from which rather cumbersome expressions for $\Delta\varepsilon_\delta$ and $\Delta\varepsilon_\alpha$ can be derived. Following a suggestion by Clark, eqn (7.34) has been studied by Attard *et al.* (1987b).

The presence of conductivity, whether AC or DC, would complicate the above equations since assumptions of uniform E or uniform D are then unphysical. The generalization of eqn (7.28) in the context of a laminar LCD has been given independently by Thurston (1984) and Clark (1984, 1985).

Disordering of the director also causes problems when determining the absolute values of $\varepsilon_\parallel$ and $\varepsilon_\perp$. These values are needed in order to quantify the interaction of the sample with an electric field. In LMM materials it is usually possible, at least for nematics, to obtain complete alignment by use of an appropriate surface treatment or external field. Techniques exist (Clark et al., 1980) for extrapolating the change in permittivity with increasing field to infinite field (and hence complete alignment). However, in an LCP, since the response to an aligning field is dominated by kinetic considerations (high viscosity) rather than static ones (stiffness constants) the development of rigorously-based procedures for extrapolating to complete alignment is less straightforward. Although useful information can obviously be obtained by studying the dependence of the measured permittivity on the field strength used during the alignment procedure described above, the difficulty of validating a purely empirical extrapolation to perfect alignment should be emphasized. Although complete disappearance of the δ relaxation can be used as an indication of complete perpendicular ($\varepsilon_\perp$) alignment, no similar indicator exists for parallel alignment.

For a polymer in which $\varepsilon_\parallel > \varepsilon_\perp$ at low frequencies, i.e. $\mu_l > \mu_t$ according to eqns (7.20) and (7.21), there normally exists a frequency within the δ relaxation at which $\varepsilon_\parallel = \varepsilon_\perp$. Provided this 'cross-over frequency' is above the range in which low-frequency conductivity effects occur (section 7.6) it follows from the fact that the δ relaxation does not affect $\varepsilon_\perp$ that, independently of the degree of misalignment, the value of the permittivity at the cross-over frequency is an accurate measure of $\varepsilon_\perp$ (Haws et al., 1987). This method is a simple and accurate one for determining $\varepsilon_\perp$ at frequencies below the α relaxation. Table 7.2 lists a number of values of $\varepsilon_\perp$ obtained by this method. These values are consistent with $\varepsilon_\perp$ values for LMM materials containing the same mesogenic groups when measurement temperatures are normalized as a fraction of the appropriate T_c. If the spherical mean $\bar{\varepsilon} = \frac{1}{3}(\varepsilon_\parallel + 2\varepsilon_\perp)$ can also be determined, either from completely randomized samples or by extrapolation from the isotropic phase, both principal permittivities can then be determined without any recourse to aligned samples.

In the empirical study of materials comparison of the physical properties of different materials is an important technique. It is essential that comparisons be performed on the proper normalized basis. This might be, for example, at the same absolute temperature or the same temperature relative to the glass transition. Ideally, one would like to uncover universal behaviour, i.e. a basis for comparisons in which data for all compounds fit on to a single universal curve. For the static aspects of dielectric

Table 7.2 Values of $\varepsilon_\perp$ obtained from permittivity measurements at the cross-over frequency

Code (Table 7.1)	Temperature (°C)	$\varepsilon_\perp$
GN3/3	38.5 (1.18T_g)	7.3
	67.4 (0.92T_c)	7.0
GN3/14	40 (0.87T_c)	7.3
GN3/19	40 (0.85T_c)	4.7
GN4/19	50 (0.92T_c)	9.6
GN4/33[†]	56.2 (1.18T_g)	13.5
	122.6 (0.92T_c)	11.5

[†]Homopolymer corresponding to GN4/17

H

behaviour ($\varepsilon_\parallel, \varepsilon_\perp$ as functions of frequency) eqns (7.20) and (7.21) suggest that $T\varepsilon_\parallel$ or $T\varepsilon_\perp$ for different compounds should be compared at equal $\overline{P_2}$ provided, as is invariably the case, the permanent dipole contributions dominate. Following mean field theory it is generally presumed that equal T/T_c, where T_c is the absolute clearing temperature, is an adequate basis for comparisons at equal $\overline{P_2}$.

Unfortunately the situation is not so simple for the truly dynamic quantity, namely the relaxation frequency f_R. For the δ relaxation it seems reasonable, building upon the arguments of Meier and Saupe (1966), to write f_R as the quotient of an underlying relaxation frequency f_D divided by a retardation factor g. The latter is reasonably taken as a function of T/T_c. Here we modify the remarks of Clark (1985) for LMM nematics by noting the weight of empirical evidence that, for both LMM and polymer mesophases, the underlying relaxation frequency f_D has Vogel–Fulcher form:

$$f_D = f_D^0 \exp[-A/(T-T_0)] \tag{7.36}$$

where T_0 is a temperature related to the glass transition temperature T_g. The prefactor f_D^0 may also be a (polynomial) function of temperature. Based on these arguments we propose the empirical form

$$f_R \approx f_D(T-T_0)/g(T/T_c) \tag{7.37}$$

The complications of performing comparisons in a two-dimensional space $(T-T_0, T/T_c)$ will be evident. Further, one often resorts to using T_g in the Vogel–Fulcher form to avoid attempting to determine the actual T_0.

However, it is worth remarking that the Vogel–Fulcher law is one of the considerations connecting dielectric behaviour to other kinetic properties such as viscosity, since these, too, tend to obey the Vogel–Fulcher form. A second similar, and possibly related (Palmer et al., 1984), connection is the Kohlrausch–Williams–Watts function, eqn (7.19), which has found application to a wide variety of relaxation phenomena. The connection with Doolittle's free-volume law is also worthy of note (Zeller, 1982). Indeed, it is productive to compare the kinetic properties of liquid crystals (polymeric or otherwise) with those of supercooled liquids (Harrison, 1976), both types of liquid being usefully thought of as liquids with rather limited free volumes. As examples of experimental studies specifically comparing different relaxation phenomena we cite Wetton et al. (1986), who compared dynamic mechanical analysis and dielectric analysis (both using thermal scanning, see section 7.1) for polymers and other materials, and Barlow and Erginsav (1974) who studied supercooled benzyl benzoate, a compound closely related to an important class of LMM mesogens. Figure 8.17 in Chapter 8 of this book compares activation energies determined from dielectric loss with those determined from ^{2}H NMR for two side chain LCPs.

In molecular terms, the above discussion concerning the scaling of $T\varepsilon(\omega)$ and f_R highlights the hybrid nature of side chain LCPs, which exhibit both a weakly first-order orientational order/disorder transition (mesophase–isotropic transition) and a second-order segmental motion freezing transition (mesophase–glass transition). Our proposition is that physical properties which are primarily determined by the orientationally anisotropic molecular interactions should be scaled by T_c, whereas properties primarily related to cooperative motions which freeze at the glass transition should be scaled relative to T_g.

Finally, it is worth noting that in real systems permanent dipoles may be located within different subunits of the macromolecule which are connected by non-rigid segments, as for example in acrylate and methacrylate LCPs. In such systems additional relaxation modes may become operative. Zentel *et al.* (1985) report five relaxation modes for polyacrylate LCPs, labelled δ, α, β, γ_1 and γ_2 in ascending order of frequency. In the case of polysiloxane LCPs, only the δ and α modes have normally been observed.

7.5 Structural influences on the dielectric properties of liquid crystalline polymers

Side chain polymers are constructed from three types of structural unit, namely the backbone, the side chain group, which may be mesogenic or non-mesogenic, and the spacer group which joins a mesogenic group to the backbone. The length of this spacer strongly influences the phase diagram (Finkelmann *et al.*, 1978; Gemmel *et al.*, 1985). A mesogenic group may be further subdivided into an anisometric core consisting of rings and bridging groups, and terminal or lateral substituents. In this section we examine the relationship between these structural features and the dielectric spectra.

7.5.1 *Mesogenic side chain group*

It is the interactions between mesogenic groups which lead to the formation of liquid crystal phases. Furthermore, the mesogenic group is a major location for permanent electric dipoles, particularly if polar substituents are attached to the core. The structure of the mesogenic group is reflected not only in long-range nematic or smectic order but also, to a varying degree, in short-range correlations. Antiparallel correlation associated with, for example, cyanobiphenyl mesogens is well understood for LMMLCs. Its effects are an apparent decrease in the molecular dipole and, as discussed in section 7.2, characteristic temperature-dependence of the mean permittivity. Haase *et al.* (1985) and Parneix *et al.* (1987) have reported data on polyacrylates, indicating that similar effects occur in LCPs. Haws *et al.* (1987) have pointed out that the core overlap of the cyanobiphenyl group in a polysiloxane LCP is identical to that in the corresponding monomeric nematic, and that such correlations, by in effect 'bridging' between chains or different parts of the same chain, may induce inter- or intra-chain correlations.

Tables 7.1 and 7.3 show examples we shall use to illustrate the influence of the mesogenic group on dielectric properties. Table 7.1 displays their structures and phase transitions. It is seen that the terminal and lateral substituents are usually rigidly attached extensions to the core and frequently are the source of much of the permanent dipole associated with the mesogenic group. Although most of the examples bear the polar cyano terminal substituent, GN3/22 carries the much less polar methoxy group. The effect of substituents on the range and type of mesophases formed is noteworthy.

In an attempt to unravel the relative importance of chain dynamics and mesogen dynamics in determining f_R – cf. section 7.4, especially eqn (7.37)—we give in Table 7.3(a) data on f_R for both $0.92T_c$ and $1.18T_g$. It is clear that, with the latter scaling, something close to universal behaviour appears, indicating that at least well below T_c (over 20°C below for all data in the table) the glass-forming dynamics dominate. The observed universality over a range of mesogenic groups, spacer lengths, and phase sequences, including both copolymers and homopolymers, suggests that these dynamics are determined by transitions between different conformations of the polymer backbone. In terms of eqns (7.36) and (7.37) the observed universality suggests, simplistically, that for each relaxation mode the parameter A in eqn (7.36) is

Table 7.3 Dielectric data illustrating the influence of the mesogenic side chain group
(a) Relaxation frequencies

		0.92T_C			1.18T_g	
Code	Temp.(°C)	$\log_{10}(f_R^\delta)$(Hz)	$\log_{10}(f_R^\alpha)$(Hz)	Temp.(°C)	$\log_{10}(f_R^\delta)$(Hz)	$\log_{10}(f_R^\alpha)$(Hz)
GN3/3	67	4.75	—	38	3.39	5.68
GN3/14	56	4.33	6.93	35	3.29	5.96
GN3/15	65	3.90	5.05	55	3.23	4.63
GN3/16	30	3.3	4.92	31	3.3	4.92
GN3/19	65	4.76	—	31	3.42	5.65
GN3/22	26	2.95	—	31	—	—
GN4/17	51	4.17	6.03	30	3.02	5.11
GN4/18	40	3.06	5.36	45	3.31	5.63
GN4/19	50	4.32	6.34	25	3.08	5.14
GN3/40	50	2.85	4.27	50	2.85	4.27

(b) Temperature-scaled dielectric decrements[†] and Arrhenius activation energies

	0.92T_c		1.18T_g		E_A^δ	E_A^α
Code	$T\Delta\varepsilon_\delta$(K)	$T\Delta\varepsilon_\alpha$(K)	$T\Delta\varepsilon_\delta$(K)	$T\Delta\varepsilon_\alpha$(K)	(kJ mol^{-1})	(kJ mol^{-1})
GN3/3	602	802	613	837	106	—
GN3/14	479	895	472	930	94	91
GN3/19	147	—	152	442	80	95
GN3/22	—	—	—	—	128	—
GN4/17	683	1478	613	1723	98	75
GN4/18	556	1085	571	1140	98	—
GN4/19	309	2003	310	1747	—	—
GN3/40	337	1870	337	1870	—	—

[†]$T\Delta\varepsilon = 2T\varepsilon''_{max}/\beta$ [see eqn (7.7)] measured for samples with no preferential alignment.

proportional to T_g, and that more than 20° below T_c the corresponding retardation factor g has much the same value for all the materials considered.

In Table 7.3(b) we list data on the dielectric decrement, see eqn (7.7), multiplied by the absolute temperature T. From eqns (7.20), (7.21) and the discussion of director misalignment in section 7.4 it will be seen that these quantities depend on the dipole components μ_l and μ_t, the director order parameter $\overline{q_{zz}}$, and the molecular order parameter $\overline{P_2}$. For a completely randomized sample ($\overline{q_{zz}} = 0$):

$$T\Delta\varepsilon_\delta = \tfrac{1}{3}(\rho h F^2/3\varepsilon_0 k_B)\mu_l^2(1 + 2\overline{P_2}) \tag{7.38}$$

$$T\Delta\varepsilon_\alpha = (\rho h F^2/3\varepsilon_0 k_B)[\tfrac{2}{3}\mu_l^2(1 - \overline{P_2}) + \mu_t^2] \tag{7.39}$$

Since the data in Table 7.3(b) refer to temperatures significantly below T_c, $\overline{P_2}$ will vary only weakly with temperature, and the quantities $T\Delta\varepsilon_\delta$ and $T\Delta\varepsilon_\alpha$ should do the same, as is seen from the table to be the case. Thus, as anticipated in section 7.4, the actual values of the permittivity $\varepsilon(\omega)$ are correlating with the molecular properties of the mesogenic moiety, rather than with T_g. Table 7.4 lists estimates of the effective dipole moments of the mesogenic groups, calculated from the sum of eqns (7.38) and (7.39) using data on $T(\Delta\varepsilon_\delta + \Delta\varepsilon_\alpha)$ from Table 7.3(b). The correlation with molecular structure is excellent. Note the impact of the bridging ester group in GN3/3 and GN3/14 when compared with GN3/19, and the effect of changing the fluorine position in GN4/18 compared with GN4/17 and GN4/19. The somewhat larger effective dipole of GN3/40,

Table 7.4 Dipole moments estimated from the dielectric decrements listed in Table 7.3(*b*)

Code (Table 7.1)	$\mu \times 10^{30}$ (C m)
GN3/3	12
GN3/14	12
GN3/19	7
GN4/17	16
GN4/18	13
GN4/19	16
GN3/40	15

relative to the copolymers GN3/3 and GN3/14, could be taken as an indication that antiparallel correlation is more hindered in the homopolymer GN3/40 which also has an Et substituted backbone and a lateral Me on the mesogen, both of which may further hinder antiparallel correlation. The absolute value of the moment obtained for GN3/19 is a little smaller than the value of $\sim 10 \times 10^{-30}$ Cm obtained from dielectric studies of monomeric alkyl cyanobiphenyls, which in turn is smaller than the $\sim 15 \times 10^{-30}$ Cm of a 'free' molecule without antiparallel correlation (Dunmur *et al.*, 1978). The values for GN3/3 and GN3/14 are distinctly smaller than the estimate of 25 $\times 10^{-30}$ Cm made for monomeric cyanophenyl benzoates by Klingbiel *et al.* (1974). For completeness we give also the Arrhenius-fit activation energies (cf. Table 7.3(*b*)).

7.5.2 Role of the spacer unit

The spacer unit is a flexible link, usually a methylene chain, which joins the mesogenic group to the polymer backbone. The length of this spacer is influential in determining the mesophase behaviour (Finkelmann *et al.*, 1978; Gemmel *et al.*, 1985). Both Zentel *et al.* (1985) and Parneix *et al.* (1987) have associated a relaxation found below T_g in polyacrylate LCPs with the spacer unit, the former calling the relaxation γ_1 and the latter β. Since the CH_2 group does not carry a significant dipole, the dielectric activity of the spacer unit evidently arises from its flexibility allowing internal reorientations between the relatively stiff, and polar, acrylate chain and the pendant mesogens, thus leading to multiple relaxation peaks (Böttcher and Bordewijk, 1978, section 74). Above the glass transition temperature increasing the number of spacer units leads, other things being equal, to loss peaks shifting to higher frequencies and activation energies decreasing, in agreement with the view that the spacer's role is to introduce flexibility. Parneix *et al.* (1987) have published useful data.

7.5.3 The polymer backbone

As with the spacer unit, the flexibility or otherwise of the polymer backbone has an important influence on the phase diagram. Consistently with our picture (section 7.1) of LCPs as mesogens strung together by polymer chains, whereas polyacrylate and polymethacrylate chains tend to favour nematic phases, the more flexible polymethysiloxane chain favours smectics. This, presumably, is because rotation about the Si—O bond gives the siloxane chain access to a wider range of conformations, allowing it to accommodate the greater order of a smectic phase more easily. The impact of chain flexibility on relaxation frequencies can be quantified by studies of the scaling relative to T_g as discussed in section 7.5.1.

Unlike the spacer, the backbone frequently carries polar groups and may therefore be dielectrically active in its own right, sometimes leading to complicated multiple peak spectra which are difficult to interpret unambiguously. In other instances spectra remain simple; for example Bormuth et al. (1987) showed that substitution of chlorine on to the backbone enhanced the δ relaxation in the polyacrylate (1) where X, R = H, Me or Cl, Bu, respectively.

$$\text{X}-\overset{\overset{\displaystyle |}{\text{CH}_2}}{\underset{|}{\text{C}}}-\text{C}\overset{O}{\underset{\text{O}-(\text{CH}_2)_6-\text{O}-}{\diagdown}}\!-\!\!\!\!\bigcirc\!\!\!\!-\text{CO}_2-\!\!\!\!\bigcirc\!\!\!\!-\text{OR} \qquad (1)$$

Substitution of groups having a larger molar volume onto the backbone (e.g. alkyl replacing H) will tend to disrupt the liquid crystallinity by disturbing the packing of side chains and diluting the mesogenic content. Zentel et al. (1985) have compared polyacrylates with polymethacrylates, while compounds GN3/15 and GN3/40 in Tables (7.2) and (7.3) provide a similar comparison for polysiloxanes.

7.5.4 Copolymers

Many of the materials already used as examples are copolymers. However, from a dielectric point of view the effect of interspersing between the mesogen-bearing silicons in the siloxane chain others bearing only methyl groups is unspectacular. Although the properties of such copolymers confer benefits in applications such as thermo-optic optical storage (McArdle et al., 1987a, b), we have in mind here another feature of side-chain LCPs thought attractive for technological applications. This is the possibility of substituting on to the chain non-mesogenic electroactive groups to confer properties such as dichroism and optical non-linearity, without the problems of phase separation and immiscibility which often arise in guest–host systems. However, the non-mesogenic substituents not only dilute the mesogenic groups but may hinder oriented packing in the fluid. Thus the cyanopropyl substituent in GN3/39 (Table 7.1) suppresses the mesophase, lowering the clearing point by over 100° relative to polymers such as GN3/15 or GN3/40 and causing it to lie below an also depressed T_g of $-23°C$.

Interestingly, GN3/39 displays two loss peaks at $1.14T_g$ even though it is isotropic at that temperature. The peak locations of $10^{2.72}$ and $10^{4.42}$ Hz are surprisingly consistent with the data in Table 7.3(a), while the Fuoss–Kirkwood β values of 0.64 and 0.27, respectively, are ordered the same as for the data in Table 7.3(a) although the β for the lower frequency peak is distinctly smaller.

The comparability of the relaxation frequencies in isotropic and liquid crystalline materials might be taken as further evidence that these frequencies are determined by the dynamics which are ultimately responsible for glass formation. However, the observation of two peaks in isotropic GN3/39 could be explicable on either of two hypotheses:

(i) Recalling (section 7.2) that mixtures of sufficiently dissimilar molecules show distinct relaxations associated with each molecule, the peaks may be assigned to relaxations of the mesogenic group and the cyanopropyl group, presumably with the latter at higher frequency. On this basis the dielectric decrements of 6.05 and 6.67, respectively, would give estimates of 14×10^{-30} Cm and 15×10^{-30} Cm for the dipole moments of the mesogenic and cyanopropyl groups respectively. These estimates are consistent with Table 7.4 and with published group moments (Gordon and Ford, 1972).

(ii) Alternatively, recalling again from section 7.2 and Figure 7.3 that if the anisotropy of the rotational diffusion tensor, eqn (7.18), is sufficiently great, two relaxations will be seen in the isotropic phase even if the two dipole-bearing side groups are behaving as a single 'average' moiety. By use of eqn (4.24) of Bone *et al.* (1984), the data on GN3/39 at $1.14\,T_\mathrm{g}$ yield an estimate of about 105 for $D_\parallel/D_\perp$. This appears unrealistically high when compared with the values of 12 and 7.5 reported by Haws *et al.* (1987) for GN3/17 and GN3/18 (Table 7.1) in their isotropic phases at $1.39\,T_\mathrm{g}$ and $1.30\,T_\mathrm{g}$ respectively, although values as high as 83 were reported for monomeric alkylated phenyl benzoate esters by Bone *et al.* (1984).

We have dwelt on the example of GN3/39 at some length because it provides an opportunity to demonstrate, more clearly than is often done in the literature, the extent to which the interpretation of dielectric loss peaks in terms of molecular processes may be ambiguous.

7.6 Low-frequency and DC conductivity phenomena

Up to this point we have discussed only the dielectric properties that result from the dynamics of permanent dipoles set within the molecules of the material. At low frequencies, and in the limit of DC fields, the dielectric properties of both monomeric and polymeric materials are dominated by a variety of conduction phenomena. In this section we first review the most significant of these effects and then, in section 7.6.2, discuss the behaviour of side chain LCP films.

The relegation of conductivity effects to the end of this chapter reflects the difficulties which beset their study both by theory and by experiment rather than their technological significance. In our experience they are in fact of greater immediate significance in LCP device physics than the effects discussed earlier. The relevance to issues of material purity and electric field alignment were mentioned in section 7.1; related to these is the question of dielectric breakdown as a possible cause of device failure. These are important issues because of the high voltages, 50 to 200 V rms across a 10 μm film (McArdle *et al.*, 1987*a*, *b*; Attard and Williams, 1986*b*), required to align LCP films. Such voltages are an order of magnitude greater than those required for LMMLCs, and can lead to problems of driver loading and film breakdown when attempting to align large areas.

7.6.1 *Conductivity effects*

The interpretation of low frequency or DC conductivity data in terms of underlying mechanisms is rarely straightforward, the observed phenomena frequently consisting of contributions from several polarization processes which, furthermore, may interact with each other. In this section we describe some of the processes that are commonly encountered in liquid and glassy dielectrics.

One of the most frequently encountered causes of low-frequency conductivity is the presence of ionic species in the material studied. Polar materials tend to induce partial dissociation of these extrinsic ionic impurities. Application of an AC field whose period is long compared with the diffusion time of the ions results in polarization due to charge motion. Charges coming into contact with the electrodes may either be removed by electrochemical reactions or, if unable to react, may form a space-charge layer at the interface. In the absence of space-charge layers, for low fields charge recombination gives rise to an ohmic relation between the applied voltage and the resulting current due to ionic charge carriers. At higher fields, charge generation by dissociation may not keep pace with the increasing voltage and hence the $I-V$ curve will exhibit a plateau.

Also at higher fields, charge injection may occur. The most commonly postulated mechanisms are Schottky emission from the electrode or Poole–Frenkel emission from impurities:

$$I(\text{Schottky}) = AT^2 \exp\left[-(\phi - \beta E^{1/2})/k_B T\right] \tag{7.40}$$

$$I(\text{Poole–Frenkel}) \sim E \exp\left[-(\phi - 2\beta E^{1/2})/k_B T\right] \tag{7.41}$$

where A is an effective Richardson constant, ϕ is the barrier height or work function, E the electric field, and β is given by

$$\beta = (e^3/4\pi\varepsilon_0\varepsilon_\infty)^{1/2} \tag{7.42}$$

e being the protonic charge and ε_∞ the high-frequency permittivity of the medium. Both processes give non-Ohmic $I-V$ curves with $\log(I)$ linear in $V^{1/2}$ to within attainable experimental accuracy. They can be distinguished by plotting $\log(I/T^2)$ against $1/T$ or by varying the electrode material.

In the absence of surface polarization effects, the presence of DC conductivity will affect only the imaginary part of the permittivity $\varepsilon''(\omega)$, causing it to increase rapidly with decreasing frequency (Figure 7.5). Empirically this increase is typically proportional to ω^{-n} where $0 < n \leqslant 1$, but usually $n \sim 1$. In practice, with materials of interest to this book, electrode polarization invariably occurs. This has the effect on the sample equivalent circuit of placing an electrode impedance Z_{el} in series with the parallel capacitance and conductance comprising the actual sample film. Thus the apparent permittivities $\varepsilon'_{\text{app}}$ and $\varepsilon''_{\text{app}}$ measured experimentally are given by (Johnson and Cole, 1951)

$$(\sigma_{\text{app}} + i\omega\varepsilon'_{\text{app}})^{-1} = (\sigma + i\omega\varepsilon')^{-1} + Z_{el} \tag{7.43}$$

where Z_{el} is taken as normalized to unit empty cell capacitance. Rearranging eqn (7.43) and taking

$$Z_{el} = Z_0(i\omega)^{-n} \quad (0 < n \leqslant 1) \tag{7.44}$$

yields

$$\sigma_{\text{app}} = \sigma - (Z_0 \cos \tfrac{1}{2}n\pi)\omega^{-n}\sigma^2 \tag{7.45}$$

$$\varepsilon'_{\text{app}} = \varepsilon' + (Z_0 \sin \tfrac{1}{2}n\pi)\omega^{-(n+1)}\sigma^2 \tag{7.46}$$

by equating real and imaginary parts and making use of the inequalities $Z_0\omega^{-n}\sigma \ll 1$ and $\omega\varepsilon' \ll \sigma$ which usually apply at sufficiently low frequencies. The value of n is typically unity for a (low-viscosity) liquid, decreasing to $\tfrac{1}{2}$ for a solid or solid-like material. Both this value and the inequalities yielding the simplified forms, eqns (7.45) and (7.46), should be checked when applying this correction to $\varepsilon'(\omega)$ data. Figure 7.6

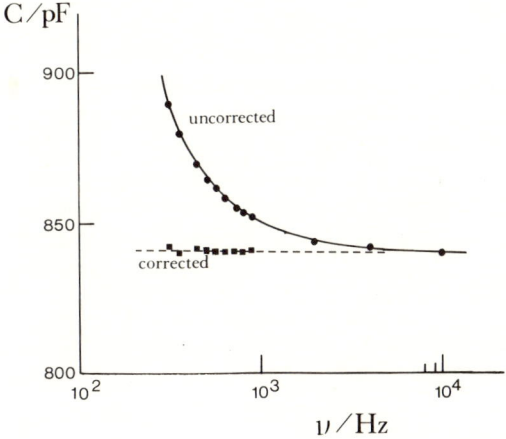

Figure 7.6 Uncorrected and corrected (broken line) values of the measured cell capacitance for polymer GN3/19 (Table 7.1) at 155°C.

shows both uncorrected and corrected (using $n = 0.5$) values of the measured cell capacitance for the polymer GN3/19 (see Table 7.1) at 155°C. Note that over most of the frequency range shown in Figure 7.6 the inequality $\omega\varepsilon' \ll \sigma$ mentioned above is violated ($\omega\varepsilon' \sim \sigma$ at 835 Hz) and the simplified forms given in eqns (7.45) and (7.46) are insufficient.

The rapid increase in $\varepsilon''(\omega)$ with decreasing frequency is a significant feature of the loss spectrum of most LCPs. With increasing temperature it can even hide dielectric features.

Under conditions such as the presence of inclusions or phase separation, where the sample contains internal interfaces between regions having different permittivities and/or conductivities, a loss peak due to the Maxwell–Wagner–Sillars (MWS) process (Böttcher and Bordewijk, 1978, section 98; Daniel, 1967) may be observed. In practice the MWS peak may occur at very low frequencies, when it will be hidden under the DC conductivity effects. However, in these circumstances eqns (7.43) plus (7.44) will no longer be sufficient to describe the measured curves $\varepsilon'_{app}(\omega)$ and $\varepsilon''_{app}(\omega)$.

The effect of a space charge layer on a charge-injecting electrode will be to enhance the local field, leading to Schottky emission. This effect can be represented by replacing E in eqn (7.40) by γE, where γ is a field enhancement factor (Sessler *et al.*, 1986). We discuss this effect for LCPs at greater length in section 7.6.2.

The preceding brief introduction to conductivity phenomena is generally applicable to isotropic media. Clearly the presence of orientational long-range order in liquid crystalline phases means that we should use tensorial quantities to describe the anisotropy in conductivity. In practice, it appears from work on LMMLCs that the anisotropy in the DC conductivity is not large and that therefore the discussion presented above is a reasonable guide to the physics of conduction in mesogenic materials. It should be stressed, however, that because electrode polarization dramatically affects dielectric behaviour, the anisotropy at the electrode-polymer interface may have important implications for the observed conductivity effects.

7.6.2 Behaviour of side chain LCPs

It is established (Blinov, 1983, section 5.1) that in LMMLCs the electrical conductivity is due to ionic carriers and is modestly anisotropic, with the sign and magnitude of the anisotropy depending on structure and phase (Jadzyn and Kedziora, 1987). Pre-transitional effects are observable. Space charge layers form at the electrodes even with the most purified of materials, although equilibration of the electrical double layer may be quite slow, with time constants of about one second (Mada and Osajima, 1986).

At sufficiently low fields, both LMMLCs and side chain LCPs show ohmic $I-V$ curves with similar parameters. Figure 7.7 shows plots of the low-field AC conductivity against $1/T$ at $1.1 T_2$ and $0.92 T_c$ (see footnote to Table 7.1) for a series of mixtures of the material GN3/14 (Table 7.1) with the commercial LMM smectic mixture S2 (BDH Ltd). All samples are dyed blue for technological reasons (McArdle et al., 1987a, b and Chapter 13) with an anthraquinone dye known to be highly stable. The data clearly suggest that the steady-state carrier transport processes in the bulk of the LMM and LCP materials are essentially identical.

Originating with the work of Heilmeier et al. (1968) on dynamic scattering, it has been known for LMMLCs that Schottky-type $I-V$ curves ($\log I \propto V^{1/2}$) can be

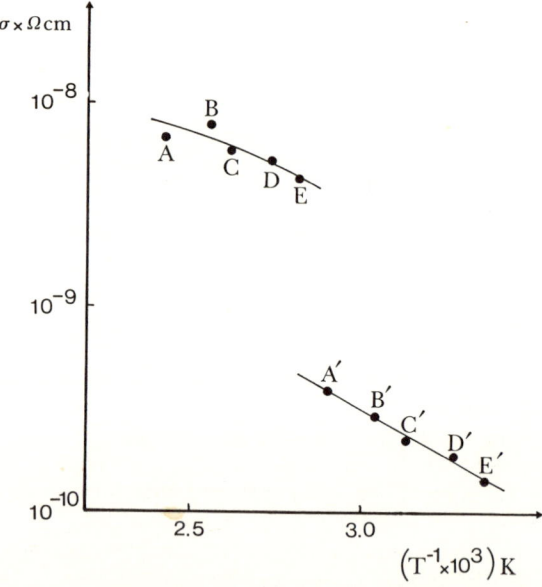

Figure 7.7 Low-field AC conductivity against $1/T$ at $1.1 T_2$ (A to E) and $0.92 T_c$ (A' to E') for a series of mixtures of polymer GN3/14 with the commercial LMM smectic mixture S2:

$A, A' = 100\%$ GN3/14	—
$B, B' = 75\%$ GN3/14	25% S2
$C, C' = 50\%$ GN3/14	50% S2
$D, D' = 25\%$ GN3/14	75% S2
$E, E' = —$	100% S2

observed at higher fields. We have verified that this is also the case for siloxane LCPs. Figure 7.8 shows an $I-V$ plot for the polymer GN3/3 (Table 7.1) at 70°C in a 25 μm cell having one ITO electrode (5 $\Omega/\square$) and one Al electrode. On application of the DC voltage large displacement currents were observed, with the time to a steady state taking from fifteen minutes to an hour depending on the voltage applied. Above a threshold of about 4 V, plots of log I against $V^{1/2}$ are very accurately linear for both the forward and reverse characteristic, although the two lines have different slopes.

Both the relatively low threshold voltage and other features, such as the different slopes, may be understood by recalling (section 7.6.1) that field enhancement effects arising from electrode polarization modify the Schottky emission formula, eqn (7.40), to read

$$I = AT^2 \exp\{-[\phi - \beta(\gamma E)^{1/2}]/k_B T\} \tag{7.47}$$

Estimates of γ may be obtained from the observed slopes by assuming a value for ε_∞ in eqn (7.42). Taking $\varepsilon_\infty = 2.8$ (an estimate of the refractive index squared) we obtain γ equal to 7.8 and 3.4 for the forward and reverse characteristics, respectively. We have confirmed this evidence that the electrode polarization is influenced by the nature of the electrode material in other (presently unpublished) studies. The dependence of the high-field $I-V$ characteristic on electrode material is indicative of a Schottky (i.e. surface) rather than a Poole–Frenkel (i.e. bulk) process.

The relatively low threshold for the onset of this Schottky emission is consistent with the large field enhancement factors observed. A feeling for the reasonableness of these

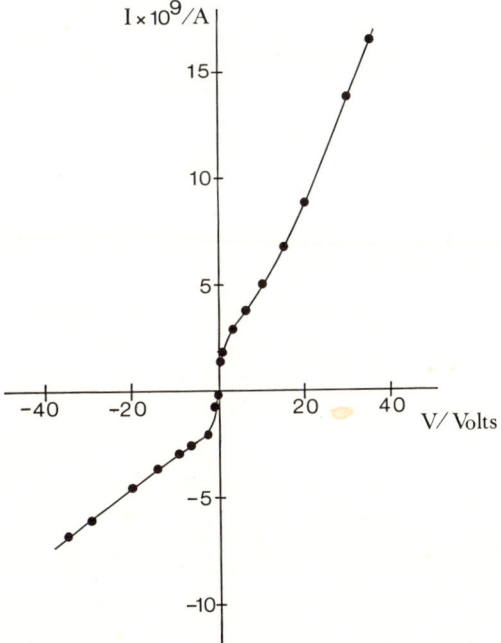

Figure 7.8 I–V characteristic of polymer GN3/3 at 70°C in a 25 μm cell with one ITO electrode and one Al electrode.

numbers may be obtained by applying a Debye–Hückel approximation to describe the electrode polarization. Assuming, for the sake of argument, a cell of thickness d with identical electrodes and filled with an insulating medium containing as an impurity a 1:1 electrolyte yielding monovalent ions, then

$$\gamma \approx \tfrac{1}{2}\lambda d \tag{7.48}$$

where

$$\lambda = (2e^2 n_0/\varepsilon_0 \varepsilon k_B T)^{1/2} \tag{7.49}$$

n_0 being the number concentration of ionized impurities and ε the permittivity of the medium. Taking, with the above experimental data on GN3/3 in mind, $d = 25\,\mu m$, $T = 343$ K, $\varepsilon = 10$, and $n_0 = (1.3 \times 10^{27})C_0 m^{-3}$, where C_0 is the number fraction of ionic carrier pairs relative to the number concentration of mesogenic groups, eqns (7.48) and (7.49) give $\gamma = 5$ for $C_0 = 10^{-9}$. This value of C_0 is, coincidentally, exactly the fraction of intrinsic carriers in germanium at room temperature. Finally, substituting eqn (7.48) into eqn (7.47) yields

$$I = AT^2 \exp\left\{-\left[\phi - \beta(\tfrac{1}{2}\lambda dE)^{1/2}\right]/k_B T\right\} \tag{7.50}$$
$$= AT^2 \exp\left\{-\left[\phi - \beta(\tfrac{1}{2}\lambda V)^{1/2}\right]/k_B T\right\} \tag{7.51}$$

from which we note that, in this model, (i) at fixed field the current density will increase with electrode spacing (see Heilmeier et al., 1968, for a different mechanism leading to qualitatively the same result), and (ii) the I–V characteristic is quantitatively independent of cell spacing, suggesting that breakdown will be determined by the voltage (rather than the field) applied to the cell.

Acknowledgements

We gratefully acknowledge the benefits of collaboration with the Department of Chemistry, Hull University, and LaserScan Ltd, funded in part by the Department of Trade and Industry under the Joint Opto-Electronics Research Scheme, and of helpful discussions with G. Williams. GSA thanks the SERC for an Advanced Fellowship.

References

Araki, K., Attard, G.S., Kosak, A., Williams, G., Gray, G., Lacey, D. and Mentor, G. (1988) *J. Chem. Soc. Faraday Trans II* **84**, 106.
Attard, G.S., Araki, K. and Williams, G. (1987a) *J. Mol. Electron.* **3**, 1.
Attard, G.S., Araki, K. and Williams, G. (1987b) *Brit. Polym. J.* **19**, 119.
Attard, G.S. and Williams, G. (1986a) *Polymer Commun.* **27**, 2.
Attard, G.S. and Williams, G. (1986b) *Polymer Commun.* **27**, 66.
Barlow, A.J. and Erginsav, A. (1974) *J. Chem. Soc. Faraday Trans. II* **70**, 885.
Barnik, M.I., Blinov, L.M., Ivashchenko, A.V. and Shtykov, N.M. (1979) *Sov. Phys. Crystallogr.* **24**, 463.
Bata, L. and Molnar, G. (1975) *Chem. Phys. Lett.* **33**, 535.
Blinov, L.M. (1983) *Electro-Optical and Magneto-Optical Properties of Liquid Crystals*. Wiley, Chichester, Chapter 4.
Block, H. (1979) *Adv. Polym. Sci.* **33**, 93.
Blythe, A.R. (1979) *Electrical Properties of Polymers*. Cambridge University Press, Cambridge, Chapter 3.
Bone, M.F., Price, A.H., Clark, M.G. and McDonnell, D.G. (1984) *Liq. Cryst. Ordered Fluids* **4**, 799.
Bormuth, F.J., Haase, W. and Zentel, R. (1987) *Mol. Cryst. Liq. Cryst.* **148**, 1.
Böttcher, C.J.F., Van Belle, O.C., Bordewijk, P. and Rip, A. (1973) *Theory of Electric Polarisation*, 2nd edn., vol. I, Elsevier, Amsterdam.
Böttcher, C.J.F. and Bordewijk, P. (1978) *Theory of Electric Polarisation*, 2nd edn., vol. II, Elsevier, Amsterdam, Chapter 9.

Boyd, R.H. (1983) *J. Polymer Sci: Polymer Phys. Edn.* **21**, 505.
Boyer, R.F. (1978) in *Molecular Basis of Transitions and Relaxations*, eds. Meier, D.J. Gordon and Breach, New York, 329–331.
Bradshaw, M.J. and Raynes, E.P. (1983) *Mol. Cryst. Liq. Cryst.* **91**, 145.
Clark, M.G., Raynes, E.P., Smith, R.A. and Tough R.J.A. (1980) *J. Phys. D: Appl. Phys.* **13**, 2151.
Clark, M.G. (1984) Plenary Lecture, 10th Int. Liquid Crystal Conf. York, UK.
Clark, M.G. (1985) *Mol. Cryst. Liq. Cryst.* **127**, 1.
Daniel, V.V. (1967) *Dielectric Relaxation.* Academic Press, London, Chapters 6, 7, and 14.
De Gennes, P.G. (1974) *The Physics of Liquid Crystals.* Clarendon Press, Oxford, section 3.1.
Deuling, H.J. (1972) *Mol. Cryst. Liq. Cryst.* **19**, 123.
Dunmur, D.A., Manterfield, M.R., Miller, W.H. and Dunleavy, J.K. (1978) *Mol. Cryst. Liq. Cryst.* **45**, 127.
Dunmur, D.A. and Miller, W.H. (1980) *Mol. Cryst. Liq. Cryst.* **60**, 281.
Durand, G. and Martinot-Lagarde, Ph. (1980) *Ferroelectrics* **24**, 89.
Edwards, D.M.F. and Madden, P.A. (1983) *Mol. Phys.* **48**, 471.
Finkelmann, H., Ringsdorf, H. and Wendorff, J.H. (1978) *Makromol. Chem. Rapid Commun.* **179**, 273.
Gemmel, P.A., Gray, G.W. and Lacey, D. (1985) *Mol. Cryst. Liq. Cryst.* **122**, 205.
Gordon, A.J. and Ford, R.A. (1972) *The Chemist's Companion.* John Wiley, New York, 124.
Haase, W., Pranoto, H. and Bormuth, F.J. (1985) *Ber. Bunsenges. Phys. Chem.* **89**, 1229.
Harrison, G. (1976) *The Dynamic Properties of Supercooled Liquids.* Academic Press, London, Chapter 6.
Hashin, Z. and Shtrikman, S. (1962) *J. Appl. Phys.* **33**, 3125.
Haws, C.M., Clark, M.G. and McArdle, C.B. (1987) *Proc Int. Conf. on Liquid Crystal Polymers, Bordeaux*, paper 6P8; (1987) *Mol. Cryst. Liq. Cryst.* **153**, 537.
Heilmeier, G., Zanoni, L.A. and Barton, L.A. (1968) *Proc. IEEE* **56**, 1162.
Jadsyn, J. and Kedziora, P. (1987) *Mol. Cryst. Liq. Cryst.* **145**, 17.
Johnson, J.F. and Cole, R.H. (1951) *J. Amer. Chem. Soc.* **73**, 4536.
Kivelson, D. and Madden, P.A. (1983) *J. Phys. Chem.* **87**, 4823.
Klafter, J. and Shlesinger, M.F. (1986) *Proc. Natl. Acad. Sci. USA* **83**, 848.
Klingbiel, R.T., Genova, D.J., Criswell, T.R. and Van Meter, J.P. (1974) *J. Amer. Chem. Soc.*, **96**, 7651.
Lamb, J. (1978) in *Molecular Basis of Transitions and Relaxations*, ed. Meier, D.J., Gordon and Breach, New York, 25–62.
Mada, H. and Osajima, K. (1986) *J. Appl. Phys.* **60**, 3111.
Madden, P.A. and Kivelson, D. (1984) *Adv. Chem. Phys.* **56**; 467.
Madhusudana, N.V. and Chandrasekhar, S. (1975) *Pramana Suppl.* No 1, 57.
Maier, W. and Meier, G. (1961) *Z. Naturforsch.* **16a**, 262.
Maxwell, J.C. (1904) *A Treatise on Electricity and Magnetism*, 3rd edn., vol. I, Clarendon Press, Oxford, 440.
McArdle, C.B., Clark, M.G., Haws, C.M., Whiltshire, M.C.K., Parker A., Nestor, G., Gray, G.W., Lacey, D. and Toyne, K.J. (1987a) *Liquid Crystals* **2**, 573.
McArdle, C.B., Clark, M.G. and Haws, C.M. (1987b) *Proc. Eurodisplay*, IOP, London, 160–3.
Meier, G. and Saupe, A. (1966) *Mol. Cryst.* **1**, 515.
Nordio, P.L. and Segre, U. (1979a) in *The Molecular Physics of Liquid Crystals*, eds. Luckhurst, G.R. and Gray, G.W., Academic Press, London, Chapter 18.
Nordio, P.L. and Segre, U. (1979b) in *The Molecular Physics of Liquid Crystals*, eds. Luckhurst, G.R. and Gray, G.W. Academic Press, London, Chapter 19.
Palmer, R.G., Stein, D.L., Abrahams, E. and Anderson, P.W. (1984) *Phys. Rev. Lett.* **53**, 958.
Parneix, J.P., Njeumo, R., Legrand, C., Le Barny, P. and Dubois, J.C. (1987) *Liquid Crystals* **2**, 167.
Pranoto, H., Bormuth, F.-J. and Hasse, W. (1986) *Makromol. Chem.* **187**, 2453.
Sessler, G.M., Hahn, B. and Yoon, D.Y. (1986) *J. Appl. Phys.* **60**, 318.
Shlesinger, M.F. (1984) *J. Stat. Phys.* **36**, 639.
Thurston, R.N. (1984) *Mol. Cryst. Liq. Cryst.* **108**, 61.
Wetton, R.E., Morton, M.R. and Rowe, A.M. (1986) *International Laboratory* (March 1986) 70–81.
Zeller, H.R. (1981) *Phys. Rev.* **A23**, 1434.
Zeller, H.R. (1982) *Phys. Rev.* **A26**, 1785.
Zentel, R., Strobl, G.R. and Ringsdorf, H. (1985) *Macromolecules* **18**, 960.

8 NMR methods for studying molecular order and motion in liquid crystalline side group polymers

C. BÖEFFEL and H.-W. SPIESS
Max-Planck-Institut für Polymerforschung, D-6500, Mainz, FRG.

8.1 Introduction

The knowledge of molecular parameters such as order and dynamics is of great interest for technical applications of polymeric liquid crystals. The phase behaviour of liquid crystalline side group polymers with the same mesogenic groups is also influenced by the type of the polymer chain and the length of the flexible spacer. Thus the development of such materials for special uses needs a detailed characterization of the mesophases and their order structure. In this chapter new NMR methods for the characterization of molecular order and dynamics of polymeric side group liquid crystals are described.

A wide variety of liquid crystalline side group polymers (Gordon and Platé, 1984; Chapoy, 1985; Shibaev and Platé, 1985) has been synthesized on the basis of the spacer concept (cf. Chapters 3, 4) (Finkelmann et al., 1978; Shibaev et al., 1979). The decoupling of the different motional and ordering tendencies of the mesogenic group and the polymer backbone is achieved by inserting a flexible spacer. Questions arise whether liquid crystalline side group polymers based on this concept are comparable with low-molar-mass systems with respect to their liquid crystalline behaviour and which kind of interactions exist between the mesogenic groups and the polymer chain. In contrast to low-molar-mass liquid crystals, the mesophase order is easily frozen in at the glass transition. These ordered glasses are a new class of materials that offers a broad field of technical applications (cf. Chapter 12, 13, 14). Consequently they have to be characterized also in the solid state with respect to their molecular order and dynamics in order to relate molecular properties to macroscopic behaviour.

Different physical methods can be applied to answer these questions. Information on the *molecular order* of the mesogenic group may be obtained by X-ray (Zugenmaier and Mügge, 1984, 1985; Zentel and Strobl, 1984; Davidson et al., 1985; Mattoussi et al., 1986; Nachaliel et al., 1987), ^{1}H-NMR (Piskunov et al., 1982; Roth and Krücke, 1986), IR- and UV-dichroism (Ringsdorf et al., 1985, 1986) as well as ESR (Wassmer et al., 1985). Information on the overall conformation of the main chain is obtained using small-angle neutron scattering (SANS) by measuring the anisotropy of the radii of gyration (Kirste and Ohm, 1985; Keller et al., 1985; Moussa et al., 1987). Information on *molecular motions* of liquid crystalline side group polymers is obtained by dielectric relaxation studies (Zentel et al., 1985; Kresse et al., 1982; Attard et al., 1986a, b, 1987a, b), but the relaxation process cannot be directly identified with a motion on a molecular level. Dynamical characteristics are also obtained from ESR (Wassmer et al., 1985) monitoring the motion of a spin label dissolved in the liquid crystalline matrix. It should be pointed out that, by this method, the molecular dynamics of the matrix is only detected indirectly via the motion of the spin label in the liquid crystalline matrix.

In order to relate the properties of these materials to their behaviour on the molecular level, molecular order and mobility of the different building blocks have to be characterized individually. ^{2}H-NMR is particularly attractive in this area, since the different parts of the molecule can be studied separately after selective labelling. The analysis of the NMR line-shapes in the frozen glassy state yields the complete orientational distribution function, as well as the type and time scale of the molecular motion (Spiess *et al.*, 1982, 1985; Boeffel *et al.*, 1983, 1986, 1988; Geib *et al.*, 1982; Pschorn *et al.*, 1986; Müller *et al.*, 1985). The molecular order of a macroscopic oriented sample is also obtained by a two-dimensional ^{13}C *magic angle spinning* experiment, where one dimension detects the chemical structure and one dimension is sensitive to molecular order (Harbison *et al.*, 1986, 1987; Blümich *et al.*, 1987). This experiment does not require isotopic labelling. From the sideband intensities in both dimensions, the orientational distribution function for the different residues can be derived, if the isotropic chemical shifts of the carbons in those sites are distinguishable.

This chapter first describes these techniques and then reviews our results on the molecular order and dynamics of polymeric side group liquid crystals, where different parameters are varied: the flexibility of the polymer chain, the length of the flexible spacer and the nature of the mesogen, i.e. rod- or disc-like. These results will be compared with SANS, X-ray, ESR and dielectric relaxation data, and finally an outlook will be given on current studies of liquid crystalline elastomers (cf. Chapter 10) and on a method to measure ultraslow director reorientations on a molecular level.

8.2 Techniques

The information on molecular order and dynamics is obtained from a line-shape analysis of the NMR spectra and spin relaxation times (Haeberlen, 1976; Mehring, 1983; Spiess, 1978). The ^{2}H-NMR spectrum is governed by the interaction of the nuclear quadrupole moment with the electric field gradient of the C^{2}H bond. For ^{13}C there are two relevant interactions: the dipole–dipole coupling between carbons and protons, and the smaller chemical shielding interaction, resulting from the coupling of the nucleus with the field of the surrounding electrons. All these are second-rank tensor interactions, and their effect on the NMR frequency depends on the orientation of the tensor with respect to the magnetic field. The orientation of this tensor as well as change in its orientation on the 'NMR time scale' are detected in the NMR spectrum. In the solid state one usually observes broad spectra due to different possible orientations of the tensors. In order to distinguish between different positions in the molecule, the sample has to be labelled selectively, or unwanted interactions have to be averaged out, e.g. by spinning the sample at the magic angle (in ^{13}C spectra) in combination with proton decoupling.

Liquid crystalline side group polymers show a high molecular order of the mesogenic group reflecting their liquid crystalline nature (Geib *et al.*, 1982) and heterogeneous motions, reflecting their polymeric properties (Haward, 1973). In addition, there is a gradient in molecular order from the mesogenic group to the polymer chain. The line-shape calculations, therefore, have to take into account the orientational distribution of the molecule as well as the type and the correlation times of the molecular motion. In the following chapter a short outline of this analysis is given; for details, the references cited above should be consulted.

8.2.1 ^2H-NMR

The ^{2}H-NMR spectrum is dominated by the quadrupole coupling of the deuterium nucleus (spin $I = 1$) with the electric field gradient of the C^2H bond. The NMR frequency in the solid state for a single deuterium is then given by (Haeberlen, 1976; Mehring, 1983; Spiess, 1978; Abragam, 1961)

$$\omega = \omega_0 \pm \delta(3\cos^2\Theta - 1 - \eta\sin^2\Theta\cos 2\Phi) \tag{8.1}$$

where ω_0 is the central frequency, $\delta = 3e^2qQ/8\hbar$ and $e^2qQ/\hbar$ in turn is the quadrupole coupling constant. The asymmetry parameter η describes the deviation from the cylindrical symmetry of the tensor around the z-axis. The orientation of the magnetic field $\vec{B}_0$ in the principal axes system of the field gradient is specified by the polar angles Θ and Φ. For deuterons in CH bonds of rigid solids $\delta/2\pi \approx 62\,\text{kHz}$, $\eta \approx 0$ and Θ is the angle between the respective CH bond and the magnetic field. Thus, molecular order and mobility is monitored via the orientation of individual CH bond directions.

The ^{2}H NMR lineshape changes in the presence of motion. If the motion is rapid on a time scale defined by the inverse width of the spectrum in absence of motion δ^{-1} (correlation times $\tau_c < 10^{-7}$ s), a spectrum due to a *averaged field gradient* is observed.

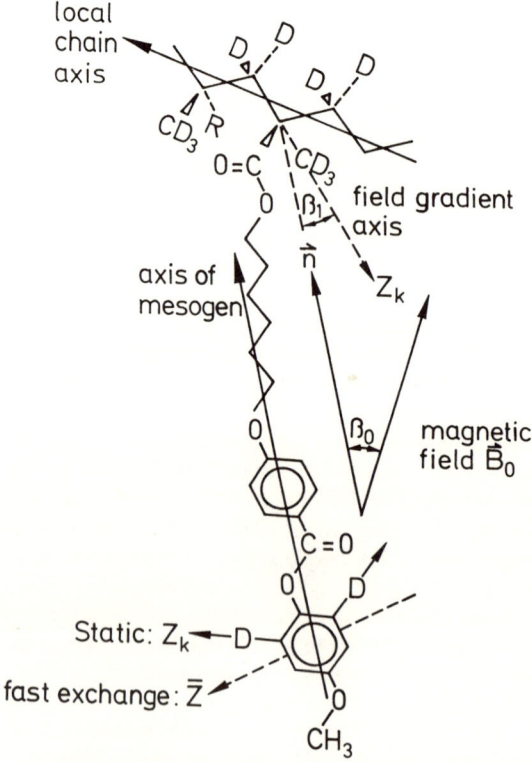

Figure 8.1 Direction of the principal axes at different sites of the molecule with respect to the order axis in a side group liquid crystal polymer.

In this rapid motional limit the NMR frequency is given by

$$\omega = \omega_0 \pm \bar{\delta}(3\cos^2\theta - 1 - \bar{\eta}\sin^2\theta\cos^2\varphi) \qquad (8.2)$$

$\bar{\delta}$ and $\bar{\eta}$ are now the quadrupole coupling constant and the asymmetry parameter, respectively, of the averaged field gradient tensor, θ and φ specify the polar angles of $\vec{B}_0$ in the principal axes system of the averaged field gradient. In Figure 8.1 the directions of the unique axes of the static and averaged field gradients for a deuteron in the phenylene ring are indicated by z_k and $\bar{z}$, respectively. The fast rotation of the methyl group around its C_3 axis leads to an averaged field gradient along the C—CH_3 bond as denoted for the methyl group in the polymer chain. In solids the averaging may either result from local motions like phenyl flips ($\bar{\delta} = \frac{5}{8}\delta$ and $\bar{\eta} = \frac{3}{5}$), from rotations of the methyl group around its C_3 axis ($\bar{\delta} = \frac{1}{3}\delta$ and $\bar{\eta} = 0$) or from conformational changes of hydrocarbon chains like kink motions ($\bar{\delta} = \frac{1}{2}\delta$ and $\bar{\eta} = 1$) (Spiess, 1983). In Figure 8.2a two spectra are shown, representing the line-shape of a rigid sample and of a sample containing a phenyl ring moving around its twofold symmetry axis in the fast motional limit. In both cases the C^2H bonds are distributed isotropically.

In the liquid crystalline phase of low molar mass systems, due to the rapid molecular motion the principal axes system of the *averaged* field gradient is associated with the liquid crystalline phase (Diehl and Khetrapal, 1969). Thus, in a uniaxial systems like a nematic or smectic A, the $\bar{z}$-axis is along the director $\vec{n}$ and $\bar{\eta}$ is zero. If biaxial order exists as in cholesterics or in smectic C, the asymmetry parameter $\bar{\eta}$ may differ from zero (Chidichimo *et al.*, 1982). The resulting quadrupole splitting in uniaxial phases is directly related to the order parameter. The situation is somewhat different in polymeric side group liquid crystals. The attachment of the mesogenic group to the polymer chain results in high viscosity of these materials, and the correlation time of the motion is slowed down, even at temperatures near the clearing point (Boeffel *et al.*, 1983). Furthermore, there is no transition from the mesophase to a crystalline state accompanied by a decrease in correlation times for reorientational motions of several orders of magnitude. Instead, they change gradually with temperature near the glass transition (Pschorn *et al.*, 1986), typical for glassy polymers. In this range the quadrupole splitting is no longer a measure of the size of the order parameter and a complete line-shape analysis, taking into account the type and the correlation time of the motion, must be performed.

At the glass transition, the motion of the polymer chain is frozen and only local motions remain. These motions are gradually frozen in with decreasing temperature. In this region, the correlation time of the motion τ_c is of the order of the inverse width of the static spectrum δ^{-1}, *intermediate exchange regime*: $10^{-7} \leqslant \tau_c \leqslant 10^{-4}$ s (Pschorn *et al.*, 1986). Then the NMR line-shapes have to be calculated, taking explicitly into account the exchange of frequencies resulting from the molecular motion during the solid echo pulse sequence, where the two RF pulses delayed by the time τ are applied (Spiess and Sillescu, 1981). In this regime the echo intensity may decrease to a few percent only of the intensity of the free induction decay (FID). So both the line-shapes and the signal intensities contain information about the type as well as the time scale of the motion.

Distribution of correlation times. In polymers, motions are usually not uniform as known from dielectric, mechanical and NMR relaxation studies as well as from photon correlation spectroscopy (Williams and Watts, 1970; Lindsey and Patterson, 1970;

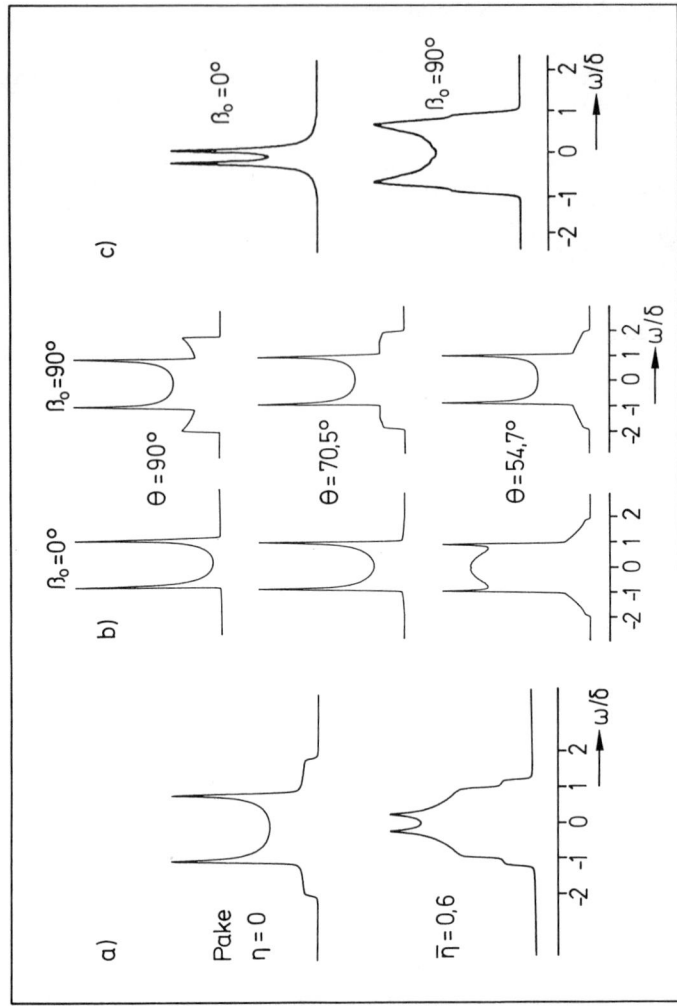

Figure 8.2 ^{2}H-NMR line shapes. (*a*) isotropic powders: rigid with $\eta = 0$ and phenyl flip with $\bar{\eta} = 0.6$ (*b*) Rigid spectra (asymmetry parameter zero) with different distributions of the CH^2 bonds with respect to the order axis as obtained by method (i), the orientational distribution function used for the calculation has a width at half height of $\pm 20°$ (*c*) Oriented (Gaussian distribution with width parameter $\bar{\beta} = \pm 20°$) including phenyl flip ($\bar{\eta} = 0.6$).

McBrierty and Douglas, 1980; Patterson, 1983). It is important to know whether the distribution of correlation times is homogeneous or heterogeneous. In the first case, every moving group shows a non-exponential relaxation described by a distribution of correlation times; in the second case, the distribution is caused by packing differences of the molecules, so that individual groups move with different correlation times. The line-shape at a given temperature can be written as a superposition of single spectra (Schmidt *et al.*, 1985):

$$\Gamma(\omega) = \int_{-\infty}^{\infty} S^{\Omega,\tau}(\omega)\, R^{\tau}(\ln\Omega)\, K(\ln\Omega)\, d\ln\Omega, \tag{8.3}$$

where $S^{\Omega,\tau}(\omega)$ described the line-shape at a fixed pulse delay τ and a fixed jump rate Ω. $R^{\tau}(\ln\Omega)$ gives the signal intensity at that jump rate. The distribution of correlation times is described by the distribution function $K(\ln\Omega)$. In all cases studied so far, it is sufficient to assume a Gaussian function with the centre at Ω_0 and a variance of σ^2:

$$K(\ln\Omega) = (\sigma\sqrt{2\pi})^{-1} \exp\left[-\frac{\left(\ln\dfrac{\Omega}{\Omega_0}\right)^2}{2\sigma^2} \right] \tag{8.4}$$

Orientational distribution. The orientational distribution function $P(\alpha,\beta,\gamma)$ describes the probability density of finding a molecule in a range between α and $\alpha + d\alpha$, β and $\beta + d\beta$ and γ and $\gamma + d\gamma$ with respect to a reference frame in the sample. Generally it can be expressed in terms of Wigner rotation matrices (Roe, 1965; McBrierty, 1974):

$$P(\alpha,\beta,\gamma) = \sum_{l}^{\infty} \sum_{k=-l}^{l} \sum_{k'=-l}^{l} P_{lkk'} D_{kk'}^{(l)}(\alpha,\beta,\gamma) \tag{8.5}$$

where α,β,γ specify the orientation of the molecule in the principal axes system, e.g. of the liquid crystal frame. If we confine ourselves to symmetric systems, straightforward ways to analyse the NMR line-shape in order to get the orientational distribution of a certain site of the molecule can be applied as follows (Spiess, 1982; Hentschel *et al.*, 1978, 1981):

(i) Expansion of the line-shape in terms of spherical functions
(ii) Expansion of the line-shape in terms of planar distributions
(iii) Expansion of the line-shape in terms of conical distributions.

First the different frames, and their relation to each other, have to be explained: the *principal axes* system, which is related to the *laboratory* system by the polar angles Θ and Φ and to the *director* frame by the angle β_1 (if it is a uniaxial phase), the relation to the *molecular* system is given by the angles θ and φ. The relation between the *molecular* frame and the *director* is given by the angle β (again in the uniaxial case) (cf. Figure 8.3).

The parameter of interest in a polymeric liquid crystal concerning molecular order is the orientation of a molecular axis, e.g. that of the mesogenic group or the polymer chain, with respect to the director. The detectable parameter in the NMR experiment on the other hand is the orientation of the C^2H bond with respect to the director, where the relation to a molecular direction is not known in many cases (cf. Figure 8.1) (Boeffel and Spiess, 1988).

In the following β denotes the angle between a molecular axis and the director,

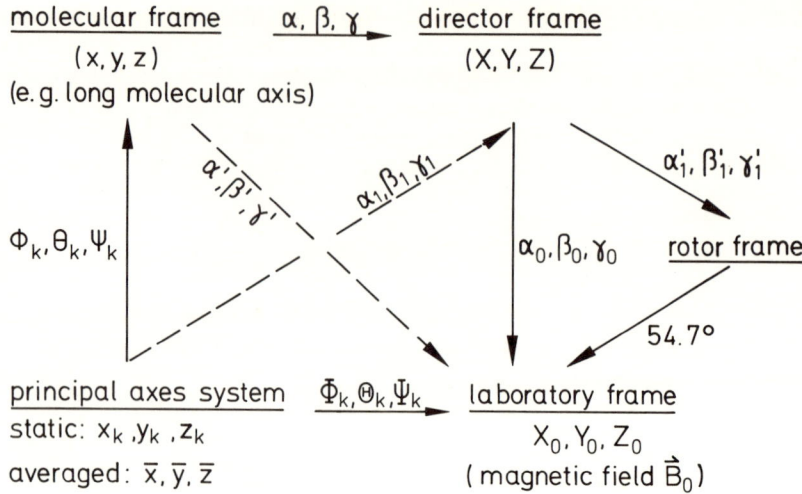

Figure 8.3 Relation between different frames: principal axes frame, director frame, molecular frame, laboratory frame and rotor frame. The transformations between the frames are described by the sets of Euler angles also given in the figure.

whereas β_1 is the angle between the electric field gradient and the director as denoted in Figures 8.1 and 8.3.

(i) *Expansion of the line-shape in terms of spherical functions.* If the molecules are uniformly distributed around the director and a molecular frame can be chosen such that the principal axes of the field gradient tensor are also uniformly distributed around the unique axis of this frame, we deal with systems of *transverse isotropy* (McBrierty, 1974). In this case the orientational distribution function can be expanded in terms of Legendre polynomials (Spiess, 1982; Hentschel *et al.*, 1981), representing the simplest form of spherical functions

$$P(\beta) = \sum_{l=0}^{\infty} P_{l00}P_l(\cos \beta)$$

$$= \sum_{l=0}^{\infty} \frac{2l+1}{8\pi^2} \langle P_l(\cos \beta) \rangle P_l(\cos \beta)$$

$$= \sum_{l=0}^{\infty} \frac{2l+1}{8\pi^2} \langle P_l \rangle P_l(\cos \beta) \qquad (8.6)$$

where P_{l00} are the moments of the orientational distribution functions and the expansion coefficients $\langle P_l(\cos \beta) \rangle = \langle P_l \rangle$ are the order parameters for a given molecular direction. The maximum of the orientational distribution function as defined in eqn (8.6) can be interpreted as the most probable angle between the director and the molecular axis. In absence of motion the NMR line-shape can easily be calculated using this expansion (Spiess, 1982; Boeffel and Spiess, 1988; McBrierty, 1974; Hentschel *et al.*, 1978). The total spectrum is then obtained by a superposition of subspectra $S_l(\omega)$

weighted by the order parameters $\langle P_l \rangle$ with even l:

$$S(\omega) = \sum_{l=0,2,4..}^{\infty} (2l+1)\langle P_l \rangle S_l(\omega) P_l(\cos \beta_0)$$

$$S(\omega) = S_0(\omega) \sum_{l=0,2,4..}^{\infty} (2l+1)\langle P_l \rangle P_l(\chi(\omega)) P_l(\cos \beta_0) \tag{8.7}$$

β_0 is the angle between the director and the magnetic field, and $S_0(\omega)$ represents the line-shape of an isotropic sample, centred at ω_0:

$$S_0(\omega) = \frac{1}{6\delta} \frac{1}{\chi(\omega)} \tag{8.8}$$

$$\chi(\omega) = \frac{1}{\sqrt{3}} \sqrt{\frac{\omega}{\delta} + 1} \tag{8.9}$$

with $-1 \leqslant (\omega/\delta) \leqslant 2$. From eqn (8.7) it is clear that the subspectra $S_l(\omega)$ are obtained by simply weighting $S_0(\omega)$ by the corresponding Legendre polynomial of order l. $S(\omega)$ describes the line-shape of a single NMR transition only. The total ^{2}H-NMR spectrum consists of a superposition of $S(\omega)$ and its mirror image with respect to $\omega = 0$ (for details see Mehring, 1983; Abragam, 1961). As mentioned before, the NMR experiment is sensitive only to the orientation of the electric field gradient with respect to the magnetic field. Therefore the order parameters $\langle P_l \rangle$ obtained from the experiment are referred to the unique axis of the field gradient, typically along the C^2H bond direction. This order parameter can easily be related to a molecular order parameter, if the conformation of that molecular site is known:

$$\langle P_l \rangle_{CH} = \langle P_l \rangle_{mol} P_l(\cos \theta), \tag{8.10}$$

where θ is the angle between the C^2H bond and the molecular axis. Orientational distributions centred at different angles with respect to the director can clearly be distinguished in the ^{2}H-NMR experiment as shown in Figure 8.2b. The relation in eqn (8.10) is valid only if there exists one defined conformation. In the presence of different conformations, the mean value $\langle P_l(\cos \theta) \rangle$ has to be considered.

The expansions based on Legendre polynomials or spherical harmonics diverge for high order. A better convergence can be achieved by an exponential expansion based on a mean field approximation of the intermolecular potential as introduced by Luckhurst (1979). In addition to relying on that specific model this method has the disadvantage that the maximum of the orientational distribution (that is, the conformation) must be known. Liquid crystalline polymers are typically sufficiently disordered, however, that the expansion in terms of Legendre polynomials is adequate in most cases. For higher ordered systems and in presence of motion the two approaches described in the following should be used.

(ii) *Expansion of the line-shape in terms of planar distributions.* Here molecules are considered having one molecular axis in common (Spiess, 1982; Hentschel et al., 1981). For deuterons in *trans* configurations of alkyl chains and for deuterons in the mesogenic group, where rapid molecular motion leads to a time-averaged field gradient reflecting the molecular symmetry, this common axis will be parallel to a principal axis of the field gradient. The sub-spectra $S_{\beta'}(\omega)$ of such an ensemble depend parametrically on the angle β' between the common axis and $\vec{B}_0$, cf. Figure 8.3. They can be calculated analytically for any value of the asymmetry parameter (that may have values $0 \leqslant \eta \leqslant 1$ in the presence of motion). In absence of motion the subspectra show two symmetric singularities for a particular β'.

The total spectrum is obtained by a weighted superposition of such subspectra $S_{\beta'}^{\Omega,\tau}(\omega)$, where the superscripts Ω and τ indicate that, in presence of motion, they may also depend on the jump frequency Ω and the pulse spacing τ of the solid echo sequence:

$$S^{\Omega,\tau}(\omega) = 2\pi \int_{-1}^{1} S_{\beta'}^{\Omega,\tau}(\omega) Q_{\beta_0}(\beta') \, d \cos \beta' \qquad (8.11)$$

The weighting factors $Q_{\beta_0}(\beta')$ in turn can be calculated from the orientation distribution function $P(\beta)$:

$$Q_{\beta_0}(\beta') = \int_{-\pi}^{\pi} P(\beta) \, d\gamma' \qquad (8.12)$$

where β is the angle between the director and the molecular axis defined above. β_0 is the angle between $\vec{B}_0$ and $\vec{n}$, and the various angles are interrelated according to

$$\cos \beta = \sin \beta_0 \sin \beta' \sin \gamma' + \cos \beta_0 \cos \beta' \qquad (8.13)$$

(for details see Spiess, 1982).

The line-shape calculation can be performed for any $P(\beta)$, but in practice a suitable functional form is chosen and then the calculated line-shape is fitted to the experimental spectrum by varying the parameters defining this function. Usually a simple Gaussian function or an expansion in Legendre polynomials is used. An example is shown in Figure 8.2c; the spectra were calculated taking into account an orientational distribution function of Gaussian shape (width parameter $\bar{\beta} = 20°$ with $\sigma = \sin \bar{\beta}$, cf. eqn 8.4) and the fast motion of the phenylene ring around $180°$.

(iii) *Expansion of the line-shape in terms of conical distributions.* This approach should be applied, if the common molecular axis is not parallel to a principal axis of the field gradient (Spiess, 1982). Even in absence of motion, a simple analytical expression can be given only if the asymmetry parameter is zero. The subspectra show up to three unsymmetrical singularities. This method allows an analysis of, for example, spectra due to deuterons in gauche bonds or in rigid phenylene rings of the mesogenic groups. In the intermediate exchange regime the subspectra can be calculated in a straightforward way as a superposition of NMR spectra for individual molecules, as in the case of planar distributions (Pschorn *et al.*, 1986).

8.2.2 ^{13}C-MAS-NMR

^{2}H-NMR is a useful method for studying molecular order in various systems, but requires isotopic labelling, an expensive and laborious step. A distinction of different chemical shifts resulting from carbons at the different sites in the molecule can be obtained in ^{13}C-MAS (Mehring, 1983), where the chemical shift anisotropy and the dipolar coupling are averaged out by spinning about the magic angle and proton decoupling. For spinning rates smaller than the frequency shift due to the chemical shift anisotropy (1 to 4 kHz compared to 5 to 15 kHz), the information about the chemical shift tensor is retained in a sideband pattern. In a 2D ^{13}C-MAS NMR experiment

(Harbison and Spiess, 1986; Harbison *et al.*, 1987) one obtains information on the chemical structure in one dimension and the information on molecular order is displayed in the second dimension. An outline of the experimental principle and the data analysis is briefly reviewed.

For an introduction to the experiment, consider a completely ordered system, such as a single crystal, in a fixed orientation with respect to the magnetic field. The NMR spectrum of such a single crystal depends on its orientation with respect to $\vec{B}_0$ at the time of measurement. This statement also applies for sample rotation periods smaller than the total data acquisition time such as in MAS. The MAS-NMR spectrum then depends on the rotor phase at the beginning of the experiment (Maricq and Waugh, 1979; Herzfeld and Berger, 1980; Munowitz and Griffin, 1982). This leads to a periodic modulation of phase and amplitude of the individual sidebands. Through a second Fourier transformation over the rotor phase this modulation is translated in the second frequency dimension as a measure of the molecular order. Thus the 2D MAS experiment requires a set of spectra at different rotor phases (Harbison and Spiess, 1986; Harbison *et al.*, 1987). The two-dimensional Fourier transformation produces a two-dimensional spectrum, which contains a sideband pattern in ω_2 (the chemical shift dimension) as well as a sideband pattern in ω_1 (the molecular order dimension), introduced via the rotor phase. Sidebands in the latter dimension are observed only for macroscopically oriented samples, while for isotropic samples only a centreband results.

The sample must be oriented in the rotor with an oblique angle with respect to the rotor axis as shown in Figure 8.4. Most efficient angles are between 30° and 70°; an orientation angle of 90° produces only even order sidebands (Harbison *et al.*, 1987). The data acquisition must be synchronized with the rotor phase in order to get angular dependent spectra. This is performed with an optical trigger as schematically shown in Figure 8.5. After a proton 90° pulse and cross-polarization to ^{13}C, the free induction decay (FID) is observed (Figure 8.5a). A simplification of the spectrum can be achieved

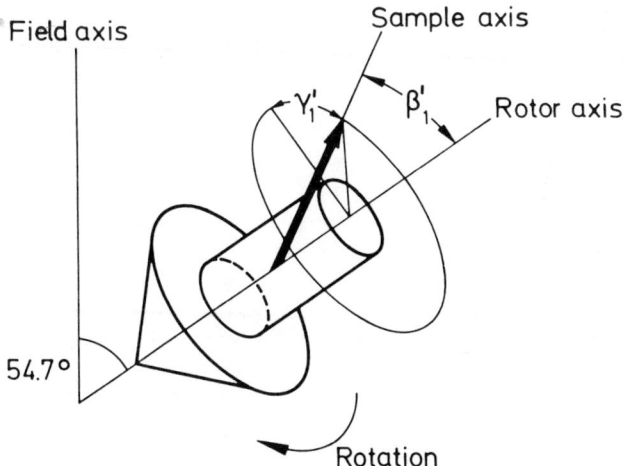

Figure 8.4 Definition of the angles β'_1 and γ'_1, which relate the sample axis to the rotor and field axis (Harbison *et al.*, 1987).

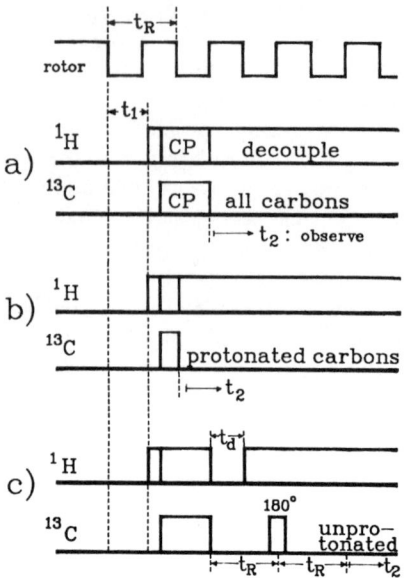

Figure 8.5 Pulse sequences used in the MAS experiments. 'Rotor' refers to the digitized output of the optical sensor used to detect the rotor position, t_R indicates one rotor cycle. Sequence (*a*) detects all carbons, by shortening the CP-time in (*b*) only protonated carbons are detected, whereas by sequence (*c*) the protonated carbons are suppressed (Blümich *et al.*, 1987).

by applying different pulse techniques (Blümich *et al.*, 1987), as shown in Figure 8.5*b* and *c*: by choosing a short cross-polarization time, the unprotonated carbons are suppressed, whereas by sequence *c* the protonated carbons can be suppressed by gated proton decoupling.

The two-dimensional spectrum is obtained by a first Fourier transformation over the acquisition time t_2 and a second one over the rotor phase t_1. For analysis, the full 2D spectrum need not be shown, but only slices through the spinning sidebands parallel to the chemical shift dimension ω_2. These slices are labelled by the sideband number $\pm M$, which are symmetric with respect to the centreband. An example on drawn polyethylene terephthalate (PET) is shown in Figure 8.6 of the centreband and of the sidebands parallel to the chemical shift dimension as obtained by the pulse sequences in Figure 8.5 in order to demonstrate also the simplifications from sequences *b* and *c*. The spectrum is recorded at a spinning speed of 3 kHz, where sidebands from the smaller chemical shift tensors of the glycolic ethylene are also present.

The data analysis closely follows the approach (i) for static samples outlined above. For systems with transverse isotropy, the orientational distribution function is again expanded in Legendre polynomials. The 2D ^{13}C spectrum is obtained by a weighted superposition of 2D sub-spectra $S_l(\omega_1, \omega_2)$. The weighting factors are again the order parameters $\langle P_l \rangle$ as introduced by eqn (8.6); the two-dimensional spectrum can therefore be written as

$$S(\omega_1, \omega_2) = \sum_{l=0}^{\infty} \langle P_l \rangle S_l(\omega_1, \omega_2), \tag{8.14}$$

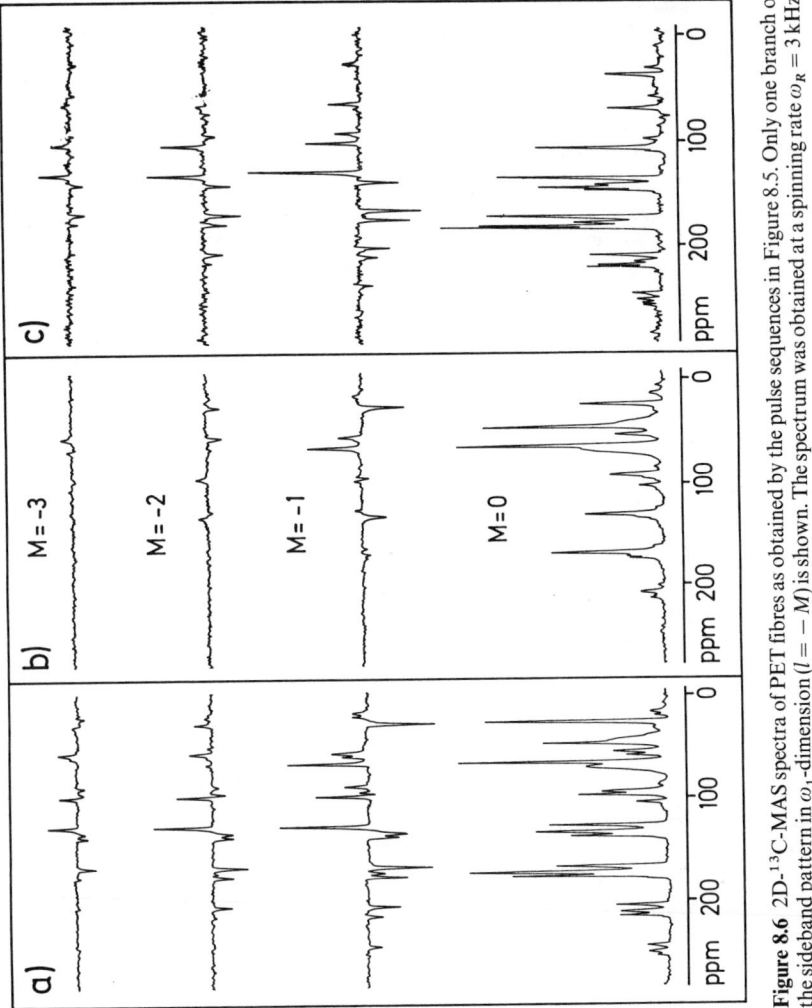

Figure 8.6 2D-^{13}C-MAS spectra of PET fibres as obtained by the pulse sequences in Figure 8.5. Only one branch of the sideband pattern in ω_1-dimension ($l = -M$) is shown. The spectrum was obtained at a spinning rate $\omega_R = 3$ kHz.

where the subspectra $S_l(\omega_1, \omega_2)$ include the dependence on the orientation of the sample in the rotor. From such a subspectral analysis the orientational distribution function is obtained in a straightforward fitting procedure (Harbison *et al.*, 1987). If the system is macroscopically oriented, but lacks transverse isotropy, e.g. due to biaxially, the orientational distribution function may be expressed in terms of spherical harmonics or in terms of Wigner rotation matrices (Harbison *et al.*, 1987; McBrierty, 1974).

As discussed previously, the expansion in Legendre polynomials is not convergent in static samples of high order. For the 2D MAS-NMR experiment, however, this analysis is generally applicable, because MAS mixes contributions from all single crystallite pattern into each sideband. This leads to an inherently lower angular resolution. Thus although the expansion of the orientational distribution function might diverge, the expansion of the NMR spectra *always* converges. The contribution of the subspectrum of order l is related to both the respective moment of the distribution function as well as to the number M of the sidebands in direction ω_1. Moments $\langle P_l \rangle$ have contributions in the sidebands up to $|M| = l$. If the sideband of order M is significant in the spectrum, then also the moment $l = |M|$ of the orientational distribution function must be considered, e.g. the PET spectra in Figure 8.6 contain sidebands up to the third order, so that at least the fourth moment is relevant for the orientational distribution function. By reducing the spinning speed, the number of detectable sidebands can be increased and 6th to 8th moments may be measured for highly ordered samples.

8.3 Experimental examples

Extensive studies were performed on *rod-like* liquid crystalline side group polymers. The systems under investigation are summarized in Figure 8.7 together with their phase behaviour. ^{2}H-NMR studies were performed on systems labelled in the mesogenic group (all systems except PMA64), in the spacer at the position next to the main chain (PA6) and in the main chain (PA6, PMA6 and PMA64). The analysis of the 2D ^{13}C-MAS spectrum will be shown on the example of PA6.

The NMR experiments were performed on Bruker CXP-300 and MSL-300 spectrometers. The samples were macroscopically oriented in the magnetic field by heating to the isotropic phase and slowly cooling through the nematic–isotropic phase transition. After annealing them for 15–30 minutes just below the phase transition, they were subsequently cooled below the glass transition, where the mesophase order is retained. Due to the positive diamagnetic susceptibility of most of the rod-like liquid crystals, they align with the director parallel to the magnetic field axis.

Angular dependent ^{2}H-NMR spectra were measured using a customized goniometer probe (Vogt, 1988), that allowed the sample to be turned around an axis perpendicular to the magnetic field. The ^{13}C-MAS spectra were taken with a Bruker double bearing probe with spinning rates between 1 and 4 kHz, chosen in such a way that an overlapping of sidebands and centrebands could be avoided. The two-dimensional data set consisted of 16 slices for the different rotor positions.

8.3.1 *Molecular motion of mesogens in the glassy state*

In polymeric side group liquid crystals, the low-temperature mesophase is frozen at the glass transition into an ordered glass. The molecular order introduced in the

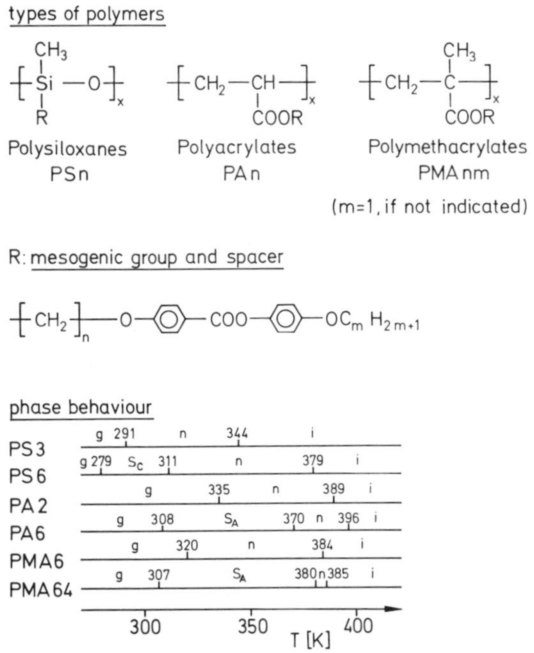

types of polymers

Polysiloxanes   Polyacrylates   Polymethacrylates
PSn   PA n   PMA nm

(m=1, if not indicated)

R: mesogenic group and spacer

phase behaviour

Figure 8.7 Systems under investigation: PA2, PA6, PS3, PS6, PMA6 and PMA64. The deuterated positions for the several systems are indicated. The mesophase ranges are given by the bars in the lower part of the figure.

mesophase is retained. At the glass transition, collective motions like the rotation around the long axis of the mesogen and translational motions of the polymer backbone are frozen in. Local motions like flips around 180° and kink motions in the spacer are still active. The local motion of the phenylene flip in the glassy state and its dependence on different parameters was studied in different mesomorphic systems: nematic, smectic A and smectic C, and in polymers with differing degrees of stiffness in the backbone: polymethacrylates, polyacrylates and polysiloxanes.

The type of motion exhibited by the phenylene rings of the mesogenic group consists of 180° jumps around the local C_2 axis (Geib et al., 1982). The time scale of this process can be monitored by observing changes in the line-shape and intensity during cooling far below the glass transition (Pschorn et al., 1986). In an oriented sample the angle between the director and the magnetic field should be chosen such, that the line-shapes in the rapid and in the slow motion limit, respectively, differ significantly. This excludes $\beta_0 = 0°$ and $\beta_0 = 90°$, because then the magnetic field axis is close to symmetry planes of the phenylene ring performing a motion between symmetry-related positions. The NMR line-shape of highly ordered samples is then largely independent of the exchange rate.

Figure 8.8 shows experimental and calculated ^{2}H-NMR spectra of the frozen nematic PA2 and the frozen smectic PA6 as a function of temperature (Pschorn, 1985). The spectra were recorded at an orientation of the director of $\beta_0 = 37°$ with respect to the magnetic field. They are plotted according to their integrated intensity, as also given

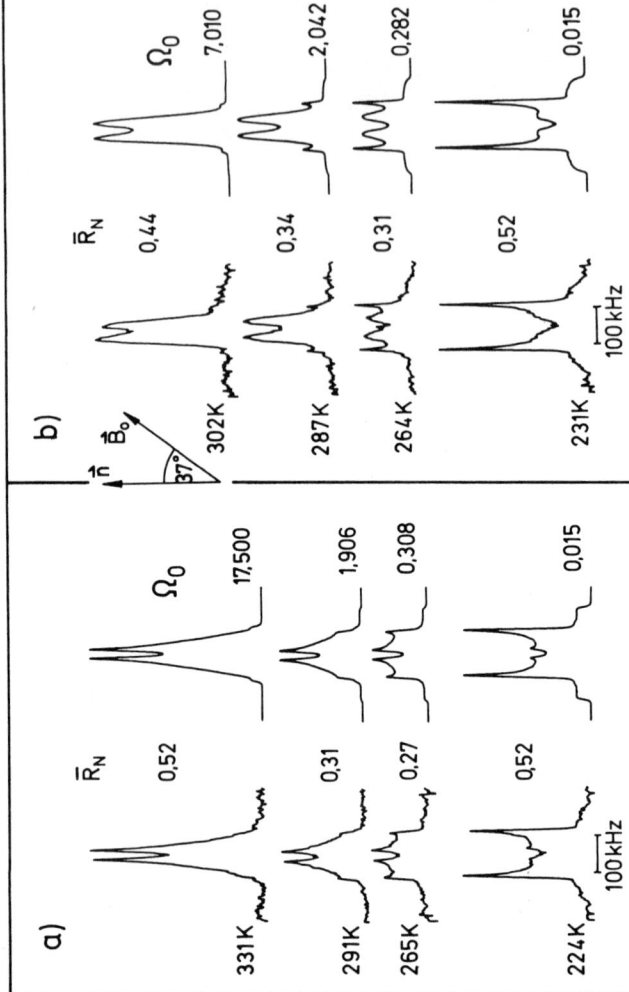

Figure 8.8 Observed and calculated ^{2}H-NMR spectra of the frozen nematic PA2 and the frozen smectic PA6 as a function of temperature. The spectra are plotted indicating their total integrated intensities, given also numerically by $\bar{R}_N$, Ω_0 (in units of δ) is the mean jump rate.

numerically by the average reduction factors $\bar{R}_N$, which are normalized to unity for both, the rapid and the slow exchange limit with $\bar{R}_N = 1$ for $\beta_0 = 0°$. Note that even at temperatures slightly below the glass transition significant intensity is lost at $\beta_0 = 37°$, whereas there is almost no change in intensity for $\beta_0 = 0$ at those temperatures. On the other hand, the major changes in line-shape are accompanied by relatively small changes in intensities, in contrast to what should be observed for a system where molecular dynamics can be described by a single correlation time. In the latter case one would expect a minimum echo intensity of 5% of the FID, whereas in these experiments the intensities in the intermediate regime are of the order of 30–40%. Moreover the line-shapes in the transition regime show significant contributions from both the slow and rapid exchange limit.

As shown in Figure 8.8, the line-shapes can be fitted both in shape and in intensity by taking into account a distribution of correlation times. The normalized reduction factor $\bar{R}_N$ reflects both the distribution of correlation times $K(\ln \Omega)$ and the reduction factors $R^\tau(\ln \Omega)$ of the spectra for single exchange rates Ω:

$$\bar{R}_N = \int_{-\infty}^{\infty} R^\tau(\ln \Omega) K(\ln \Omega) \, d \ln \Omega \qquad (8.15)$$

where $K(\ln \Omega)$ is a function of temperature. The distribution of correlation times could

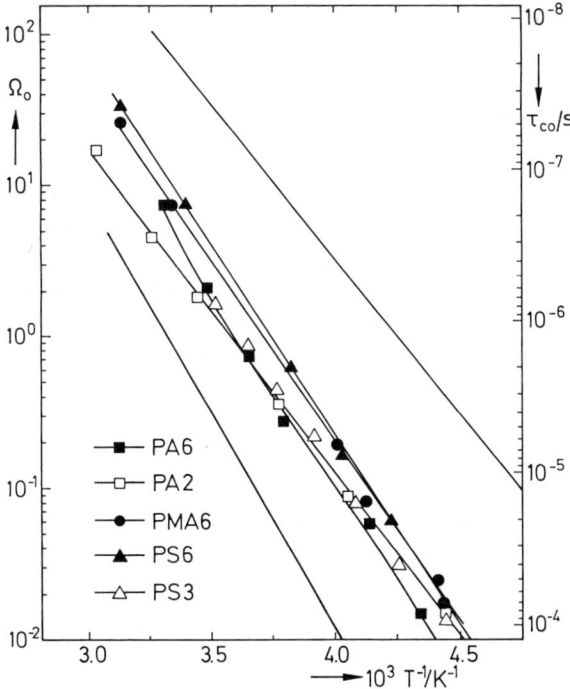

Figure 8.9 Temperature dependence of the mean jump rate for the systems under investigation: PS3, PS6, PA2, PA6 and PMA6. The width of the mean distribution of correlation times is indicated by the solid lines.

be identified as heterogeneous in nature ascribed to differences in molecular packing represented by a log-Gaussian function, where the temperature dependence of the mean jump rate in the centre of the distribution is given by

$$\Omega_0 = \Omega_{00} \exp - \left[\frac{\bar{E}_A}{RT} \right] \tag{8.16}$$

and can be described by a mean activation energy $\bar{E}_A$.

An activation diagram containing the results of all systems is shown in Figure 8.9, and all data are summarized in Table 8.1. The mean activation energy for the different systems is in a range between 42 and 48 kJ/Mol, and the distribution of correlation times is between 2.2 and 2.6 decades. The variation in activation energies and in widths of the distribution of correlation times is within experimental error. This means that the motional variety due to differences in the molecular packing within one system is more pronounced as the difference between systems having different mesophases.

In refinements of the line-shape analysis, an inclination between the local C_2 axis of the phenylene ring and the long axis of the mesogenic group of about 10°, cf. Figure 8.1, as well as a Gaussian distribution of jump angles around 180° with a half width of $\pm 10°$, were explicitly taken into account. Both possibilities resulted in only minor changes in the total spectra (Pschorn, 1985). Thus these NMR experiments cannot distinguish whether the 180° jumps occur about the long axis of the mesogenic group or about the local C_2 axis (Pschorn et al., 1986). The comparison of the data for different systems shows that the phenylene motion is a thermally activated process, which takes place also at higher temperatures in the liquid crystalline phase as was suggested long ago for low molar mass liquid crystals (Luz et al., 1974). In the liquid crystalline phase, this motion is augmented by diffusive rotation about the axis of the mesogen by arbitrary angles, as evidenced by the substantial narrowing of the NMR spectra (Geib et al., 1982). These two mechanisms can thus be separated by cooling below the glass transition, where the diffusive motion is frozen, whereas the local motion persists.

This type of phenylene motion is a common process also observed in other amorphous polymeric materials such as polystyrene and polycarbonate (Spiess, 1983).

Table 8.1 Molecular order and motion of one phenylene unit in the mesogenic group in the glassy state

Polymer	PA2 nematic	PA6 smectic A	PS3 nematic	PS6 smectic C	PMA6 nematic
Order parameter S	0.65	0.88	0.65	0.85	0.65
Orientational distribution function					
Shape	—	—	Gaussian	—	—
Width parameter $\bar{\beta}$	18°	10.5°	19°	11.5°	18.5°
Correlation time distribution function					
Shape	—	—	Gaussian	—	—
Width 2σ	2.2	2.5	2.6	2.4	2.3
Mean activation energy E_A of the 180° jump in kJ mol^{-1}	42	47	43	48	46

There the process has a similar activation energy; also, the width of the distribution of correlation times is comparable, and is mainly influenced by the free volume in the system and can be related to its mechanical properties. In particular it is hindered by additives, which suppress the mechanical relaxation (Wehrle *et al.*, 1987). Compared to these amorphous materials, it may be surprising that the highly ordered liquid crystals show such similar motional behaviour. Apparently the overall packing differences between the different mesophases are too small, compared to the packing differences within each phase itself, to introduce different motional behaviour in the respective phases. In accord with this observation, the intermolecular distance between the mesogenic groups that give rise to a diffuse wide-angle scattering X-ray reflection at about 0.45 nm was found to be essentially the same in all systems.

8.3.2 *Molecular order of mesogen, spacer and polymer backbone*

Temperature dependence of the order parameter in the liquid crystalline phase. In the liquid crystalline phase, the order parameter is directly related to the quadrupole splitting, provided the molecular motions are in the fast exchange limit and if the phase is uniaxial, according to (Diehl and Khetrapal, 1969):

$$\Delta\omega_Q = \bar{\delta}S = \bar{\delta}\tfrac{1}{2}\langle 3\cos^2\beta - 1\rangle \qquad (8.17)$$

The temperature dependence of the quadrupole splittings for the polymers PMA6, PA6 and PS6 is shown in Figure 8.10. The quadrupole splittings in the isotropic phase are zero due to the isotropic motion of the molecules in the liquid phase corresponding to an order parameter $S = 0$. At the transition from the isotropic to the nematic phase, the order parameter jumps to $S = 0.2$–0.3, values that are smaller than for the corresponding low molar mass liquid crystals (Boeffel *et al.*, 1983). For comparison of systems with

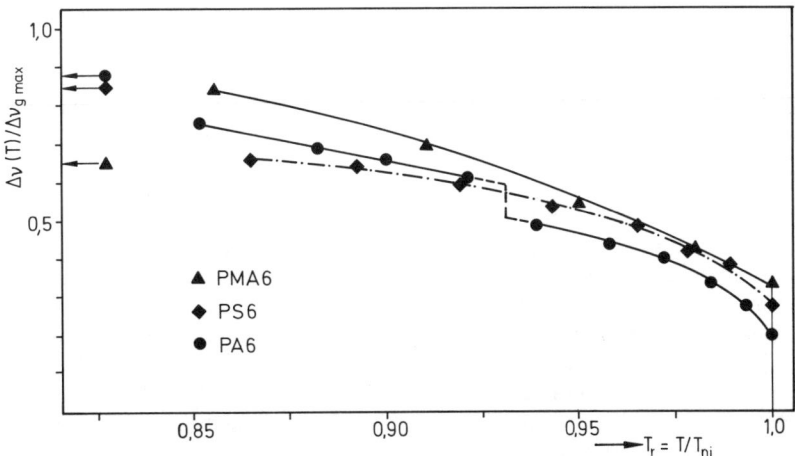

Figure 8.10 Temperature dependence of the quadrupole splittings for three systems with equal spacer length but different main chain structure. The dashed line marks the nematic–smectic phase transitions for PA6. The arrows indicate the order parameter in the glassy state as obtained from the line-shape analysis.

different mesophase ranges, the order parameters are plotted versus reduced temperatures T_r. PA6 has a transition from the nematic to the smectic A phase at $T_r = 0.932$, which is indicated by a jump in the order parameter due to its first order nature. The transition from nematic to smectic C for PS6 at $T_r = 0.821$ is not in the range shown here. The general temperature dependence of the order parameter is similar to that observed in low molar mass liquid crystals near the nematic–isotropic transition. At a reduced temperature below 0.85 (even higher for PMA6), the quadrupole splitting increases due to the slowing down of the correlation time for the rotation around the long molecular axis. Thus the quadrupole splitting can be larger than the order parameter. This fact is due to the polymer nature of the samples; it depends on the molecular mass as well as on the decoupling between the polymer chain and the mesogenic group via the flexible spacer. The polymethacrylate PMA6 has the highest molecular mass ($M_n = 143\,000$), and those of PA6 and PS6 are $40\,000$ and $27\,000$, respectively. The bulk viscosity of a polymer increases with the 3.4th power of the molecular mass (Ferry, 1980; De Gennes, 1979), so that in PMA6 the molecular motion is slowed down at higher temperatures due to its higher viscosity. In addition, the polymethacrylate chain is less flexible than the polyacrylate or the polysiloxane chain. This induces a stronger coupling between the polymer chain and the mesogenic group (cf. results on the polymer chains below). The coupling between the mesogenic group and the polymer chain also depends on the length of the flexible spacer. An anomalous increase of the quadrupole splitting not due to an increase of order in the nematic phase is also observed for PA2 and PS3 at temperatures 10° to 15° below the clearing point (Boeffel et al., 1983; Boeffel, 1987). For PS3, the motional averaging is not even complete at the clearing point at 346 K. Thus the mobility of the side group is influenced by the viscosity of the sample depending on the molecular mass as well as on the temperature.

The temperature dependence of the order parameter can be interpreted in a molecular field theory as introduced by Maier and Saupe (1958, 1960) and refined by Luckhurst (1979). Unlike low molar mass analogues of the same type of mesogenic groups, deviations of the intermolecular potential in the polymers from cylindrical symmetry can be neglected (Boeffel et al., 1983). Furthermore, the order parameter at the nematic–isotropic transition is not as high as predicted in the Maier–Saupe theory, indicating that the potential is not as steep to begin with.

The order parameter of the spacer and the polymer chain cannot be derived from the NMR data in the same way as for the mesogenic group, because of an incomplete averaging of molecular motions in the liquid crystalline phases. In the nematic phase, the order parameter measured for the spacer is half of that for the mesogenic group, indicating the increase in gauche bonds at that position resulting from kink motions. In the smectic phase, there is a dramatic increase in the quadrupole splitting due to the slowing down of the correlation times of molecular motions. For the polymer chain, even at the clearing point the averaging of the molecular motion is incomplete. To get the order parameters in this regime, the line-shapes have to be analysed, taking into account the orientational distribution function, the type and the correlation time of the diffusive motion (Müller et al., 1985).

Order parameter in the glassy state. At the glass transition, molecular motions are frozen except local ones as described before, whereas the molecular order introduced in the liquid crystalline phase is retained. By cooling below the glass transition, the molecular order of the different parts of the molecule can be determined by ^{2}H-NMR after selective labelling or by ^{13}C-MAS-NMR. Results from both methods are shown in Figures 8.11 and 8.12 for PA6. In Figure 8.11, ^{2}H-NMR spectra are plotted for the mesogenic group, the spacer and the main chain. They are obtained at angles of the director with respect to the magnetic field of 0° and 90°. The decrease of order from the mesogenic group to the polymer chain is reflected in the decreasing angular dependence of the spectra. The ordering tendencies in all parts of this molecule are the same: the mesogenic group, the spacer and the main chain are aligned preferentially parallel to the director (the results from the main chain will be discussed explicitly later). The order parameters ($\langle P_2 \rangle$) are 0.88, 0.52 and 0.25, respectively (Boeffel et al., 1986). The order parameters $\langle P_l \rangle$ as obtained by the NMR line-shape analysis are summarized in Table 8.2.

Figure 8.12 shows the centreslice and the sidebands labelled by $-M$ parallel to the ω_1-dimension of the two-dimensional ^{13}C-MAS sideband spectrum of PA6. The first-order sideband is enlarged to show the assignment of the different carbons used in the data analysis: the quaternary carbon on the outer phenylene ring next to the OCH$_3$-group (4), the spacer carbons (2) and the acrylic carbonyl (3). As shown before, the

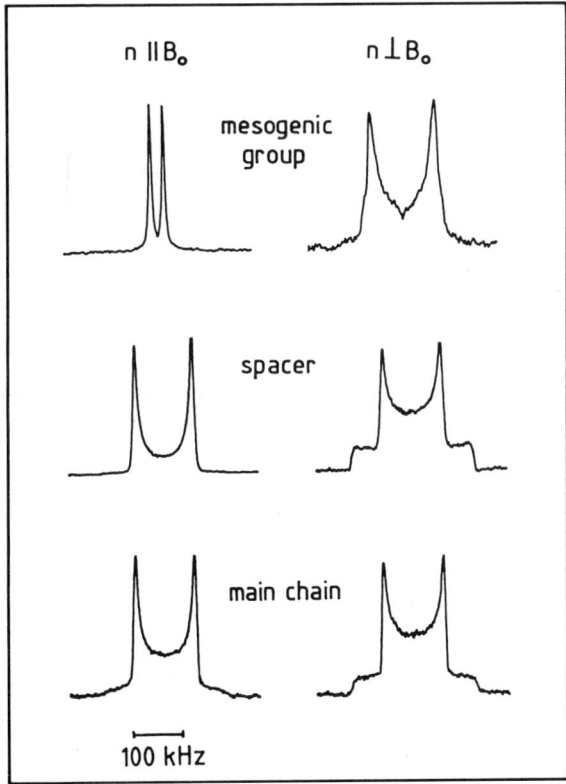

Figure 8.11 Orientation-dependent ^{2}H-NMR spectra of PA6 at different sites of the molecule in the frozen smectic state.

J

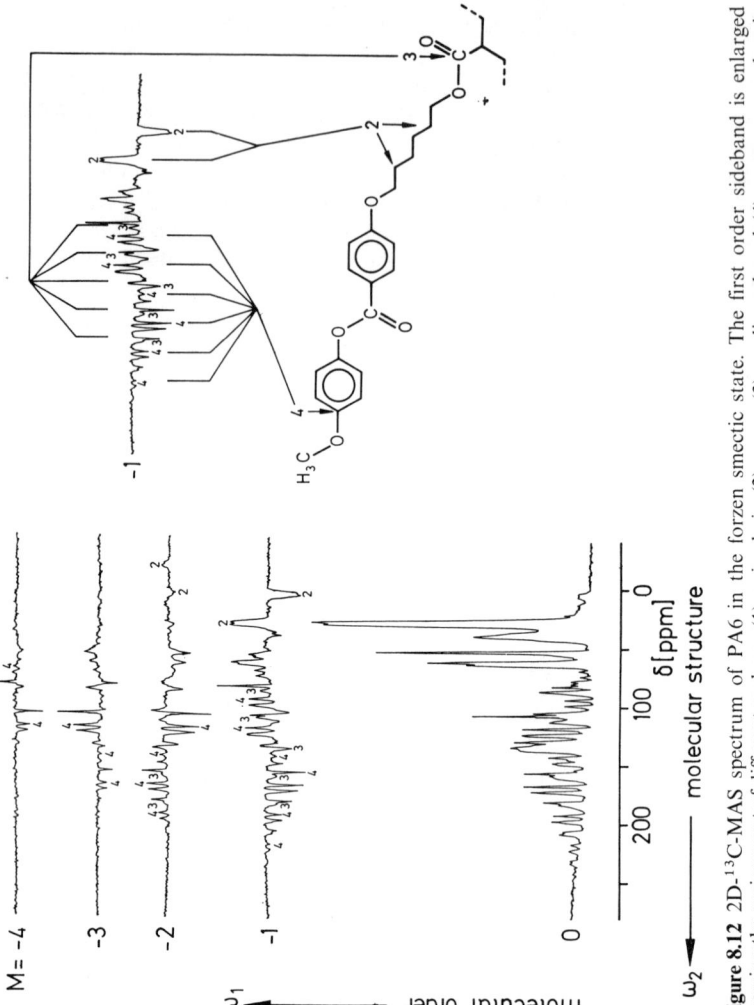

Figure 8.12 2D-^{13}C-MAS spectrum of PA6 in the forzen smectic state. The first order sideband is enlarged showing the assignment of different carbons: (1) main chain, (2) spacer, (3) acrylic carbonyl, (4) quaternary carbon in the outer phenylene ring connected to the OCH$_3$-group. Only the sidebands labelled by -M are shown, cf. Figure 8.6. The spectrum was obtained at a spinning rate $\omega_R = 2.1$ kHz.

Table 8.2 Order parameters $\langle P_l \rangle$ for different sites in PA6 as
obtained from ^{2}H- and 2D- ^{13}C MAS-NMR data

| | 1 | $\langle P_l \rangle_{\text{mol}}$ | |
		^{2}H-NMR	^{13}C-NMR
Mesogenic group	2	0.90	0.92
	4	0.72	0.78
	6	0.48	0.54
	8	0.35	0.40
Spacer	2	0.52	0.52
	4	0.19	0.19
Acrylic	2	—	0.45
Carbonyl	4	—	0.10
Polymer	2	0.25	—
Chain	4	0.05	—

number of the observed sidebands in the first dimension reflects the order at that site of
the molecule. For the mesogenic group, sidebands up to the fifth order (shown here up
to the fourth only) are observed, whereas the lower-ordered spacer and acrylic carbonyl
show sidebands to the second order only. The results (Blümich *et al.*, 1987) are
summarized in Table 8.2 together with those of the ^{2}H-NMR data. They are in good
agreement for the mesogenic group and the spacer. Sidebands of the polymer chain
(position 1) do not show up in the ^{13}C spectrum because of an orientation of the carbon
tensor at an angle close to 54° with respect to the rotor axis. No gradient in the order
parameter is observed for the spacer carbons, which is different to the ordering
behaviour of the alkyl chains in low molar mass liquid crystals (Boeffel *et al.*, 1983;
Boden *et al.*, 1981). The gradient from the spacer to the acrylic carbon next to the
polymer chain is not very significant, whereas there is a strong decrease in order from
that position to the chain through only one bond, suggesting the importance of
rotation about that bond for the decoupling (see also the following section).

Orientation of the polymer chain. According to the spacer model of side group liquid
crystal polymers, the polymer backbone should retain its random coil conformation if
the orientational decoupling with the mesogenic groups is effective. This requires that
there are no orientation-dependent inter- and intramolecular interactions between
these two molecular sites. In order to tackle this question, the molecular order was
studied in systems with the same mesophase structure but different chain flexibility, in
order to obtain information on how the polymer chain itself influences its ordering
structure. Investigations were also performed in systems with the same chain but
different mesophase structure in order to obtain information on the influence of the
anisotropic interactions resulting from the mesophase.

Figure 8.13 shows angular dependent spectra of the smectic polyacrylate and
polymethacrylate (Boeffel *et al.*, 1986; Boeffel and Spiess, 1988). The spectra were taken
at temperatures where molecular motions of both the polymer chain and the side
groups are frozen in.

For the polymethacrylate, the orientation of the C—C^{2}H$_3$ bond direction is
considered. The fast rotation of the methyl group around its C$_3$ axis leads to an

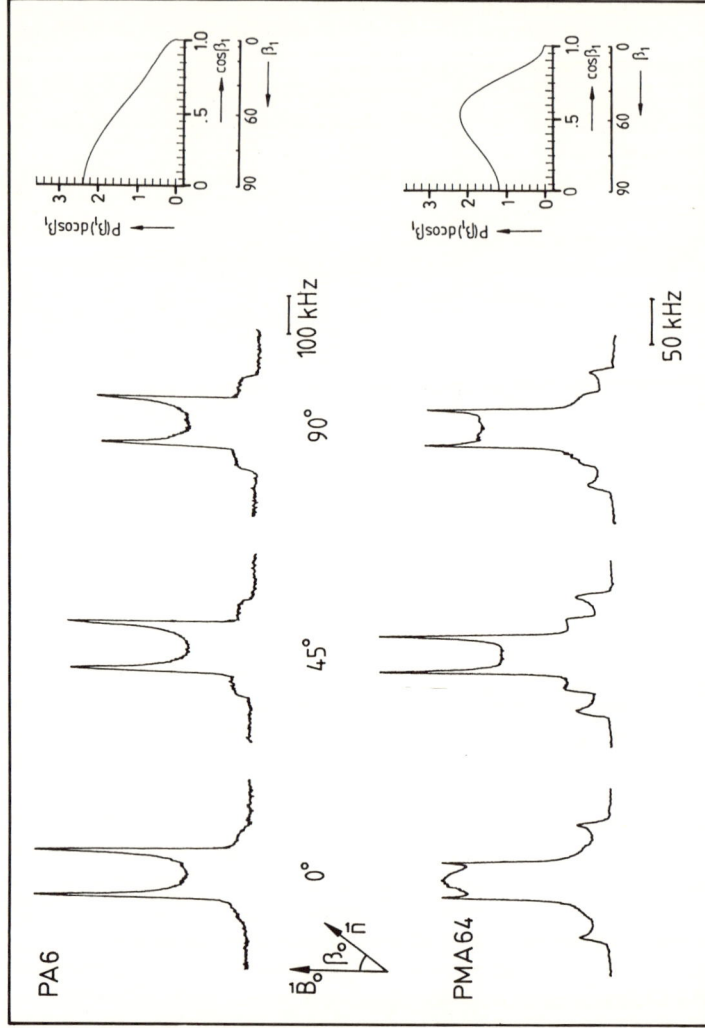

Figure 8.13 Orientation-dependent ^{2}H-NMR spectra of PA6 and PMA64, deuterated in the main chain, in their frozen smectic phase and on the right-hand side the orientational distribution functions as obtained from the line-shape analysis.

averaged field gradient with its mean value oriented along this axis (cf. Figure 8.1). Therefore the polymethacrylate spectrum consists of a superposition of a line-shape resulting from the time-averaged field gradient of the methyl group and a line-shape due to the rigid methylene group. The results discussed for these systems are only based on the methyl spectra, because the methylene spectrum lacks sufficient resolution to warrant detailed analysis.

The line-shapes of the spectra of the two smectics show remarkable differences resulting from clearly distinguishable orientational distributions of equivalent directions in the monomer unit as plotted on the right-hand side of Figure 8.13. The maximum of this function is at 90° for the polyacrylate but at 54° for the polymethacrylate. This maximum displays the most probable orientation of the interaction tensor with respect to the director. The preferential orientation of the C^2H bonds at an angle of 90° with respect to the director in the polyacrylate is only consistent with an elongation of the chain parallel to it. The width of the orientational distribution ($\pm 50°$ at half height) indicates that the order of the polymer chain is rather low. The coil is only slightly elongated, it can be interpreted as a rotational ellipsoid with its long axis parallel to the director. For the polymethacrylate the preferential orientation of the $C—C^2H_3$ bonds at 54° with respect to the director, on the other hand, is only consistent with an orientation of the chain perpendicular to it. The width of only $\pm 20°$ at half height of the orientational distribution function indicates the high conformational order at that molecular site. Assuming a Gaussian shape of the orientational distribution function and analysing it by expansion in Legendre polynomials, an molecular order parameter $\langle P_2 \rangle = 0.6$ results for the bisector of the $O_2C—C—CH_3$ moiety, cf. Figure 8.1. The order parameter of the mesogenic group is $\langle P_2 \rangle = 0.85$, therefore the polymer chain must be highly elongated perpendicular to the director.

The comparison of two systems with the same polymer chain but different mesophase structure shows, that this specific conformation is present already in the nematic phase (Boeffel and Spiess, 1988). In Figure 8.14 the orientational distribution functions of the $C—C^2H_3$ bonds for the frozen nematic PMA6 and the frozen smectic PMA64 are compared. In both cases the maximum is at 54°, but it is much less pronounced for the nematic polymer indicating that there the random coil conformation of the polymer backbone is only slightly distorted.

The main chain orientation also depends on the length of the flexible spacer, as observed in liquid crystalline elastomers based on a polysiloxane backbone (Finkelmann et al., 1984) (cf. Chapter 10). Elastomers of the same chemical type as the linear systems studied in this work have shown a similar tendency of order. In strained elastomers the *polymer chain* is oriented *parallel* to the axis of stress; the *mesogenic groups* are oriented *parallel* and *perpendicular* to it in polyacrylates and in polymethacrylates, respectively (Zentel and Benalia, 1987). The only difference between the polyacrylate and the polymethacrylate chain consists in the replacement of the hydrogen by a methyl group on the carbon that is linked to the mesogenic group. As mentioned above, the rotational freedom around the bond connecting the carboxyl group to the chain apparently plays an important role in decoupling of the mesogens from the polymer backbone. The steric difference between the chains introduced by this methyl group, therefore, is probably responsible for the different behaviour of these systems. Studies on the even more flexible polysiloxane backbones should lead to a

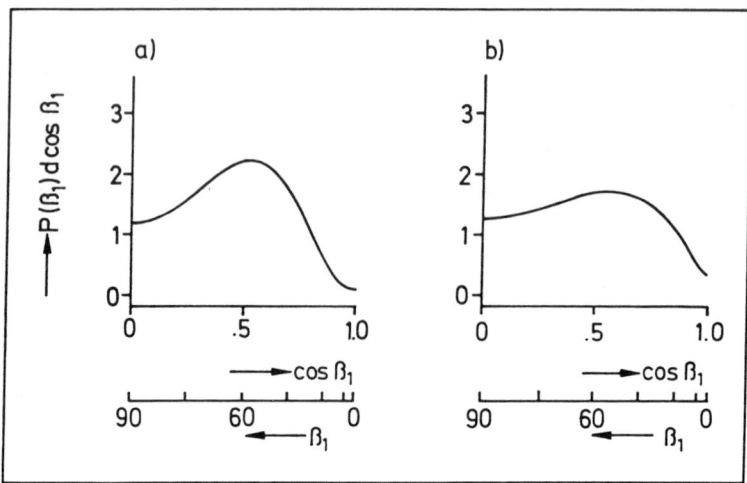

Figure 8.14 Orientational distribution functions of frozen smectic (*a*) and frozen nematic (*b*) polymethacrylates PMA64 and PMA6 (Boeffel and Spiess, 1988).

better understanding of the delicate coupling between the mesogenic groups and the polymer chain.

^{2}H-NMR and SANS studies on polymers *dissolved* in a nematic matrix have shown (Dubault *et al.*, 1985) that the anisotropic intermolecular forces through the mesogenic group introduced only little alignment of the polymer chain in the liquid crystalline matrix. The coupling between the polymer chain and the mesogenic group via covalent bonds is apparently necessary in order to achieve such high alignment in the polymer backbone, as observed in the systems studied here.

Vasilenko *et al.* (1985) calculated order parameters for the different sites of the molecule as a function of chain stiffness, spacer length and length of the mesogenic group. Our experimental results in the polyacrylates show lower chain order parameters and higher spacer order parameters than predicted in this theory. The theories of Warner *et al.* (cf. Wang and Warner, 1987; Renz and Warner, 1988 and Chapter 2) can also model liquid crystalline side group polymers. The different behaviour is interpreted on the basis of the interaction between the polymer chain, the spacer and the mesogenic group. Thus our results on the polyacrylate are consistent with their N_{III} phase, where nematic-like interactions between the spacer and the polymer chain exist because of local entropic forces and a parallel orientation of these two sites is preferred. For the polymethacrylates, the repulsive forces between the spacer and the polymer chain are dominant, the perpendicular alignment of the backbone and the mesogenic group is consistent with the N_I phase of Warner *et al.* It should be mentioned that the polymer chain in this type of comb-like polymer has no mesogenic character as predicted in the theory (Wang and Warner, 1987). More detailed calculations are in progress, where the interaction parameters are calculated on the basis of local entropic forces (Renz and Boeffel, 1988).

8.3.3 Polymers with discotic mesogens

The previous studies considered only systems with rod-like mesogens in the side group. A new class of liqiud crystals is formed by disc-shaped molecules (Chandrasekhar et al., 1977), where the discs are stacked into columns, which in turn form a two-dimensional structure. The arrangement of the molecules within the columns and of the columns themselves leads to new types of mesophases (Destrade et al., 1981). The synthesis of polymers having disc-shaped molecules either in the side group or in the main chain has been described by Kreuder et al. (1984, 1985, 1986), and Wenz (1985), but the phase characterization is difficult because of the extremely high viscosities in the discotic liquid crystalline phases. An important criterion for the existence of a mesophase is its ability to orient in the magnetic field. Thus ^{2}H-NMR can help to characterize the phase behaviour in discotic liquid crystals, in particular by comparing the polymers with low molar mass analogues in both their dynamics and their order structure.

^{2}H-NMR studies were performed on a homologous series of monomer, dimer and trimer as model compounds for main chain polymers and on a main chain polymer as well as on a side group polymer. The systems are based on a triphenylene core with either pentyloxy or heptyloxy side chains. The dimer, trimer and main chain polymer were synthesized by condensation of the respective functionalized monomers with hexadecanoic acid, discotic side groups were fixed to a polysiloxane chain via a spacer of eleven methylene units. The systems were deuterated in the aromatic core and in the α- and γ-position of the side chains (Kranig, 1987; Hueser, 1987).

Compared to monomers, the molecular motion in the liquid crystalline phase is restricted in the dimer, trimer and the polymers. Because of the linkage of the discs either to each other (dimer, trimer, main chain polymer) or to the polysiloxane chain, the motion around the sixfold axis perpendicular to the molecular plane is quenched (Kranig, 1987; Hueser, 1987; Hueser and Spiess, 1988), whereas this is a typical process in monomeric systems (Goldfarb et al., 1981; Luz et al., 1985). This means that backfolding of the polymer within a discotic column (Wenz, 1985) is not dominant in those systems. For all the systems, a macroscopic orientation in the magnetic field was observed (Hueser and Spiess, 1988), as shown in Figure 8.15 and Figure 8.16. Because of the negative diamagnetic susceptibility, discotic phases orient with the director distributed in a plane perpendicular to the magnetic field, as schematically drawn in Figure 8.15. Figure 8.16 shows the difference spectra between orientations 0° and 90° and 0° and 45°, respectively, of the normal of the director plane and the magnetic field, for the dimer and the main chain polymer, deuterated in the aromatic core or in the α-position of the side chains. The results for the side group polymer are comparable to those of the main chain polymer. For the simulation of the spectra of the aromatic core, an asymmetry parameter $\eta = 0.07$ was taken into account. The macroscopic orientation was achieved by cooling the samples from the isotropic phase slowly to 20° below the phase transition at 2 K h^{-1} and then faster below the glass transition. Because of the high viscosity of the polymers themselves and of the mesophases, the alignment time is longer than for the rod-like side group systems. The type of order introduced in the systems is the same in all samples. From the monomer, it is known that it forms an ordered discotic hexagonal phase (Destrade et al., 1981). By analogy, we conclude that the dimer and the higher homologues as well as the side group polymer likewise from discotic hexagonal phases (Hueser et al., 1989). This conclusion is also proved

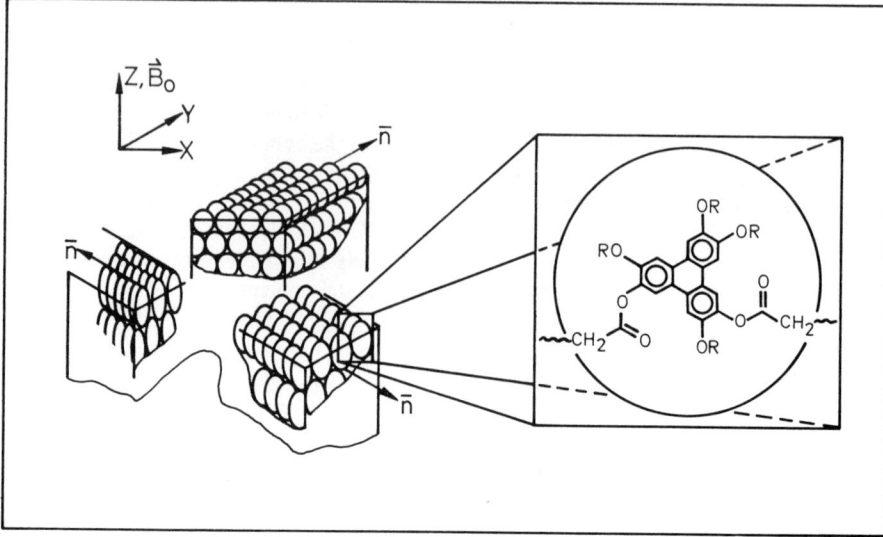

Figure 8.15 Orientation behaviour of discotic phases in the magnetic field (Goldfarb *et al.*, 1981).

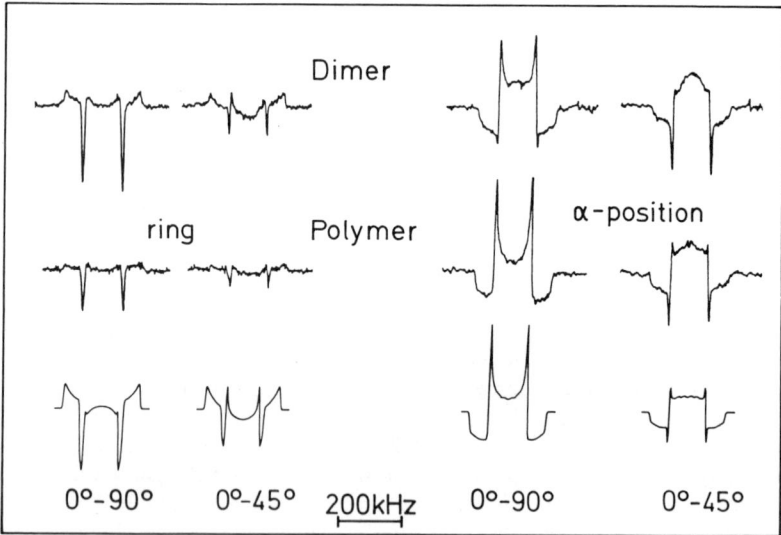

Figure 8.16 Discotic liquid crystals: difference spectra for the dimer (upper row) and the main chain polymer (middle) with pentyloxy side chains deuterated in the aromatic core (left) and the α-position (right). The difference spectra are obtained by subtraction of spectra measured at angles of 0°, 45° and 90° of the normal of the director plane with respect to the magnetic field. The bottom line shows simulated spectra with widths of the orientational distribution functions of the column axes of $\pm 12°$ for the aromatic core and $\pm 15°$ for the α-position (Hueser and Spiess, 1988).

by X-ray analysis (Herrmann-Schönherr *et al.*, 1986; Hueser *et al.*, 1989). The order parameter for the mesogenic core decreases from 90% to 80% from the dimer to the main chain polymer. The side group polymer is even less ordered by the magnetic field—the transition temperature to the isotropic phase is only 345 K, so that the mobility of the molecules is not sufficient to turn them in the magnetic field at that temperature. At temperatures well below T_g the alkyl side groups are oriented in the plane of the core and are not tilted with respect to the central unit. The order in the side groups decreases from the α- to the γ-position as expected for the high degree of motional freedom for this molecular site.

It was possible from the NMR experiments to determine the order at the different sites of the molecule—aromatic core, alkyl side group and spacer—as well as the conformation at different positions in the spacer and the alkyl chain (Hueser and Spiess, 1988). This demonstrates that ^{2}H-NMR is a useful method even for the characterization of complicated systems such as discotic liquid crystal polymers.

8.4 Comparison with other methods: neutron scattering, X-ray scattering, ESR spectroscopy, dielectric relaxation

Neutron scattering. Information on the conformation of the polymer chain can be obtained by *small-angle neutron scattering* (SANS), using mixtures of protonated and chain deuterated samples (Cotton *et al.*, 1974). The radii of gyration parallel and perpendicular to the director are obtained from macroscopically aligned samples by using a two-dimensional detection method. Neutron scattering probes the overall chain dimensions, but it does not give information about the conformation at individual sites which can be probed by NMR. The two methods thus nicely complement each other.

Studies on the polymethacrylates were performed by Kirste *et al.* (1985) (PMA6) and Keller *et al.* (1985) (PMA64). The ratio of the radii of gyration in the nematic phase 1:1.25, with the longer dimension perpendicular to the director. This small anisotropy is consistent with the broad orientational distribution found for the orientation of the C—C^{2}H$_3$ bonds with respect to the director from the ^{2}H-NMR data. For PMA64, a drastic increase in the anisotropy of the radii of gyration was observed at the transition from the nematic to the smectic phase reaching a ratio of 1:4 parallel and perpendicular to the director, respectively, in the latter. This result also fits our ^{2}H-NMR data well (cf. Figure 8.14) showing that this drastic change in the anisotropy of the radii of gyration is reflected in the respective widths of the orientational distribution functions, which indicate high chain orientation in the smectic phase. This fact has been taken as evidence for largely 'phase-separated' polymer chains between the smectic layers (Davidson *et al.*, 1985). The size of the radius of gyration parallel to the mesogenic groups (2.2 nm), however, is comparable with the layer spacing (2.95 nm). Thus considerable interpenetration between the smectic layers and the polymer chain must exist rather than a true phase separation (Keller *et al.*, 1985).

The anisotropy of the radii of gyration found in the frozen smectic phase of the polyacrylate (PA6) is much smaller than for the polymethacrylates, but of the opposite nature, with the ratio of approximately 1.1:1 (Boeffel and Jung, unpubl.). From this data the angular dependence of the spectra in the ^{2}H-NMR experiment should be even smaller than the observed one. As noted above, this angular distribution is consistent with a most probable orientation of the C^2H bonds in a plane perpendicular to the order axis as found for deuterons in *trans* conformations with the local chain axis parallel to the director. For *gauche* conformations, one deuteron of the CH_2 group is also in a plane perpendicular to the chain axis, whereas the second one lies on a cone at an angle of 35.2°. The latter conformation does not contribute significantly to the angular dependence of the NMR spectra, because the anisotropic contribution to the line-shape due to these two orientations largely cancel each other. NMR and SANS are consistent with a model assuming that the probability for *trans* conformations is high parallel to the director, whereas perpendicular to it, where backfolding of the chain is present, the probability for *gauche* is high. This type of chain arrangement leads to a nearly isotropic coil, whereas locally chain segments can be oriented considerably. After completion of this review Keller *et al.* (1989) concluded from a SANS study that the polymer chain in the polyacrylate is oriented perpendicular to the director in the smectic phase.

These comparisons of the ^2H-NMR data with SANS demonstrate that both methods provide complementary information, which is necessary for the understanding of the chain order behaviour in different liquid crystalline polymers.

X-ray scattering. X-ray scattering techniques provide information on the structure of liquid crystalline polymers, where the scattering results from both short- and long-range order of the mesogenic groups. The polymer chain gives rise to sufficient scattering intensity only if it is labelled with heavy atoms like chlorine, but this also modifies the chemical behaviour of the system. Thus X-ray is a suitable method for the characterization of the mesophase, but details about the chain conformation are only obtained indirectly.

X-ray studies yielding information on the layer structure and periodicity as well as on tilt angles and their change with temperature, have been performed on several systems also studied in this work (Zugenmaier and Mügge, 1984, 1985; Zentel and Strobl, 1984; Mattoussi *et al.*, 1986; Nachaliel *et al.*, 1987). Undulations were observed by Davidson *et al.* (1985) in magnetically aligned smectic polymethacrylates, parallel to the smectic layers. They were related to the polymer chain, the spacer and the tail group. From these results, it was also concluded that the polymer chain is preferentially oriented perpendicular to the mesogenic groups. Further information on the coupling between the mesogenic groups and the polymer chain was obtained from fibre patterns. In a fibre the polymer chain is expected to be oriented along the fibre axis. The orientation of the mesogenic group then contains information about the coupling of the two molecular sites. Results on PS6 from Zugenmaier *et al.* (1984), (1985) have shown, that the mesogenic groups tend to align perpendicular to the fibre axis and hence to the polymer chain. Evidence for a different ordering behaviour of the polysiloxane chain was obtained from SAXS measurements of a similar polymer differing in the spacer length, which was dissolved in a low molar mass nematic solvent (Mattoussi *et al.*, 1986). In the nematic phase an anisotropy of the radii of gyration was determined with a ratio of 1.4:1 with its long axis parallel to the director. X-ray studies on fibres of polyacrylates and polymethacrylates gave contradictory results, with orientation of the mesogenic group either parallel or perpendicular to the draw direction depending on the spacer length and the drawing conditions (Zentel and Strobl, 1984; Davidson *et al.*, 1985). The difficulty in producing well-defined fibres seems to be a limiting factor for these studies.

Fibres of reproducible quality were obtained from discotic liquid crystal polymers (Herrmann-Schönherr et al., 1986; Hueser et al., 1989) after annealing in the mesophase. For the main chain polymers the column axes were found to be oriented perpendicular to the draw direction, with a planar director distribution as also obtained after the alignment in the magnetic field (cf. Figure 8.15). The mesogens in the side group polymers, however, orient with the column axes parallel to the fibre axis (Hueser et al., 1989) and the mesogenic groups are highly oriented within the columns. This high order is already introduced by drawing, and cannot be increased by annealing in the mesophase as observed for the main chain polymers. This demonstrates that the decoupling between the siloxane polymer chain and the mesogenic groups must be efficient in the discotic side group liquid crystals, and different macroscopically aligned materials can be obtained by orientation in the magnetic field or through mechanical forces.

ESR spectrocopy. ESR studies were performed on the polyacrylates PA2 and PA6 by Wassmer et al. (1985). The molecular dynamics was studied in systems having different molecular masses and spacer lengths, and was also compared to a low molar mass analogue. The limitation of the ESR technique is the fact, that the system under investigation is not detected directly at a specific site, as in NMR, through a tensor at that molecular position, but its behaviour is studied via a spin label, which is dissolved in the matrix. The use of the low concentrations of the spin label usually assures that the behaviour of the matrix is monitored.

In these studies the cholestane spin probe was used, which reflects the behaviour of the mesogenic unit due to the rigid molecular structure of the probe. The order parameters, determined from the ESR data, do not show a molecular mass dependence for the polymers, they are also in good agreement with those obtained for the mesogenic group by NMR. The slowing down of the correlation times, as observed for systems with increasing molecular mass as well as for PA2 compared to PA6 due to the stronger coupling to the polymer backbone through the short spacer, was also detected. The advantage of the ESR experiment is the higher sensitivity of the line-shapes for shorter correlation times than in ^{2}H-NMR. More detailed information could be obtained from ESR, if the spin label reflects the chemical structure of a specific site of the molecule and if it is covalently linked to the polymer, so that also the translational degree of freedom is the same for the spin label and for the matrix. Such studies would extend the dynamic time scale towards shorter correlation times as compared to ^{2}H-NMR, and can yield information about the behaviour of solutes in liquid crystalline polymers.

Dielectric relaxation. A series of liquid crystalline side group polymers has been investigated by dielectric relaxation studies (Zentel et al., 1985; Kresse et al., 1982; Haase et al., 1985; Attard et al., 1986, 1986a, b, 1987a, b, cf. Chapter 7). This method is sensitive to reorientations of the dipole moments in the molecule. The molecular dynamics can be studied over a wide temperature and frequency range. Due to the complex structure of liquid crystalline side group polymers, typically several relaxation processes are observed and ascribed to different degrees of freedom: (i) reorientations of the whole side chain at high temperatures in the isotropic and in the mesophase; (ii) the motions of the polymer chain at the dynamical glass transition; and (iii) local processes below the glass transition, depending on the chemical structure corresponding to

motions in the spacer, in the terminal group or in the mesogenic group, if there is an
ester linkage between two phenyl rings (denoted as the β_1-relaxation following the
notation in Zentel et al. (1985)). The β_1-process occurs in the same temperature and
frequency range, where in the ^{2}H-NMR the phenylene motion is observed (cf. section
8.3.2). A combined activation diagram is shown in Figure 8.17 for PA2 (a) and PA6 (b)
(Zentel et al., 1985; Vallerien et al., 1989). The activation energies determined from the
dielectric loss curves of the β_1-relaxation are similar to those determined from the ^{2}H-
NMR for the phenylene motion. The measured dielectric data were fitted using the
Havriliak–Negami function (Havriliak and Negami, 1966), which describes asymmetric relaxation time distributions. This delivers directly the temperature-
dependent relaxation rate (Vallerien et al., 1989). The curves from the dielectric
relaxation data are slightly shifted to lower rates as compared to the NMR data. This
reflects the use of the asymmetric distribution function in the data analysis, which has a
longer tail to higher frequencies, whereas for the data evaluation of the ^{2}H-NMR
spectra, symmetric Gaussian functions were used to describe the distribution of
correlation times.

For a completely localized phenylene motion, however, the state of symmetry is
conserved and no change is induced in the dipole moment of the adjacent ester group.
This leads to the conclusion that both techniques probe coupled reorientations of the
mesogenic group. Such a statement could in principle be proved by a two-dimensional
experiment, where during the preparation time a dielectric relaxation•experiment is
performed, which after the evolution is detected by NMR. The two-dimensional
spectrum then contains the dielectric relaxation in one dimension and the NMR
response in the second dimension. The correlation of the two experiments should

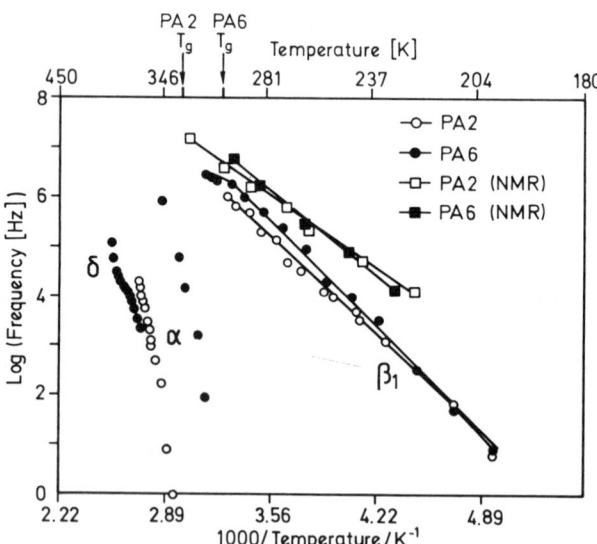

Figure 8.17 Combined activation diagram of dielectric relaxation and ^{2}H-NMR data for PA2
and PA6 (Vallerien et al., 1989). γ denotes the reorientation of the side chain at high temperatures,
α the motion associated with the glass transition, and β_1 the local motion in the glassy state.

contain an insight about the molecular process, responsible for the dielectric relaxation, which is not obtained directly from the experiment.

For a detailed treatment of dielectric relaxation spectroscopy of side chain LCPs, see Chapter 7.

8.5 New developments: LC elastomers, director reorientation measured on a molecular level

LC elastomers. The interaction between the polymer chain and the mesogenic group discussed above strongly influences the behaviour of liquid crystalline elastomers. In particular, high orientation of the mesogens can be achieved by relatively small deformations of the network (Finkelmann *et al.*, 1984). The order is introduced by mechanical forces in the liquid crystalline phase and retained by cooling below the glass transition, preserving any information stored in the mesophase.

First, ^{2}H-NMR studies were performed on the cross-linked analogue of the linear PS3 (Dames, 1987). The amount of cross-linking agent in the reaction mixture was 12%, which corresponds to one link per five monomer units. This results in a highly cross-linked network, which can only be deformed to a maximum elongation of 25%. This system is far from an ideal network, in particular it does not relax back to its original size even after heating to the isotropic phase.

In agreement with the results of Finkelmann *et al.* (1984, 1987) the mesogenic groups orient by applying mechanical forces perpendicular to the axis of stress and perpendicular to the surface of the sample. This effect is supported by the magnetic field, if the sample is stretched in the NMR probe with the orientation of the surface perpendicular to the field. In this geometry, the tendencies of inducing order are parallel for the mechanical forces and the magnetic field. In distinction to the behaviour of the nematic linear polymer at the same reduced temperature, the director in the cross-linked system remains largely parallel to the field when rotating the sample by 90°. This observation indicates that such high degrees of cross-linking influence the liquid crystalline behaviour.

Director reorientation measured at the molecular level. Up to now we have concentrated on local molecular properties, which can be studied by NMR methods. NMR can also be used, however, to investigate the *collective* behaviour of liquid crystalline phases. If a liquid crystal is rotated about an axis perpendicular to the magnetic field, it experiences two torques due to the magnetic field and due to the sample rotation, respectively (Leslie *et al.*, 1972). If the former predominates (low spinning speeds, low viscosity) the director remains essentially parallel to $\vec{B}_0$, if the latter predominates (high spinning speeds, high viscosity), it follows the sample rotation. For an accurate determination of the rotational viscosity γ_1 a spinning speed ω_R should be chosen such, that the two torques are comparable in magnitude. According to the continuum theory of liquid crystals, the director is predicted to align at an angle θ_∞ to the field given by (Leslie *et al.*, 1972):

$$\sin 2\theta_\infty = \frac{2\omega_R \, \gamma_1}{H_0 B_0 \Delta\chi} \quad \text{where} \quad \theta \leqslant 45° \tag{8.18}$$

$\Delta\chi$ is the anisotropy of diamagnetic susceptibility, H_0 is the magnetic field of the inductance B_0. The orientation of the director can be either measured macroscopically (Heppke and Schneider, 1974; Bock et al., 1986), or through the resulting quadrupole splitting in ^{2}H-NMR (Emsley et al., 1981). The rotational viscosity γ_1 can be measured more accurately, if the time dependence of the director at the beginning of the sample rotation is followed. If the rotation is started at time $t = 0$, the angle θ between the director at time t and $\vec{B}_0$ follows the equation (Bost, 1987):

$$\tan \theta =$$

$$\frac{(\sin 2\theta_\infty - \tan \theta_0)\left(1 - \exp\left[-\frac{t}{\tau}\sin 2\theta_\infty\right]\right) + \cos 2\theta_\infty \tan \theta_0\left(1 + \exp\left[-\frac{t}{\tau}\sin 2\theta_\infty\right]\right)}{(1 - \sin 2\theta_\infty \tan \theta_0)\left(1 - \exp\left[-\frac{t}{\tau}\sin 2\theta_\infty\right]\right) + \cos 2\theta_\infty\left(1 + \exp\left[-\frac{t}{\tau}\sin 2\theta_\infty\right]\right)}$$

$$(8.19)$$

where

$$\tau = \frac{\gamma_1}{H_0 B_0 \Delta\chi}. \tag{8.20}$$

The angle θ can be measured from the quadrupole splitting, if the data acquisition is synchronized with the sample rotation. In polymeric liquid crystals, values of ω_R below $10^{-2}\,\mathrm{rad\,s^{-1}}$ are of interest and the synchronization can be achieved easily. A related experiment has been published by Martins et al. (1986), where the director reorientation back to equilibrium is followed after a sudden rotation through a fixed angle α. By following the time-dependence of the ^{2}H-NMR spectra explicitly, deviations from the simple hydrodynamic treatment based on a single rotational viscosity can be detected.

8.6 Summary and outlook

The application of NMR to liquid crystalline polymers based on rod- and disc-shaped mesogens demonstrates the utility of this technique in studying *molecular order* and *dynamics*. ^{2}H-NMR is especially suitable in this field, because after selective labelling the different sites of the molecule can be monitored separately. 2D ^{13}C MAS-NMR does not require isotopic labelling and therefore opens a new field for investigating the molecular order of liquid crystalline polymers in a routine way.

Studies of this type have confirmed the spacer concept of liquid crystal side group polymers, but have also shown its limitations: whereas the dynamical and orientational behaviour of the mesogenic groups in the glassy state are largely independent of the polymer chain, the mesogenic groups can transfer considerable order to the polymer chain via the flexible spacer. This coupling depends on such different parameters as spacer length and flexibility of the polymer backbone (see also Percec and Pugh, Chapter 3).

An even deeper insight into the relationship between molecular parameters and macroscopic behaviour will be obtained by combination of methods which complement each other, such as ^{2}H-NMR and SANS, or NMR and dielectric relaxation experiments. The design of new two-dimensional experiments, where the NMR experiment is used as a detection method after an optical, electrical or mechanical excitation, may lead to a better understanding of these complex materials, because processes following these excitations are monitored directly on a molecular level.

Acknowledgements

This review is based on projects on polymeric liquid crystals in our group. Our work has benefited greatly from collaboration with the group of Professor Ringsdorf at the University of Mainz, which is highly appreciated. The main part of the review was written while one of us (CB) was on a postdoctoral fellowship at the Weizmann Institute of Science in Rehovot, Israel. It is a pleasure to thank Professor Zeev Luz for his stimulating discussions on this work. The continuous financial support of this project by the Deutsche Forschungsgemeinschaft (SFB 41) is greatly acknowledged.

References

Abragam, A. (1961) *The Principles of Nuclear Magnetism*, Oxford University Press, Oxford.

Araki, K. and Attard, G. (1986) *Liquid Crystals* 1, 301.

Attard, G. (1986) *Mol. Phys.* 58, 1087.

Attard, G. and Williams, G. (1986) *Liquid Crystals* 1, 253.

Attard, G., Araki, K. and Williams, G. (1987a) *Brit. Polym. J.* 19, 119.

Attard, G., Moura-Ramos, J. and Williams, G. (1987b) *J. Polym. Sci., Polym. Phys. Ed.* 25, 1099.

Blümich, B., Boeffel, C., Harbison, G., Yang, Y. and Spiess, H. (1987) *Ber. Bunsenges. Phys. Chem.* 91, 1100.

Bock, F., Kneppe, H. and Schneider, F. (1986) *Liquid Crystals* 1, 239.

Boden, N., Clark, L., Bushby, R., Emsley, J., Luckhurst, G. and Stockley, C. (1981) *Mol. Phys.* 42, 565.

Boeffel, C. (1987) PhD thesis, University of Mainz.

Boeffel, C. and Spiess, H. (1988) *Macromolecules* 21, 1626.

Boeffel, C., Hisgen, B., Pschorn, U., Ringsdorf, H. and Spiess, H. (1983) *Israel J. Chem.* 23, 388.

Boeffel, C., Spiess, H., Hisgen, B., Ringsdorf, H., Ohm, H. and Kirste, R. (1986) *Makromol. Chem., Rapid Commun.*, 7, 777.

Bost, R. (1987) Diploma thesis, University of Mainz.

Chandrasekhar, S., Sadashiva, B. and Suresh, K. (1977) *Pramana* 9, 471.

Chapoy, L. (ed.) (1985) *Recent Advances in Liquid Crystalline Polymers*. Elsevier Applied Science, London.

Chidichimo, G., Yaniv. Z., Vaz, N. and Doane, J. (1982) *Phys. Rev. A* 25, 1077.

Cotton, J., Decker, D., Benoit, H., Franoux, B., Higgins, J., Jannink, G., Ober, R., Picot, C. and des Cloiseaux, J. (1974) *Macromolecules* 7, 863.

Dames, B. (1987) Diploma thesis, University of Mainz.

Davidson, P., Keller, P. and Levelut, A. (1985) *J. Physique* 46, 939.

DeGennes, P. (1979) *Scaling Concepts in Polymer Physics*. Cornell University Press, Ithaca.

Destrade, C., Huu Tinh, N., Gasparoux, H., Malthete, J. and Levelut, A. *Mol. Cryst., Liq. Cryst.* 71, III.

Diehl, P. and Khetrapal, C. (1969) in *NMR Basic Principles and Progress*, eds. Diehl, P., Fluck, E. and Kosfield, R., vol. 1, Springer, Heidelberg.

Dubault, A., Ober, R., Veyssie, M. and Cabane, B. (1985) *J. Physique* 46, 1227.

Emsley, J., Khoo, S., Lindon, J. and Luckhurst, G., (1981) *Chem. Phys. Lett.* 77, 609.

Ferry, J. (1980) *Viscoelastic Properties of Polymers*, 3rd edn., Wiley, New York.

Finkelmann. H., Ringsdorf, H. and Wendorff, J. (1978) *Makromol. Chem.* 179, 273.

Finkelmann, H., Kock, H., Gleim, W. and Rehage, G. (1984) *Makromol. Chem., Rapid Commun.* 5, 287.

Geib, H., Hisgen, B., Pschorn, U., Ringsdorf, H. and Spiess, H. (1982) *J. Amer. Chem. Soc.* 104, 917.

Gleim, W. and Finkelmann, H. (1987) *Makromol. Chem.* 188, 1489.

Goldfarb, D., Luz, Z. and Zimmerman, H. (1981) *J. Physique* 42, 1303.

Gordon, M. and Platé, N. (eds.) (1984) *Advances in Polymer Science*, Springer, Berlin, 59–61.

Haase, W., Pranoto, H., and Bormuth, F. (1985) *Ber. Bunsenges. Phys. Chem.* 89, 1229.

Haeberlen, U. (1976) in *High Resolution NMR in Solids: Selective Averaging*, (*Adv. Magn. Reson. Suppl.* 1), ed. Waugh, J., Academic Press, New York.

Harbison, G. and Spiess, H. (1986) *Chem. Phys. Lett.* 124, 128.

Harbison, G. Vogt, V. and Spiess, H. (1987) *J. Chem. Phys*, 86, 1206.

Havriliak, S. and Negami, S. (1966) *J. Polym. Sci.* C14, 99.

Haward, R. (ed.) (1973) *The Physics of Glassy Polymers.* Applied Science, London.
Hentschel, R., Schlitter, J., Sillescu, H. and Spiess, H. (1978) *J. Chem. Phys.* **68**, 56.
Hentschel, R., Sillescu, H. and Spiess, H. (1981) *Polymer* **22**, 1516.
Heppke, G. and Schneider, F. (1974) *Z. Naturforsch.* **29A**, 1356.
Herrmann-Schonherr, O., Wendorff, J., Kreuder, W. and Ringsdorf, H. (1986) *Makromol. Chem., Rapid Commun.* **7**, 97.
Herzfeld, J. and Berger, A. (1980) *J. Chem. Phys.* **73**, 6021.
Hueser, B. (1987) PhD thesis, University of Mainz.
Hueser, B. and Spiess, H. (1988) *Makromol. Chem., Rapid Commun.* (in press).
Hueser, B., Pakula, T. and Spiess, H. (1989) *Macromolecules* **22** (in press).
Keller, P., Carvalho, B., Cotton, J., Lambert, M., Moussa, F. and Pepy, G. (1975) *J. Physique Lett.* **46**, L-1065.
Keller, P., Cotton, J., Moussa, F, Noirez, L., Pepy, G., Hardouin, F., Richard, H. and Strazielle, C. (1989) *Macromolecules* (in press).
Kirste, R. and Ohm, H. (1985) *Makromol. Chem., Rapid Commun.* **6**, 179.
Kranig, W. (1987) Diploma thesis, University of Mainz.
Kresse, H., Kostromin, S. and Shibaev, V. (1982) *Makromol. Chem. Rapid Commun.* **3**, 509.
Kreuder, W. (1986) PhD thesis, University of Mainz.
Kreuder, W. and Ringsdorf, H. (1984) *Makromol. Chem. Rapid. Commun.* **4**, 807.
Kreuder, W., Ringsdorf, H. and Tschirner, P. (1985) *Makromol. Chem. Rapid Commun.* **6**, 367.
Leslie, F., Luckhurst, G. and Smith, H. (1972) *Chem. Phys. Lett.* **13**, 368.
Lindsey, C. and Patterson, G. (1970) *J. Chem. Phys.* **73**, 3348.
Luckhurst, G. (1979) in *The Molecular Physics of Liquid Crystals*, eds. Luckhurst, G. and Gray, G., Academic Press, London.
Luz, Z., Hewitt, R. and Meiboom, S. (1974) *J. Chem. Phys.* **61**, 1758.
Luz, Z., Goldfarb, D. and Zimmerman, H. (1985) in *NMR of Liquid Crystals*, ed. Emsley, J. Reidel, Dordrecht.
Maier, W. and Saupe, A. (1958) *Z. Naturforsch.* **13a**, 546.
Maier, W. and Saupe, A. (1959) *Z. Naturforsch.* **14a**, 882.
Maier, W. and Saupe, A. (1960) *Z. Naturforsch.* **15a**, 287.
Maricq, M. and Waugh, J. (1979) *J. Chem. Phys.* **70**, 3300.
Martins, A., Esnault, P. and Volino, F. (1986) *Phys. Rev. Lett.* **57**, 1745.
Mattoussi, H., Ober, R., Veyssie, M. and Finkelmann, H. (1986) *Europhys. Lett.* **2**, 233.
McBrierty, V. (1974) *J. Chem. Phys.* **61**, 872.
McBrierty, V. and Douglas, D. (1980) *Phys. Rep.* **64**, 61.
Mehring, M. (1983) *High Resolution NMR Spectroscopy in Solids*, 2nd ed., Springer, Berlin.
Moussa, F., Cotton, J., Hardouin, F., Keller, P., Lambert, M., Pepy, G., Maussac, M. and Richard, H. (1987) *J. Physique* **48**, 1079.
Muller, K., Meier, P. and Kothe, G. (1985) in *Progress in NMR Spectroscopy*, eds. Emslly, J., Freeney, J. and Sutcliffe, L., Pergamon, Oxford.
Munowitz, M. and Griffin, R. (1982) *J. Chem. Phys.* **76**, 2848.
Nachaliel, E., Keller, P., Davidov, D., Zimmerman, H. and Deutsch, M. (1987) *Phys. Rev. Lett.* **58**, 896.
Patterson, G. (1983) *Adv. Polym. Sci.* **48**, 124.
Piskunov, M., Kostromin, S., Stroganov, L., Shibaev, V. and Platé, N. (1982) *Makromol. Chem. Rapid Commun.* **3**, 443.
Pschorn, U. (1985) PhD thesis, University of Mainz.
Pschorn, U., Spiess, H., Hisgen, B. and Ringsdorf, H. (1986) *Makromol. Chem.* **187**, 2711.
Renz, W. and Warner, M. (1988) *Proc. R. Soc. London* (in press).
Ringsdorf, H., Schmidt, H., Baur, G. and Kiefer, R. (1985) in *Recent Advances in Liquid Crystalline Polymers*, ed. Chapoy, L., Elsevier Applied Science, London.
Ringsdorf, H., Schmidt, H., Baur, G., Kiefer, R. and Windscheid, F. (1986) *Liquid Crystals* **1**, 319.
Roe, J. (1965) *J. Appl. Phys.* **36**, 2004.
Roth, H. and Krucke, B. (1986) *Makromol. Chem.* **187**, 2655.
Schmidt, C., Kuhn, K. and Spiess, H. (1985) *Progr. Colloid. Polym. Sci.* **71**, 71.
Shibaev, V. and Platé, N. (1985) *Pure Appl. Chem.* **57**, 1589.
Shibaev, V., Platé, N. and Freidzon, Y. (1979) *Polym. Sci., Polym. Chem. Ed.* **17**, 1655.

Simon, R. and Coles, H. (1986) *Liquid Crystals* **1**, 281.

Spiess, H. (1978) in *NMR Basic Principles and Progress*, eds. Fluck, E. and Kosfeld, R., Springer, Berlin.

Spiess, H. (1982) in *Developments in Oriented Polymers–1*. ed. Ward, I., Applied Science, London.

Spiess, H. (1983) *Colloid. Polym. Sci.* **261**, 193.

Spiess, H. (1985) in *Advances in Polymer Science* **66**, 23 eds. Kausch, H. and Zachmann, H., Springer, Berlin.

Spiess, H. and Sillescu, H. (1981) *J. Magn. Reson.* **42**, 381.

Vallerien, S., Kremer, F. and Boeffel, C. (1988) *Liquid Crystals* (in press).

Vasilenko, S., Shibaev, V. and Khoklov, A. (1985) *Makromol. Chem.* **186**, 1951.

Vogt, V. (1988) PhD thesis, University of Mainz.

Wang, X. and Warner, M. (1987) *J. Phys.* **A20**, 713.

Wassmer, K., Ohmes, E., Portugall, M., Ringsdorf, H. and Kothe, G. (1985) *J. Amer. Chem. Soc.* **107**, 1511.

Wehrle, M., Hellmann, G. and Spiess, H. (1987) *Colloid. Polym. Sci.* **265**, 815.

Wenz, G. (1985) *Makromol. Chem. Rapid Commun.* **6**, 577.

Williams, G. and Watts, D. (1970) *Trans, Faraday Soc.* **66**, 80.

Zentel, R. and Strobl, G. (1984) *Makromol. Chem.* **185**, 2669.

Zentel, R. and Benalia, M. (1987) *Makromol. Chem.* **188**, 665.

Zentel, R., Strobl, G. and Ringsdorf, H. (1985) *Macromolecules* **18**, 960.

Zugenmaier, P. and Mügge, J. (1984) *Makromol. Chem., Rapid Commun.* **5**, 11.

Zugenmaier, P. and Mügge, J. (1985) in *Recent Advances in Liquid Crystalline Polymers*, ed. Chapoy, L., Elsevier Applied Science, London.

9 Cholesteric polymers with side mesogenic groups: structure, optical properties and intramolecular mobility

V.P. SHIBAEV and Ya.S. FREIDZON Chemistry Department, M.V. Lomonosov Moscow State University, Moscow 119899, USSR

9.1 Introduction

Among the many various types of liquid crystalline (LC) compounds, the cholesteric liquid crystals attract particular interest on account of their unusual helical supermolecular structure, which determines a number of unique optical properties, such as selective reflection of light in different wavelength ranges, high sensitivity of the selective light reflection to temperature, and an extreme optical activity (Belyakov and Sonin, 1982). These properties provide an extensive application of the cholesteric liquid crystals as highly sensitive thermoindicators, IR and super-high-frequency radiation visualizers, thermographic materials for medical diagnoses and non-perturbing quality control of electronics devices.

It has been only the cholesteric type of mesophase that was discovered *for the first time* both in the low molar mass systems (cholesteryl benzoate) and for polymeric systems (lyotropic solutions of poly-γ-benzyl-L-glutamate) (Elliott and Ambrose, 1950; Robinson, 1956). Moreover, early attempts to synthesize thermotropic LC polymers with the side mesogenic groups were initiated by the desire to create cholesteric polymers via incorporation of chiral* cholesterol molecules as the side groups into acrylic and methacrylic polymers.[†]

9.2 General considerations on structure and peculiarities of optical properties of cholesterics

First of all it should be emphasized that the cholesteric mesophase is formed only by chiral molecules, and this specifies the appearance of their helical twisting (Belyakov and Sonin, 1982). The character of molecular ordering at short distances in the cholesteric mesophase is identical to that of a nematic mesophase, i.e. there is no long-range order in the arrangement of centres of mass of molecules, and rod-like molecules are oriented mainly along the director **n**. As can be seen in Figure 9.1, the molecules lie as if in the quasi-nematic layers characterized by definite orientations of the director **n**.

---

*Chirality denotes that a property of an object is incongruent with its reflection in an ideal plane mirror; chirality alone determines optical activity of molecules, and chiral compounds exist usually as a pair of isomers, a lefthanded and a righthanded one.
[†]Poly(cholesteryl acrylates) (PChA) and poly(cholesteryl methacrylates) (PChM), synthesized by polymerization of the corresponding monomers in the works cited, however, proved to possess no LC properties.

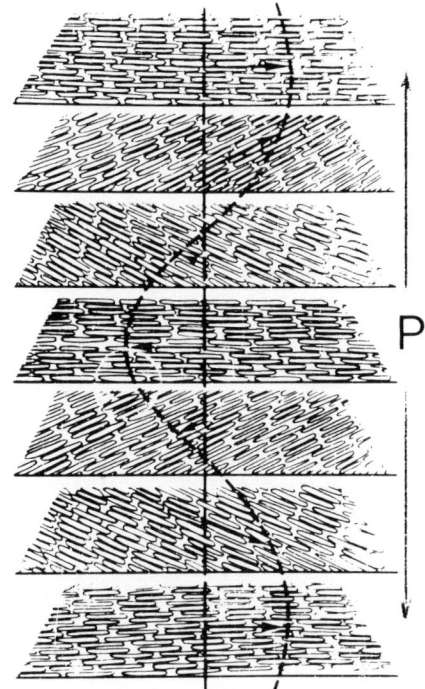

P

Figure 9.1 Schematic representation of helical twisting of molecules in the cholesteric mesophase. Arrows show the direction of the long axes of molecules.

In going from one layer to the next, the director rotates by a small angle φ, so that on the whole a supermolecular structure with some helical twisting of molecules occurs, describing a helix with pitch P. When the helix axis is oriented along the axis Z, the director has the following components:

$$n_X = \cos \varphi \quad n_Y = \sin \varphi \quad n_Z = 0.$$

Such a structure possesses periodicity along the Z axis. The pitch of the helix, which is determined by the nature of the molecule, corresponds to a rotation of the director by an angle of 2π, and the optical property periodicity is $d = P/2$ due to the nonpolar character of the structure (n and $-n$ are equivalent). Hence, $P = 2\pi d/\varphi$, where φ is the twist angle.

A specific helical supermolecular structure of the cholesteric mesophase leads to an appearance of unique optical properties, which are summarized very briefly as follows.

(i) One of the remarkable peculiarities of the cholesteric structure is an anomalously high specific optical rotation, as much as 10^3–10^4 deg mm^{-1}, being a hundred times as great as that resulting merely from the chirality of molecules.

(ii) A periodic cholesteric structure reflects incident light as a normal diffraction lattice. The wavelength of reflection λ_R is related to the pitch P ($P = 2d$) of the cholesteric phase by

$$2\bar{n}d \sin \theta = \lambda_R \qquad (9.1)$$

where $\bar{n}$ is the average refractive index, d is the lattice period, and θ is an angle between the incident beam and the cholesteric plane.

For the right angle of incidence ($\theta = 90°$) eqn (9.1) becomes

$$2\bar{n}d = \lambda_R \quad \text{or} \quad \bar{n}P = \lambda_R \qquad (9.2)$$

The refractive indices of ordinary low molar mass LCs are similar to those of polymeric cholesterics, $c.$ 1.5. The selective reflection of visible light occurs if the pitch of the cholesteric helix is around 360–500 nm and a thin layer of the cholesteric appears coloured.

(iii) The cholesteric structure displays circular dichroism, i.e. the light selectively reflected by a cholesteric liquid crystal is circularly polarized, the direction of circular polarization being coincident with the direction of twist of the cholesteric helix. Light of the same wavelength as that reflected but having an opposite circular polarization is not affected by the cholesteric liquid crystal.

(iv) In many cases the pitch of the cholesteric helix depends very critically on temperature. For the majority of cholesteric liquid crystals, the wavelength of selective reflection decreases with increasing temperature. However, there are a number of cholesteric systems, including polymeric ones, whose helical pitch is usually increased when the temperature is raised.

Many molecular-statistical theories and models have been suggested to account for the temperature dependence of the pitch of the helix, each elucidating some of the dependencies. Most of the theoretical relationships are valid for the cholesteric systems at temperatures far from the region of pre-transition phenomena.

However, no theoretical relationships relating the temperature dependence of P near the pre-transition region of temperatures to parameters of the molecular and supermolecular structure are currently available.

9.3 Molecular structure of cholesteric polymers with side mesogenic groups

Depending on the character of arrangement of chiral centres within macromolecules, all cholesteric polymers with side groups can be roughly divided into two groups.

The first group consists of rigid-chain or semi-rigid-chain polymers with various side groups, containing chiral centres in their backbones. These are, for example, branched polypeptides, such as poly-γ-alkylglutamate (Uematsu and Uematsu, 1984), or hydroxypropylcellulose and its derivatives (Werbowy and Gray, 1976; Kulitshikhin and Golva, 1985; Shibaev and Ekaeva, 1987). The helical form of macromolecules of these polymers, which can be modelled by rigid rods, determines the formation of the lyotropic, and in some cases of the thermotropic, cholesteric mesophase.

The second group involves homopolymers and copolymers with mesogenic substituents with chiral centres in side branchings. These polymers are usually synthesized via (i) homopolymerization of chiral mesogenic monomers; (ii) copolymerization of nematogenic monomers (i.e. the monomers which produce homopolymers forming the nematic mesophase) with chiral mesogenic monomers; (iii) copolymerization of nematogenic monomers with chiral nonmesogenic monomers; or (iv) by attachment of chiral mesogenic low molar mass substances or their mixtures with nematogenic compounds to the polymer chain by means of polymer analogous reactions (for details see Chapters 3, 4).

Methods (ii)–(iv) are based on inducing the cholesteric helix in the nematic mesophase under the effect of chiral mesogenic or nonmesogenic additives, an approach widely used in the case of low molar mass liquid crystals. Here it should be emphasized that in all the cases (i)–(iv) the formation of the cholesteric mesophase originates from the ordering of the mesogenic fragments but not of the backbones of macromolecules, as in polymers of the first group. Precisely for this reason the term 'polymers with *side groups*' usually means 'polymers with *side mesogenic groups*', which contribute to physicochemical behaviour of macromolecules. That is why much emphasis in the present review is put on this most extensive and, perhaps, most promising group of polymers.

Derivatives of cholesterol are most widely used as chiral mesogenic groups for such cholesteric side chain polymers; recently other chiral fragments, such as chiral alkyl derivatives of phenyl benzoates have also been employed for this purpose. To date, a great number of cholesterol-containing homo- and copolymers with acrylic, methacrylic, and siloxane backbones have been synthesized.

A peculiar structural feature of the cholesteric polymers is the presence of a spacer, i.e. of an aliphatic (or oxymethylene) fragment between the mesogenic groups and the backbone. As pointed out in the Introduction (see footnote), the attachment of a mesogenic fragment such as cholesterol directly to the backbone did not lead to the formation of the LC phase, due to the steric hindrance applied by the main chain to the packing of the bulky cholesteric groups in PChA and PChMA.

The idea of the autonomy of the mesogenic groups was realized for the first time in comb-shaped polymers which have been used as convenient 'matrices' (supports) for constructing LC polymers (Freidzon *et al.*, 1974; Shibaev *et al.*, 1975, 1976, 1977, 1978). If mesogenic groups are added to the end groups of the side chains of comb-shaped polymers, i.e. if they are moved some distance from the backbone, it will be possible to decrease the above-mentioned steric hindrances. Using this concept, a few hundred LC comb-shaped polymers not only of cholesteric types but the nematic and smectic types have already been synthesized (see Chapters 3 and 4).

However, even the presence of a flexible spacer does not necessarily ensure the formation of the cholesteric mesophase only by incorporating side mesogenic chiral groups. The smectic mesophase is most typical for the comb-shaped cholesterol-containing polymers, which in fact seems to relate to the presence of the backbone (or very long length of spacer) predetermining the tendency to pack mesogenic groups in layers. So, for example, only smectic-type mesophases were found for the cholesterol-containing polymers (1)–(4).

$$-CH_2-C(R)-$$
$$|$$
$$CONH(CH_2)_n-COO-Chol \ ,$$

$R = H, CH_3$ (1)
$n = 2 - 11$

(Freidzon *et al.*, 1974, 1980; Shibaev *et al.*, 1976, 1977, 1978)

$$\begin{array}{c} CH_3 \\ | \\ -CH_2-C- \\ | \\ COO-\bigcirc-(CH_2)_n-COO-Chol \ , \end{array}$$

$n = 2, 6, 12$ (2)

(Finkelmann *et al.*, 1978a).

Chol =

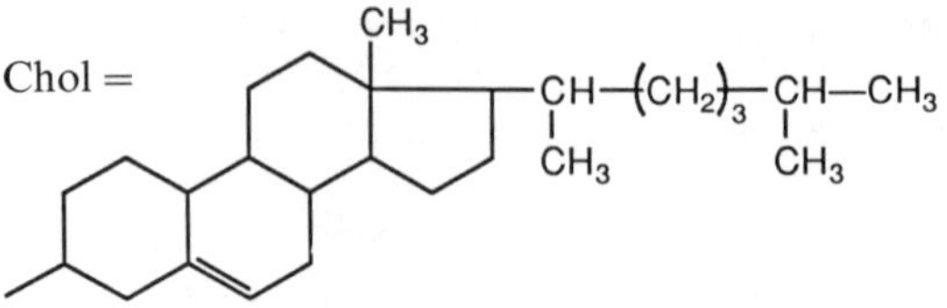

$$
\begin{array}{c}
CH_3 \\
| \\
-O-Si- \\
| \\
(CH_2)_3-COO-Chol
\end{array}
$$

Phase transitions: g45°C S115°C i°C   (3)
(Finkelmann *et al.* 1980)

$$
\begin{array}{c}
R \\
| \\
-CH_2-C- \\
| \\
COO-(CH_2)_n-COO-Chol
\end{array}
$$

$R = H, CH_3$ (4)
$n = 5, 14$
(Freidzon *et al.*, 1980)

A similar situation was also observed for the siloxane homopolymer with chiral phenyl benzoate mesogenic groups (5).

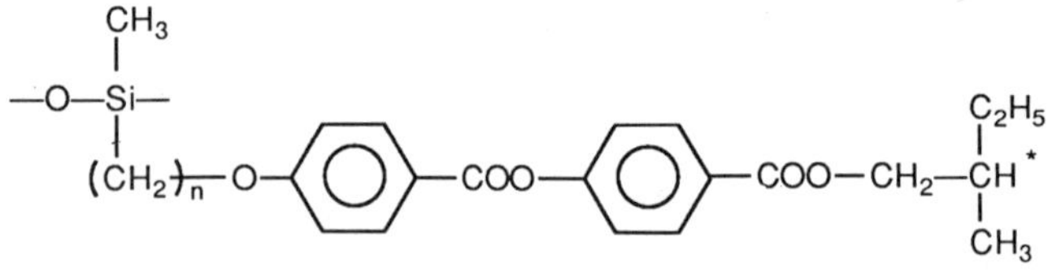

(5)

$$
\begin{array}{c}
CH_3 \\
| \\
-O-Si- \\
| \\
(CH_2)_n-O-\!\!\!\bigcirc\!\!\!-COO-\!\!\!\bigcirc\!\!\!-COO-CH_2-\overset{*}{C}H
\end{array}
$$

C₂H₅   Phase transition:
$n = 5$ g $- 5°C$ S 20°C i
(Finkelmann *et al.*,
CH₃   1982)

At the same time, for the cholesterol-containing homopolymers (6a), (6b) there was observed the formation of the cholesteric mesophase on melting of the smectic phase. The cholesteric mesophase is also inherent for some polymers of series (5):

$n = 3$, phase transitions: g 2°C Ch 13°C i   (5a)
$n = 4$, phase transitions $g - 3°C$ S 17°C Ch 37.5°C i   (5b)

$$
\begin{array}{c}
-CH_2-C(R)- \\
| \\
COO-(CH_2)_{10}-COO-Chol
\end{array}
$$

(Tropsha *et al.*, 1985)

(6)

$R = H$ $\bar{M}_w = 2 \times 10^5$, phase transitions S_A 153°C Ch 157°C i   (6a)
$R = CH_3$, $\bar{M}_w = 3.5 \times 10^5$, phase transitions S_A 125°C Ch 171°C i   (6b)

From the viewpoint of molecular structure, the formation of the cholesteric mesophase in homopolymers should be ensured by sufficiently flexible back bones, and flexible and long spacers, which provides autonomic behaviour of bulky cholesterol (or any other chiral mesogenic) groups, necessary for the realization of their helical twisting.

In this respect these possibilities are best realized in the copolymers consisting of chiral and mesogenic units, rather than in the homopolymers.

This approach has successfully been developed in the works of Soviet (Mousa *et al.*, 1982; Freidzon *et al.*, 1985, 1986*a, b*; Wedler *et al.*, 1987) and West German (Finkelmann *et al.*, 1978*b*, 1980, 1984) researchers, who have synthesized a large number of copolymers having nematogenic and chiral units. Some of the polymers are presented in Table 9.1.

Particular attention should be focused on series (3) shown in the table, i.e. the copolymers of the nematogenic monomer with cholesteryl acrylate (ChA). The homopolymerization of ChA is known to produce an amorphous polymer. However, the copolymers containing up to 33 mol% of the ChA units do form the cholesteric mesophase that selectively reflects light in the visible part of the spectrum. Further increase of the ChA units results in a gradual disturbance of the LC structure, and the copolymers containing more than 50 mol% of ChA are altogether amorphous (Freidzon *et al.*, 1986*a*).

These important results show that the copolymerization of a nematogenic monomer with small amounts of a chiral monomer promotes the cholesteric mesophase even in those cases where the homopolymer of the chiral monomer is amorphous.

Similar results have also been obtained for the copolymers of series (1) (Table 9.1). The homopolymer A forms smectic and nematic mesophases, the homopolymer B is amorphous. The copolymers containing up to 25 mol% of chiral units B form the cholesteric mesophase.

Copolymerization of chiral mesogenic monomers with nematogenic monomers thus enables the preparation of cholesteric copolymers with a very broad range of compositions.

9.4 Structure of cholesteric polymers

A 'comb-shaped' structure of macromolecules with side mesogenic groups specifies their tendency to form layered structures. In this connection two questions arise: in which way is the helical supermolecular structure formed in this system, and what peculiarities are there, as compared to the cholesteric structure of low molar mass liquid crystals? These questions were resolved in the works of Freidzon *et al.* (1986*b*, 1987) and Platé *et al.* (1987), where the structure of cholesteric mesophase for homopolymers and copolymers has been studied.

Now let us consider the *structure of the homopolymers* 6, which form cholesteric and smectic mesophases.

At room temperature these polymers form the smectic S_A mesophase, in which cholesterol groups are packed in the layers with a liquid-like intralayered order, the longer axes of these groups being perpendicular to the layer planes (Freidzon *et al.*, 1987). Two types of packing of side groups were observed to coexist in the structure of layers. One of the types corresponds to the antiparallel packing, so that the cholesterol groups of one macromolecule are surrounded by the methylene chains of the

Table 9.1 Cholesteric copolymers

Series	Structure of copolymer		Content of chiral units (mol %)	T_g(°C)	T_{cl}(°C)	λ_R at $T^* = 0.99 T_{cl}$	Ref.
1	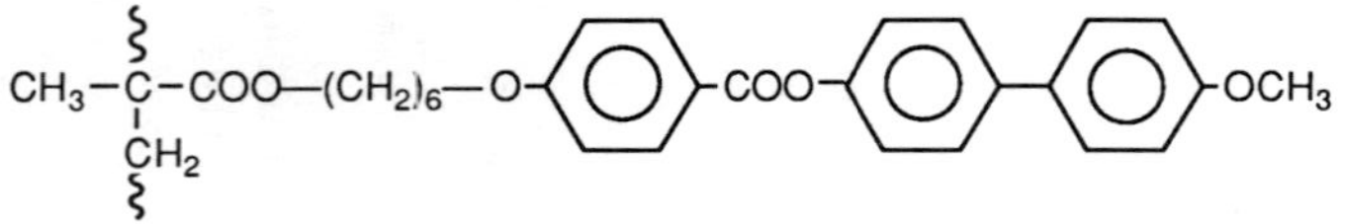	A	11 17 21	70 73 77	247 229 216	1260 712 562	Finkelmann et al. (1978)
	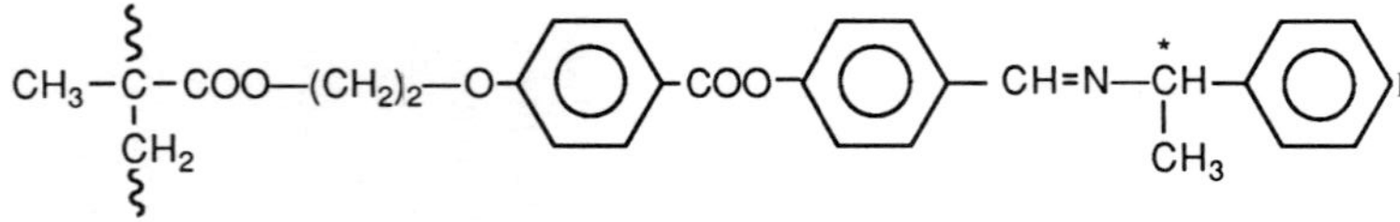	B	25	80	203	467	
2	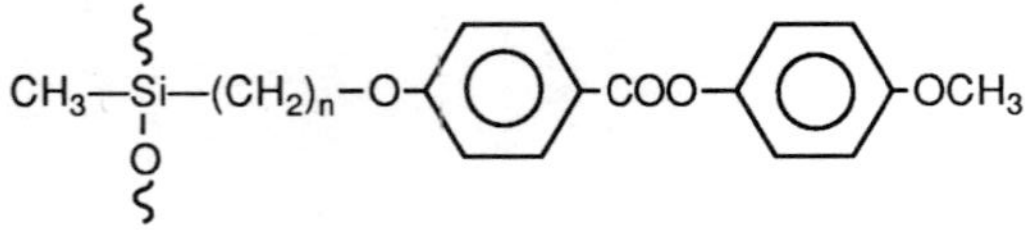	(2.1) $n = 3$	3 7 15	— — —	— — —	1700 800 380	Finkelmann and. Rehage (1980)
	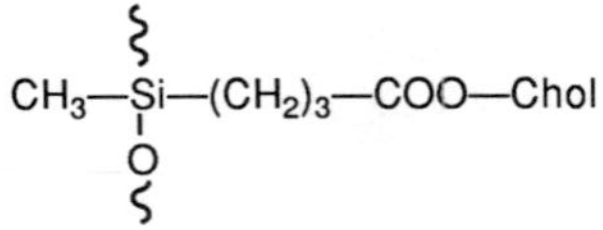	(2.2) $n = 6$	5 10 15	— — —	— — —	1500 800 590	

No.	Structure	Designation					Reference
3	~CH–COO–(CH₂)₅–COO–〇–COO–〇–OCH₃ / –CH₂~		15	20	120	650	Freidzon et al. (1985, 1986a)
			33	60	95	500	
	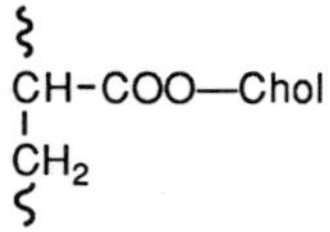 ~CH–COO—Chol / –CH₂~		40	80	90	440	
			52	100	Amorphous		
4	~CH–COO–(CH₂)₅–COO–〇–COO–〇–OCH₃ / –CH₂~	(4.1) R = H, n = 5	20	25	115	550	Freidzon et al. (1985, 1986a)
			35	35	103	500	
		(4.2) R = H, n = 10	45	40	110	400	
			17	20	120	770	
	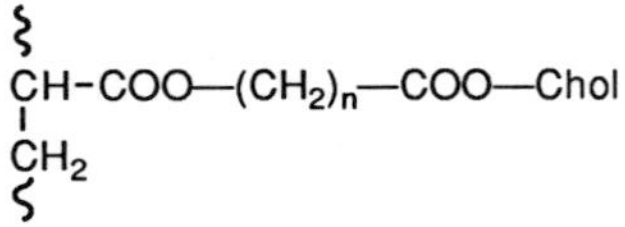 ~CH–COO–(CH₂)ₙ–COO—Chol / –CH₂~	(4.3) R = CH₃, n = 10	21	25	110	666	
			28	25	115	560	
5	~CH–COO–(CH₂)ₙ–O–〇–〇–CN / –CH₂~	(5.1) n = 4	39	30	112	460	Wedler et al. (1987)
			19	20	116	740	
			22	25	118	640	
			33	35	120	515	
	~CH–COO–(CH₂)₅–COO—Chol / –CH₂~	(5.2) n = 5	21	50	106	765	Freidzon et al. (1986b)
			27	50	89	680	
			33	55	106	575	
			28	50	102	660	
			36	50	105	600	
			52	55	150	540	

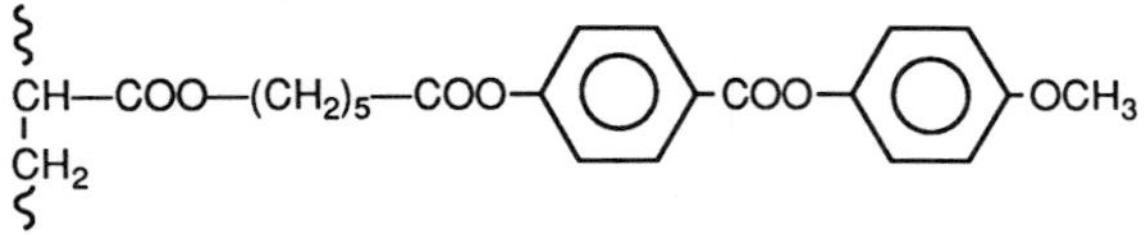

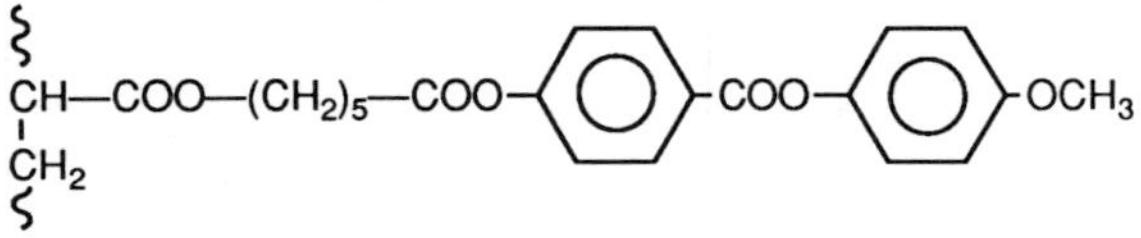

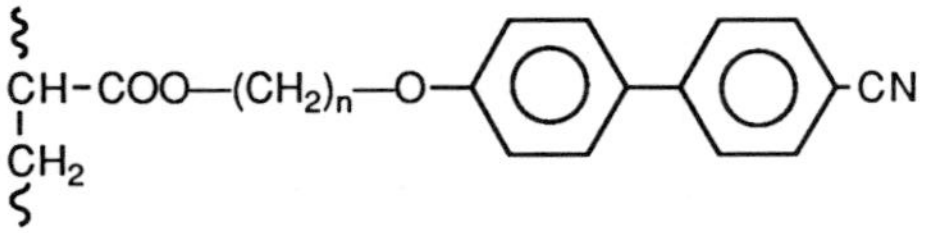

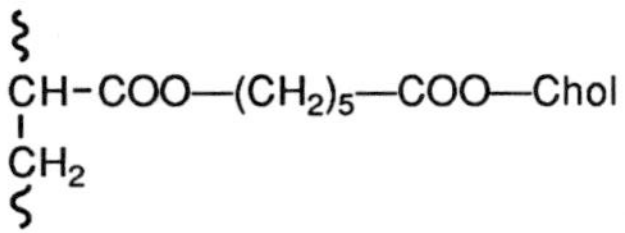

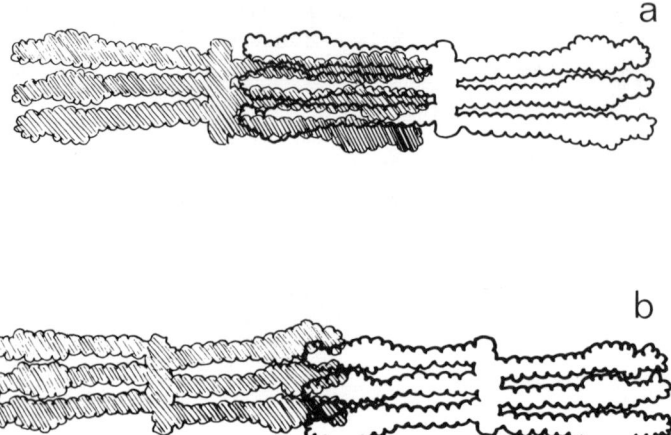

Figure 9.2 Schematic representation of the packing of macromolecules of the cholesteric homopolymers: (*a*) single-layer antiparallel packing; (*b*) intermediate packing with partial overlapping of cholesterol groups. The shaded molecules lie in the plane parallel to the plane of the figure (Freidzon *et al.*, 1987).

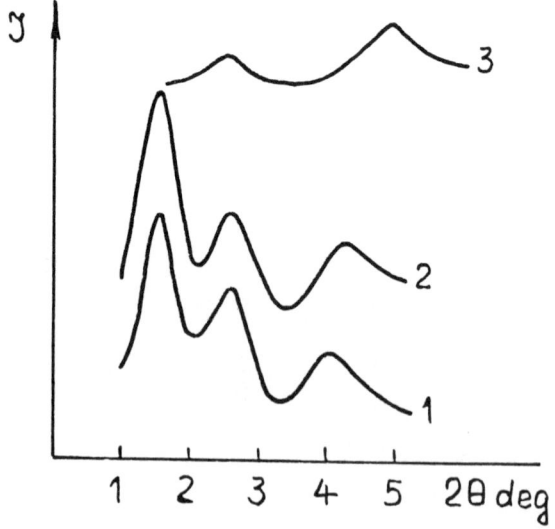

Figure 9.3 Small-angle X-ray diffraction patterns for the polymer of formula (**6b**) in S_A (1, 2) and Ch(3) phases at 20°C (1), 90° (2), and 140°C (3) (Freidzon *et al.*, 1987).

neighbouring macromolecules (single-layer packing) (Figure 9.2a). The other type of packing involves the overlapping of alkyl 'tails' of the cholesterol groups (Figure 9.2b).

The transition to the cholesteric mesophase changes the character of the X-ray diffraction pattern in comparison with the smectic mesophase, however, the presence of low-angle maxima shows that the layer order in the arrangement of side groups is retained (Figure 9.3).

The cholesteric mesophase of low molar mass liquid crystals is usually considered as a twisted nematic mesophase. The cholesteric mesophase of the polymer just considered possesses a pronounced layered order, and hence this mesophase cannot be regarded as a twisted nematic one. The cholesteric mesophase of polymers in this case can be considered as follows (Figure 9.4). At low temperatures a smectic mesophase type S_A is formed; in the region of the $S_A \rightarrow Ch$ transition there occurs a twisting of the smectic layers in such a way that each section perpendicular to the axis of twisting represents a structure typical of the S_A phase.

The structure of the cholesteric mesophase of copolymers depends very critically on the nature of the nematic matrix and on the concentration of cholesterol units.

Incorporation of small amounts of the cholesterol units into the nematic polymer, thus inducing the helical supermolecular structure formation, does not change the arrangement of side mesogenic groups. When the portion of cholesterol units is higher than 30 mol %, they tend to form layered structures, which manifest themselves as low-angle maxima on the X-ray diffraction patterns of these copolymers. That is, the copolymers form a cholesteric structure with elements of layered order, where side mesogenic groups are arranged perpendicular to the layer planes.

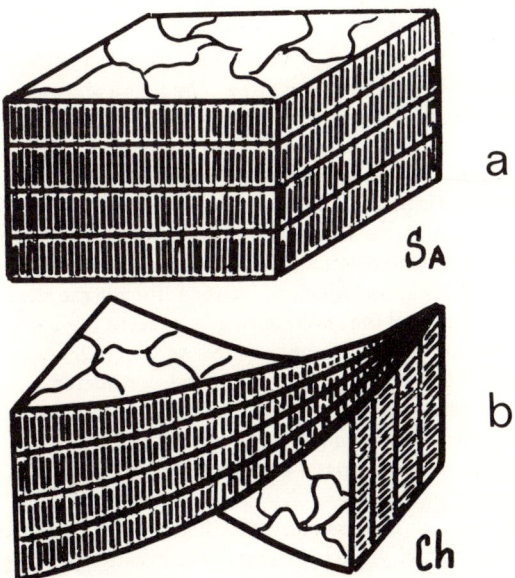

Figure 9.4 Schemes of packing of mesogenic groups in S_A (a) and cholesteric (b) phases of side chain LC polymers (Freidzon et al., 1987).

9.5 Optical properties of cholesteric polymers

Cholesteric polymers containing cholesterol moieties as the chiral units form a lefthand cholesteric helix, and, hence, the selectively reflected light possesses lefthand circular polarization.

9.5.1 *Effect of chemical structure of cholesteric copolymers on selective reflection of light: helical twisting power*

In studying optical properties of the induced cholesteric mesophase in low molar mass nemato-cholesteric mixtures, the notion of the twisting force of a chiral additive, or the 'helical twisting power' (A), is widely used; the latter is expressed by the equation

$$A = \frac{dP^{-1}}{dx_{Ch}} \quad \text{for } x_{Ch} \ll 1 \tag{9.3}$$

where x_{Ch} is the molar fraction of the chiral additive. Taking into account the relationship (9.2), the above equation is usually rewritten as

$$A = n\frac{d\lambda_R^{-1}}{dx_{Ch}}, \tag{9.4}$$

i.e. the helical twisting power is determined by the slope of a linear region of the dependence of λ_R^{-1} on x_{Ch}. The helical twisting power characterizes both the capability of the nematic matrix to be twisted into a helix under the influence of chiral additives, and the twisting force of the chiral additive.

For all cholesteric copolymers, the helical pitch decreases on increasing the proportion of cholesterol-containing units, and at a certain concentration there is observed the selective reflection of the visible light.

In Figure 9.5 the dependence of λ_R^{-1} on the molar fraction of cholesterol-containing units x_{Ch} for different polymers* is presented. The figure shows that the copolymers based on the same nematic monomer and different cholesterol-containing monomers are described by the same dependence of λ_R^{-1} on x_{Ch} (points 1–4). The helical twisting power for these copolymers is about $10\,\mu m^{-1}\,mol^{-1}$. These results show that the capability of the polymeric nematic matrix to be twisted into a helix on introduction of cholesterol-containing units is identical for all cholesterol-containing monomers independent of their chemical structure. The geometrical shape of the cholesterol group and the molar optical rotation, which is identical for all the monomers (since it is determined by the chirality of the cholesterol group) seem to be the decisive factors in this case and for low molar mass liquid crystals as well (Chilaya, 1985).

The value of the helical twisting power for the considered series of copolymers coincides with the corresponding value for a mixture of low molar mass liquid crystals of p-butoxybenzylidene-p-butylaniline and cholesteryl propionate, where no specific interaction is present in the nematic matrix and the chirality is provided by the cholesterol fragment.

Replacement of a nematic matrix changes the helical twisting power. In matrices that display certain smectogeneity, such as copolymers with cyanobiphenyl mesogenic groups (series 5.2, Table 9.1) the helical twisting power is lower, $A = 8\,\mu m^{-1}\,mol^{-1}$.

---

*Optical properties of different copolymers have been compared at the temperature $T^* = 0.99T_{Cl}$.

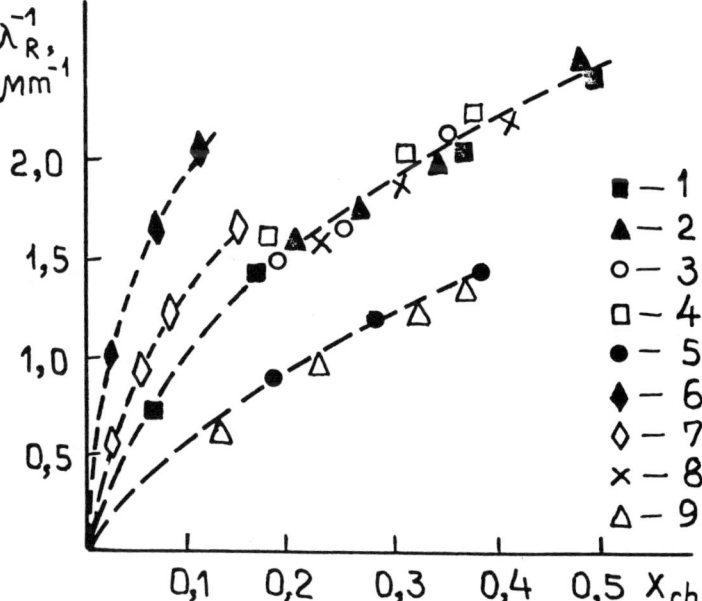

Figure 9.5 Dependence of λ_R^{-1} on the molar fraction of cholesterol-containing component for the copolymers in Table 9.1: 4.1(1), 4.2(2), 4.3(3), 3(4), 5.2(5), 2.1(6), 2.2(7), and for low molar mass mixtures of cholesteryl propionate p-butoxybenzylidene-p-butylaniline (8) and cholesteryl propionate p-butoxybenzylidene-p-aminobenzonitrile (9) (Finkelmann *et al.*, 1980; Freidzon *et al.*, 1985a, 1986a, b).

These data show that strong 'lateral interactions' between the mesogenic groups, which are responsible for smectogeneity, hinder the helical twisting in these systems. The dependence of λ_R^{-1} on x_{Ch} for mixtures of cholesteryl propionate with the cyano-containing nematic liquid crystal, p-butoxybenzylidene-p-aminobenzonitrile, given in Figure 9.5, shows that this system and the copolymers of series (5.2) (points 5 and 9) are characterized by identical values of A.

All this indicates that the mechanism of formation of the induced cholesteric mesophase in copolymers is the same as in low molar mass liquid crystals. This allows us to anticipate that the different mesogenic groups, promoting the formation of the nematic mesophase, and the different chiral additives (mesogenic and non-mesogenic), widely used in low molar mass nemato-chiral mixtures, can be employed in the synthesis of cholesteric polymers. This, in turn, permits us to control the optical, electrical and other properties of cholesteric polymers within very broad limits.

Theoretical models accounting for the appearance of the chiral supermolecular structure in the nematic mesophase, built up of chiral molecules, show that the angle of relative rotation of molecules (α) depends on two order parameters, S and D.

If the preferential arrangement of molecules is the parallel packing and the angle between their long axes, and the director of the mesophase is θ, then

$$S = 3/2(\cos^2 \theta - 1/3) \qquad (9.5)$$

The consideration of the order parameter for the system of chiral molecules should take into account the possibility of rotation of molecules around their long axes. If ψ is an angle of rotation of a chiral molecule around its long axis, then

$$D = \cos 2\psi \qquad (9.6)$$

When the order parameters S and D become larger, the twist angle of the molecule also increases, i.e. the pitch of the helix decreases.

When the cholesteric mesophase is formed by the mesogenic groups attached to the backbone of the polymer, the order parameter S is somewhat lower than in low molar mass liquid crystals (Piskunov et al., 1982), and this should lead to some increase in the helix pitch. At the same time, the rotation of the side mesogenic groups around their long axes is hindered in comparison with the free molecules. This should result in an increase of the order parameter D and a decrease of the helix pitch. In the copolymers considered above these two effects probably compensate for each other. If the freedom of rotation of side mesogenic groups is restricted further, the increase of the order parameter D should reduce the helix pitch considerably. This was demonstrated by Finkelmann and Rehage (1980) on the copolymers of the series (2) (Table 9.1). The shortening of the spacer between the mesogenic phenyl benzoate group and the backbone from 6 to 3 methylene units hinders the rotation of side groups, and thus promotes a larger twisting of the mesogenic groups in the copolymer (Figure 9.5, points 6 and 7).

Let us now consider the character of textures formed in films of the copolymers. The selective reflection of light is typical for the planar texture of cholesteric liquid crystals, in which long axes of molecules lie in the planes parallel to the surface of the liquid crystal layer, whereas the axis of the cholesteric helix is perpendicular to them. The maximum intensity of the selectively reflected light is observed for the ideal planar texture of the so-called 'cholesteric single crystal'. Appearance of imperfections in the texture reduces the intensity of the reflected light. In particular, if the texture has regions of planar texture, in which the axis of the cholesteric helix makes up an angle φ with the normal, then along with the reflection of light with the wavelength $\lambda_R = \bar{n}P$ there will be observed the reflection of light with the wavelength $\lambda = \bar{n}P \cdot \sin \varphi$, i.e. the

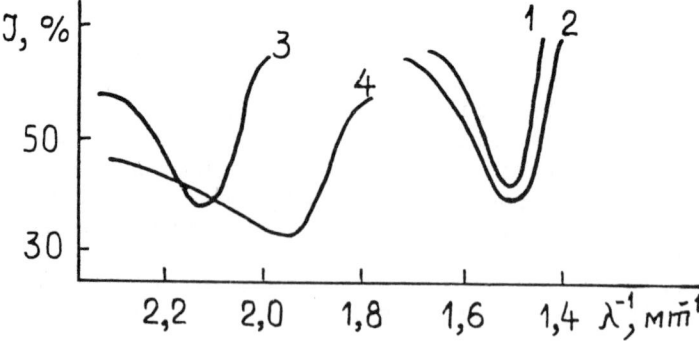

Figure 9.6 Transmittance spectra for the copolymers of the series 4.2 (Table 9.1) containing 21 (1, 2) and 39 (3, 4) mole % of cholesterol units, at temperatures 105° (1, 3), 20° (2), and 85°C (4) (Freidzon et al., 1986a).

reflection peak will be broadened towards the short wavelength region. Appearance of elements of the focal-conic texture, which intensively scatters light, should also lead to a similar broadening of the reflection peak.

For the copolymers forming the cholesteric mesophase without elements of the layered order (Figure 9.6, spectra 1, 2 and 3) the reflection peaks are symmetric, sufficiently intensive and have small widths (25–40 nm). In the copolymers forming the cholesteric mesophase with a pronounced layered order the planar texture easily transforms to the focal-conic one, and the reflection peak (spectrum 4, Figure 9.6) is asymmetric, i.e. a non-selective light scattering takes place, especially in the short wavelength region.

9.5.2 Temperature dependence of helix pitch

One of the most important characteristics of cholesteric liquid crystals is the temperature dependence of the helix pitch which can be exploited in the practical use of cholesterics for recording thermal fields of different origin. For polymers this characteristic becomes even more significant, since it is now possible to retain (to freeze) the cholesteric helical structure in the solid polymers by cooling polymeric films below the glass transition temperature.

When the cholesteric structure is induced in the nematic mesophase by small additives of cholesterol-containing units, the helical pitch is independent of temperature within the whole range of existence of the mesophase (Figure 9.7, curves 1–3). This pitch may be retained by cooling below T_g, thereby allowing the production of solid polymeric glasses, which selectively reflect the circularly-polarized light of a designated wavelength.

For the copolymers containing more than 30 mol % of cholesterol units the other temperature dependence of the helix pitch is observed. As already pointed out, the X-ray diffraction patterns for such copolymers contain some low-angle reflections caused by elements of layered order. The formation and perfection of the layered order in the copolymers with decreasing temperature results in an untwisting of the cholesteric

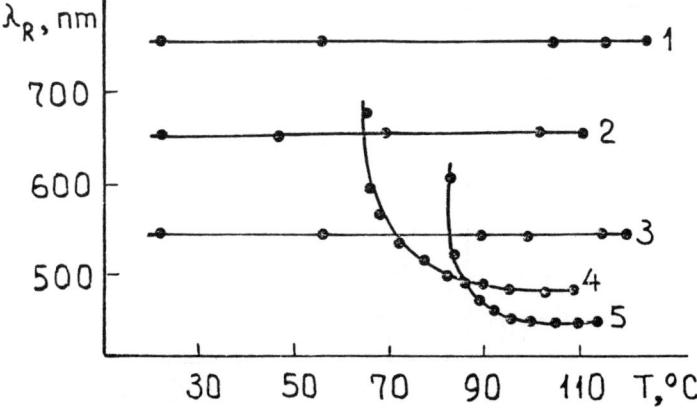

Figure 9.7 Temperature dependence of λ_R for copolymers of the series 4.2 (Table 9.1) containing 17 (1), 21 (2), 28 (3), 36 (4), and 39 (5) mole % of cholesterol units. (Freidzon et al., 1986a).

helix and a shift of the selective light reflection to the long wavelength region. With increasing numbers of cholesterol-containing units, the smectogeneity becomes ever more pronounced. As a result, firstly the untwisting of the helix begins at higher temperatures, and secondly the sharp increase in λ_R takes place in the region of 70–90°C (Figure 9.7, curves 4 and 5). It should be pointed out that no transitions to the smectic mesophase were detected for all the copolymers of the series 4 (Table 9.1). The

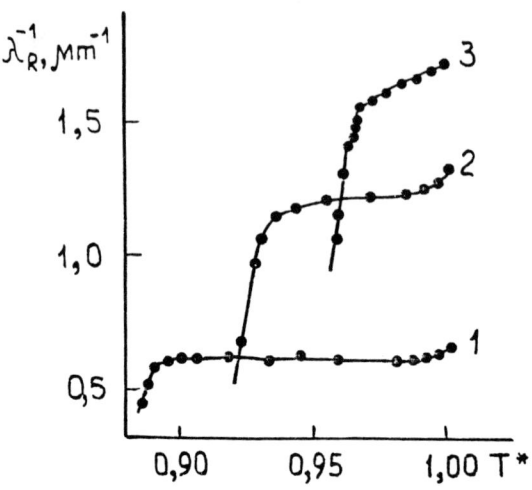

Figure 9.8 Temperature dependence of λ_R^{-1} on $T^* = T/T_{cl}$ for copolymers of the series 2.2 containing 5 (1), 10 (2), and 15 (3) mole % of cholesterol units. (Finkelmann and Rehage, 1980).

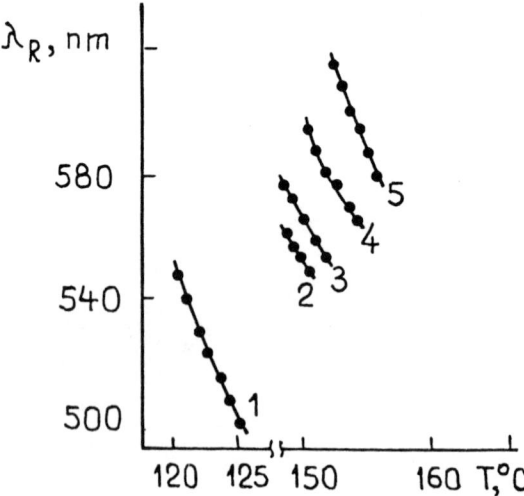

Figure 9.9 Temperature dependence of λ_R for the fractions of polymer (**6a**): (1) $\bar{M}_w = 1 \times 10^4$; (2) $\bar{M}_w = 6 \times 10^4$; (3) $\bar{M}_w = 1.2 \times 10^5$; (4) $\bar{M}_w = 1.4 \times 10^5$; (5) $\bar{M}_w = 2.0 \times 10^5$ (Freidzon *et al.*, 1985b).

colour preservation of the copolymeric films on cooling to room temperature also indicates that no transition to the smectic mesophase takes place.

In the copolymers, for which the cholesteric mesophase is preceded by the smectic one, in the vicinity of the transition Ch → S there occurs an abrupt untwisting of the cholesteric helix, leading to an increase of the selective light reflection wavelengths (Figure 9.8). At temperatures far enough from the region of transition to the smectic phase the helix pitch is practically independent of temperature and determined only by the copolymer composition.

The effect of untwisting of the cholesteric helix near the transition to the smectic phase with decreasing temperature manifests itself most clearly in the cholesteric homopolymer (6a), for which the temperature range of the cholesteric mesophase does not exceed 5°C. For this polymer the temperatures of the transitions $S_A \to Ch$ and $Ch \to I$ rise with increasing molar mass (Freidzon et al., 1985a), so the intervals of existence of the cholesteric mesophase for different fractions are shifted along the temperature scale. As a consequence, the pitch of the cholesteric helices for fractions of different molar mass varies considerably (Figure 9.9).

9.5.3 Compositions of cholesteric polymers with low molar mass liquid crystals

One of the methods of controlling the temperature interval of existence of the cholesteric mesophase and optical properties of cholesteric polymers is to create compositions of cholesteric polymers with low molar mass cholesteric or nematic liquid crystals.

Figure 9.10 shows the temperature dependencies of the wavelengths of selective light reflection for the compositions of one of the same cholesteric copolymers (series 4, Table 9.1) with different low molar mass liquid crystals (Freidzon et al., 1985c). Addition of small amounts of a low molar mass liquid crystal (8) to the copolymer

$$CH_3O-\langle\bigcirc\rangle-COO-\langle\bigcirc\rangle-OC_6H_{13} \qquad (8)$$

reduces a relative content of cholesterol units, shifting the selective light reflection towards the long wavelength region. As can be seen from Figure 9.10, the introduction of 30 mol% of the nematic molecules shifts the selective light reflection about 100 nm.

A stronger concentration dependence of the helix pitch is observed for the mixtures of a lefthanded copolymer and righthanded cholesteryl chloride. The introduction of 20 mol% of cholesteryl chloride increases the wavelength of selective light reflection by almost 200 nm. If the lefthanded liquid crystal, such as cholesteryl pelargonate, is added to the cholesteric copolymer, the helix pitch of the composition is essentially less than that of the copolymer (Figure 9.10, curves 6 and 7).

Figure 9.10 (curves 1–5) shows that for the copolymer and its compositions with a low molar mass nematic (8) and cholesteryl chloride the pitch of the helix is independent of temperature (a small increase of the pitch for the composition containing 20 mol% of cholesteryl chloride results from a reduction of the order parameter S); in contrast, the helix pitch of the compositions containing smectogenic cholesteryl pelargonate sharply increases with decreasing temperature, and this relates to the formation of elements of smectic structure in the cholesteric mesophase.

The formation of the helical structure in compositions of cholesteric polymers with low molar mass nematics and cholesterics obeys the same regularities as for

K

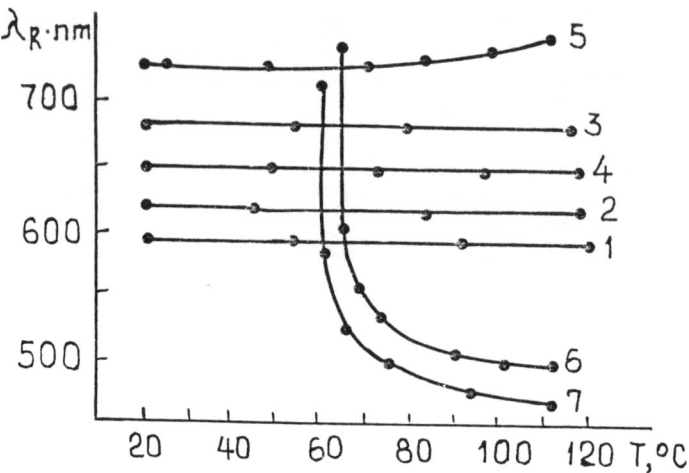

Figure 9.10 Temperature dependence of λ_R for copolymer of the series 4.2 (Table 9.1) containing 24 mole % of cholesterol units, curve (1) and its mixtures with low molar mass liquid crystals: nematic (8)–[10, curve (2) and 30, curve (3) mole %]; cholesteryl chloride–[10, curve (4) and 20, curve (5) mole %]; cholesteryl pelargonate–[20, curve (6) and 30, curve (7) mole %]. (Freidzon et al., 1985c).

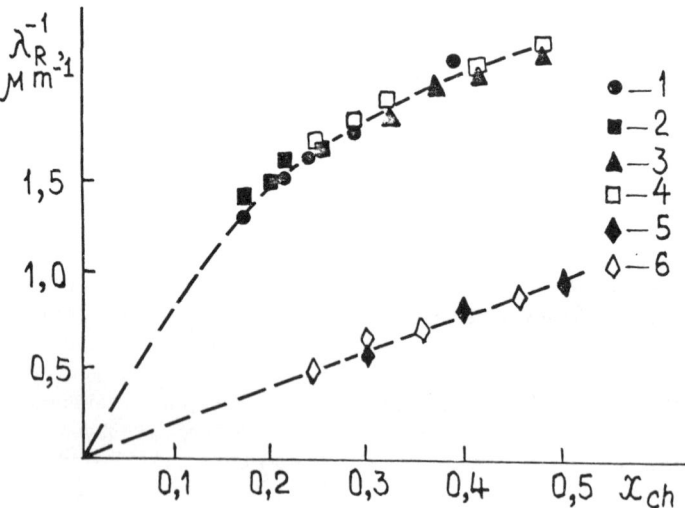

Figure 9.11 Dependence of λ_R^{-1} on the molar fraction of a chiral component for the copolymers of the series 4.2 (Table 9.1), curve (1), their mixtures with nematic (8), curve (2) cholesteryl pelargonate, curve (3), and cholesteryl acetate, curve (4), for the copolymers (9), curve (5) and for a mixture of the nematic homopolymer (10) with a chiral compound (9), curve (6) (Freidzon et al., 1985c; Finkelmann and Rehage, 1984).

copolymers (Freidzon *et al.*, 1985c). Figure 9.11 shows that the dependence of λ_R^{-1} on the molar fraction of the cholesterol-containing component is described by one and the same curve for both copolymers and their compositions with low molar mass liquid crystals (Freidzon *et al.*, 1985c).

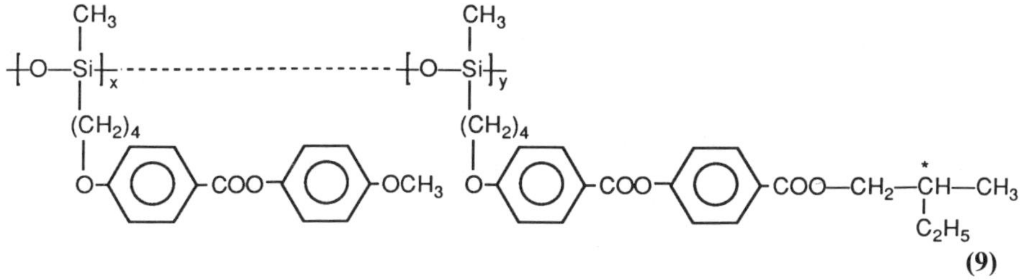

(9)

Similar results have been obtained in the work of Finkelmann and Rehage (1984) in which the copolymers containing phenyl benzoate chiral and achiral units (9) as well as the compositions of a nematic homopolymer (10) with a low molar mass chiral compound (11) have been studied.

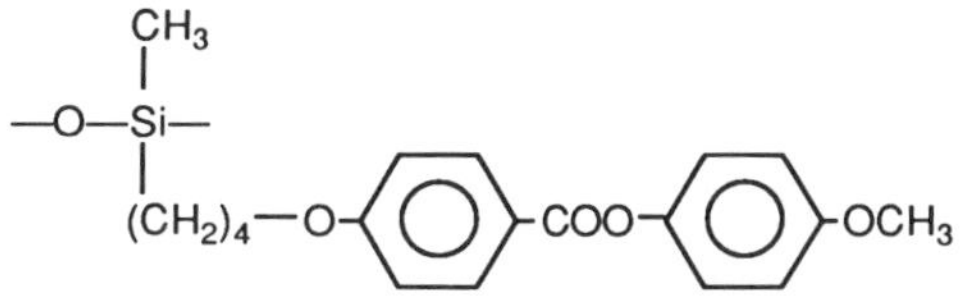

(10)

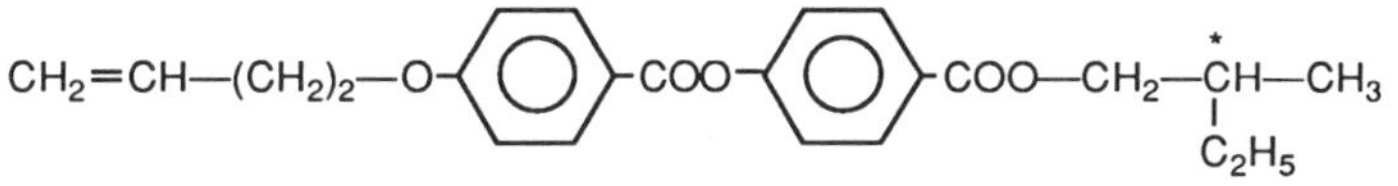

(11)

The same dependence of λ_R^{-1} on x_{Ch} (Figure 9.11) was observed both for the copolymers (9) and for the compositions of the nematic homopolymer (10) with the low molar mass chiral compound (11).

An interesting method for the preparation of compositions from cholesteric polymers and low molar mass cholesteric liquid crystals has been proposed by Shannon (1984). This method employed cholesterol-containing monomers which are able to form a cholesteric mesophase selectively reflecting the visible light.

A rapid photopolymerization of a mixture of monomers in the cholesteric phase leads to formation of polymeric films preserving the optical characteristics of the starting monomers. These films, representing polymer-monomeric compositions and containing 20–30% of monomers, do not change their properties even after one year.

9.6 Behaviour of polymeric cholesterics in electric fields

An effective method of governing the structure and optical properties of cholesteric polymers is by way of an electric field. The main result of the effect of electric field on a layer of cholesteric polymer, which is characterized by a large positive value of anisotropy of dielectric permeability $\Delta\varepsilon$ (these properties are typical of copolymers containing cyanobiphenyl units), is the transformation of a helical planar structure to an optically active homeotropically oriented structure. Analysis of the dependence of the optical transmittance and wavelength of selectively reflected light (λ_R) on the value of applied voltage shows up two stages in this process (Figure 9.12) (Talroze *et al.*, 1986).

In the first stage (I) the axes of the cholesteric helices are tilted with respect to the film plane, λ_R is shifted into the blue region of the spectrum and the transition from the planar texture to the focal-conic turbid texture takes place. Intensity of light passing through a sample sharply drops.

The second stage (II) is the untwisting of the cholesteric helix, leading to a homeotropically oriented structure, in which side groups of macromolecules are oriented along the applied electric field, normal to the film's plane. The value of λ_R drastically increases and the transparency also increases.

Similar structural changes are known also for low molar mass nemato-cholesteric mixtures, where the untwisting of the helix is considered as a field-induced cholesteric → nematic phase transition. However, owing to the above-mentioned peculiarities of the cholesteric mesophase of comb-like polymers, where side groups have a tendency to form a layered structure, the untwisting of the helix is accompanied by the formation, not of a nematic but rather a smectic phase (S_A). This fact is proved by the presence of small-angle equatorial reflections on the X-ray diffraction patterns of homeotropically oriented films of the copolymers (Figure 9.12).

The threshold character of the untwisting of the cholesteric helix enables an estimation of the twist elastic constant K_{22} for the copolymers, using the formula

$$U_2 = \frac{\pi^2 d}{2P} \sqrt{\frac{4K_{22}}{\Delta\varepsilon}} \qquad (9.7)$$

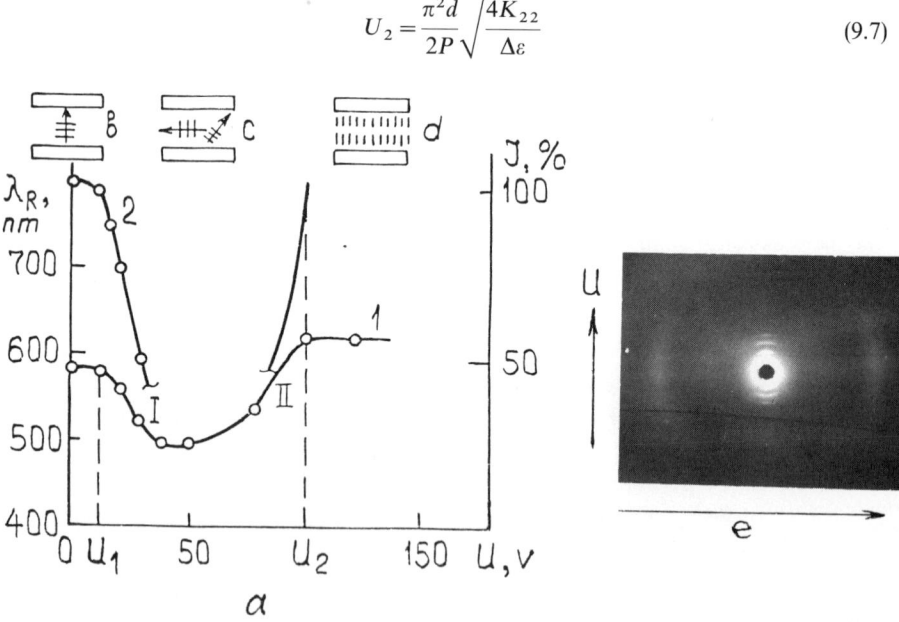

Figure 9.12 Dependence of optical transparency (1) and λ_R (2) on voltage for the copolymer of the series 5.2 (Table 9.1) containing 36 mole % of cholesterol units at $T = 102°C$ (*a*). Schematic picture of planar (*b*), focal-conic (*c*), and homeotropic (*d*) textures of a polymeric film; an arrow shows the direction of the cholesteric helix axis. X-ray diffraction pattern of this polymer oriented by an electric field (*e*) (arrow shows direction of electric field) (Talroze *et al.*, 1986).

where U_2 is the threshold voltage for the untwisting of the helix, $\Delta\varepsilon$ is the dielectric anisotropy, and d is the sample thickness.

The values of K_{22} estimated from eqn (9.7) lie, depending on the copolymer composition, in the interval $4-7 \times 10^{-7}$ dyne, which are of nearly the same magnitude as low molar mass liquid crystals. This fact, together with the independence of K_{22} of the length of the polymeric chain (at least within a range of the degree of polymerization between 50 and 160), indicate that the elasticity of the cholesteric polymeric melts is defined by the interaction of the side mesogenic groups, and in the first approximation has nothing to do with the chain structure of macromolecules (Talroze et al., 1986).

At the same time, the dynamic properties of the polymeric cholesterics are highly sensitive to the degree of polymerization; this is clearly manifested in the kinetics of the untwisting of the cholesteric helix (Figure 9.13) (Talroze et al., 1987).

In Figure 9.13, optical transparency as a function of time is shown. Here two stages of the orientational process are also seen. The values of orientation time in the first stage τ_I are practically independent of the degree of polymerization (DP) whereas the values of τ_{II}, corresponding to the untwisting time of the helix, increase with increasing DP (Talroze et al., 1987), as is clearly seen from experimental data—values of orientation time and effective activation energy (E_A) for the orientation process for cholesteric copolymer series (5.2) (Table 9.1), containing 21 mole % of cholesterol units ($U = 130\,\text{V}$, $T = 82°\text{C}$).

DP	50	80	160
$\tau_I(s)$	1.0	1.5	1.0
$\tau_{II}(s)$	16	97	138
$E_A, \text{kJ mol}^{-1}$ (for τ_{II})	240	210	210

These values of E_A, of about $210-240\,\text{kJ mol}^{-1}$, are very high and comparable with the values of E_A for viscous flow controlled by mobility of the macromolecule segments. Thus the untwisting of the cholesteric helix probably involves individual segments of

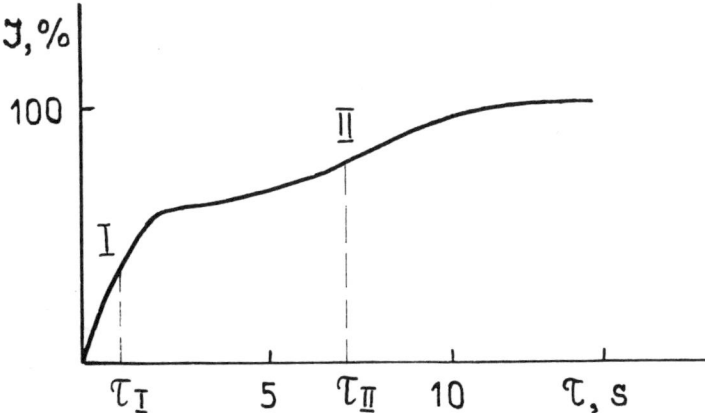

Figure 9.13 Kinetics of optical transparency of the copolymer film of the series 5.2 (Table 9.1), containing 28 mole % of cholesterol units at $T = 101°\text{C}$ and $U = 140\,\text{V}$ (Talroze et al., 1986).

the chains. At the same time, the increase of the untwisting time τ_{II} with increasing DP shows that the macromolecule as a whole also contributes to this process.

It should be noted that the process considered above is a reversible one, and after an electric field has been switched off the polymeric system relaxes to an initial state. The curve of the transparency change shown in Figure 9.14 reflects the course of the relaxation process (Talroze et al., 1987). The S-shaped curve indicates that the first stage of relaxation proceeds as a nucleation-type process. During the induction period, τ_0, structure defects arise in the homeotropically oriented layer of the smectic liquid crystal, and the consequent twisting of the cholesteric helix begins at these defects. The τ_0 value essentially depends on the electric field voltage at which the untwisting of the helix has proceeded. It appears that, at small voltages, macroscopically oriented samples preserve some regions with helical structures, and in this case the relaxation process takes place without an induction period. If the voltage increases, the number of these regions diminishes and this results in an increase in τ_0. At sufficiently high voltage, the defects in the oriented sample vanish, and hence the τ_0 value becomes constant

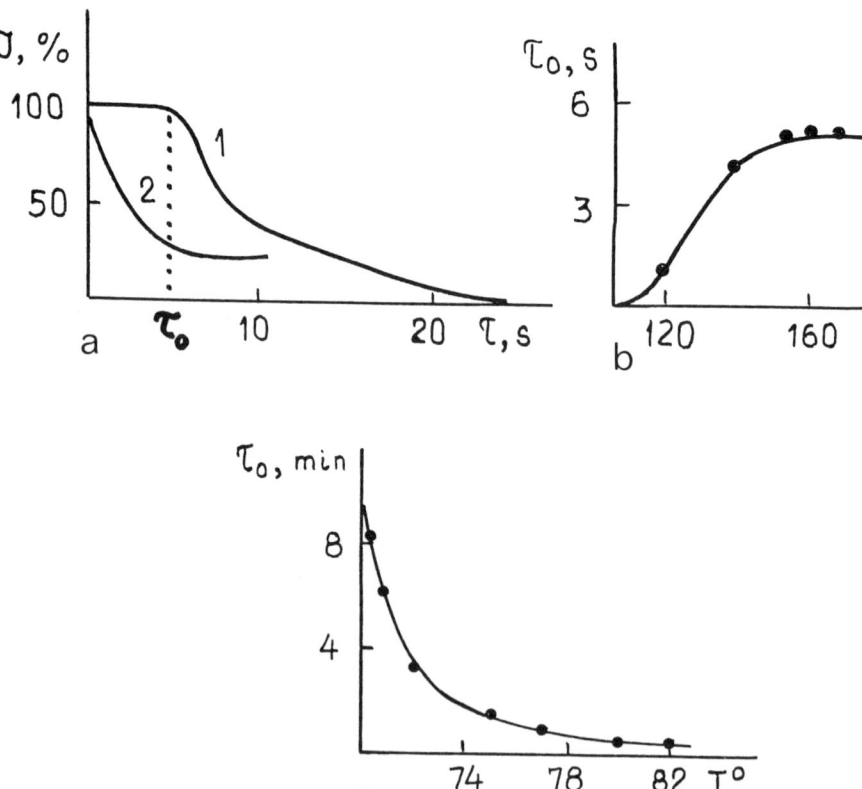

Figure 9.14 The change of optical transparency of the copolymer film (see Figure 9.13) during relaxation for initial orienting voltages 130 (1) and 100 V (a) at 100°C. Dependence of the induction period on initial voltage of the initial field (b) and temperature (c) (Talroze et al., 1987).

(Figure 9.14*b*). The induction period depends very strongly on temperature (Figure 9.14*c*). This enables an increase in memory time by cooling the sample below the glass transition temperature, and memorizes the structure created by the field in the polymeric film.

9.7 Formation of intramolecular structures in cholesterol-containing polymers in dilute solutions

The chain nature of macromolecules of LC polymers gives rise to the formation of a distinct new type of structure, the intramolecular mesophase formed by a single macromolecule. This is displayed most clearly in cholesterol-containing polymers (Freidzon *et al.*, 1981; Anufrieva *et al.*, 1982, 1984).

$$\begin{array}{c} CH_3 \\ | \\ -CH_2-C- \\ | \\ COO-(CH_2)_n-COO-Chol \end{array} \qquad (12)$$

or (PChM–*n*), with $n = 5, 10, 14$.

Figure 9.15 presents the temperature dependence of the radius of gyration $(\bar{R}^2)^{1/2}$,

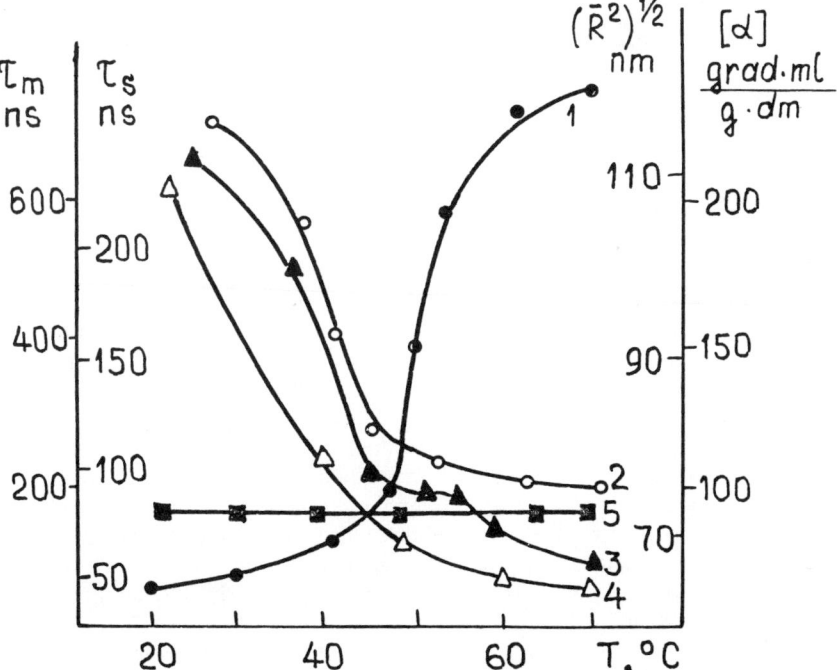

Figure 9.15 Temperature dependence of $(\bar{R}^2)^{1/2}$ (1); [α] (2); τ_m (3); τ_s (4) for solutions of the polymer (12) (PChM-10) $n = 10$) and [α] for the corresponding monomer (5) in heptane ($\lambda = 350$ nm) (Freidzon *et al.*, Anufrieva *et al.*, 1982).

the specific optical rotation $[\alpha]$, and relaxation times τ_m and τ_s, characterizing the mobility of the main chains and side groups, respectively, for a cholesteric polymer (**12**), (PChM–10), in heptane solution. These curves show that a sharp change of all these parameters takes place in the temperature region 40–60°C.

The decrease in $(\bar{R}^2)^{1/2}$ results from the formation of a compact intramolecular structure. A decrease of the intramolecular mobility (an increase of the relaxation time τ_m) is also indicative of such a process. The values of $(\bar{R}^2)^{1/2}$ and τ_m are evidence that at temperatures below 50°C the macromolecular coil transforms into a dense globule having very low intramolecular mobility. The data about the internal structure of the globule can be obtained from the results of optical rotation measurements.

At temperatures below 40°C, $[\alpha]$ values are seen to be larger than those caused by the proper molecular optical activity of the cholesterol group (the figure gives, for comparison, values for the monomer (Figure 9.15, curve 5), which are seen to be unchanging within the wide temperature range). A high optical activity (of about the same order of magnitude as observed for the polymer at 20°C) is typical of cholesteric liquid crystals. All these results indicate unambiguously that the liquid crystalline globules formed at temperatures below 40°C possess the cholesteric type structure.

Analysis of the relationships shown in Figure 9.15 highlights two stages in the observed process. In the first stage some decrease in size and reduction of the intramolecular mobility takes place. The side cholesterol groups do not contribute to the interaction. This is evidenced by relatively low values of τ_s and $[\alpha]$. This stage is a transition between a 'coil' and an 'isotropic liquid globule', in which side mesogenic groups are distributed randomly. In the second stage (below 45°C) the side mesogenic groups become involved in the interaction, and the subsequent insignificant decrease in size (Figure 9.15, curve 2) is accompanied by a sharp drop of the intramolecular mobility and a sharp increase of optical activity (Figure 9.15, curves 2–4). At this stage the transition from an 'isotropic liquid globule' to a 'liquid crystalline globule of the cholesteric type' occurs in which the mesogenic cholesterol groups are packed in a dense helical structure.

The formation of the liquid crystalline globule is a result of the interaction between side mesogenic groups, and hence, the process of globule formation and its structure depend essentially on the length of the spacer connecting the cholesterol group to the backbone. An increase of the spacer length shifts the temperature interval of the transition from a 'coil' to a 'globule' towards low temperatures (Figure 9.16). Macromolecules, having a short spacer of five methylene units, exist in the globular conformation within the whole temperature region studied (20–70°C); however, the density of the globules is considerably lower than that of globules of macromolecules having the longer spacers. The intramolecular liquid crystalline structure of PChM-5 globules is also less perfect, as can be seen by comparing the $[\alpha]$ values for the globules of polymers having spacers of different length (Figure 9.16). These results show that the 'remoteness' of mesogenic groups from the backbone secures sufficient autonomy for the side chains from the backbone and assists their cooperative interaction, giving a more ordered structure.

The mechanism of formation of the globule with intramolecular cholesteric structure is well manifested in the study of polymeric fractions of different molar mass (Figure 9.17). The isotropic liquid globule is formed within exactly the same temperature range (45–55°C) regardless of the molar mass. The temperature region of formation of the LC globule essentially depends on the molar mass of the polymer: the

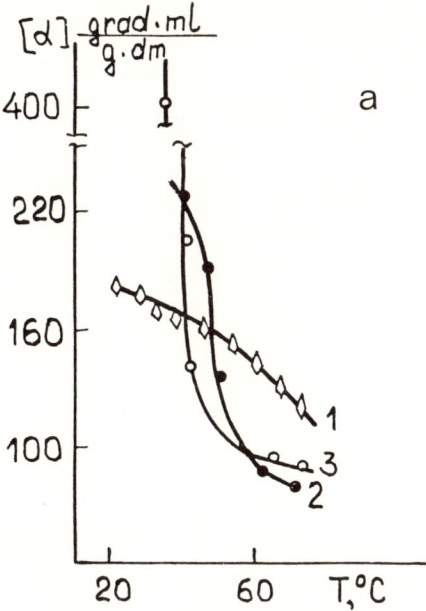

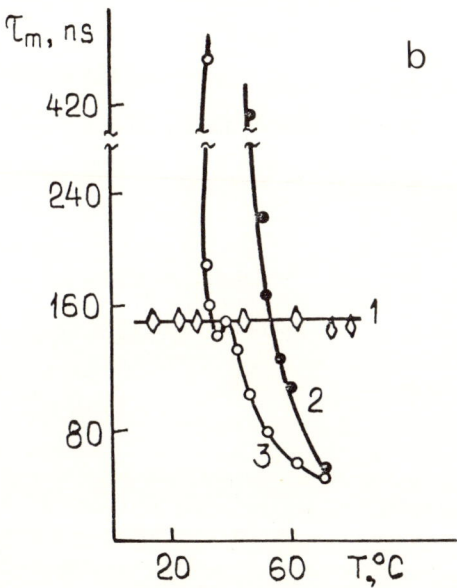

Figure 9.16 Temperature dependence of $[\alpha]$ (*a*) and τ_m (*b*) for solutions of polymer (**12**): PChM-5 (1); PChM−10 (2), and PChM−14 (3) in heptane. (Freidzon *et al.*, 1981).

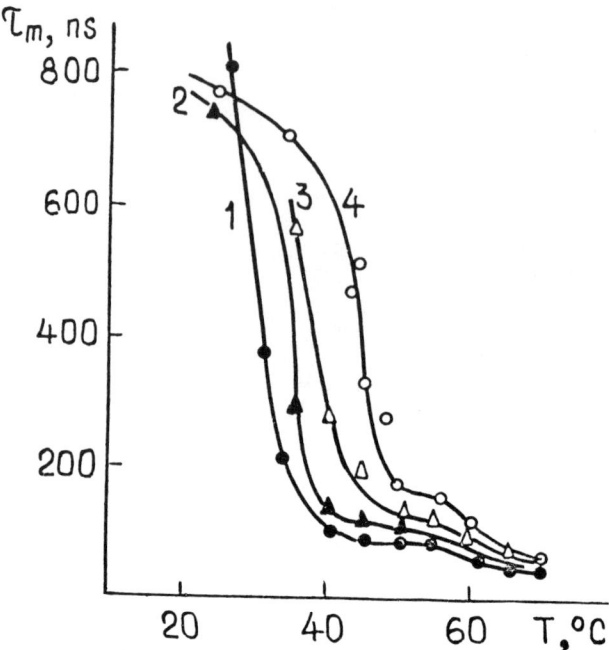

Figure 9.17 Temperature dependence of τ_m for solutions of fractions of polymer (**12**) (PChM-10) in heptane; $\bar{M}_w = 0.35 \times 10^6$ (1); 0.46×10^6 (2); 0.72×10^6 (3); 6.6×10^6 (4) (Anufrieva *et al.*, 1984).

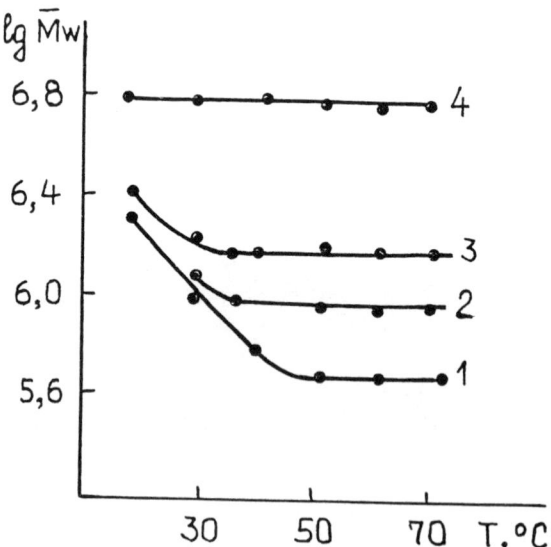

Figure 9.18 Temperature dependence of the mass of the polymeric coil for solutions of the polymer (**12**)(PChM-10) in heptane: $\bar{M}_w = 0.35 \times 10^6$ (1); 0.72×10^6 (2); 1.4×10^6 (3); 6.6×10^6 (4) (Anufrieva *et al.*, 1984).

higher the molar mass, the higher the temperature at which the formation of the LC globule begins to take place. Such a dependence can be accounted for by analysing the variation of the molar mass of the polymer with decreasing temperature.

For the high molar mass sample ($M_w = 6.6 \times 10^6$) the molar mass does not change within the whole temperature range, i.e. the 'LC globule' involves a single macromolecule (Figure 9.18). The formation of the 'LC globule' for macromolecules with molar mass less than $M_w < 1.5 \times 10^6$ is accompanied by an increase of the effective mass, i.e. an association of the macromolecules occurs. These results show that a certain critical content of mesogenic groups is necessary for the formation of the intramolecular mesophase; for the polymer considered this content is about 3000.

The above experimental data on the formation of compact globular structures and coil → globule transitions are in good agreement with theoretical studies (Grosberg, 1980).

9.8 Conclusion

It may be concluded from the data above that cholesteric polymers with side mesogenic groups behave in many respects like their low molar mass liquid-crystal counterparts, i.e. they selectively reflect light in the IR, UV or visible part of the spectrum, display circular dichroism, and are susceptible to the action of electric field with untwisting of the helix.

At the same time, the specific macromolecular nature of cholesteric polymers predetermines the unique possibility of creating the 'frozen cholesteric structure' in bulk solids, films, fibres and overcoats. This feature alone of cholesteric polymers opens up wide perspectives for obtaining spectrum filters and reflectors, circular dichroic optical filters, colour-controlled optical elements, thermoindicators, and coloured decorative and technical overcoatings possessing all the advantages of polymeric materials.

References

Anufrieva, Ye., Pautov, V., Freidzon, Ya., Shibaev, V. and Platé, N. (1982) *Vysokomol. Soedin.* **A-24**, 825.

Anufrieva, Ye., Pautov, V., Friedzon, Ya., Shibaev, V. and Platé, N. (1984) *Dokl. AN USSR* **278**, 383.

Belyakov, V. and Sonin, A. (1982) *Optics of Cholesteric Liquid Crystals* (in Russian) Nauka, Moscow, pp. 360.

Chilaya, G. (1985) *Physical Properties and Application of Liquid Crystals with Induced Helical Structure* (in Russian) Metsniereba, Tbilisi, pp. 87.

De Visser, A., Feyen, J., De Groot, K. and Bantjes, A. (1970) *J. Polym. Sci.* **B-8**, 805.

Elliot, A. and Ambrose, E. (1950) *Disc. Faraday Soc.*, **9**, 246.

Finkelmann, H., Ringsdorf, H., Siol, W. and Wendorff, J. (1978a) *Makromol. Chem.* **179**, 829.

Finkelmann, H., Koldehoff, J. and Ringsdorf, H. (1987b) *Angew. Chem.* **B90**, 92.

Finkelmann, H. and Rehage, G. (1980) *Makromol. Chem., Rapid Commun.*, *1*, 733.

Finkelmann, H. and Rehage, G. (1982) *Makromol. Chem., Rapid Commun.* **3**, 859.

Finkelmann, H. and Rehage, G. (1984) *Adv. Polym. Sci.* **60/61**, 99.

Freidzon, Ya., Shibaev, V. and Platé, N. (1974) *3rd All-Union Conf. on Liquid Crystals, Abstracts*, Ivanovo, 214.

Freidzon, Ya., Shibaev, V. and Platé, N. (1980) in *Advances in Liquid Crystal Research and Application*. ed. Bata, L., Pergamon, Oxford; Akademiai Kiado, Budapest, 899.

Freidzon, Ya., Shibaev, V., Pautov, V., Bronitsh, T., Shelukhina, G., Kasaikin, V. and Platé, N. (1981) *Dokl. AN USSR* **256**, 1435.

Freidzon, Ya., Boiko, N., Shibaev, V. and Platé, N. (1985a) in *Polymeric Liquid Crystals*, ed. Blumstein, A., Plenum, New York, 303.

Freidzon, Ya., Tropsha, Ye., Shibaev, V. and Platé, N. (1985b) *Makromol. Chem., Rapid Commun.* **6**, 625.

Freidzon, Ya., Boiko, N., Tropsha, Ye. and Shibaev, V. (1985c) in *XXII Conf. on High Molecular Compounds, Alma-Ata, Abstracts*, 118.

Freidzon, Ya., Boiko, N., Shibaev, V. and Platé, N. (1986a) *Eur. Polym. J.* **22**, 13.

Freidzon, Ya., Kostromin, S. and Shibaev, V. (1986b) *Vysokomol. Soedin.* **B-28**, 686.

Freidzon, Ya., Tropsha, Ye., Tsukruk, V., Shilov, V., Shibaev, V. and Lipatov., Yu. (1987) *Vysokomol. Soedin.* **A-29**, 1371.

Grosberg, A. (1980) *Vysokomol. Soedin.* **A-22**, 96.

Kulitchikhin, V. and Golva, L. (1985) *Khimiya Drevesiny* **N3**, 9.

Mousa, A., Freidzon, Ya., Shibaev, V. and Platé, N. (1982) *Polym. Bull.* **6**, 485.

Pishkunov, M., Kostromin, S., Stroganov, L., Shibaev, V. and Platé, N. (1982) *Makromol. Chem. Rapid Commun.*, **3**, 443.

Platé, N., Talroze, R., Freidzon, Ya. and Shibaev, V. (1987) *Polymer J.*, **19**, 135.

Robinson, C. (1956) *Trans. Faraday Soc.* **52**, 571.

Saeki, H., Iimura, K. and Takeda, M. (1972) *Polymer J.* **3**, 414.

Shannon, P. (1984) *Macromolecules* **17**, 1873.

Shibaev, V. and Platé, N. (1977) *Vysokomol. Soedin.* **A-19**, 923.

Shibaev, V. and Ekaeva, I. (1987) *Vysokomol. Soedin.* **A-29**, 2681.

Shibaev, V., Freidzon, Ya. and Platé, N. (1975) *11th. Mendeleev Congr. on General and Applied Chemistry, Moscow, Abstracts*, **2**, Nauka, 164.

Shibaev, V., Freidzon, Ya. and Platé, N. (1976) *Dokl. AN USSR* **227**, 1412.

Shibaev, V., Platé, N. and Freidzon, Ya. (1978) in *Mesomorphic Order in Polymers and Polymerization in Liquid Crystalline Media*, ACS Symp. Ser. **74**, ed. Blumstein, A., American Chemical Society, Washington DC, 33.

Talroze, R., Korobeynikova, I., Shibaev, V. and Platé, N. (1986) *Dokl. AN USSR* **290**, 1164.

Talroze, R., Korobeynikova, I., Shibaev, V. and Platé, N. (1987) *Vysokomol. Soedin.* **A-29**, 1037.

Toth, W. and Tobolsky, A. (1970) *J. Polym. Sci.* **B-8**, 289.

Tropsha, Ye., Freidzon, Ya. and Shibaev, V. (1985) *5th All-Union Conf. on Liquid Crystals, Abstracts*, Ivanovo, **2**, 117.

Uematsu, J. and Uematsu, Y. (1984) *Adv. Polym. Sci.* **59**, 37.

Wedler, W., Talroze, R., Korobeynikova, I., Freidzon, Ya., Shibaev, V. and Platé, N. (1987) *Krystallographiya* **32**, 1222.

Werbowy, R. and Gray, D. (1976) *Mol. Cryst. Liq. Cryst. Letts.* **34**, 97.

10 Side chain liquid crystalline elastomers

W. GLEIM and H. FINKELMANN Institut für Makromolekulare Chemie,
Universität Freiburg, Stefan-Meier-Str. 31, 7800 Freiburg, FRG

10.1 Introduction

The preceding chapters of this book have shown that the liquid crystalline state is realizable with linear main chain and side chain polymers. The phase behaviour of these liquid crystalline (LC) polymers is analogous to their low molar mass counterparts. Hence the nematic and cholesteric state as well as the smectic polymorphism is observed with these LC polymers.

Owing to their chemical structure, LC polymers are characterized by a combination of liquid crystalline order with polymer properties. As result of their structural constitution the physical properties of LC polymers are determined by a subtle interaction of anisotropic alignment and chain coiling.

Conventional polymers cross-linked to networks are of particular interest due to their elasticity and stability of shape. It is an obvious question whether in a similar way LC polymers may be transferred to LC elastomers combining liquid crystalline order and network properties.

The starting point for consideration of the realization of LC elastomers is the fact that cross-linking prevents only the 'macro-Brownian' motion of polymer chains. The 'micro-Brownian' motion, which is the mobility of the chain segments, is however not impaired except close to the network junctions. Thus if the contour length between adjacent netpoints is sufficiently high, cross-linking of linear LC polymers should yield elastomers with liquid crystalline properties. Based on these assumptions, side chain LC elastomers were prepared for the first time with a siloxane polymer backbone (Finkelmann et al., 1981).

In the meantime, LC elastomers with a different chemical constitution of the main chain as well as mesogenic side groups have been synthesized. The intention of this chapter is to demonstrate some theoretical and experimental features of this new type of elastomeric material.

After a brief insight into synthesis and characterization of side chain LC elastomers, some of their principal physical properties are described, for example their phase transitions, director orientation, mechanical and photoelastic behaviour, and some possible future developments and applications of LC elastomers are demonstrated.

10.2 Synthesis and phase behaviour

Formation of networks by cross-linking of linear polymer chains is well known. The usual methods can be transferred to LC polymers if adequate chemical methods are applied.

The most convenient way to synthesize elastomers is to start from well-characterized

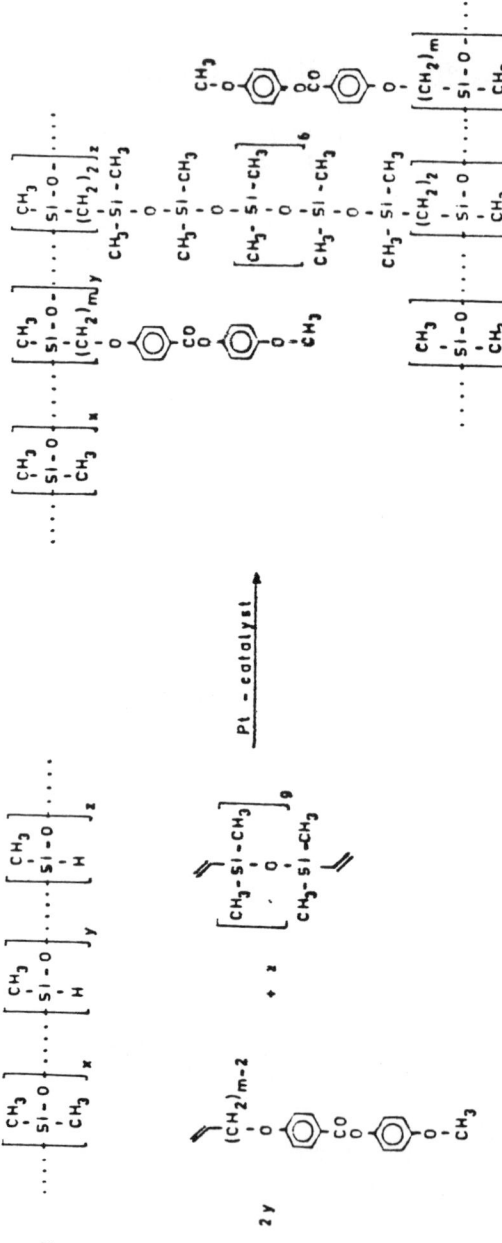

Figure 10.1 Schematic synthesis route for siloxane LC elastomers with end-on fixed mesogenic side chains.

linear polymers carrying reactive centres. By means of a suitable reaction or a cross-linking agent, the reactive centres are connected. The first liquid crystal side chain elastomers were formed in this way based on a polysiloxane main chain (Finkelmann *et al.*, 1981). The route of synthesis is sketched in Figure 10.1. By a single-step reaction, mesogenic side groups and cross-linker are added to the linear poly(methylhydrogensiloxane) chain. The phase behaviour of these elastomers is given in Table 10.1. In Figure 10.2 the DSC curves of a linear and the corresponding cross-linked polymer (samples 1f and 2f, Table 10.1) are shown. Owing to the flexible cross-linker which carries no mesogenic groups, the glass transition and the liquid crystalline → isotropic phase transformation of the elastomer is lowered. The extent of the liquid crystalline phase, however, remains constant.

In a similar way, LC elastomers with a polyacrylate and polymethacrylate main chain are formed (Zentel and Benalia, 1987). The first step of this synthesis involves the copolymerization of the mesogenic monomer with a certain amount of OH-functionated monomer. In the second step the linear chains are linked via the OH-groups by reaction with a diisocyanate component. Smectic and nematic elastomers are obtained. In analogy to the siloxane networks, short spacers lead to nematic phases whereas smectic phases are preferred with long spacers. As the mesogens are linked with one end of the polymer backbone, this type is termed 'end-on fixed'. By lateral fixing of mesogens, polymers are obtained which offer different new perspectives (Hessel *et al.*, 1987). Cross-linking of these side-on fixed LC polymers leads to nematic elastomers,

Table 10.1 Phase transition temperatures of linear and cross-linked siloxane LC polymers with end-on fixed mesogenic side chains.

Linear polymers

Homopolymers	x	y	m	Phase transition temperatures (K)
1a	0	120	3	g 288 n 334 i
b	0	120	4	g 288 n 368 i
c	0	120	6	g 278 s 319 n 385 i
Copolymers				
d	60	60	3	g 276 n 294 i
e	60	60	4	g 267 s 323 i
f	60	60	6	g 263 s 350 i

Cross-linked polymers

Homopolymers	x	y	z	m	Phase transition temperatures (K)
2a	0	108	12	3	g 273 n 335 i
b	0	108	12	4	g 274 n 312 i
Copolymers					
d	60	48	12	3	g 263 n 283 i
e	60	48	12	4	g 258 s 305 i
f	60	48	12	6	g 253 s 332 i

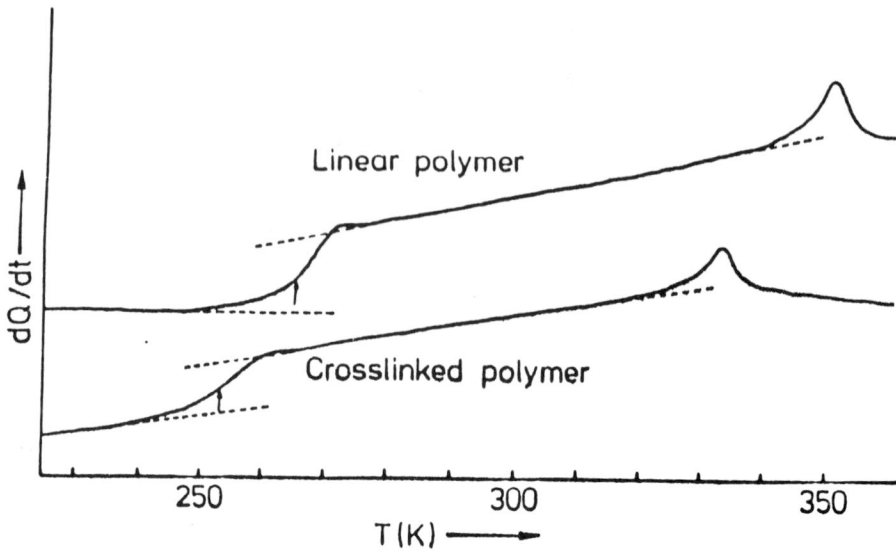

Figure 10.2 DSC traces of the linear and cross-linked polymers 1f and 2f of Table 10.1.

which are distinguished by a small biphasic gap at the phase transition. In Scheme A, a side-on fixed elastomer is characterized.

By reaction of mesogenic monomers with monomers having a functionality greater than 2, polymerization and cross-linking are made to occur simultaneously. The phase behaviour of the elastomers obtained in this way does not differ from that of the corresponding linear polymers (Davis *et al.*, 1986). Liquid crystalline phases are observed up to 10 mol % content of the cross-linking agent. At a further increase of network density, the liquid crystalline phase drops, due to the disturbing influence of the cross-linker.

10.3 Physical properties of LC elastomers

10.3.1 *Phase transitions*

The LC elastomers show phase behaviour analogous to the corresponding linear polymers, as shown in section 10.2. By thermodynamic investigations, information about phase transitions and phase stability relations of LC elastomers is obtained. These elastomers are characterized by a combination of liquid crystalline and network properties. It is therefore of interest whether the phase behaviour of LC elastomers resembles that of conventional networks. The statements below refer to the Ehrenfest classification as well as to the Landau–de Gennes treatment of phase transformations.

The Ehrenfest scheme describes an nth-order phase transformation by a discontinuity of the nth derivative of the free enthalpy at the transformation point (Ehrenfest, 1933). According to this classification, the liquid crystalline → isotropic transformation is first-order, and this applies to low molar mass liquid crystals as well as to side chain LC polymers (Frenzel and Rehage, 1980). The latent heat and the change of volume,

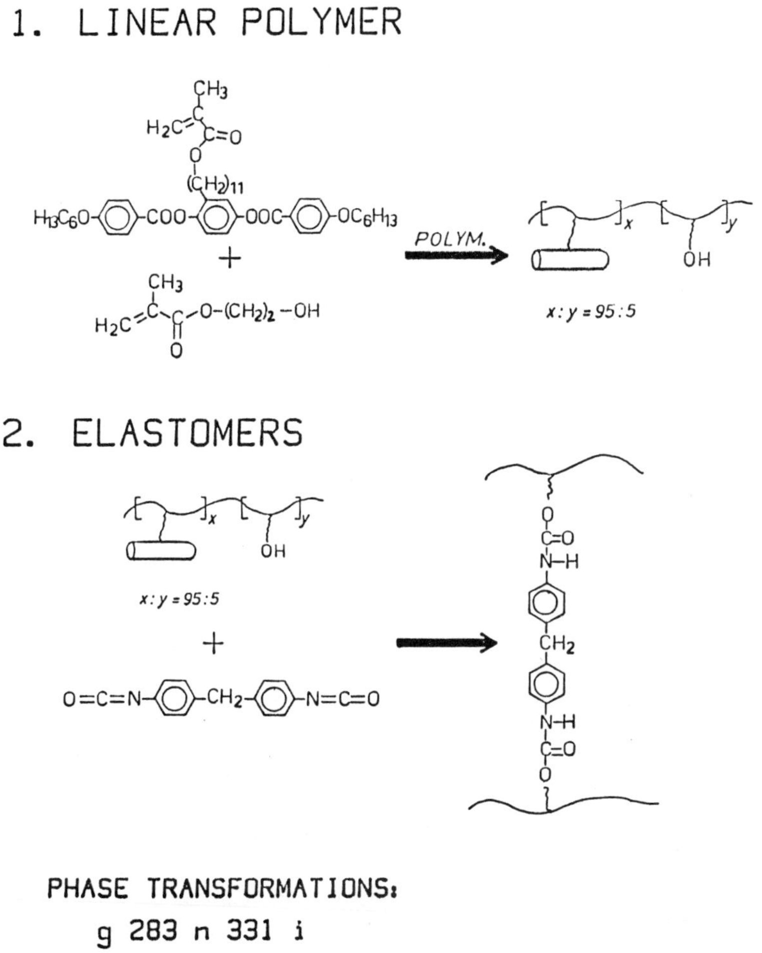

1. LINEAR POLYMER

$x:y = 95:5$

2. ELASTOMERS

$x:y = 95:5$

PHASE TRANSFORMATIONS:

g 283 n 331 i

Scheme A Synthesis and phase behaviour of a methacrylic LC elastomer with side-on fixed mesogenic side chains.

however, are small compared to other first-order transformations such as crystal → isotropic melt and liquid → gas. The nematic → isotropic transformation especially is accompanied by rather small changes in enthalpy and volume. This only weak first-order character of the nematic → isotropic transformation leads to interesting phenomena displayed in the pre-transitional region. This will be discussed in detail in section 10.3.4, where the photoelastic properties of LC elastomers are demonstrated.

It should be mentioned that, besides the first-order liquid crystalline → isotropic transformation, additional phase transformations of second-order are observed for some transitions between different smectic modifications (Anisimov et al., 1985; Thoen et al., 1982). Likewise, a cholesteric → smectic transformation has been proved to be of second order (Pollmann et al., 1988).

As the phase behaviour of LC elastomers is similar to that of non-cross-linked polymers, the order of the phase transformation is not changed by cross-linking. Below the clearing temperature T_c, a homogenous liquid crystalline phase is present, whereas above T_c the isotropic elastomer exists. At the clearing temperature T_c the liquid crystalline elastomer coexists with the isotropic elastomer. In case of a strained network, under conditions of phase equilibrium two anisotropic phases coexist, one of which is the oriented melt. The other coexisting phase is a mechanically ordered liquid crystalline phase, provided network anisotropy and liquid crystalline order are coupled.

The thermodynamic description of a mechanically loaded conventional elastomer undergoing a first-order phase transformation from isotropic to crystalline is given by an analogue of the Clausius–Clapeyron equation (Mandelkern, 1964):

$$\frac{\mathrm{d}f}{\mathrm{d}T} = \frac{f}{t} - \frac{\Delta H}{T \Delta L} \qquad (10.1)$$

where ΔH is the transformation enthalpy, f the retractive force, ΔL the change of the sample length and $(\mathrm{d}f/\mathrm{d}T)$ the slope of the coexistence line in the f–T diagram. At the transformation temperature the complete elastomer passes into the crystalline state. Similarly, at the clearing temperature T_c a first-order phase transformation converts the LC elastomer into a homogeneous liquid crystalline phase. In this case, by means of eqn (10.1) the latent heat ΔH of the isotropic → liquid crystalline transformation can be determined by thermoelastic measurements.

Crystallization of conventional networks under strain is a well-known phenomenon. The state of order of the chain segments is influenced by mechanical stress. Due to the improved alignment of network chains, the crystalline → isotropic phase transformation is shifted towards higher temperatures. If in the case of LC elastomers the network anisotropy is coupled to the state of order of the mesogenic groups, mechanical strain improves this state of order. Consequently the liquid crystalline state is stabilized by deformation too.

Phase transitions of mechanically strained elastomers are similar to field effects of low molar mass liquid crystals. In this case, the state of order of the mesogenic molecules is influenced by magnetic and electric fields. Thus the phase transformation temperatures are shifted as shown, both theoretically and experimentally for the nematic → isotropic transformation in magnetic and electric fields (Helfrich, 1970; Rosenblatt, 1981). The effects are very small, however, and high field strengths are required.

Magnetic and electric field effects as well as strain-induced phase transformations belong to the general context of phase transitions in the presence of external fields. They can be treated in a very elegant way by the Landau theory of phase transitions (Landau, 1937). Based on a power-law expansion of the free energy in terms of an order parameter, this was originally developed for second-order phase transformations. Later, the Landau theory was extended by de Gennes to the nematic → isotropic transformation of low molar mass liquid crystals (de Gennes, 1971). In this form, the theory covers the phenomena observed in the vicinity of the phase transformation such as electric and magnetic birefringence and pretransitional light scattering. This applies to low molar mass nematic liquid crystals as well as to nematic side chain polymers (Ullrich and Wendorff, 1985). Furthermore, the theory clearly shows the field-dependent shift of the transformation temperature, leading respectively to electrical and magnetic critical point at sufficient high field strength (Gramsbergen et al., 1986).

This is demonstrated in Figure 10.3, where the nematic order parameter S has been plotted against a measure of temperature for different magnetic field strengths. The detection of a magnetic or electric critical point, however, seems to be beyond any experimental verification, because of the extreme field strengths required.

For LC elastomers, the Landau–de Gennes theory must be expanded by an additional term considering the network free energy (Warner *et al.*, 1987; cf. Chapter 2). With this extension, the mechanical critical point is recovered; this was originally suggested for main chain LC elastomers (de Gennes, 1975). This means that by mechanical strain the first-order nematic → isotropic phase transformation is shifted up to a critical temperature, where it becomes continuous. Owing to the better orientation of LC networks achieved by strain, the probability for detecting the mechanical critical point seems to be higher than that for detecting the magnetic and electric critical points respectively.

Of the different possibilities for detecting the mechanical critical point of LC elastomers, two methods are now discussed. One of these is the investigation of the state of order at the nematic → isotropic transformation. The other makes use of the photoelastic properties of LC elastomers above the nematic → isotropic transformation.

Owing to the first-order nature of the nematic → isotropic phase transformation, the

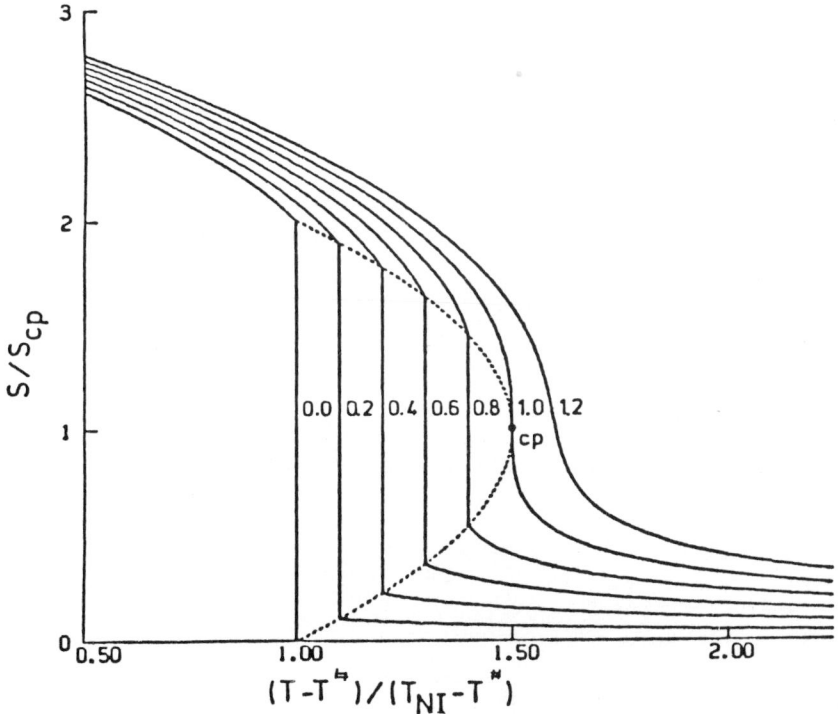

Figure 10.3 Order parameter S/S_{cp} v. temperature $T - T^{\#}$ for different values of the field variable h/h_{cp}. The dashed line is nematic–isotropic coexistence line, cp is the critical point.

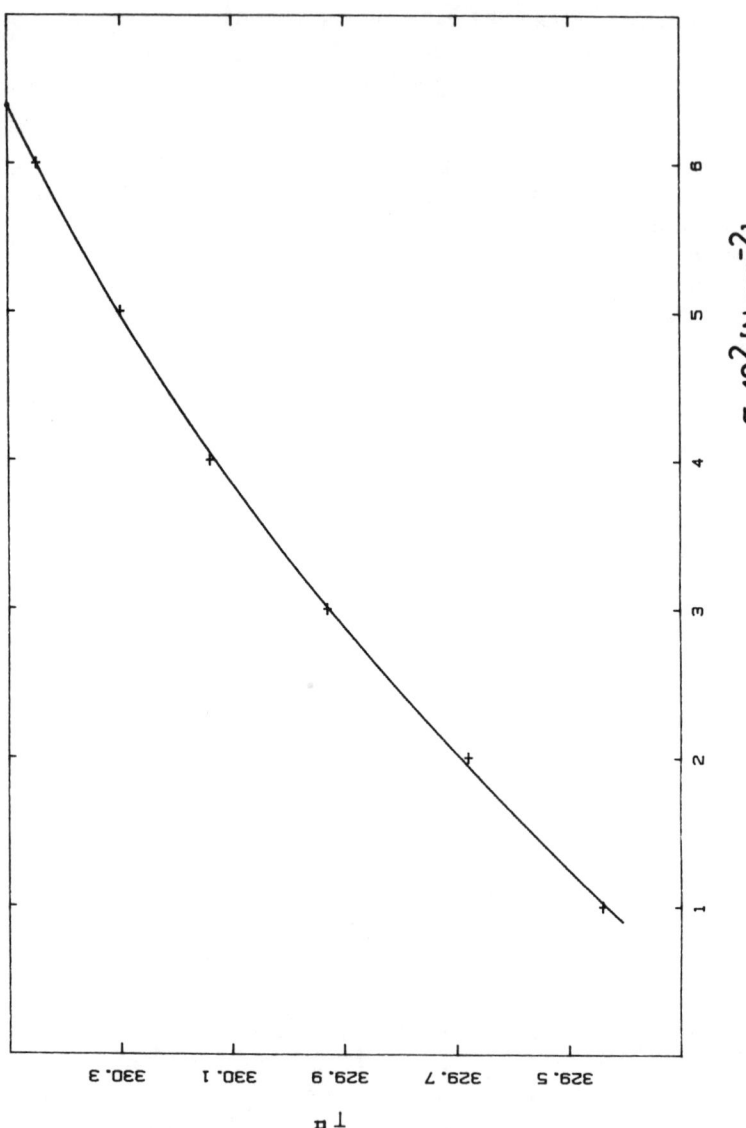

Figure 10.4 Dependence of $T^{\#}$ on stress σ for the elastomer of in Scheme A.

state of order of LC elastomers, characterized by an order parameter S, suffers a discontinuous jump at the transformation temperature T_{ni}. This can be detected by IR dichroism measurements of adequately functionalized mesogenic side groups (Schätzle and Finkelmann, 1987). The Landau–de Gennes theory shows that the jump of the order parameter S at the nematic $\rightarrow$ isotropic transformation is directly related to the applied stress σ. It decreases with increasing stress until the critical value σ_{cr} is reached, where there is no further jump in S. For stresses greater than σ_{cr} there is no phase transformation and nematic phase and oriented melt are indistinguishable.

Discontinuous jumps in thermodynamic parameters at first-order phase transformations are a consequence of the existence of a spinodal. The Landau-de Gennes theory allows for extrapolation to the stability limit by photoelastic measurements. The stress birefringence diverges at a temperature $T^{\#}$, where the deformed elastomer becomes completely unstable with respect to the nematic phase. $T^{\#}$ is found to be 1–2° below the nematic $\rightarrow$ isotropic transformation temperature (Schätzle, 1988). This is discussed in more detail in section 10.3.4. With increasing stress, $T^{\#}$ and T_{ni} are shifted towards higher temperatures. In Figure 10.4 the dependence of $T^{\#}$ on stress is shown for a methacrylate elastomer (example from scheme A), directly reflecting the theoretical predictions. It must be noted that this clear shift of $T^{\#}$ or T_{ni} respectively cannot be observed as a result of electric or magnetic field effects. According to Figure 10.3, the difference between T_{ni} and $T^{\#}$ is diminished with increasing stress. At the critical point, T_{ni} and $T^{\#}$ are identical.

The first experiments performed with side chain LC elastomers clearly show the strain dependence of the nematic $\rightarrow$ isotropic phase transformation. It remains an open question whether the mechanical critical point can be reached with these elastomers. The perspective for detecting the critical point seems realistic for LC elastomers with a strong coupling of network anisotropy and liquid crystalline order. Of course, this is best for main chain LC elastomers. But the side chain LC elastomers also offer many possibilities for improved coupling of anisotropic order and chain properties.

Besides the influence on the phase behaviour, structural modification of LC elastomers also affects the orientational properties of these networks. In the following sections some relations between chemical structure and orientational behaviour are pointed out.

10.3.2 *Director orientation by mechanical deformation*

Depending on the degree of coupling of liquid crystalline order to network properties, a mutual interrelation of network anisotropy and orientation of mesogenic groups in LC elastomers exists. A detailed theoretical analysis shows that several different energetic and entropic influences determine the orientational behaviour of side chain LC polymers (Wang and Warner, 1987, and Chapter 2). Preferred parallel or perpendicular orientation of mesogenic groups with respect to the main chain follows as a result of competition between anisotropic order and conformational disorder. The balance of energetic and entropic influences is very sensitive to structural modifications. Therefore, flexibility of the main chain and the spacer as well as chemical constitution of the mesogens should be reflected by the orientation behaviour.

Another approach to the explanation of the orientation properties of side chain LC networks is a model considering different domains as present in the elastomer

(Schwarz, 1986). Depending on the spacer length, a particular kind of domain occurs with greater frequency, determining the director orientation with respect to the stress axis. The model, however, *a priori* introduces some unlikely polar coupling between the mesogenic moieties and does not reflect the experimental findings (Schätzle and Finkelmann, 1987).

By means of X-ray investigations or IR-dichroism measurements of strained elastomers, the orientation of mesogenic side groups is easily determined. For siloxane networks with identical end-on fixed mesogens, the orientation depends on the length of the alkyl spacer (Schätzle and Finkelmann, 1987). Short spacers with three methylene units lead to perpendicular orientation, whereas four methylene units effect a parallel alignment of mesogenic side groups. As the chain flexibility is improved by increasing spacer length, parallel orientation seems to be attributed to sufficient flexible main chains. Consequently, a spacer with five methylene units should yield parallel orientation. However, perpendicular orientation of mesogenic groups is observed in this case. That chain flexibility influences the orientation behaviour of mesogenic side groups is demonstrated by investigations on polyacrylate and polymethacrylate networks with identical side chains (Zentel and Benalia, 1987). The change from the more flexible polyacrylate to the stiffer polymethacrylate networks changes the orientation from parallel to perpendicular. On the other hand, restrictions on the chain flexibility by very short spacers with two methylene units have no influence on the orientation behaviour of polyacrylate networks (Mitchell *et al.*, 1987). A parallel alignment is still observed in this case, *gauche* conformations of a considerable number of coupling chains being adopted.

The preceding results have shown that the orientation mechanism of the side chain LC elastomers is not quite clear. Detailed investigations on the different factors influencing orientation of mesogenic side groups are still required, especially conformation statistics of the main chain and the flexible spacer.

The orientation behaviour of LC elastomers is surely a very important factor which contributes to the mechanical properties of these materials. In the next section some principles of rubber elasticity are discussed.

10.3.3 Mechanical behaviour

The elastic properties of LC elastomers are of particular importance. Firstly, a brief survey of the physical factors contributing to the mechanical behaviour of conventional elastomers is given. Subsequently, some aspects of the elastic properties of LC elastomers in the liquid crystalline state are discussed.

The elastic behaviour of conventional networks is described by the statistical theory of rubber elasticity (Treloar, 1975). In its simplest form this is based on Gaussian chain statistics and affine deformation. The change of thermodynamic variables upon deformation is attributed to intramolecular reasons only. As long as the volume remains constant, the non-conformative part of the partition function does not change. The free energy of deformation is calculated, from which the stress–strain relations for various kinds of deformation are evaluated. For uniaxial extension this is given by

$$\sigma = vkT \frac{\langle r^2 \rangle}{\langle r^2 \rangle_0} (\lambda^2 - \lambda^{-2}) \tag{10.2}$$

Statistical theory gives a very simple equation of state, relating the stress σ (retractive force per cross-section of the strained network) to the strain ratio λ. The formula contains the number v of chains per unit volume, the temperature T, Boltzmann constant k and the front factor $\langle r^2 \rangle / \langle r^2 \rangle_0$, which considers different rotational isomers with different energy. Here $\langle r^2 \rangle_0$ is the mean square end-to-end distance of unstrained free chains, and $\langle r^2 \rangle$ is the same quantity for the cross-linked chains.

Further developments of the theory account for different aspects neglected in the simple form. Eigenvolume and limited extensibility of real network chains lead to non-Gaussian chain statistics (Treloar, 1975; Chompff, 1977). On the other hand non-affine deformation and fluctuating netpoints are considered (Ronca and Allegra, 1975; Erman and Flory, 1982). Thus the functionality, that is, the number of chains emanating from a cross-link, is incorporated (Edwards, 1971). Furthermore, chain entanglements fixed by chemical cross-linking are supposed to be elastically effective (Langley and Ferry, 1986; Graessley, 1974). A different approach is a van der Waals-like equation of state with two parameters related to finite chain extensibility and global interactions (Kilian, 1981).

The elastic behaviour of liquid crystalline elastomers in the isotropic state is also governed by these different influences. In the liquid crystalline state, the coupling of liquid crystalline order to network properties is the decisive factor which determines the elastic behaviour. Due to the aligning potential of the liquid crystalline order, the factors contributing to the elastic properties work differently in different directions. Consequently the elastic moduli in the liquid crystalline state depend on the direction of deformation, thus adopting tensorial character.

In the isotropic state, stress–strain relations for LC elastomers are derived from statistical theory as for conventional rubbers. Because of the coupling of liquid crystalline order to network properties in the liquid crystalline state, the calculation of deformation free energy starts from anisotropic networks. Therefore the conformational properties of the chains as well as energetic interactions have to be regarded as dependent on direction. This implies anisotropic chain statistics and an anisotropic contribution of netpoint fluctuation and chain entanglements to elastic properties. Entanglement slippage is favoured parallel to a preferred direction, while restricted in the perpendicular direction, as shown by de Gennes for main chain LC polymers (de Gennes, 1982). This applies also to side chain LC elastomers. In this case, the influence of bulky side groups to chain slipping has also to be considered. The appearance of entanglements is related to the degree of polymerization N_e, which is about 300 for conventional elastomers as calculated from the plateau modulus of the corresponding linear polymer (Graessley, 1974). For LC polymers in the liquid crystalline state N_e is unknown.

As for conventional elastomers, the elastic behaviour of LC networks in the isotropic state is determined by the conformational properties of the network chains. Hence retractive forces are mainly entropic in origin. Energetic contributions may arise from non-isoenergetic rotational isomers and distortion of bond angles. In the liquid crystalline state, the conformational properties also play an important role in the elastic behaviour. In addition, coupling of network anisotropy to the state of order of the liquid crystalline phase probably makes a considerable contribution of energetic sources to restoring forces. This should be reflected by the different temperature dependence of the elastic moduli as compared to the isotropic state.

The experimental verification of these different aspects contributing to the elastic properties of LC elastomers presents an open field to be claimed. Experiments are described below which give a first insight into some features of the mechanical behaviour of side chain LC elastomers.

The temperature-dependent state of order of mesogenic side groups is mirrored by the thermoelastic behaviour (Gleim, and Finkelmann, 1987). This is shown in Figure 10.5 for a siloxane elastomer (sample no. 2d of Table 10.1). The nominal stress σ^0 (retractive force per initial cross-sectional area) of a sample fixed at constant length decreases rapidly with decreasing temperature below the phase transformation (Figure 10.5b). At lower temperature, the tensile stress changes into a compressive force. Simultaneously, the length of the unloaded sample increases, becoming larger than the initial length achieved by stretching (Figure 10.5a). The reorientation process of the mesogens, as indicated by the linear thermal expansion coefficient and stress behaviour, converts the isotropic network into a macroscopically anisotropic sample. Hence the elastic response to mechanical deformation not only depends on the polymer backbone but also on the orientation of the mesogens with respect to the direction of deformation.

Above the phase transformation, the usual thermoelastic behaviour of an elastomer below the thermoelastic inversion is observed. Due to the small deformation of the sample, the entropic force is dominated by the thermal expansion. At higher elongations, of course, an increase of tensile stress with increasing temperature would be observed. The linear dependence of the nominal stress on temperature shows that above the phase transformation the elastic behaviour is predominantly dependent on network properties covered by the statistical theory of rubber elasticity. Short-range order, as in formation of clusters, does not seem to have a major effect on the thermoelastic properties. The deviations seen close to the indicated phase transformation in Figure 10.5 are likely to be due to a broad biphasic region of this elastomer.

The experiments shown so far in this chapter describe the elastic response to a static deformation. Information about mobility of network chains and mesogenic side groups can be obtained by dynamic mechanical measurements (Oppermann et al., 1982). The considerable change of mobility of the mesogens at the phase transformation should be reflected in the dynamic mechanical behaviour. Figure 10.6 shows the temperature dependence of both storage and loss modulus (G' and G'') of a siloxane network (sample no. 2d of Table 10.1). Besides a β-relaxation process at low temperatures, the phase transformation occurs as a discontinuous jump within the tail of the glass transition. The change in G' and G'' at the clearing temperature is small. Thus the viscoelastic behaviour is mainly attributed to the properties of the main chain. The liquid crystalline → isotropic phase transformation represents only a small disturbance.

The elastic properties of the LC elastomers in the liquid crystalline state are strongly time-dependent. The time scale on which strain-induced reorientation of the mesogenic groups occurs is revealed by the frequency-dependence of the elastic moduli (Mitchell et al., 1987). The time required for reorientation by mechanical strain is about an order of magnitude smaller than the release of orientation when the stress is switched off. This is analogous to electric and magnetic field effects. Continuing the analogy, an increase of the applied stress should diminish the switching time of reorientation. However, the properties of the main chain, especially the glass transformation, have to be kept in mind when dealing with the time-dependence of molecular reorientations of mesogenic side groups.

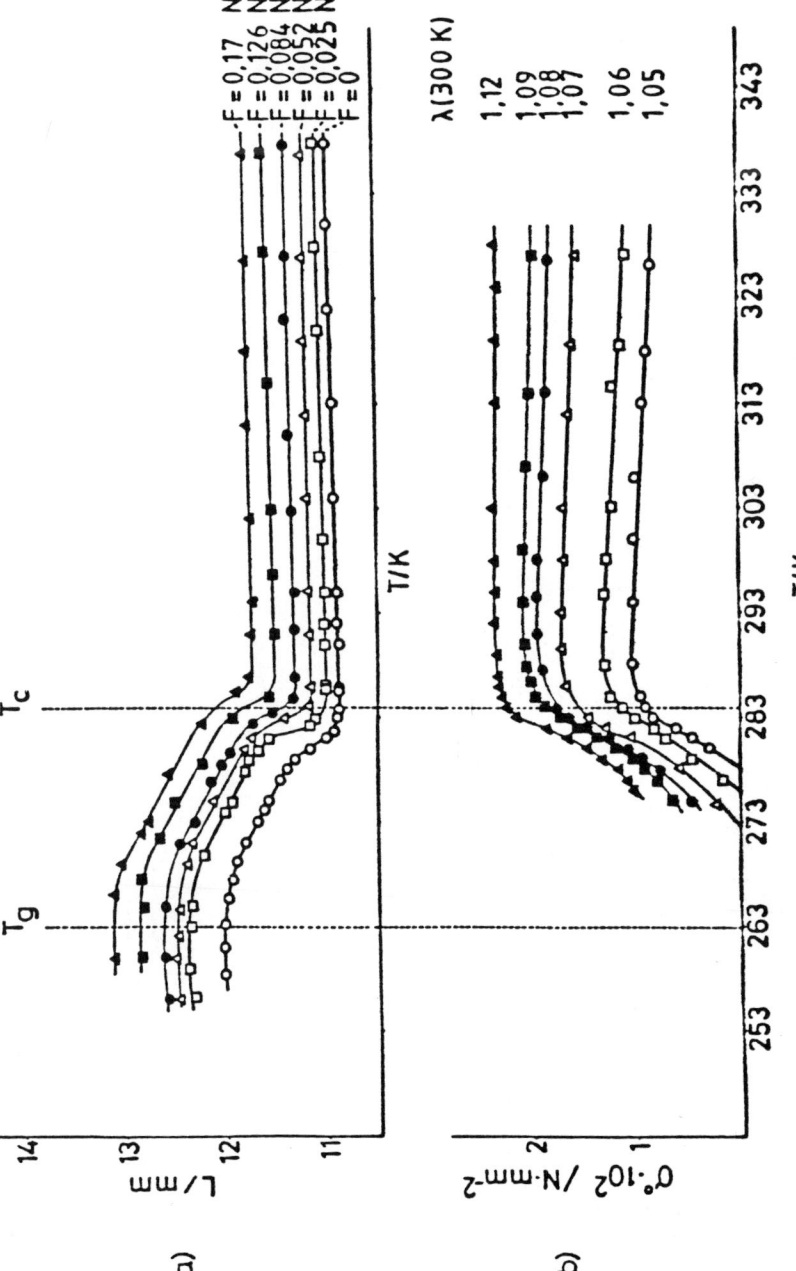

Figure 10.5 Thermoelastic behaviour of elastomer 2d of Table 10.1 (*a*) Temperature dependence of sample length *L* at constant load F as indicated; (*b*) temperature dependence of nominal stress σ^0 at constant sample length; the indicated strain ratios refer to 300 K.

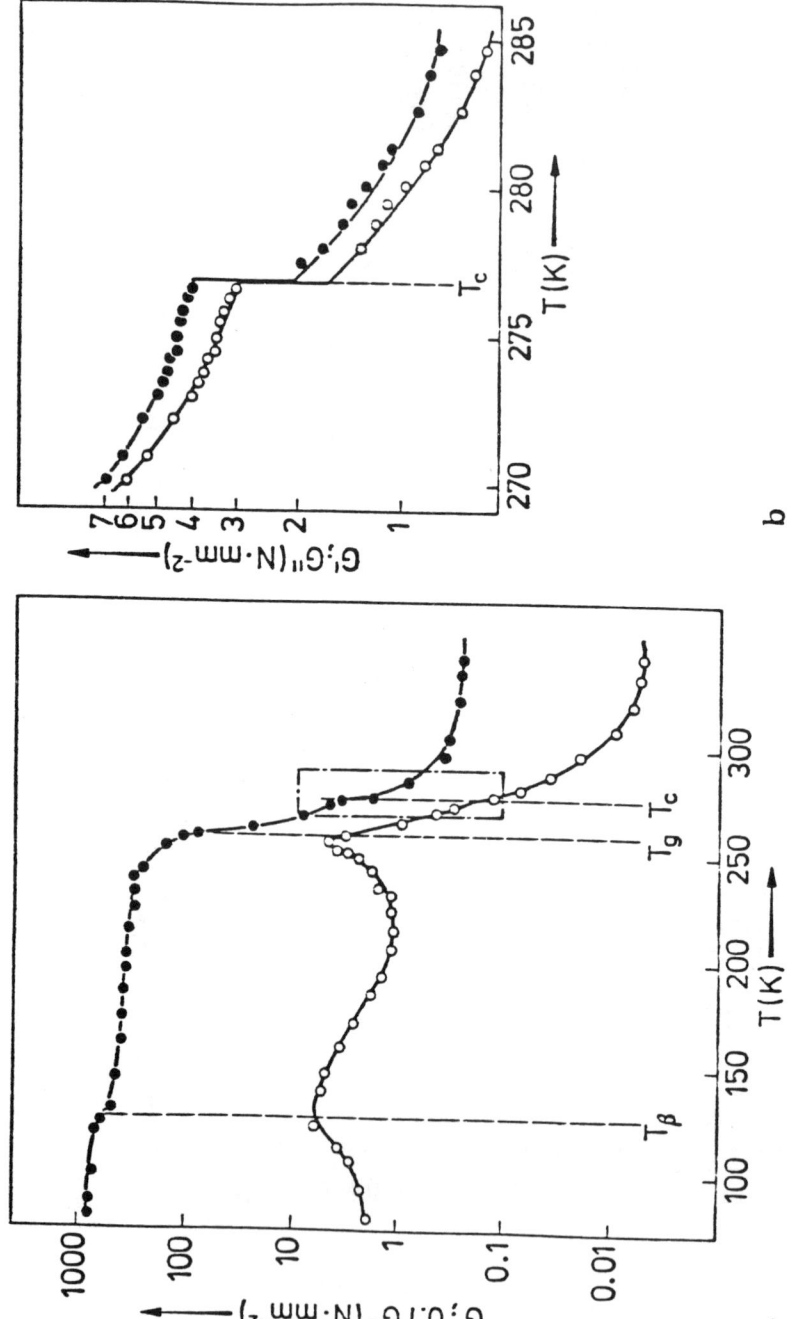

Figure 10.6(*a*) Temperature dependence of the storage modulus *G'* (filled circles) and loss modulus *G''* (open circles), measured at 0.5 Hz, for elastomer 2d of Table 10.1; (*b*) Detailed section of Figure 10.6*a*.

10.3.4 *Photoelasticity*

The stress optical properties of conventional elastomers are closely related to their mechanical behaviour. This applies equally well to the LC elastomers in the isotropic state. In the liquid crystalline state, due to the coupling of anisotropic order to network properties, unusual photoelastic properties are expected. Some features of photoelasticity of conventional elastomers are demonstrated below.

The theory of stress birefringence for conventional networks was developed by Kuhn and Grün on the basis of the statistical segment model (Kuhn and Grün, 1942). Upon deformation, a network becomes birefringent because the optical anisotropic chain segments are oriented, which clearly means an intramolecular origin of stress birefringence. Assuming the validity of the Lorentz–Lorenz equation, the relation of birefringence and strain for a uniaxially stretched network is given by

$$\Delta n = \frac{2\pi}{45}\frac{(\bar{n}^2 + 2)^2}{\bar{n}}\frac{\langle r^2 \rangle}{\langle r^2 \rangle_0}\nu(\alpha_1 - \alpha_2)(\lambda^2 - \lambda^{-2}) \tag{10.3}$$

Here Δn is the difference between the main refractive indices n_1, parallel, and n_2, perpendicular to the direction of stress, α_1 and α_2 are the corresponding main polarizabilities of the statistical segment, and $\bar{n}$ is the mean refractive index. A comparison with eqn (10.2) shows that the stress-optical coefficient C is given by

$$C = \frac{\Delta n}{\sigma}\frac{2\pi}{45kT}\frac{(\bar{n}^2 + 2)^2}{\bar{n}}(\alpha_1 - \alpha_2) \tag{10.4}$$

thus following the well-known proportionality of stress and birefringence. Insofar as the optical anisotropy $(\alpha_1 - \alpha_2)$ does not depend on temperature, the stress-optical coefficient should be inversely proportional to temperature. If different rotational isomers are not energetically equivalent, a slight temperature dependence of $(\alpha_1 - \alpha_2)$ follows (Stein, 1976). According to eqns (10.3) and (10.4), the signs of birefringence and stress-optical coefficient respectively are correlated to the sign of the optical anisotropy of the statistical chain segment. This has important implications for the relation between mesogenic side-group orientation and photoelastic behaviour.

In the following section, stress-optical properties of conventional and LC elastomers are compared. It is shown how the intrinsic tendency of the mesogenic side groups to form liquid crystalline phases leads to characteristic deviations in photoelastic behaviour. Furthermore, information about orientation of mesogenic groups is extracted from stress-optical measurements and compared to X-ray investigations. Firstly, however, some principles of photoelasticity in the liquid crystalline state are discussed.

10.3.4.1 *Photoelastic behaviour in the anisotropic state.*
The stress-optical behaviour of LC elastomers in the liquid crystalline state is determined mainly by the optical properties of the liquid crystalline phase. Network properties are of subordinate importance; thus the Kuhn–Grün theory of stress birefringence is not applicable in the liquid crystalline state. At present, no adequate theory of photoelasticity in the anisotropic state exists. Hence a correlation of stress-optical properties and molecular quantities is not possible.

The photoelastic behaviour of LC elastomers in the liquid crystalline state depends on the director orientation of the mesogenic groups with respect to the axis of deformation. It has been shown by X-ray measurements that the director orientation of mesogenic groups with respect to the strained network chains is independent whether the elastomer is under compression or elongation (Hammerschmidt and Finkelmann, 1988). Thus, if the state of deformation changes from extension to compression, the director orientation of the mesogenic groups is rotated by 90°. Clearly a drastic change of photoelastic properties follows.

Stress-optical measurements in the liquid crystalline state present some difficulties due to insufficient transparency of the samples. Photoelastic investigations are more convenient in the isotropic state where the samples are translucent. The following section deals with results of photoelastic measurements in the isotropic state.

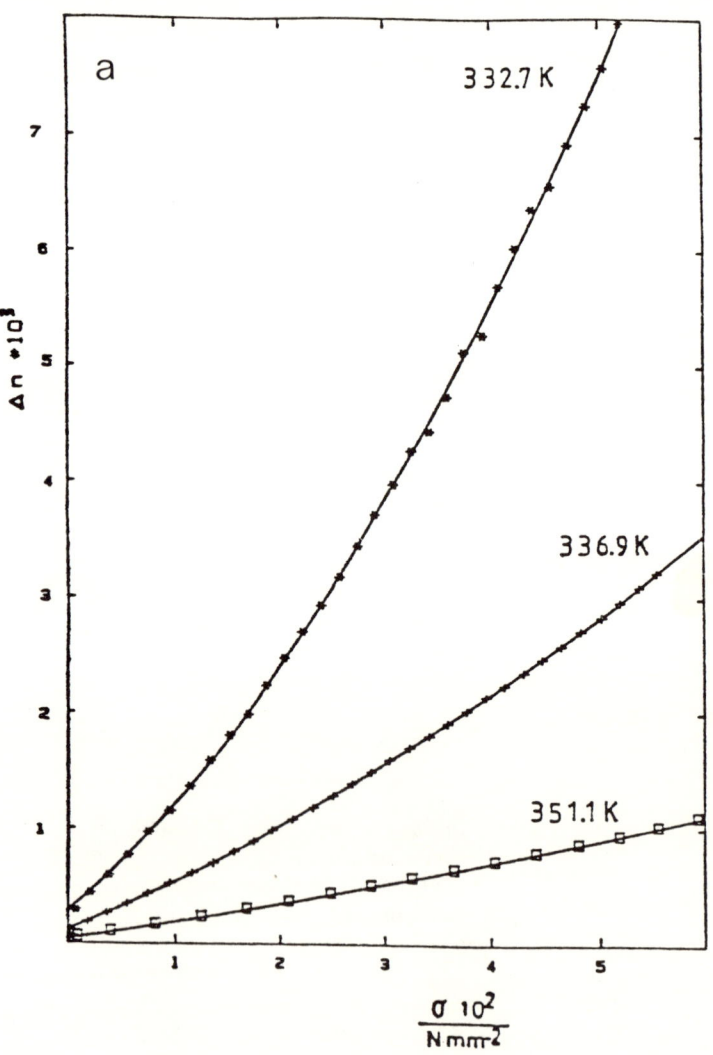

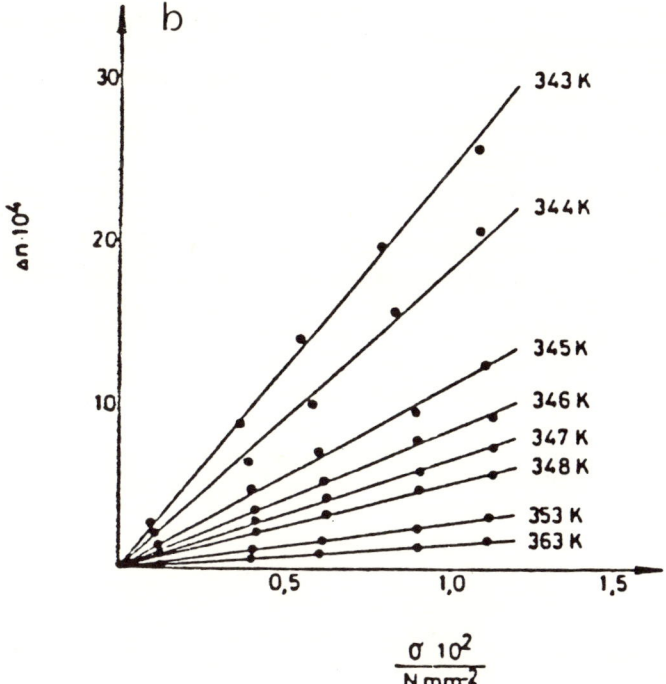

Figure 10.7 Dependence of birefringence Δn on applied stress σ. (a) Elastomer of Scheme A; (b) elastomer 2b of Table 10.1.

10.3.4.2 *Photoelastic behaviour in the isotropic state.* In this section, some principles of photoelasticity of LC elastomers in the isotropic state are demonstrated by very recent experiments.

It is well known that the physical properties of liquid crystals near the nematic → isotropic phase transformation are strongly influenced by pretransitional effects (de Gennes, 1974). Some degrees above the clearing temperature, nematic clusters are formed in the isotropic melt (Porter and Johnson, 1963). These clusters are oriented by external electric and magnetic fields, causing birefringence, the value of which is related to the number and size of clusters and their state of order with respect to the direction of the external field. This relation has already been shown in Figure 10.3 where, in the isotropic state, the order parameter S is directly related to the birefringence.

In the LC elastomers these clusters are formed as well, and are oriented by mechanical strain, leading to marked effects displayed by the photoelastic properties. Figure 10.7a shows a plot of birefringence and stress for a methacrylic network (example from Scheme A) above the nematic → isotropic transformation (Schätzle, 1988). No proportionality of stress and birefringence over the whole range of deformation is observed in this case, in contradiction to eqn (10.4). The shape of the curves suggests that the applied strain not only affects orientation but also increases the intrinsic anisotropy of the network. Close to the phase transformation, the stress–

birefringence curves do not extrapolate to zero, which means that the unloaded network is macroscopically ordered in this case.

In Figure 10.7*b* the same plot is shown for a siloxane elastomer (example no. 2b of Table 10.1); this time, birefringence and stress are proportional (Gleim and Finkelmann, 1987). This corresponds to conventional elastomers free of cluster formation. However, the observed proportionality is more probably due to the small deformation range ($< 10\%$) and the flexible cross-linking chain carrying no mesogenic groups, than to the absence of cluster formation.

Nematic clusters are also formed in the siloxane networks. This is clearly demonstrated by the strong temperature dependence of CT, the product of stress-optical coefficient and temperature. According to eqn (10.4), only a smooth temperature dependence of CT should be observed. On approaching the phase transformation, however, CT increases rapidly, as seen in Figure 10.8. This corresponds to the large increase of electric and magnetic birefringence observed in the pretransitional region (Stinson and Litster, 1982).

The pseudocritical behaviour due to cluster formation is described by the Landau–de Gennes theory. According to this, the electric and magnetic birefringence diverge at $T^{\#}$

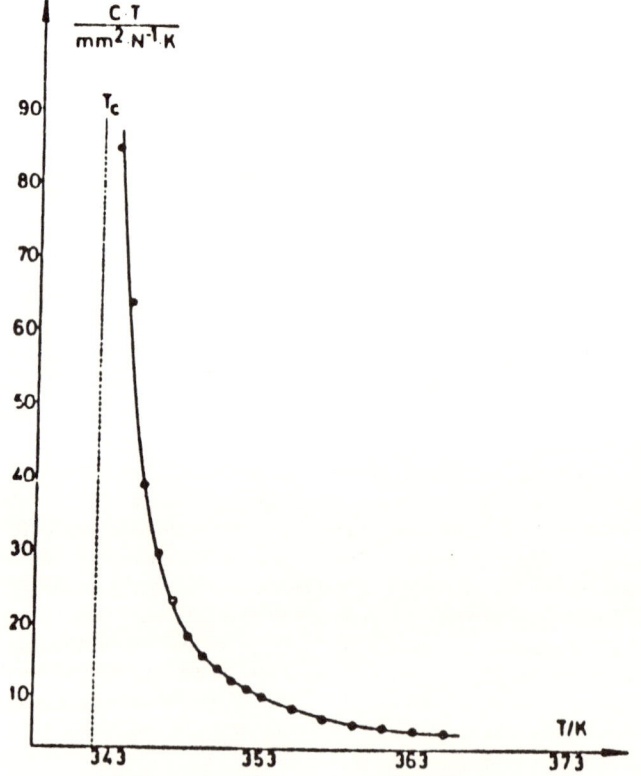

Figure 10.8 Product of stress-optical-coefficient C and temperature T *v.* temperature T for elastomer 2b of Table 10.1.

according to the subsequent proportionality:

$$\Delta n(E,H) \sim \frac{1}{(T - T^{\#})^{\gamma}} \qquad (10.5)$$

Within the Landau-de Gennes theory, the critical exponent of this divergency is unity. Thus $T^{\#}$ is obtained by extrapolation of $\Delta n^{-1/\gamma}$ to zero.

The theory applies also to strained elastomers. This is shown in Figure 10.9 by a linear relation between $\Delta n^{-1/\gamma}$ and temperature for a siloxane network (Figure 10.9a) (Gleim and Finkelmann, 1987) and a methacrylic network (Figure 10.9b) (Schätzle, 1988). The critical exponent found by these measurements is close to unity. The extrapolation leads to a temperature $T^{\#}$, which is about 2°C below the transformation temperature T_{ni}.

These experiments clearly show the correlation of photoelastic and thermodynamic properties linked by the Landau–de Gennes theory of the nematic → isotropic phase transformation. The divergency of stress birefringence directly indicates the appearance of a nematic phase at lower temperature for these elastomers.

As has been stated in section 10.3.4, stress-optical measurements provide information about orientation of the mesogenic side groups. If these highly polarizable molecules are attached to the polymer backbone, the anisotropy of the statistical chain segments depends on the orientation of this side group with respect to the segment axis. The anisotropy of the main chain can be neglected, because it is small compared to that of the mesogenic group.

As the greatest polarizability of the mesogen lies within its long axis, the positive birefringence of the siloxane and methacrylic networks (Figure 10.7a, b) indicates a preferred parallel alignment of the mesogenic groups with respect to the segment axis. The view is supported by additional X-ray investigations which show that the orientation is not significantly changed upon passing the isotropic → liquid crystalline transformation (Gleim and Finkelmann, 1987). It should be noted that parallel alignment of mesogenic side groups is not generally the case. Perpendicular orientations are also observed. Here the chemical constitution of the mesogen and main chain play an important role (Schätzle and Finkelmann, 1987). Even the type of deformation (compressional or extensional) must be considered (Hammerschmidt and Finkelmann, unpubl.).

10.4 Conclusion

The combination of liquid crystalline order and network properties is realized by linking low molar mass liquid crystals as side groups to a polymer network. The physical properties of these side chain LC elastomers are related to the interaction of network anisotropy and state of order of the liquid crystalline phase.

The LC elastomers in the liquid crystalline state are anisotropic systems irrespective of mechanical loading. Consequently their physical properties are dependent on direction, adopting tensorial character in the liquid crystalline state.

By mechanical deformation, a uniform orientation of the director is achieved. Hence strained LC elastomers in the liquid crystalline state behave optically as single crystals of the same dimension. This property offers a wide field of possible applications. In this connection, among other possibilities, their use for optical devices as well as for membranes should be mentioned.

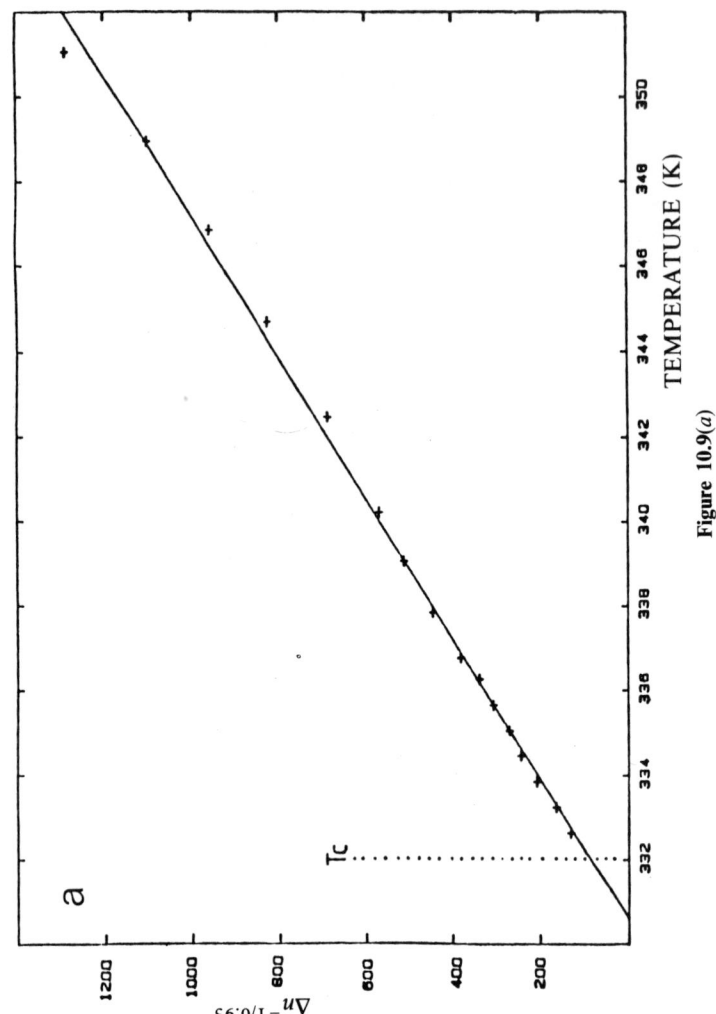

Figure 10.9(a)

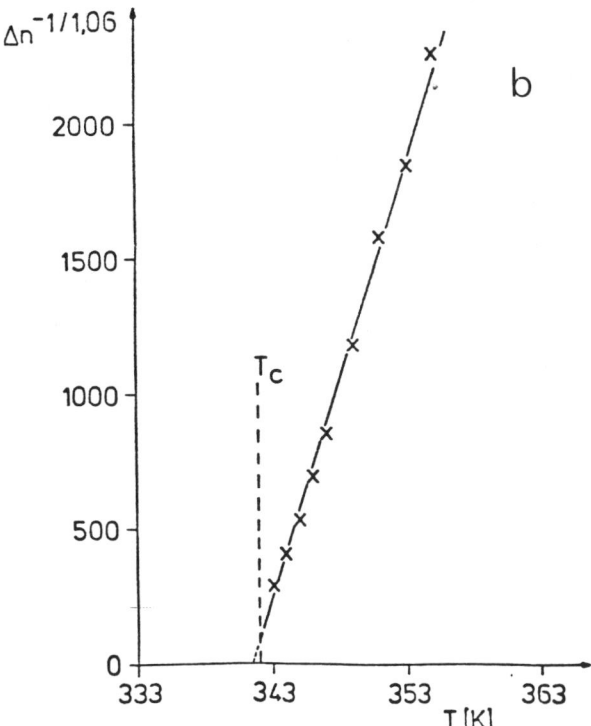

Figure 10.9 Temperature dependence of $\Delta n^{-1/\gamma}$ (a) Elastomer 2b of Table 10.1; (b) Elastomer of Scheme A.

On the theoretical side, LC elastomers present a great challenge. Many questions concerning network coupling to anisotropic alignment of mesogenic groups and its results on the physical properties are still open.

References

Anisimov, M.A., Voronov, V.P., Kulkov, A.O. and Kholmurodov, F. (1985) *J. Physique* **46**, 2137.
Chompff, A.J. (1977) in *Chemistry and Properties of Crosslinked Polymers*, ed. Labana, S.S., Academic Press, New York.
Davis, F.J., Gilbert, A., Mann, J. and Mitchell, G.R. (1986) *J. Chem. Soc., Chem. Commun.* 1333.
de Gennes, P.G. (1971) *Mol. Cryst. Liq. Cryst.* **12**, 193.
de Gennes, P.G. (1974) *The Physics of Liquid Crystals*, Clarendon Press, Oxford.
de Gennes, P.G. (1975) *C.R. Acad. Sci. Paris* **281B**, 101.
de Gennes, P.G. (1982) in *Polymer Liquid Crystals* eds. Ciferri, A., Krigbaum, W.R. and Meyer, R.B., Academic Press, New York.
Edwards, S.F. (1971) in *Polymer Networks, Structural and Mechanical Properties*, eds. Chompff, A.J. and Newman, S., Plenum, New York.
Ehrenfest, P. (1933) *Proc. K. Ned. Akad. Wet, Amsterdam* **36**, 153.
Erman, B. and Flory, P.J. (1982a) *Macromolecules* **15**, 800.
Erman, B. and Flory, P.J. (1982b) *Macromolecules* **15**, 806.
Finkelmann, H., Kock, H.J, and Rehage, G. (1981) *Makromol. Chem., Rapid Commun.* **2**, 317.
Frenzel, J. and Rehage, G. (1983) *Makromol. Chem.* **184**, 1685.

Gleim, W. and Finkelmann, H. (1987) *Makromol. Chem.*, **188**, 1489.

Graessley, W.W. (1974) *Adv. Polym. Sci.* **16**, 1.

Gramsbergen, E.F., Longa, L. and de Jeu, W.H. (1986) *Phys. Rep.* **135(4)** 195.

Helfrich, W. (1970) *Phys. Rev. Lett.* **24**, 201.

Hessel, F., Herr, R.P. and Finkelmann, H. (1987) *Makromol. Chem.* **188**, 1579.

Kilian, H.H. (1981) *Polymer* **22**, 209.

Kuhn, W. and Grün F. (1942) *Kolloid Z.* **101**, 248.

Landau, L.D. (1937) *Phys. Z. Sowjetunion* **11**, 26.

Langley, N.R. and Ferry, J.D. (1968) *Macromolecules* **1**, 353.

Mandelkern, L. (1964) *Crystallization of Polymers*, McGraw-Hill, New York.

Mitchell, G.R., Davis, F.J. and Ashman, A. (1987) *Polymer* **28**, 639.

Oppermann, W., Braatz, K., Finkelmann, H., Gleim, W., Kock, H.J. and Rehage, G., (1982) *Rheol. Acta* **21**, 423.

Pollmann, P., Wiege, B. and Rothert, A., (1988) *Liquid Crystals* **3(2)**, 225.

Porter, R.S. and Johnson, J.F. (1963) *J. Appl. Phys.* **34**, 51.

Ronca, G. and Allegra, G. (1975) *J. Chem. Phys.* **63**, 4990.

Rosenblatt, R. (1981) *Phys. Rev.* **A24**, 2236.

Schätzle, J. and Finkelmann, H. (1987) *Mol. Cryst. Liq. Cryst.* **142**, 85.

Schätzle, J. (1988) Thesis, University of Freiburg.

Schwarz, J. (1986) *Makromol. Chem. Rapid. Commun.* **7**, 216.

Stein, R.S. (1976) *Rubber Chem. Technol.* **49**, 458.

Stinson, T.W. and Litster, J.D. (1982) *Phys. Rev. Lett.* **25**, 503.

Theon, J., Marynissen, H. and van Dael, W. (1982) *Phys. Rev. A* **26(5)**, 2886.

Treloar, L.R.G. (1975) *The Physics of Rubber Elasticity*, Oxford University Press, Oxford.

Ullrich, K. and Wendorff, J.H. (1985) *Mol. Cryst. Liq. Cryst.* **131**, 361.

Wang, X.J. and Warner, M. (1987) *J. Phys. A: Math. Gen.* **20**, 713.

Zentel, R. and Benalia, M. (1987) *Makromol. Chem.* **188**, 665.

11 Field-induced effects in side chain liquid crystal polymers

WOLFGANG HAASE, Institut für Physikalische Chemie, Petersenstr. 20, Technische Hochschule, D-6100 Darmstadt, FRG

11.1 Introduction

Side chain liquid crystalline polymers (LCPs) are a new type of substance combining liquid crystalline properties with the properties of polymers. These compounds have attracted much attention from a scientific and an applications point of view. For most applications, such as display devices or optical storage, a macroscopically oriented domain over a large area has to be prepared. Maintaining this, preoriented samples are then exposed to external electric or magnetic fields, in order to obtain the desired effects. In this case the energy density of the applied external field must overcome the elastic energy terms. Much information concerning the properties of the side chain LCPs can be obtained by considering the coupling strength between the external forces and molecular forces.

In this chapter information regarding the behaviour of side chain liquid crystalline polymers under the influence of an external electric or magnetic field will be discussed, including the closely related surface alignment effects. A point for discussion will be whether field induced properties are determined to a first approximation by the liquid crystalline properties alone, or to what extent, if any, polymer specific features like spacer length or main chain structures are involved.

11.2 Alignment and dynamics of LCP cells

11.2.1 General considerations

The properties of low molecular mass nematic liquid crystals composed of rod-shaped molecules under the influence of external fields are well documented (e.g. de Gennes, 1974; Blinov. 1983) and will be referred to here only briefly. In the side chain liquid crystalline polymers, the influence of the non-mesogenic part, the flexible spacer and the polymeric backbone have also to be considered.

If the external field strength exceeds a threshold field, induced reorientation takes place as a consequence of the given anisotropy of the dielectric or diamagnetic permittivity of the liquid crystal samples. If the external field is taken away, the liquid crystal molecules relax back to their initial state.

In general, in the presence of an electric field $\vec{E}$, the free energy density G of the nematic liquid crystal is

$$dG = dG_{\text{elastic}} + dG_{\text{dielectric}} \qquad (11.1)$$

The director $\vec{n}$ is defined in the z direction by

$$G_{\text{elastic}} = \tfrac{1}{2} \int_V \{k_{11}(\text{div}\, \vec{n})^2 + k_{22}(\vec{n} \cdot \text{rot}\, \vec{n})^2 + k_{33}(\vec{n} \cdot \text{rot}\, \vec{n})^2\} dv \qquad (11.2)$$

G_{elastic} is the Frank free energy density, and k_{11}, k_{22} and k_{33} are the splay, twist and bend elastic constants respectively. v is the space coordinate of the volume V. The elastic constants are of the order of $10^{-11} N$ for polymeric as well as for low molecular mass liquid crystals (see section 11.5.5).

It should be mentioned here that the model used is only valid for bulk properties, neglecting conductivity contributions or surface effects. The dielectric contribution is

$$G_{\text{dielectric}} = \tfrac{1}{2} \int_V \vec{D} \cdot \vec{E}\, dv \qquad (11.3)$$

The dielectric displacement $\vec{D}$ is related to the electric field $\vec{E}$ by

$$\vec{D} = \langle \varepsilon \rangle \cdot \vec{E} \qquad (11.4)$$

The dielectric permittivity tensor ε_{ij} is given by

$$\varepsilon_{ij} = \varepsilon_\perp \delta_{ij} + (\varepsilon_\parallel - \varepsilon_\perp) n_i \cdot n_j; \quad i,j = 1, 2, 3 \qquad (11.5)$$

where δ_{ij} is the Kronecker symbol.

The dielectric anisotropy is given by

$$\Delta \varepsilon = \varepsilon_\parallel - \varepsilon_\perp \qquad (11.6)$$

$\varepsilon_\parallel$ and $\varepsilon_\perp$ are the dielectric constants parallel and perpendicular to the optical axis, respectively, described by the director $\vec{n}$.

With some assumptions for the planar cell (isotropic approximation for eqn 11.2), the elastic free energy can be written as

$$G_{\text{elastic}} = \tfrac{1}{2} \int_0^d k(\text{div}\, \vec{n})^2\, dz = \tfrac{1}{2} \int_0^d k\left(\frac{\partial \varphi}{\partial z}\right)^2 dz \qquad (11.7)$$

where d is the cell thickness and $k = k_{11}$.

For the dielectric contribution, eqn (11.8) is satisfied:

$$G_{\text{dielectric}} = \int_0^d - \varepsilon_0 \Delta \varepsilon (\vec{E} \cdot \vec{n})^2\, dz = \int_0^d - \varepsilon_0 \Delta \varepsilon \sin^2 \varphi\, dz \qquad (11.8)$$

where ε_0 is the vacuum permittivity, and φ is the angle between the director $\vec{n}$ and the initial orientation.

Now the density of free energy is:

$$G = \tfrac{1}{2} \int_0^d \left[k\left(\frac{\partial \varphi}{\partial z}\right)^2 - \varepsilon_0 \Delta \varepsilon E^2 \sin^2 \varphi \right] dz \qquad (11.9)$$

By setting electric energy equal to field energy and solving the differential equation (Euler's second-order differential equation), we get, for a small deviation φ_m of the director $\vec{n}$ in the middle of the cell,

$$E_0 = \frac{\pi}{d}\left(\frac{k}{\varepsilon_0 \Delta \varepsilon}\right)^{1/2} \qquad (11.10)$$

Here E_0 is the threshold field strength. With $U_0 = E_0 \cdot d$, the threshold voltage U_0 is given by

$$U_0 = \pi \left(\frac{k}{\varepsilon_0 \Delta \varepsilon} \right)^{1/2} \tag{11.11}$$

In this simple model, the threshold voltage U_0 depends on $k = k_{11}$, the splay elastic constant for the planar cell and on $\Delta \varepsilon$, but not on the thickness d of the cell. The model used here is valid for the Frederikz transition by deforming the planar cell. The same basic theory holds for the Frederikz deformation of a homeotropic arrangement as shown in Figure 11.1.

For the deformation of a homeotropic cell, $k = k_{33}$ is valid. The appropriate elastic constant for twist effect is given by $k = [k_{11} + \frac{1}{4}(k_{33} - 2k_{22})]$.

In the case where an external magnetic field is applied with the same level of approximation, eqn (11.3) must be substituted by eqn (11.12):

$$G_{\text{diamagnetic}} = \frac{1}{2} \int \vec{B} \cdot \vec{H} \, dv \tag{11.12}$$

assuming $\vec{B}$, the magnetic induction vector, parallel to the magnetic field vector $\vec{H}$. The magnetic threshold field strength H_0, analogous to eqn (11.10), can be derived in the

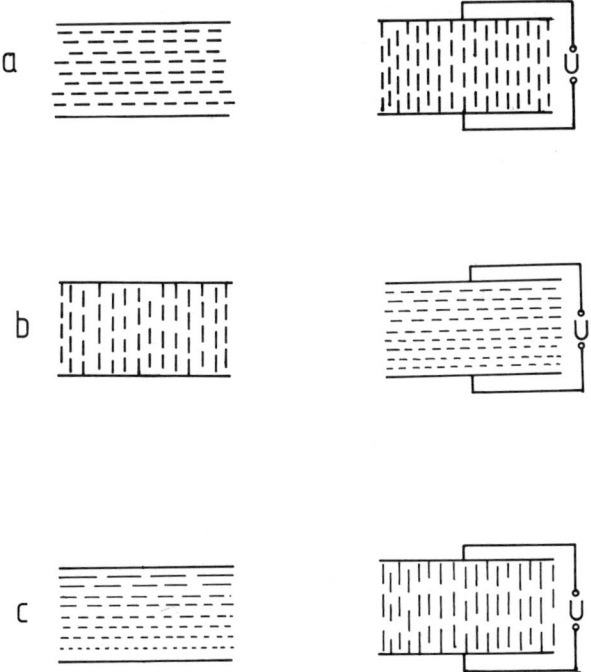

Figure 11.1(a) Frederikz transition from homogeneous to homeotropic; (b) Frederikz transition from homeotropic to homogeneous; (c) Transition from twisted-nematic to homeotropic, U, applied voltage.

same way:

$$H_0 = \frac{\pi}{d}\left(\frac{k}{\Delta\chi}\right)^{1/2} \tag{11.13}$$

$\Delta\chi = \chi_\parallel - \chi_\perp$ is the diamagnetic anisotropy parallel and perpendicular, respectively, to the field $\vec{H}$. Using eqns (11.10) and (11.13), a relationship between the electric and magnetic threshold results:

$$E_0 = \left(\frac{\Delta\chi}{\varepsilon_0\Delta\varepsilon}\right)^{1/2}\cdot H_0 \tag{11.14}$$

In case of dual field interaction, for example where the magnetic field and the electric field are parallel or perpendicular, eqn (11.3) is to be replaced by eqn (11.15):

$$G_{\text{dielectric}} + G_{\text{diamagnetic}} = \tfrac{1}{2}\int(\vec{D}\cdot\vec{E} + \vec{B}\cdot\vec{H})\,d\tau \tag{11.15}$$

The dynamics of the Frederikz transition (see Figure 11.1) are described by:

$$k_{11}\left(\frac{\partial\varphi}{\partial z}\right)^2 - \varepsilon_0\Delta\varepsilon E^2 = \eta_1\frac{\partial\varphi}{\partial t} \tag{11.16}$$

Here $\eta_1 = \alpha_3 - \alpha_2$ is a viscosity coefficient, where α_2, α_3 are Leslie coefficients. Considering the dependence of φ on z and time t, eqn (11.17) holds as a good approximation.

$$\varphi(z, t) = \varphi_m(t)\cos\left(\frac{\pi z}{d}\right) \tag{11.17}$$

$\varphi_m(t)$ tends exponentially towards $\varphi_m(\infty)$ with a time constant τ_r:

$$\varphi_m^2(t) = \frac{\varphi_m^2(\infty)}{1 + a^2\exp(-2t/\tau_r)} \tag{11.18}$$

where a is an integration constant, and τ_r is the so-called rise time:

$$\frac{1}{\tau_r} = \frac{k_{11}}{\eta_1}\left(\frac{\pi}{d}\right)^2\left[\left(\frac{U}{U_0}\right)^2 - 1\right] \tag{11.19}$$

for applied voltage U.

The passive decay time τ_d^0 for switching off the electric field is:

$$\frac{1}{\tau_d^0} = \frac{k_{11}}{\eta_1}\left(\frac{\pi}{d}\right)^2 \tag{11.20}$$

The passive decay time is dependent only on the elastic coefficient k_{11} and the viscosity coefficient η_1 at a given layer thickness d.

Combining eqns (11.19) and (11.20), the rise time τ_r is obtained as

$$\frac{1}{\tau_r} = \frac{1}{\tau_d^0}\left[\left(\frac{U}{U_0}\right)^2 - 1\right] \tag{11.21}$$

The surface effect and also the conductivity contribution will not be described here. The latter effect gives rise to the so-called Williams domains.

11.2.2 *E-field effects*

The dielectric anisotropy is mainly determined by the dipolar properties of the mesogenic group. The flexible spacer is, in most cases, a nonpolar $(CH_2)_n$ group; backbones like polysiloxane or polyacrylate can contribute to $\Delta\varepsilon$ if a chain ordering takes place through the polar group on, or as part of the chain. In general, the contribution of the main chain can be assumed as isotropic (see also Chapter 7).

At a certain voltage $U > U_0$ the outer electric field will align the mesogenic parts, depending on their dipolar components. This can be described by the well-known Maier–Meier equation, which can be shortened to:

$$\Delta\varepsilon \sim -\frac{\mu^2}{2kT}(1 - 3\cos^2\beta)S \qquad (11.22)$$

It shows $\Delta\varepsilon$ tends to be negative $(\beta > 54.7°)$ or positive $(\beta \leqslant 54.7°)$ with β the angle between the dipole moment μ and the director $\vec{n}$. S is the order parameter, k the Boltzmann constant and T the temperature.

As an example we can consider the polymers of principal structure

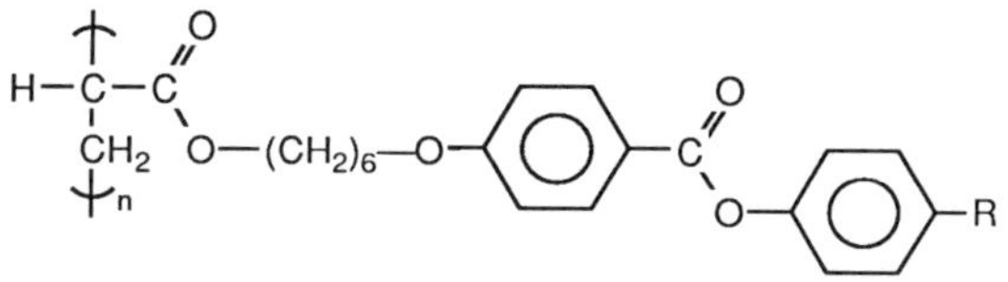

$$\text{with } R = CN \quad \textbf{(2)}$$
$$R = OCH_3 \ \textbf{(2a)}$$

Using the dipole components (Bormuth, 1988) and the vector addition model, $\mu_\parallel$ and $\mu_\perp$ can be estimated as follows:

(2)	$\mu_\parallel$	$\mu_\perp$
	5.47	2.08

(2a)	$\mu_\parallel$	$\mu_\perp$
	0.88	2.41

where the symbols $\parallel$ and $\perp$ mean parallel and perpendicular to the molecular long axis. It follows that compounds such as **(2)** will be oriented homeotropically and compounds such as **(2a)** homogeneously, on applying the voltage perpendicular to the glass plates.

A cell used for orientation is shown in Figure 11.2. The two glass plates coated with In_2O_3/SnO_2 are separated by a spacer. Normally the cell is filled by capillary action, while the sample is kept near the clearing point for several hours. The cell filling procedure is in some cases critically dependent on the viscosity of the material. Quite a good preorientation is given on the conducting side by a polyimide coating. In principle, the cell can also be configured having the conducting material on the outer side. To get a pretilt angle, coating can be performed by evaporation of SiO at a defined angle. Normally AC voltage is used in order to avoid the electrohydrodynamic disturbance due to charge transport.

The cell thickness can be varied using different spacers. For thin cells, the glass plates must be optically flat. The field strength which can be used is up to 15–50 kV cm^{-1}. At higher field strength the cell may be electrically shorted. In order to

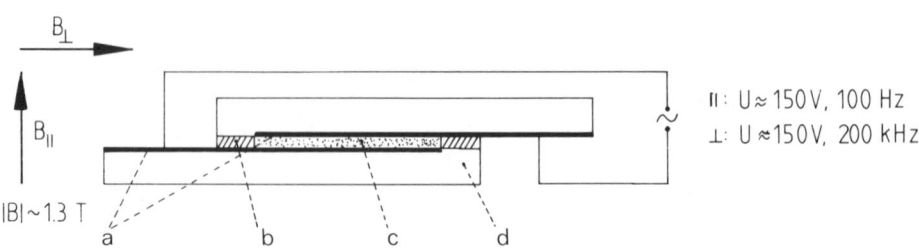

Figure 11.2 Construction of the cell. Included are the arrangements of magnetic and electric fields used for alignment. (*a*) electrodes, (*b*) Kapton spacer, (*c*) sample, (*d*) glass substrate.

measure the anisotropy effects, the electric field can be applied parallel in the plane of the glass plates with special electrode geometries. In this case, compounds like (**2**) can be aligned homogeneously, and compounds like (**2a**) homeotropically. For the latter orientation procedure, thicker cells are normally needed. Another method of changing the orientation, but using the same experimental arrangement, is to raise the frequency of the electrical field to such a high value that $\Delta \varepsilon$ changes its sign from plus to minus. This can be done if the sign reversal takes place at a normally accessible frequency ($\sim 1\,\mathrm{kHz}$) (see dual-frequency addressing, section 11.3.3 and Chapter 7).

The quality of orientation can be checked by the polarizing microscope. Also, any anisotropic property can be checked through time-dependent saturation, e.g., the capacitance (Bormuth and Haase, 1988). An interesting method for calculating the degree of orientation S_d is described by Attard and Williams (1986).

11.2.3 H-field effects

A certain magnetic field strength H with $H > H_0$ (eqn 11.13) orients the molecules with their long axis parallel to the field direction, if the diamagnetic susceptibility anisotropy $\Delta \chi = \chi_\parallel - \chi_\perp$ is positive ($\parallel$ and $\perp$ representing parallel and perpendicular, respectively, to the director).

The question arises, which part of the side chain liquid crystalline polymer molecules contributes significantly to $\Delta \chi_M$, the diamagnetic molecular anisotropy? To a first approximation the spacer and the main chain can be assumed as being random. The anisotropies of the mesogenic part of compounds like (**2**) and (**2a**) are mainly determined through the anisotropy of the two phenyl rings, estimated for the given axis definition to be $\Delta \chi_M \sim 55 \times 10^{-6}\,\mathrm{cm^3\,mole^{-1}}$ (Ibrahim and Haase, 1979). Assuming group anisotropies for polar groups e.g.—$C \equiv N$ and —O—R as $-8.8\,\mathrm{cm^3\,mole^{-1}}$ and $-6.5\,\mathrm{cm^3\,mole^{-1}}$, the anisotropies of compounds like (**2**) and (**2a**) are assumed as $\Delta \chi_M \sim 50 \times 10^{-6}\,\mathrm{cm^3\,mole^{-1}}$. This means that in compounds like (**2**) and (**2a**), polar groups make a contribution to $\Delta \chi_M$ only in the order of 0–10% of the whole molecular anisotropy. It thus follows, as consequence of this, that compounds like (**2**) and (**2a**) always orient parallel to the external field, whereas in the electrical field the orientation is different, (**2**) parallel and (**2a**) perpendicular to the external field.

In comparison with the E-field alignment, the H-field alignment shows some advantages. An important point is that there are no disturbances through hydrodynamic effects, so that the H-field can be maintained during the measurements. The homogeneous alignment is simple to arrange. Also, the tilt angle can be selected.

However, E-field alignment shows important advantages with regard to achievable, field strength. Using the relations for diamagnetic compounds with $1\,kV\,cm^{-1} \sim 1T$, valid for compounds with higher $\Delta\varepsilon$ (BDH-E7: $\Delta\varepsilon \sim 13.8$ at 10 kHz (Wu, 1985) or $1\,kV/cm \sim 0.1\,T$ valid for compounds with small $\Delta\varepsilon$ ($\Delta\varepsilon \sim 0.1$ (de Gennes, 1974)), the electrical field strength can be higher than the magnetic field strength. The diamagnetic anisotropy of the liquid crystalline phase is assumed as $\Delta\chi \sim 10^{-7}\,cm^3\,g^{-1}$.

The relations show that compounds like (2) with higher $\Delta\varepsilon$ are more easily aligned in the electrical field, whereas, for compounds such as (2a), the orientation in the magnetic field (up to 7 T) is in some cases better. The normal breakdown for the cells is about $15-50\,kV\,cm^{-1}$; on the other hand the field strength of the normal electromagnets or permanent magnets is only 2 T maximum before saturation sets in. For higher H-fields specialized magnets and/or pulse techniques are required. It should be mentioned here that the magnetic threshold H_0 can be determined by changing of the directing magnetic field.

From well-aligned monodomains the macroscopic diamagnetic anisotropy can be measured. Assuming models for calculating $\Delta\chi_M$ the diamagnetic molecular anisotropy, the order parameter S may also be extracted.

11.2.4 *Surface effects*

Various different materials have been used as the alignment coating or 'alignment layer' for LCs. These include both organic and inorganic substances and mixtures of the two. The techniques used to apply these materials on surfaces such as glass, ceramics and organic polymers include dip coating, spraying, roller coating, high-vacuum sublimation (evaporation) and even chemical reaction at the surface. In addition, a number of techniques have been used to create preferred molecular directionality and/or tilt of LC molecules on these alignment layer surfaces, including rubbing, high-speed buffing, mechanical abrasion, chemical milling and high-vacuum sublimation at oblique angles. The types of materials used to create these latter effects include such substances as polyesters, cellulose, nylon, alumina, silicon monoxide and magnesium fluoride.

Recently it has been reported by Noël *et al.* (1984) that certain polymeric liquid crystal molecules can be aligned by surface treatment. They have achieved a homeotropic alignment by simple treatment of glass slides with boiling chromic sulphuric acid, acetone and methanol (sequentially interspersed with water rinses) and rinsing with hot distilled water. Films thus prepared appear completely dark when viewed vertically between crossed polarizers. They show no birefringence, which implies that the optical axis of the molecules lies in the observation direction, perpendicular to the supporting surfaces. Noël *et al.* (1984) have also achieved homogeneous alignment when samples were prepared on coated glass plates treated with a solution of polyimide in N-methylpyrrolidone, heating to 150°C, then rubbing in a fixed direction and heating again. However, Krigbaum *et al.* (1980, 1981) and Corazza *et al.* (1984) reported contradictory results for a poly(ethylene-terephthalate-co-1,4-benzoate) containing 60 mol. % *p*-oxybenzoyl units, a polyester synthesized from 4,4'-dihydroxy-α-methylstilbene and adipic acid and a polyester prepared from 4,4'-iso-propenylene diphenol and adipic acid.

No systematic investigations are known concerning the field response properties of thinner LCP layers ($\sim 1\,\mu m$) preoriented by very strong surface alignment forces. However, thin spin-coated LCP samples have been prepared by various workers (Pinsl *et al.*, 1987) but electrical access to the samples was not required.

11.3 Some aspects to follow switching

11.3.1 *Definitions of response times*

The response time is a basic property regarding potential applications of liquid crystals in devices. In principle it is the time needed to switch from one orientation to another, e.g. from on state to off state. Unfortunately, differences in the definition are found among different research groups. The three most common cell arrangements are shown in Figure 11.1. For the response time in the TN cell, we have used the well-known definition (Figure 11.3): the time elapsing between an intensity change of 10 to 90% or 90 to 10% of maximum intensity from on to off state.

For the Frederikz cell, the intensity I that can be detected through crossed polarizers is described by

$$I = I_p \sin^2 (2\phi) \sin^2 (\delta/2) \tag{11.23}$$

where I_p is the intensity of the incident light and ϕ the angle between the optical axis and the E-vector of the incident, linearly polarized light prior to switching. Eqn 11.23 implies a $\sin^2$-shaped run of the intensity-time curve during the switching process.

$\delta/2$, the phase difference, is defined by

$$\delta/2 = \frac{\pi d}{\lambda} \Delta n \langle \cos^2 \phi \rangle_d \tag{11.24}$$

where d is the cell thickness, λ the wavelength of the incident light and $\Delta n = n_\parallel - n_\perp$ is the birefringence with $\parallel$ and $\perp$ parallel and perpendicular, respectively, to the director. The importance of the angle φ is demonstrated by Baur *et al.* (1975). For a thin Frederikz cell, a response time can be defined in a similar fashion to the TN- cell. This response time includes for example the change of intensity from $t = 0$ until the time where $I = 0.1 I_p$ (90%) or $I = 0$ (100% change of intensity). For a cell with greater thickness, the $\sin^2 (\delta/2)$-shaped curve of the change of intensity can be used for definition as well:

$$\bar{\varphi} = \arccos \sqrt{\langle \cos^2 \varphi \rangle_d} \tag{11.25}$$

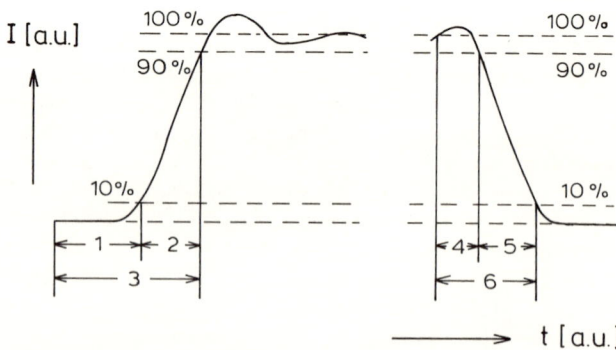

Figure 11.3 Definition of the switching times in TN-cell. 1, turn-on decay; 2, rise time; 3, turn-on time; 4, turn-off decay; 5, fall time; 6, turn-off time.

This leads to an approximately 10% change of intensity for $\bar{\varphi} = 0.2$ rad and 90% for $\bar{\varphi} = 1.0$ rad. The time needed for reorientation from $\bar{\varphi} = 0.2$ to $\bar{\varphi} = 1.0$ for a Frederikz transition can be compared with the rise time or the fall time of a TN cell, although the deformations of the layers and consequently the elastic constants are different.

So by comparing different experimental results one has to be careful if the 10/90% definition is used and, furthermore, the quality of alignment has to be checked. However, in some cases the 10/90% definition is impracticable, especially in the case of long response times. Sometimes other orientation processes instead of Frederikz or Twist-effects are investigated. When comparing the data, however, one has to be careful.

It has already been mentioned (Pranoto, 1984; Pranoto and Haase, 1988), that the relation.

$$\bar{\varphi} = \varphi_0 \exp(t/\tau) \tag{11.26}$$

with φ_0 and τ constants, respectively, is invalid for higher $\bar{\varphi}$ values, and also for the rise and decay process.

A relation

$$\bar{\varphi} = \varphi_{max}(1 - \exp(-t/\tau_r)) \tag{11.27a}$$

for the rise and another

$$\bar{\varphi} = c't + c'' \tag{11.27b}$$

with c', c'' constants for the active decay process, has to be suggested for higher values in $\bar{\varphi}$.

11.3.2 Experimental methods

Our experimental set-up (Figure 11.4) used with Frederikz planar cells as well as with the TN cells for studying the influence of external field is described elsewhere (Pranoto

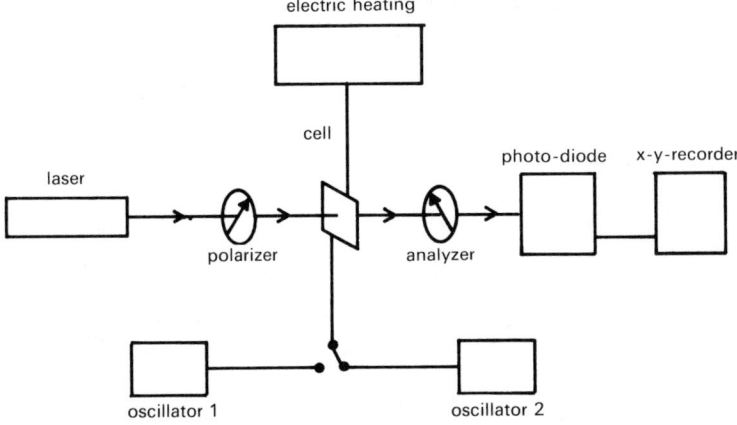

Figure 11.4 Experimental set-up for switching-time measurements.

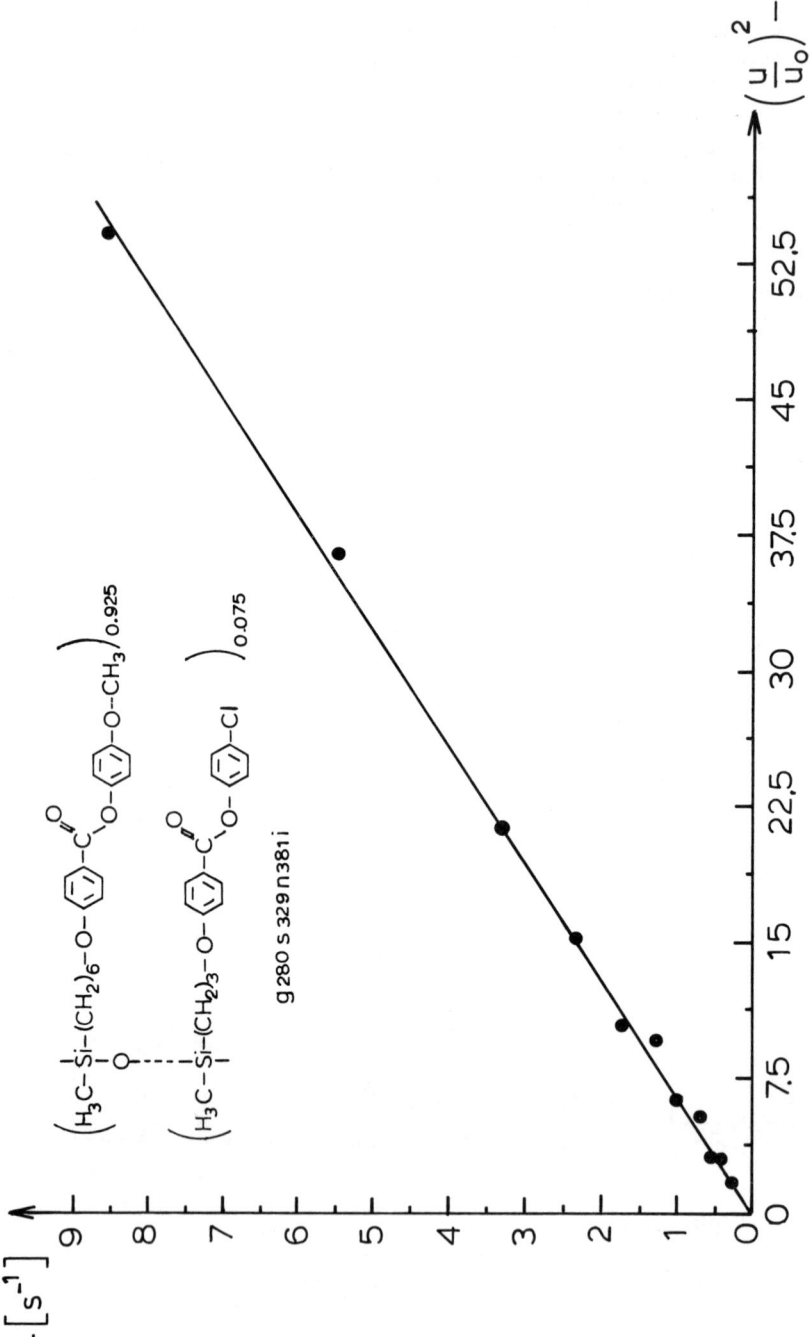

Figure 11.5 Rise time of the inserted polysiloxane SiCl. The validity of eqn (11.21) is demonstrated.

and Haase, 1983; Haase and Pranoto, 1984, 1985; Pranoto, 1984). Using a polarizing microscope the quality of the oriented layer can be checked. Besides this, the polarization quality of the incident light is increased.

The cell thickness was determined by applying an interference method using a spectrometer (Cary 17). The birefringence Δn was obtained from $n_{\parallel}$ and $n_{\perp}$ by applying a Leitz–Jelley microrefractometer, or directly using a compensator. The threshold voltage was determined in an indirect way using eqn (11.21). The direct method, though well suited to low molecular mass liquid crystals, is a little inconvenient for the polymers due to the broad range of the response time, especially for smaller voltages $(U \rightarrow U_0)$. The passive decay time τ_d^0 was determined experimentally.

The results for a polysiloxane copolymer known as SiCl (Pranoto, 1984) are depicted in Figure 11.5 for the TN cell. Linear plots of this type are a good indication of the applicability of models valid for low molecular mass liquid crystals to side chain liquid crystalline polymers, especially for the compounds investigated. In fact, for side chain polymeric liquid crystals the response behaviour in external fields is determined by the liquid crystal properties at temperatures far from the glass temperature T_g. This is an important point for understanding the properties of these mesomorphic materials. Also

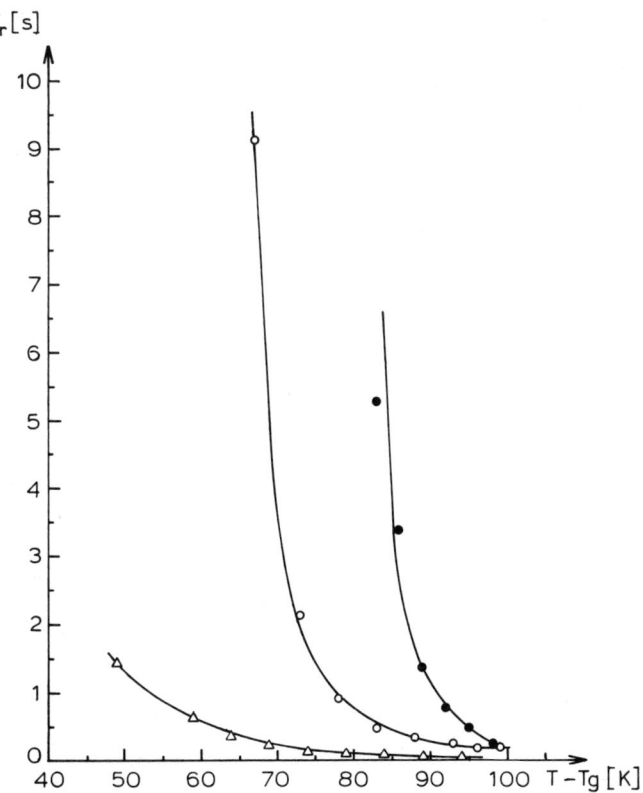

Figure 11.6 Rise times for SiCl ($\bullet$), SiCN ($\circ$) and compound 2($\triangle$) v, $T - T_g$ using TN cell.

the threshold voltage U_0 and the elasticity constants are of the same order of magnitude as the low molecular mass LCs.

The data given in Figure 11.6 for nematic phases are reduced to the same thickness $d = 10 \, \mu m$. The differences between SiCl, SiCN (SiCN is equivalent to SiCl when Cl is substituted by CN, cf. Table 11.1) and the polymeric compound (2) as regards the switching times are a consequence of the dielectric properties of the compounds investigated ($U_0 \sim \Delta\varepsilon^{-1/2}$, see eqn 11.11). Response properties for polyacrylates and polymethacrylates were investigated by Ringsdorf and Zentel (1983). Recently, Wendorff et al. (1988) determined different response properties for such compounds.

Finkelmann et al. (1983) described some response properties and threshold voltages for some siloxane copolymers, using the TN technique. Recently, Goozner and Finkelmann (1985) reported, for the first time, response properties on side chain liquid crystalline polymethacrylates having negative dielectric anisotropy in a homeotropically aligned cell. Furthermore, Williams domains were observed in such polymers.

In a series of papers the Moscow group (Platé et al., 1984; Shibaev and Platé, 1985) described different properties of thermotropic liquid crystalline polymers in electric and magnetic fields. Some data for elastic constants and response time were presented. Their dual frequency techniques will be referred to in section 11.3.3, the thermorecording aspects in section 11.5.1 (and Chapter 13).

Coles (1985), Coles and Simon (1985a, b) and Simon and Coles (1986) have described the properties of polysiloxanes or copolysiloxanes in the electric field. These compounds with high $\Delta\varepsilon$ values are heated in the S_A phase in order to study potential thermorecording. Their rise time is defined as the time for the light transmission to drop to 50% of its initial value on application of an electric field across the system.

Very few reports are available concerning the response properties of LCPs in the

Table 11.1 Some compounds investigated using Frederikz or TN cells.

Polyacrylate, polymethacrylate

$$R_1 - \underset{\underset{n}{\overset{\displaystyle |}{\downarrow}}}{\overset{\overset{\displaystyle \uparrow}{\overset{\displaystyle |}{CH_2}}}{\underset{\displaystyle |}{C}}} - COO - (CH_2)_m - O - C_6H_4 - Y - C_6H_4 - R_2$$

Compound	R_1	m	Y	R_2	Reference
(1)	H	6	—	CN	Bormuth (1988)
					Wendorff et al. (1988)
(2)	H	2,6	COO	CN	Ringsdorf and Zentel (1983)
	H	6	COO	CN	Pranoto and Haase (1983)
					Haase and Pranoto (1984, 1985)
					Pranoto (1984)
					Wendorff et al. (1988)
(3)	H	6	N=N	CN	Wendorff et al. (1987)
(4)	H	2,6	COO	C_6H_4CN	Ringsdorf and Zentel (1983)
	H, CH$_3$	2,6	COO	C_6H_4CN	Ujiie et al. (1987)
	CH$_3$	2	COO	C_6H_4CN	Goozner and Finkelmann (1985)
(2)–(3)	Copolymer				Ringsdorf and Schmidt (1984)

(**Table 11.1** *Continues on p. 324*)

magnetic field. Casagrande *et al.* (1983) studied the magnetic field response of polysiloxanes in planar cell oriented perpendicularly to the wall in order to measure different properties (see eqn 11.13). Besides these other measurements were done on non-side chain liquid crystals in order to determine elastic constants and the magnetic threshold field strength H_0 (Fernandes and Dupré, 1981) or to investigate the orientation process (Noël *et al.*, 1981). Recently Roth and coworkers (1986) have studied the effects of H-fields and temperature programs on LC polysiloxanes.

11.3.3 *Dual-frequency addressing*

If the lowest frequency relaxation in $\varepsilon'_\parallel$, the real part of the dielectric permittivity parallel to the director $\vec{n}$, is in the audio frequency region, such materials can be utilized for 'dual-frequency addressing'. Here the dielectric anisotropy is positive for lower frequencies and negative for higher frequencies. Therefore it is possible to convert an originally homeotropically oriented liquid crystal domain into a homogeneous orientation by quickly switching from low to higher frequency. The formula for the active decay time τ_d is given by

$$1/\tau_d = (1/\tau_d^0) \times ((U/U_0^+) + 1) \qquad (11.28)$$

where U_0^+ is the corresponding threshold voltage.

The principal effect is demonstrated in Figure 11.7. f_0, the optical cross-over frequency, is relevant for devices. f_R is the dielectric relaxation frequency (see

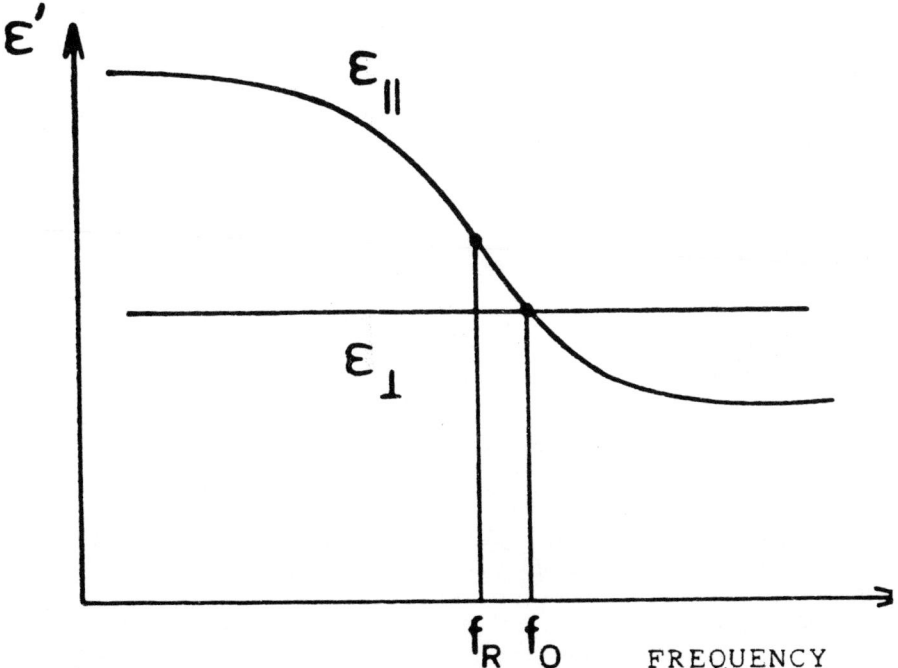

Figure 11.7 Principal frequency dependence of $\varepsilon_\parallel$ and $\varepsilon_\perp$ and definition of f_0 and f_R.

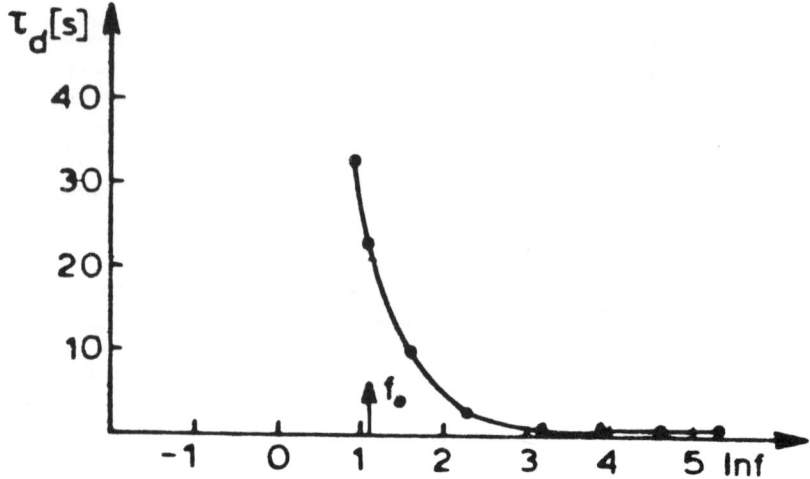

Figure 11.8 Determination of f_0 from the decay time.

Chapter 7). The optical determination of f_0 is known for low molecular mass liquid crystals (Pötzsch and Haase, 1976; Haase and Pötzsch, 1977) and sometimes used also for side chain LCPs (Pranoto and Haase, 1983; Haase and Pranoto, 1984, 1985). It should be mentioned that the frequencies f_0 and f_R are normally not equal. This is also shown in Figure 11.7.

The set-up we used for dual frequency addressing experiments is also shown in Figure 11.4. The frequency f_0 can be obtained by measuring the active decay time τ_d as a function of the frequency of the external field (see Figure 11.8). U_0^+ is determined using eqn (11.28). For $\Delta\varepsilon \to 0$ following $U/U_0^+ \to 0$ and $\tau_d \to \tau_d^0$.

The dual-frequency technique accelerates decay; usually the active decay time is about one order of magnitude smaller than the passive decay time for low molecular mass liquid crystals. The optimization of different parameters is an aspect of molecular engineering (Clark, 1985). This is also valid for polymers.

Electrohydrodynamic instabilities can also be produced in polymers under the influence of external fields. At a given temperature one can change the regime of homeotropic orientation to an electrohydrodynamically distorted orientation by varying the frequency of the electric field (Platé et al., 1984).

11.4 Dyed LCPs in guest–host arrangements

11.4.1 Guest–host effect

Liquid crystalline phases (low molecular mass phases as well as polymeric ones) can be used as solvents for organic molecules. Even those solutes possessing no liquid crystalline character can acquire orientational order. This effect was called the guest–host effect by Heilmeier and Zanoni (1968). There are many examples of this effect for applications. The best-known examples are the possibility of measuring anisotropic spectra of the solutes, and dissolving of dye molecules resulting in dyed displays with a potential use for colour switching. The latter point will be the centre of the discussion following.

11.4.2 Dyes dissolved in LCPs

Normally it is possible to dissolve 1–2% by weight of dyestuff in a polymeric liquid crystal matrix. The resulting mixtures are polymeric guest–host systems, whose characteristics are similar to low molecular mass systems. For example, the orientation of such system with the known techniques (see section 11.2) is analogous to the low molecular mass LC systems. No change in the polymer mesophase structure is observed (Platé et al., 1984) and, contrary to low molecular weight guest–host systems, the order parameter S_{dye} seems to be constant (Quotschalla and Haase, 1987) in this concentration region. The magnitude of the order parameters of equivalent dyes in low molecular mass systems at great dilution and in polymeric systems are always nearly the same.

However, a great difference is seen in the electrooptic response behaviour of guest–host systems. From low molecular mass system it is known that addition of dye molecules leads to a (strong) increase of the viscosity. The result is an increase of the response time with growing amount of dye. This effect could not be observed in polymeric systems. Coles and Simon (1985a) showed that there is no difference between the undyed system and the system containing some % per weight of dyes.

Dyed LCPs have been exploited in device applications other than displays. For example, photoisomerizable azo dyes which are inherent in the LCP structure have been used by Eich (1987) in novel experiments with holography. McArdle et al. (1987) have used dyed LCPs to demonstrate high information content data laser-written on to smectic samples. In this work the dye was used to couple laser energy into the LCP matrix (see Chapter 13). Dye doped LCPs or inherently coloured LCPs have also made an impact on non-linear optics (e.g. see Meredith et al., 1981, and Chapter 12). Photochromic LCPs have already been alluded to by Percec and Pugh in Chapter 3, and photoconductive LCPs are under investigation by Lux et al. (1987) and Chapoy and co-workers (1983).

11.4.3 Dyed copolymers

The great advantage of these systems in comparison to systems with dissolved dyes is that temperature-dependent solubility does not occur and a much greater amount of dye can be accommodated (up to 60% per weight and more, Ringsdorf et al. (1985; Ringsdorf and Schmidt, 1984). The orientation of such systems under the influence of external electrical fields is comparable with that of dissolved dye molecules (Quotschalla and Haase, 1988).

The problem, however, for polymeric systems, is as always the order of the response time for normal cells. Using mixtures with familiar low molecular weight liquid crystals leads to response times up to three times better. Further investigations in this area may lead to a good black dyestuff (copolymer containing three different dyes over the visible region) possessing no temperature-dependent solubility. Also, laser addressing and hole burning are potential areas for application, as the dyed copolymers showed some prospects in this area (Schmidt, 1984).

11.5 Switching-related properties of LCPs

11.5.1 Types of mesogenic groups and phases

The different experimental methods referred to in sections 11.3 and 11.4 have resulted in relatively little information reported in the literature concerning the field-induced effect in LCPs. Furthermore, mesogenic groups investigated to date as side chains are only rod-like molecules and discotic groups or combined LC polymers have not been investigated. Mesogenic groups investigated are biphenyl, phenylbenzoate, azobenzene, benzilidene and biphenylbenzoate groups. The reported literature is summarized in Table 11.1.

The question regarding whether side chain liquid crystalline polymers have potential as materials for display devices analogous to conventional liquid crystals can be simply answered in the negative; the reason is that the switching times are too long. Yet, and this is the important point, the side chain liquid crystalline polymers show response effects under the influence of external fields, and therefore these materials have potential for all applications where the long timescale does not contradict this. Therefore it is of great importance to learn as much as possible about the dynamics of these materials.

First, returning to the relations given in section 11.2.1 (eqns 11.19, 11.20, 11.21) and section 11.3.2, (eqn 11.26) one can see, for a given LCP and assuming the validity of the models underlying these equations, some special proportionalities to experimental and substantial parameters exist. τ_r, the rise time, and τ_d, the active decay time, remain proportional to the square of the cell thickness d and inversely proportional to the square of the applied voltage. Even in thinner cells ($\sim 10\,\mu m$) at higher voltages (up to 50 V), the response process is comparably short. Important parameters are the liquid crystalline temperature range, as well as the elastic constants (section 11.5.5), viscosity (section 11.5.6) and the threshold field strength, as well the dielectric and diamagnetic anisotropy constants in the given electric and magnetic fields, respectively.

Due to differing experimental conditions, the definitions and physical meaning of the desired data may differ in the literature. For applications, the most important time is the decay time. In Table 11.2, a comparison of data of Pranoto and Haase (1983), Haase and Pranoto (1984, 1985) and Bormuth (1988) is given; in this case, the experimental

Table 11.1 (Contd. from p. 320)

Polymethylsiloxanes as polymers and copolymers

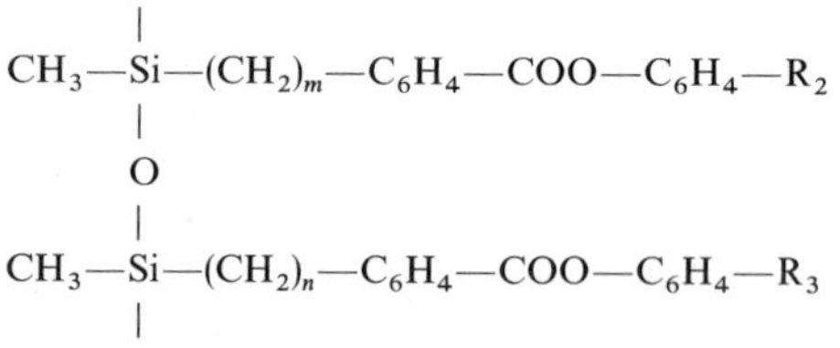

$$
\begin{array}{c}
| \\
CH_3-Si-(CH_2)_m-C_6H_4-COO-C_6H_4-R_2 \\
| \\
O \\
| \\
CH_3-Si-(CH_2)_n-C_6H_4-COO-C_6H_4-R_3 \\
|
\end{array}
$$

Compound	m, n	R_2, R_3	Reference
(5)	$m, n = 3, 4, 5, 6$	$R_{2,3} = Cl, F, NO_2, OCH_3$	Finkelmann et al. (1983)
(6a), **(6b)**	$m = 6, n = 3$	$R_2 = OCH_3, R_3 = Cl$ **(6a)**	Pranoto and Haase (1983)
		$R_2 = OCH_3, R_3 = CN$ **(6b)**	Haase and Pranoto (1984, 85)
			Pranoto (1984)
(7)	$m, n = 4, 6$	$R_2 = OCH_3$	Casagrande et al. (1983)

(6a) = SiCl, **(6b)** = SiCN

Table 11.2 Response times for compounds investigated (for compound number see Table 11.1). $d = 10\,\mu m$,
$U = 50\,V$, $T^* = (T/T_{ni}) = 0.940$; rise times (τ_r), some
passive decay times (τ_d^0) and active decay times (τ_d) are
given.

	(6a)	(6b)	(2)	(1)
τ_r	14.2	0.5	0.2	2.7
τ_d^0	9.7	10.4	56.3	154
τ_d	0.8	0.2	1.5	—

The degree of polymerization for (6a), (6b) is 95 (Haase
and Pranoto, 1985); for compound (2), 50 (Haase and
Pranoto, 1984, 1985) and for (1), 13 (Bormuth, 1988).

technique and definition used are completely comparable. The data in Table 11.2 are
scaled to the same reduced temperature (0.940; some data were derived by extrapolation from reported data), the same cell thickness $(10\,\mu m)$ and the same voltage
(50 V rms).

The comparably high values for (1) can only be interpreted as a consequence of the
strong antiparallel coupling of the mesogenic groups via cyano–phenyl interactions,
which is evidenced from dielectric anisotropy data (Haase and Pranoto, 1985). Also, for
(1) no active switch-off could be realized.

Table 11.2 shows very clearly that the active decay τ_d is considerably shorter than
the passive decay τ_d^0. The switching properties can be directly related to the so-called
δ-relaxation process (see Chapter 7). This relaxation process is comparable to the
analogous process for low molecular mass liquid crystals, but has much longer
characteristic times. Also, this process is more highly thermally activated.

In the transition from nematic to smectic phase, no change in related response
properties was found.

In some electro-optical experiments, described by Coles (1985), McArdle et al. (1987)
and Shibaev and Platé (1984), materials were heated from ambient to the required
temperature, applying a pulsed AC field or using laser heating. These effects are referred
to in Chapter 13.

11.5.2 Spacer length

The function of the spacer is to decouple the motion of the main chain from that of the
side chain. It is widely accepted that spacer length $—(CH_2)_n—$ with $n \geqslant 6$ no longer
affects the dynamic properties. Also, the most investigated compounds have a spacer
length with $n = 6$. Casagrande et al. (1983) showed drastic differences in magnetic field
experiments of compounds with $n < 6$. The compounds with $n = 4$ showed long
orientation times and a type of saturation. This was explained by a polymer backbone
influence. There is evidence that, for example, the elastic constants should vary with the
spacer length. Also, Ujiie et al. (1987) demonstrated significant differences to longer
time scales going from $n = 6$ to $n = 2$.

11.5.3 The main chain influence

We can outline at least three main chain influences:

(i) Incomplete decoupling between main chain and side chain, as was discussed
above.

(ii) Stiffening of the main chain. As can be seen in Table 11.1, the main chains investigated are polyacrylates, polymethacrylates and polymethylsiloxanes. The data in Table 11.2 reveal shorter response times for polymethylsiloxanes in comparison with polyacrylates. The reason is that the polymethylsiloxane backbone is more mobile. Also, the polymethacrylate shows higher stiffness in comparison to the polyacrylate. Another point, related to the stiffness of the main chain, is the lower orientation degree through external fields having a backbone with higher stiffness.

(iii) Polymerization degree, P. The longer the chain, the higher the response time. Some authors have reported on these properties, e.g. Shibaev and Platé (1984), Ujiie *et al.* (1985) and Bormuth (1988). Indeed, for application, polymers (= oligomers) with lower polymerization degree seem to be better candidates in comparison with polymers having long chain length. It should be emphasized that the preparation of compounds with different chain length is a part of the chemical ⋅ technique.

11.5.4 *Glass temperature*

It was pointed out in section 11.5.1 that the response properties in the external field are related to the δ process. This process is determined by the liquid crystalline properties. On decreasing the temperature, vitrification takes place. Talroze *et al.* (1981) suggested a linear relationship between the difference $T - T_g$ and the rise time, where T_g is the glass temperature and T the given temperature. This relation demonstrates the validity of the well-known Vogel–Fulcher equation for glass-forming liquids. The viscosity at T_g is always the same for polymers, and according to Rehage and Borchard (1973) has a certain value of $\sim 10^{15}$ cP. Following this, the difference shown in Figure 11.6 between the three compounds presented can be directly related to elasticity and viscosity. The problems discussed clearly demonstrated the necessity of having compounds with lower T_g.

11.5.5 *Elastic constants and threshold*

To a first approximation, the order of magnitude of the elastic constants of side chain liquid crystalline polymers is equivalent to the low molecular mass liquid crystals. This is a consequence of the fact that a longer spacer decouples side chain and backbone from each other. As was pointed out (section 11.3.2), the elastic constants k_{11} and k_{33}

Table 11.3 Elastic constants and viscosity coefficients for some compounds investigated (Pranoto, 1984). For compound structure see Table 11.1

Compound	$T[°C]$	Cell	k/k_{11} $(10^{12}\,N)$	$\eta(10^4\,\text{cp})$
(2)	105	TN	10	20
		Fre	6.7	70
(6a)	102	TN	16	0.8
		Fre	5.9	1
(6b)	103	TN	10	1
		Fre	6.7	2

can be obtained in an external electric or magnetic field studying the Frederikz transition. Also, the threshold voltage and the magnetic threshold strength H_0 can be obtained by the same methods. Elastic constants k_{11} and $k = k_{11} + \frac{1}{4}(k_{33} - 2k_{22})$ are presented in Table 11.3 for the compounds (2), (6a) and (6b) (Pranoto, 1984).

Threshold voltages U_0 for these compounds are given by Pranoto and Haase (1983), Haase and Pranoto (1984, 1985) and Bormuth (1988). After Casagrande et al. (1983), for compound (7) ($n = 6$) the magnetic threshold is $\sim 1\,\mathrm{kG}$ at $T/T_c = 0.95$.

Hopwood and Coles (1985) added some polymers to low molecular mass liquid crystals up to 40% w/w. The elastic constants k_{11} and k_{33} decreased with increasing polymer concentration, but remain principally in the same order of magnitude ($\sim 10^{-12}\,\mathrm{N}$ to $\sim 10^{-11}\,\mathrm{N}$) as for low molecular mass liquid crystals.

11.5.6 Viscosity

For switchable compounds, the viscosity for LCPs is up to three orders of magnitude higher than the analogous values of low molecular mass liquid crystals. The viscosity coefficients for compounds (2), (6a) and (6b) (Pranoto, 1984) are given in Table 11.3, and for compound (1) (Bormuth, 1988) in Table 11.4.

The temperature dependence of the viscosity is given by the Arrhenius equation. Pranoto (1984) obtained values between 76 and $123\,\mathrm{kJ\,mol^{-1}}$ for compounds (2), (6a) and (6b).

Indeed, by decreasing the temperature, the data tend to $10^{15}\,\mathrm{cP}$ at T_g. So for non-switchable compounds under the influence of external fields, the viscosity coefficients are between $\sim 10^7$ and $10^{15}\,\mathrm{cP}$.

11.5.7 Dyes

As described in section 11.4, a small amount of dye dissolved in polymers normally does not change the viscosity drastically; also, the viscosity coefficients for dyed copolymers are comparable to the data valid for the undyed polymer or assumed to be a little bit higher. From this it follows that by including dyes in polymers the response time can be increased. In case of application of dyes in polymers for thermorecording, e.g. with higher-voltage AC fields or with laser heating, this effect can be neglected.

Table 11.4 Elastic constants and viscosity coefficients for compound (1) at various temperatures (see Table 11.1) using a TN cell (Bormuth, 1988)

T/T_{ni}	$k[10^{12}N]$	$\eta(10^6\,\mathrm{cP})$
0.982	17	0.05
0.969	13	0.3
0.953	130	1.6
0.931	224	3.8

T_{ni}, clearing point.

Acknowledgement

This work was supported by the Deutsche Forschungsgemeinschaft. The author thanks colleagues and group members for stimulating discussions.

References

Attard, G.S. and Williams, G.W. (1986) *Liquid Crystals* **1**, 53.

Baur, G., Stieb, A. and Meier, G. (1975) *Appl. Physics* **6**, 309.

Blinov, L.M. (1983) *Electro-Optical and Magneto-Optical Properties of Liquid Crystals*. Wiley-Interscience, New York.

Bormuth, F.J. (1988) Unpublished PhD thesis, Technische Hochschule, Darmstadt.

Bormuth, F.J. and Haase, W. (1988) *Liquid Crystals* **3**, 881.

Casagrande, C., Veyssie, M., Weill, C. and Finkelmann, H. (1983) *Mol. Cryst. Liq. Cryst. Lett.* **92**, 49.

Chapoy, L., Biddle, D., Halstrøm, J., Kovács, K., Brunfeldt, K., Qasim, M. and Christensen, T. (1983) *Macromolecules* **16**, 181.

Clark, M.G. (1985) *Mol. Cryst. Liq. Cryst.* **17**, 127.

Coles, H.J. (1985) *Faraday Disc. Chem. Soc.* **79**, 201.

Coles, H.J. and Simon, R. (1985a) *Polymer* **6**, 1801.

Coles, H.J. and Simon, R. (1985b) in *Polymeric Liquid Crystals*, ed. A. Blumstein, Plenum, New York, 351.

Corazza, P., Sartirana, M.L. and Valenti, B. (1984) *Makromol. Chem.* **183**, 847.

de Gennes, P.G. (1974) *The Physics of Liquid Crystals*, Clarendon Press, Oxford.

Eich, M. (1987) Unpublished PhD thesis, Technische Hochschule, Darmstadt.

Eich, M. and Wendorff, J. (1987) *Makromol. Chem. Rap. Commun.* **8**, 467.

Fernandes, J.R. and DuPré, D.B. (1981) *Mol. Cryst. Liq. Cryst. Lett.* **7**, 67.

Finkelmann, H., Kiechle, U. and Rehage, G. (1983) *Mol. Cryst. Liq. Cryst.* **94**, 343.

Goozner, R.E. and Finkelmann, H. (1985) *Makromol. Chem.* **182**, 371.

Haase, W. and Pötzsch, D. (1977) *Mol. Cryst. Liq. Cryst.* **38**, 77.

Haase, W. and Pranoto, H. (1984) *Progr. Colloid & Polym. Sci.* **69**, 139.

Haase, W. and Pranoto, H. (1985) in *Polymeric Liquid Crystals*, ed. A. Blumstein, Plenum, New York, 313.

Haase, W., Pranoto, H. and Bormuth, F.J. (1985) *Ber. Bunsenges., Phys. Chem.* **89**, 199.

Heilmeier, G.H. and Zanoni, L.A. (1968) *Appl. Phys. Lett.* **13**, 91.

Hopwood, A.I. and Coles, H.J. (1985) *Polymer* **26**, 1312.

Ibrahim, I.H. and Haase, W. (1979) *J. de Physique* **40**, C-164, and references therein.

Krigbaum, W.R., Lader, H.J. and Ciferri, A. (1980) *Macromolecules* **13**, 554.

Krigbaum, W.R., Granthan, C.E. and Toriumi, H. (1981) *Macromolecules* **15**, 59.

Lux, M., Strohriegl, P., Höcker, H. (1987) *Makromole Chem.*, **188**, 811.

McArdle, C.B., Clark, M.G., Haws, C.M., Wiltshire, M.C.K., Nestor, G., Gray, G.W., Lacey, D. and Toyne, K.J. (1987) *Liquid Crystals* **2**, 573.

Meredith, G.R., Van Dusen, J.G. and Williams, D.J. (1981) *Macromolecules* **15**, 1385.

Noël, C., Billard, J., Bosio, L., Friedrich, C., Lauprêtre, P. and Strazielle, C. (1984) *Polymer* **5**, 63.

Noël, C., Monnerie, L., Achard, M.F., Hardouin, F., Sigaud, G. and Gasparoux, H. (1981) *Polymer* **22**, 578.

Patel, J.S., Leslie, J.M. and Goodby, J.W. (1984) *Ferroelectrics* **59**, 137.

Pinsl, J., Braüche, Chr., Kreuzer, F. (1987) *J. Molec. Electron.* **3**, 9.

Platé, N.R., Talrose, R.V. and Shibaev, V.P. (1984) *Pure & Appl. Chem.* **56**, 403.

Pötzsch, D. and Haase, W. (1976) *Phys. Lett.* **57A**, 343.

Pranoto, H. (1984) Unpublished PhD thesis, Technische Hochschule, Darmstadt.

Pranoto, H. and Haase, W. (1983) *Mol. Cryst. Liq. Cryst.* **98**, 99.

Pranoto, H. and Haase, W. (1988) *Z. Naturforsch. A* (in press).

Quotschalla, U., and Haase, W. (1987) *Mol. Cryst. Liq. Cryst.* **153**, 83.

Quotschalla, U. and Haase, W. (1988) *Mol. Cryst. Liq. Cryst.* **157**, 355.

Rehage, G. and Borchard, W. (1973) *The Physics of Glassy Polymers*, ed. R.N. Haward, Applied Science, London, 54.

Ringsdorf, H. and Zentel, R. (1983) *Makromol. Chem.* **183**, 145.

Ringsdorf, H., Schmidt, H.W., Baur, G. and Kiefer, R. (1985) *Recent Advances in Liquid Crystalline Polymers*, ed. L.L. Chapoy, Elsevier, New York, 253.

Ringsdorf, H. and Schmidt, H.W. (1984) *Makromol. Chem.* **185**, 37.

Roth, H. and Krucke, H. (1986) *Makromol. Chem.* **187**, 2655, and references therein.

Schmidt, H.W. (1984) Dissertation Doktor der Naturwissenschaften, Universität Mainz, 72.

Shibaev, V.P. and Platé, N.A. (1985) *Pure & Appl. Chem.* **57**, 1589 and references therein.

Shibaev, V.P. and Platé, N.A. (1984) *Advances in Polymer Science.* **60/61**, Springer, Berlin.

Simon, R. and Coles, H.J. (1986) *Polymer* **7**, 811 and references therein.

Talroze, R.V., Kostromin, S.G., Shibaev, V.P., Platé, N.A., Kresse, H., Saher, K. and Demus, D. (1981) *Makromol. Chem., Rapid Commun.* **2**, 305.

Ujiie, S.I., Koide, N. and Iimura, A. (1987) *Bordeaux Conf. on Polymeric Liquid Crystals.*

Wu, S.-T. (1985) *Proc. SPIE, Advances in Materials for Active Optics* **567**, 74.

12 Side chain liquid crystal polymers as optically nonlinear media

G.R. MÖHLMANN and C.P.J.M. van der VORST, Akzo Research Laboratories, Corporate Research, Applied Physics Department Postbox 9300, 6800 SB Arnhem, The Netherlands

12.1 Introduction

Optically nonlinear materials are currently the subject of many research programmes, owing to their potential application in integrated electro-optic systems. Such systems can be used to influence and control the temporal, spatial and frequency properties of propagating light beams and are, therefore, important for optical processing purposes.

In the class of organic materials, side chain liquid crystal polymers, among others, can be applied as optically nonlinear media if, they fulfil a number of requirements. In this chapter, these requirements will be discussed, and it is shown that the unique properties of polymers play an important role in the development of optically nonlinear materials and devices thereof.

In order to achieve second-order nonlinear effects it is necessary to give the system a uniaxial (polar) ordering; this can be done by electric field poling. In this chapter, the electric field-induced ordering of initially isotropic and liquid crystalline systems will be extensively discussed, applying theoretical models.

12.2 Origin of optically nonlinear effects

Generally, if fields (electric, magnetic, optical) act on materials, the building blocks of the materials (atoms, ions, electrons) will experience forces and be displaced through translation or rotation by those forces. The material is said to be *polarized* by the action of the fields. For small applied field strengths, the lateral displacements of the relevant building blocks (polarization) are often linearly proportional to those field strengths (linear optics).

Optically nonlinear materials are substances exhibiting nonlinear polarizations when subjected to electric, magnetic or electromagnetic (optical) fields. The induced displacements in such materials do not depend linearly on the applied field strengths any more, but now show contributions proportional to higher orders of the field strength. The amplitude of the applied fields may be constant (DC field) or be a temporally varying function (AC fields and optical fields). Generally, the applied field strengths must be strong in order to observe the occurrence of such nonlinearities.

Nonlinear electric polarizations in materials can give rise to a number of optically nonlinear phenomena, such as the linear electro-optic (Pockels) effect, the quadratic electro-optic (Kerr) effect, frequency doubling, frequency tripling or four-wave mixing. Optical fields can undergo phase shifts, frequency shifts, polarization rotation or intensity changes, if the proper conditions for the applied field (strength, direction, power), are fulfilled.

The above-mentioned nonlinear effects may be used to control the propagation of optical waves (signals) and thus may find applications in the fields of optical data communication, optical processing and computing. Therefore, it is essential to understand the optically linear and nonlinear phenomena of materials in order to develop optimized substances for use in electro-optical devices and related applications.

Besides electric or electro-magnetic fields, the propagation of light can be influenced by, for instance, magnetic and acoustic (sound) fields, leading to the Faraday and acousto-optic effects respectively. However, these effects will not be dealt with in this chapter; only electrically-induced polarizations will be discussed. A compilation of the effects with which the propagation of optical waves in materials can be influenced, mainly by changing the refractive index of the medium, can be found in the *Handbook of Lasers* (1971).

12.3 Mathematical description of nonlinear electric polarizations

Generally, the induced electric polarization component P_I in a material under the influence of an electric or optical field can be written as (summation over identical indices)

$$P_I = \chi^{(1)}_{IJ} \cdot E_J + \chi^{(2)}_{IJK} \cdot E_J \cdot E_K + \chi^{(3)}_{IJKL} \cdot E_J \cdot E_K \cdot E_L + \cdots \qquad (12.1)$$

in which $\chi^{(1)}_{IJ}$ is the linear susceptibility, $\chi^{(2)}_{IJK}$ the second order (nonlinear) susceptibility, etc.; the terms E_J, etc., are the relevant components of the electric field strength. By taking the term E_J apart, eqn (12.1) can be rewritten as

$$P_I = [\chi^{(1)}_{IJ} + \chi^{(2)}_{IJK} \cdot E_K + \chi^{(3)}_{IJKL} \cdot E_K \cdot E_L + \cdots]E_J \qquad (12.2)$$

In linear optics, the term $\chi^{(1)}_{IJ}$ is connected to the refractive index n_{IJ}, via the relation $\chi^{(1)}_{IJ} = n^2_{IJ} - 1$. The terms $\chi^{(2)}_{IJK}E_K$ and $\chi^{(3)}_{IJKL}E_K E_L$, having the same dimension as $\chi^{(1)}$, are thus also connected to the refractive index of the material. The conclusion is that materials exhibiting a nonzero value of $\chi^{(2)}$ or $\chi^{(3)}$ undergo a change in refractive index if placed in an electric field. The electric field can also be an optical field. As a consequence, for $\chi^{(3)}$-materials, the nonlinear effect is proportional to the intensity of light; the intensity of light being quadratically proportional to the amplitude of the corresponding electric field vector: $I \propto E^2$. A general, but more extensive, treatment of nonlinear polarization phenomena can be found in specialized literature (Shen, 1984).

Optically nonlinear effects connected to $\chi^{(2)}$ are the linear electro-optical (Pockels) effect, frequency doubling, frequency mixing, optical rectification, etc. The term with $\chi^{(3)}$ allows processes like the quadratic electro-optical (Kerr) effect, frequency tripling, frequency (four-wave) mixing, self-focusing, stimulated scattering processes (Raman, Brillouin, etc.), Coherent Anti-Stokes Raman Scattering (CARS), etc.

The coefficients $\chi^{(n)}$ are tensors of rank $n + 1$, relating the induced polarization vector components P_I to the appropriate field components E_J, E_K, etc. However, for the sake of simplicity, we will omit the use of the exact tensor representation and of the corresponding indices, where possible. Also, the frequencies of the fields will generally not be shown in the formula; only in specific experimental cases such as frequency doubling will they be mentioned.

It should be noted that an important symmetry constraint exists for some of the nonlinear effects to occur. For odd-ranked tensors such as $\chi^{(2)}_{IJK}$ with rank 3 not to be equal to zero, no centrosymmetry is allowed in the system. So systems exhibiting an inversion centre (centrosymmetry) do not possess even-order (second, fourth, etc.) nonlinear processes. This restriction will play an important role in the development of second-order optically nonlinear (polymeric) materials, as will be shown further on. For third- and other odd-order nonlinearities (even-ranked tensors), no such symmetry constraints exist.

This chapter will mainly deal with second- and third-order optically nonlinear effects, based on translational displacements of charge carriers (electrons, ions) by external fields. The orientational, displacements, like rotations of dipoles and of anisotropically polarizable molecules, will not be discussed extensively with respect to their resulting optically nonlinear effects; a treatment of this orientational 'giant optical nonlinearity' can be found in the work of Tabiryan et al. (1986). However, orientational displacements are important in achieving a desired order and thus optimum performance of second- and third-order nonlinearities; these nonlinearities themselves are based on lateral charge displacements in organic molecules.

Pure orientational displacements such as rotations, can be considered as third-order phenomena, because the magnitude of the induced effects (birefringence) depends quadratically on the applied field strengths.

12.4 Origin of optical nonlinearity in inorganic materials

The traditional materials which have been studied for their optically nonlinear properties are inorganic single crystals. Examples are quartz, potassium dihydrogen phosphate (KDP), zinc oxide, lithium niobate and barium titanate.

For inorganic materials and relatively slowly varying field strengths, mainly the ions in the crystal will be displaced under the influence of the field. However, ions have relatively large masses and are restricted in their lateral displacements by the presence of neighbouring ions. Therefore, the nonlinear coefficients will generally be small. For higher frequencies (> GHz) of the applied fields, the heavy ions cannot any longer easily follow the correspondingly rapid amplitude changes. At optical frequencies, it is assumed that the ions do not follow the amplitude variations of the field at all; the contribution of ions to the (nonlinear) polarizations in the optical limit approaches zero; only (small) electronic displacements then contribute. The following experimental results support this view.

(i) It has been observed that the dielectric constants (and thus the refractive index) of single-crystalline inorganic materials are considerably higher at low frequencies than at very high (optical) frequencies.
(ii) It has been found that the $\chi^{(2)}$ value for the electro-optical (Pockels) effect (measured at low frequencies) is much larger than the $\chi^{(2)}$ value for frequency doubling (measured at the high optical frequencies, up to 5×10^{14} Hz).

Clearly, the ions cannot follow the high optical frequencies in the case of frequency doubling. For frequency doubling, it is believed that electrons in the crystal are the dominant contributors. Some data, illustrating the difference between the $\chi^{(2)}$ values of the Pockels effect ($\chi_{EO}^{(2)}$) and of frequency doubling ($\chi_{FD}^{(2)}$), are given in Table 12.1 for some inorganic ionic single-crystalline materials, and for an organic molecular single crystal.

Table 12.1 $\chi^{(2)}_{EO,FD}$ values for some inorganic and organic crystals for the tensor elements I, J, K in units $10^{-12}\,mV^{-1}$

Compound	I, J, K	$\chi^{(2)}_{EO}$	$\chi^{(2)}_{FD}$
Quartz[a]	1, 1, 1	1.37	0.8
KDP[a]	3, 2, 1	21.6	0.94
Zinc Oxide[a]	3, 3, 3	26.3	14
Lithium niobate	3, 3, 3	410	82
MNA[b]	3, 3, 3	540	500

[a] *Handbook of Lasers* (1971)
[b] Garito (1983)

It is clearly seen in Table 12.1 that for the inorganics, the susceptibilities for the Pockels effect are larger than the corresponding ones for frequency doubling, supporting the fact of ionic displacements at low frequencies, and the relatively small contribution of electrons at all frequencies in such materials.

The compound MNA is an organic single crystal for which it is believed that the polarizations at all frequencies are dominated by lateral electronic displacements, as supported by the equality of the magnitudes for the two effects.

12.5 Origin of optical nonlinearity in organic materials

12.5.1 Macroscopic nonlinear properties

In order to obtain materials with larger nonlinear coefficients (at optical frequencies) than found in inorganic single crystals, one should search for systems in which the displaceable charge carriers have very small masses and are not strongly confined to a particular region in the system. Delocalized electrons in conjugated π-electron systems fulfil, in principle, these requirements.

Conjugated π-electron systems are found in organic molecules containing alternating multiple and single bonds in linear or cyclic structures. It has been found, indeed, that particular organic single crystals show a much larger second- or third-order nonlinear coefficient than inorganic materials.

A typical example of an organic (noncentrosymmetric) single crystal in which a second-order effect is found, is 2-methyl-4-nitro-aniline (MNA) (Figure 12.1). Second-order effects are always due to lateral displacements in organic molecules.

As with second-order optical nonlinearities, third-order effects in organic single crystals (e.g. polydiacetylenes) can be caused by lateral electronic displacements, too. In liquids or liquid crystals, lateral electronic as well as molecular rotational displacements can contribute to third-order effects. Third-order materials may possess centrosymmetry.

There are some important differences between inorganic and organic materials with respect to their optically nonlinear character based on lateral displacements; a summary is given in Table 12.2.

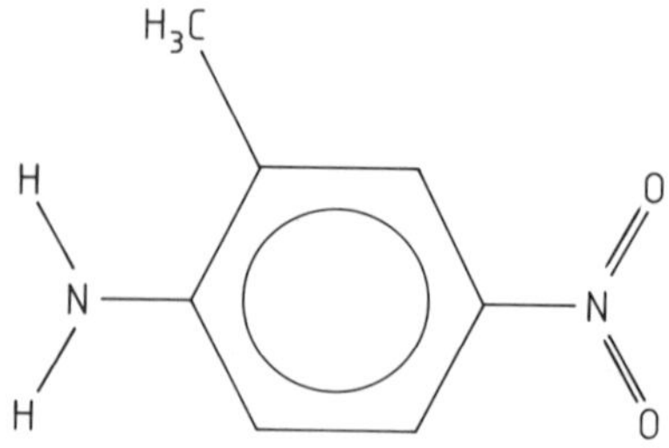

Figure 12.1 Molecular structure of the organic compound 2-methyl-4-nitroaniline (MNA).

12.5.2 Molecular nonlinear properties

In the case of organic materials with electrically neutral building blocks, a molecular analogue to the induced macroscopic polarization exists. For (single crystalline) *inorganic* materials, such a microscopic approach is not possible; the smallest particles left after breaking apart the crystalline structure, by, dissolution for instance, do not have the same composition as the bulk material. For noncontaminated *organic* systems, the smallest molecular units still represent the overall elemental composition.

In the case of organic molecules, the induced polarization is often referred to as the induced dipole moment μ_{ind}, and can be expressed as:

$$\mu_{ind} = \alpha E + \beta E \cdot E + \gamma E \cdot E \cdot E + \cdots \tag{12.3}$$

in which α is the linear polarizability, β the hyperpolarizability, γ the second hyperpolarizability, etc. Also here, the coefficients α, β, γ, etc., are represented by tensors, as in the case of macroscopic systems; however, for simplicity, we again omit the tensor notation and the corresponding indices.

Similarly as in the macroscopic case, a symmetry constraint exists on the molecular level with respect to the occurrence of even-order nonlinear optical effects. For the second-order nonlinear coefficient β not to be equal to zero, no centrosymmetry is allowed in the molecule. For example, the molecule *trans*-stilbene (see Figure 12.2*a*) exhibits an inversion centre, and will, therefore, not possess a β unequal to zero. However, by asymmetric substitution of this molecule with suitable groups, the centrosymmetry can be removed and a non-zero second-order nonlinear molecular coefficient will result. In Figure 12.2*b* an asymmetrically substituted molecule, 4-

Table 12.2 Important differences between optically nonlinear inorganic and organic materials

Inorganic	Organic
Building blocks are relatively heavy ions	Building blocks are electrically neutral molecules
Induced polarization is due to ionic (and electronic) displacements	Induced polarization is due to π-electron displacement in conjugated systems
Restricted displacement of heavy (slow) ions	Large displacement possible of light (fast) electrons

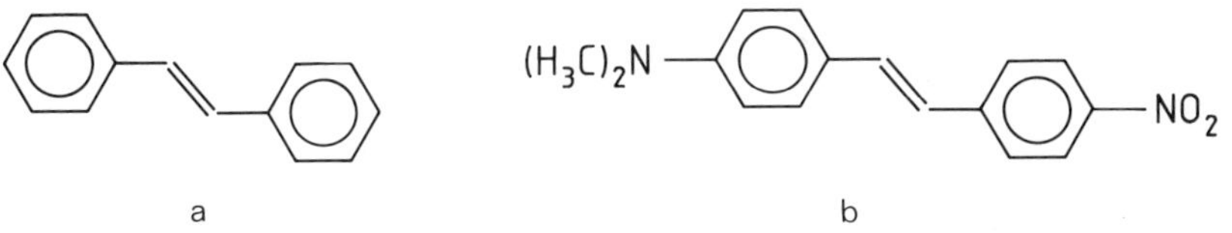

Figure 12.2(*a*) The centrosymmetric molecule *trans*-stilbene; (*b*) The noncentrosymmetric molecule DANS.

dimethylamino-4′-nitrostilbene (DANS), is shown. For third- and other uneven-order nonlinear effects, no such symmetry constraints exist.

One can easily understand why the even-order contributions to the nonlinearity disappear if centrosymmetry is present. For centrosymmetric molecules like benzene, *trans*-stilbene, etc., the magnitude of the induced dipole moment is independent of the direction (sign) of the applied electric field. The following equation thus must, therefore, hold in such case:

$$\mu_{ind}(+E) = -\mu_{ind}(-E). \tag{12.4}$$

By inserting $+E$ in eqn (12.4), one obtains for the induced dipole moment

$$\mu_{ind}(+E) = +\alpha E + \beta E \cdot E + \gamma E \cdot E \cdot E + \cdots \tag{12.5}$$

In the case of reverse (negative sign) field, one obtains:

$$\mu_{ind}(-E) = -\alpha E + \beta E \cdot E - \gamma E \cdot E \cdot E + \cdots \tag{12.6}$$

Owing to the fact that $E \cdot E$ is always positive, irrespective of the sign of the field E, relation (12.4) can only be fulfilled for centrosymmetric molecules, if β is equal to zero, or if the field strength E is equal to zero.

For asymmetrically substituted molecules, not possessing centrosymmetry, relation (12.4) need not be fulfilled.

12.5.3 *Experimental determination of the molecular second-order nonlinear coefficient*

It is possible to relate the molecular hyperpolarizability β to some spectroscopic properties of the molecule. In a simplified quantum mechanical model, taking only the lowest-lying relevant $\pi - \pi^*$ electronic transition into account (two-level model), the value for β can be expressed as:

$$\beta = \Delta\mu_{eg} \times M_{eg}^2 / H \tag{12.7}$$

in which $\Delta\mu_{eg}$ is the difference between dipole moments of the excited (e) and the ground (g) states, respectively. M_{eg} is related to the transition moment of the corresponding excitation. The factor H depends on whether the electro-optic (EO) or frequency doubling (FD) effect is considered; the complete expressions contain, among others, terms such as $(E_{eg} - E_{ph})$ and $(E_{eg} - 2E_{ph})$, respectively, where E_{eg} is the corresponding transition energy and E_{ph} is the energy of the photons under study. References to the complete expressions for H are given by Oudar and Chemla (1977), Oudar (1977), and Garito (1983).

As can be seen from eqn (12.7), molecules exhibiting large changes in dipole moment upon excitation must show a high β-value. Molecules fulfilling this condition are the so-

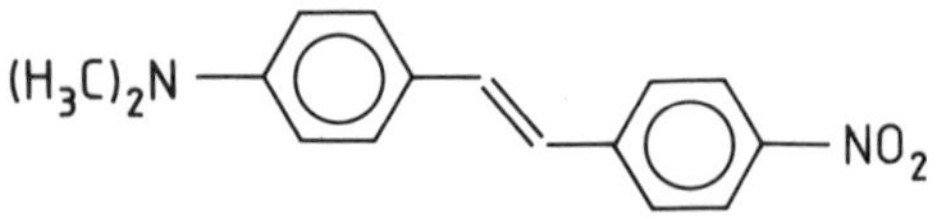

Figure 12.3.(a) Electron charge distribution in the ground state of the molecule DANS; (b) Electron charge distribution in the first CT-state of the molecule DANS.

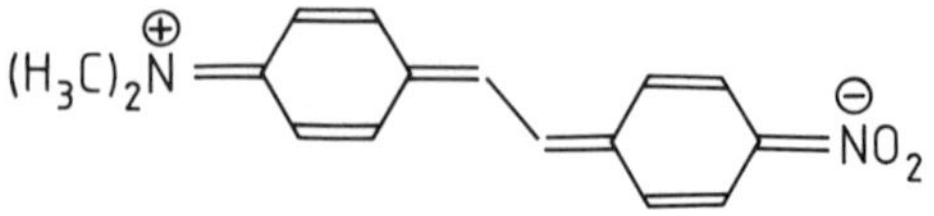

called Charge Transfer (CT) molecules. The molecule DANS is a typical example of such a CT-molecule; the charge transfer character is shown in Figures 12.3a, b, in which the displacement of the π-electron cloud, going from the ground state to the first excited CT-state, is shown. For DANS, the $\Delta\mu$-value is equal to about 18 Debye, and can be considered as a large change. For fluorescing molecules, the value for $\Delta\mu_{eg}$ can be obtained by solvatochromic spectroscopy in which the shift of absorption and fluorescence spectra as a function of solvent polarity is measured.

As can be seen in eqn (12.7), a large transition moment (corresponding to a higher molar extinction coefficient ε) is favourable for a large β-value. The ε in the case of the DANS molecule is equal to about 25 000 litre Mole^{-1} cm^{-1}.

The denominator H of the expression for the β-value (see eqn 12.7), contains differences between the excitation energy and the energy of the applied photons. By decreasing this difference, the magnitude of β increases. In the case that the difference in the denominator approaches zero (resonance excitation), the value for β rises to infinity. Evidently, β shows dispersion. Therefore, if a β-value is presented for a particular molecule, one must know under which experimental circumstances the value has been obtained (photon energy, solvent, etc.).

Sometimes, a so-called static β_0-value is given, indicating the value for zero field frequency (infinitely long wavelength). It can be derived from a quantum mechanical model that the static β_0-value for the electro-optical (Pockels) effect is twice as large as the one for frequency doubling. However, owing to the occurrence of factors such as $E_{eg} - 2 \times E_{ph}$ in the corresponding equation for frequency doubling (instead of factors like $E_{eg} - E_{ph}$ for the Pockels effect) the denominator for frequency doubling approaches zero more rapidly, and the β-value for frequency doubling thus will overtake that of the Pockels effect at about the point where E_{ph}/E_{eg} is 0.3–0.4. Schematic representation of the dispersion curves for β_{EO} and β_{ED} are presented in Figure 12.4.

It may appear from Figure 12.4 that it is advantageous to bring the transition energy (absorption peak) of the molecule towards that of the applied photons to increase the effective β-value. However, there will be a trade-off in doing so, between the increasing β-value on the one hand and the increasing absorption losses in devices applying these materials on the other hand.

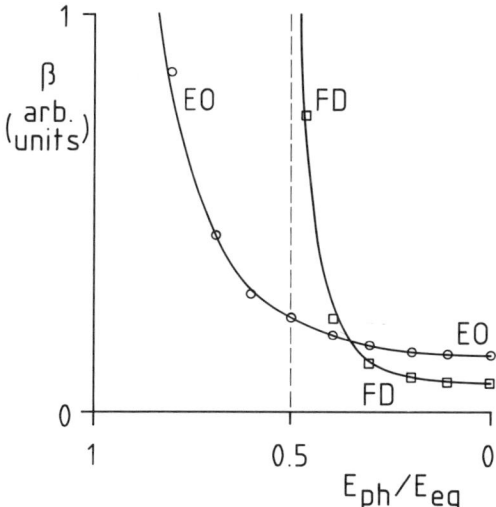

Figure 12.4 Variation of the molecular β-value as a function of the ratio of the applied photon energy and the excitation energy to a relevant CT-state.

Also, another method of obtaining the molecular β-value for frequency doubling exists: the so called Electric Field Induced Second Harmonic (EFISH) generation method (also known by the name dc-SHG). In this method, solutions $(10^{-4} - 10^{-2}$ Molar) of organic molecules are placed in a strong electric field to remove the centrosymmetry originally present in the liquid system by ordering of the dipole moments in the field. While the electric field is on, a pulsed laser beam is directed through the solution and frequency doubling will take place. The power conversion of the incident radiation into the second harmonic is then measured, from which the β-value can be derived. This method yields the nonlinear value without the model approximations made in the solvatochromic experiment. However, in the EFISH method, a secondary standard for the calibration is necessary in order to obtain 'absolute' values for β.

Comparison of the coefficients obtained experimentally in bulk samples for the Pockels effect $(\chi_{EO}^{(2)})$ with that for frequency doubling $(\chi_{FD}^{(2)})$ in the case of MNA crystals, shows that these values are almost the same. This supports the idea that the Pockels effect as well as the frequency doubling originate from the same type of displacement, namely, that of electrons.

In Table 12.3, measured β_{FD}-values for some organic molecules are presented. It is clearly seen that the molecules exhibiting the larger conjugated π-electron systems show the larger values for β_{FD}.

12.5.4 *Occurrence and determination of the molecular third-order nonlinear coefficient*

Besides lateral charge displacements leading to third-order effects, the rotational displacements of complete molecular building blocks will also give rise to such effects.

Table 12.3 Measured β_{FD}-values ($10^{-30}\,\text{cm}^5\,\text{esu}^{-1}$) for some organic molecules

Compound	β_{FD}	Remarks
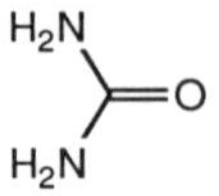 urea	0.4^a	$1.064\,\mu m$
 para-nitroailine (pNA)	15.6^b	$1.064\,\mu m$
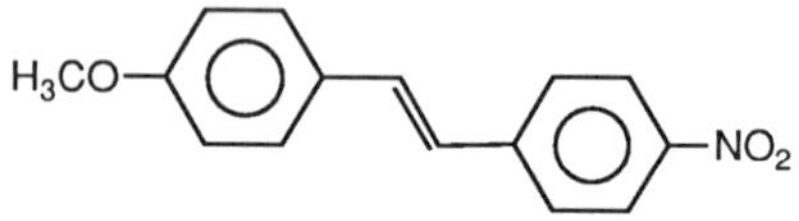 para-methoxy-nitrostilbene (MONS)	60^b	$1.064\,\mu m$
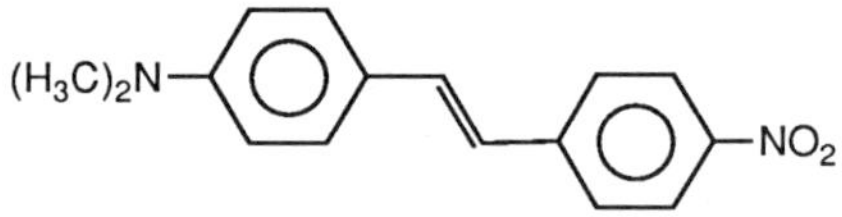 para-dimethylamino-nitrostilbene (DANS)	274^b	$1.064\,\mu m$

[a] Halbout and Tang (1987)
[b] Values measured in dioxane, by Huijts (1988, unpublished).

In the case of orientational displacements, the linear electric field dependence of the induced polar ordering is used to remove the centrosymmetry in order to achieve $\chi^{(2)}$-materials. The quadratic field dependence of induced axial order can be used to improve the electronic $\chi^{(3)}$. For third-order effects, in principle, axial alignment instead of polar ordering already suffices.

The experimental determination of γ (cf. eqn 12.3) can be carried out by measuring third harmonic generation (THG) or other third-order phenomena such as four-wave mixing, stimulated scattering, self-focusing, etc. Some of the corresponding methods are described in the work of Chemla and Zyss, Vol. 2 (1987).

12.5.5 Relation between molecular and macroscopic properties

In order to obtain optically nonlinear materials with large (macroscopic) nonlinear susceptibilities, one has to know the relation between the macroscopic $\chi^{(n)}$-values and the corresponding molecular hyperpolarizabilities.

For the electro-optic effect and frequency doubling to occur, the sample must have a second-order nonlinear susceptibility $\chi^{(2)}$.

The relation between the second-order susceptibility $\chi^{(2)}$ and the molecular β can be approximated by:

$$\chi^{(2)}_{IJK} = NF\beta_{ijk}\langle \cos\theta_{Ii}\cos\theta_{Jj}\cos\theta_{Kk}\rangle \tag{12.8}$$

in which N denotes the number of molecules per unit volume, and F the local field factor. Upper-case subscripts refer to the macroscopic frame; lower-case subscripts to the molecular frame. The cosines are projection factors, the brackets denote an averaging over molecular orientations; such averaged projection factors are polar ordering parameters.

As has been indicated before, for second-order nonlinearities no centrosymmetry is allowed. Samples containing isotropically oriented molecules do not therefore, show any second-order effect (unallowed inversion centre present). For isotropic systems, the polar ordering parameters in eqn (12.8) are equal to zero, resulting in a zero value for $\chi^{(2)}$. In order to show the desired second-order nonlinear effect, a bulk material must have some degree of polar (uniaxial) ordering; axial ordering only is not sufficient in this case.

After having obtained organic molecules with large β-values, one has to find a method for ordering the individual molecules in the macroscopic sample such that a net polar ordering results. Several methods exist for doing so, such as:

(i) Growth of noncentrosymmetric single crystals
(ii) Deposition of highly uniaxially ordered layers by the Langmuir–Blodgett technique
(iii) Electric field poling of polar optically nonlinear molecules, contained in suitable matrices.

Many polar organic molecules with promising second-order polarizabilities crystallize centrosymmetrically owing to the large dipole–dipole interaction energy. The thus resulting bulk materials do not show the desired nonlinear effects. However, a number of organic molecules (e.g. MNA) do crystallize in the favoured noncentrosymmetric way and have shown sizeable second-order nonlinear phenomena such as the electro-optic (Pockels) effect and frequency doubling.

The Langmuir–Blodgett deposition technique yields good results as long as a few monolayers are concerned. It has not yet been possible, however, to deposit many (thousand) layers to achieve sufficiently thick films for optical waveguiding and manipulation, while maintaining the required net polar ordering. Also, the procedure is time-consuming and films often contain domain walls which enhance scatter; molecular rearrangement within L–B films is also known.

Electric field poling of optically nonlinear molecules in polymeric systems, through interaction of their permanent molecular dipole moments with the field, leads to the occurrence of the desired net polar ordering. This provides a means of producing second-order nonlinear ($\chi^{(2)}$) polymers, as will be explained a little later.

In order to obtain useful second-order optically nonlinear materials by the electric field poling procedure, the induced uniaxial ordering must be maintained after the poling procedure has finished. The system containing the optically nonlinear molecules should exhibit such properties that the induced polar ordering is permanent under field-free conditions, after poling.

In the case of the EFISH measurement, the centrosymmetry of the dissolved molecular system is only temporarily removed by the (often pulsed) strong electric field. However, after switching off the electric field, the induced polar ordering rapidly disappears via orientational relaxation and the system has no second-order nonlinearities left. Solutions and other low-viscosity media tend to the situation of isotropic ordering again.

To obtain a permanent non-zero $\chi^{(2)}$-value in bulk materials via electric field poling,

M

a matrix is required, permitting sufficient molecular mobility during poling, and being sufficiently rigid after poling to prevent molecular relaxation of the induced ordering. Polymers are such materials.

In the case of third-order materials, not net ordering of any sort is strictly required. If the third-order effect is due to lateral charge displacements, the optically nonlinear effects are optimal if the electric vector of the optical field coincides with the direction of the largest polarizability. By inducing axial ordering, such overlap of directions can be maximized. The electric field induced axial ordering itself is also an optically third-order phenomenon; the magnitude of the induced birefringence varies quadratically with the applied field strength.

12.6 Optically nonlinear polymers

12.6.1 Introduction

Polymers are almost ideal systems for inducing optical nonlinearities by electrical field alignment or poling. Many polymers show a so-called glass transition temperature (T_g) Below T_g, the polymer is in the glassy state in which the polymeric matrix is rather rigid, severely restricting particular molecular segmental motions. Above T_g, the polymer is in the rubbery state in which the mobility of molecular segments is some orders of magnitude higher than below T_g.

By heating the polymer containing the optically nonlinear molecules to above the T_g and applying the correct electric fields, a net macroscopic order is induced. For optically nonlinear molecules exhibiting a permanent dipole moment, a net polar ordering will be induced, leading to noncentrosymmetric systems possessing second- and higher-order nonlinear effects. For polarizable molecules without a permanent dipole moment, axial alignment can be induced through the interaction of DC or AC electric fields with the anisotropically polarizable molecule; here, only the third- and other odd-order optical properties will be present after alignment.

Subsequently, after axial and/or polar ordering has been induced, the polymer is cooled to below T_g while keeping the external fields on. At sufficiently low (room) temperature, the electric field can be switched off; the induced order remains frozen in the now rigid polymer matrix. The bulk material should then show optical nonlinearities.

Several possibilities exist when applying polymers as matrix materials for optically nonlinear systems:

(i) Solid solutions of optically nonlinear molecules in an amorphous polymeric host

(ii) Solid solutions of optically nonlinear molecules in a liquid crystalline polymer host, for enhanced ordering

(iii) Main chain (liquid crystalline) polymers, the optically nonlinear molecules are incorporated in the backbone of the polymer

(iv) Side chain polymers where the optically nonlinear chromophores are attached exclusively as side groups to a polymer backbone.

(v) Liquid crystalline side chain copolymers where the optically nonlinear molecules are attached to a polymerizable moiety and then copolymerized with mesogenic monomers

(vi) Liquid crystalline side chain homopolymers wherein the optically nonlinear and polymerizable moieties do possess strong enough mesogenic properties themselves

(vii) Lyotropic liquid crystalline polymers wherein polymeric helicies can be oriented to produce highly polarized systems.

In the rest of this chapter we will mainly deal with polar ordering phenomena leading to materials with second-order nonlinear effects.

12.6.2 Solid solutions

Liquid solutions containing the azo-dye Disperse Red 1 (DR-1; 4-[N-ethyl-N-2(hydroxyethyl)]amino-4'-nitroazobenzene) and PMMA in a solvent have been prepared as a starting material for the production of solid solutions (Singer et al., 1986). The solutions were spun coated on to indium tin oxide (ITO) coated glass. A thin film of about 4 μm resulted, on which a gold electrode was deposited. The sample was then heated above the T_g of PMMA (about 100°C) and then poled, applying electric fields in the range 0.2–0.6 MV cm^{-1}. The $\chi_{FD}^{(2)}$ value, measured via frequency doubling in the poled film, was equal to 5 pm V^{-1}, which is about five times the value of the inorganic crystal KDP. Inserting the known experimental values into eqn (12.8), a value of 0.2 for the relevant polar order parameter can be estimated for these experimental conditions.

The solid solution of the above-mentioned azo dye DR-1 in PMMA is amorphous before and after the poling procedure. Simple statistical models exist, predicting the degree of polar ordering being achieved via poling of initially isotropic or liquid crystalline samples. For the relevant polar order parameter, ⟨cosines⟩, the following expression can be applied to a first approximation (see also section 12.7 below):

$$\langle \text{cosines} \rangle = \mu_0 \times E/(k \times T \times m) \tag{12.9}$$

in which μ_0 is the permanent dipole moment, E the electric field strength, k the Boltzmann constant and T the temperature in Kelvin; the factor m depends on the state of the system and has the value 5 for isotropic systems, and 1 for liquid crystalline systems with perfect axial order.

By inserting typical molecular values in eqn (12.9), one is able to estimate the degree of induced polar ordering in a dilute system of molecular dipoles; the result can be used to estimate the nonlinear coefficient of a material by applying eqn (12.8). For example, taking the values $\mu_0 = 7$ Debye ($= 2.335 \times 10^{-29}$ Cm), $E = 1 \times 10^8$ V m^{-1}, $k = 1.38 \times 10^{-23}$ J K^{-1}, $T = 373$ K and $n = 5$ (isotropic matrix), the value for ⟨cosines⟩ is equal to 0.091. Insertion of $n = 1$ (ideal liquid crystalline matrix) in eqn (12.9) leads to the value of 0.454 for the polar ordering. The maximum value for ⟨cosines⟩ is one; eqn (12.9) is only valid for relatively small electric field strengths because saturation effects at higher field strengths are not taken into account in these simple models.

Because one can gain in net polar ordering by starting from an already liquid crystalline system before poling, preparations of solid solutions of DANS in copolyacrylates, containing mesogenic side groups as shown in Figure 12.5, have been carried out (Meredith et al., 1982).

A disadvantage of solid solutions is the limited solubility of the active nonlinear molecules in such matrices; concentrations up to a few percent only are possible. If much higher concentrations are applied, the nonlinear molecules tend to segregate before or after the poling procedure, forming (centrosymmetric) microcrystallites. These particulates do not contribute to the nonlinear effect but may give rise to

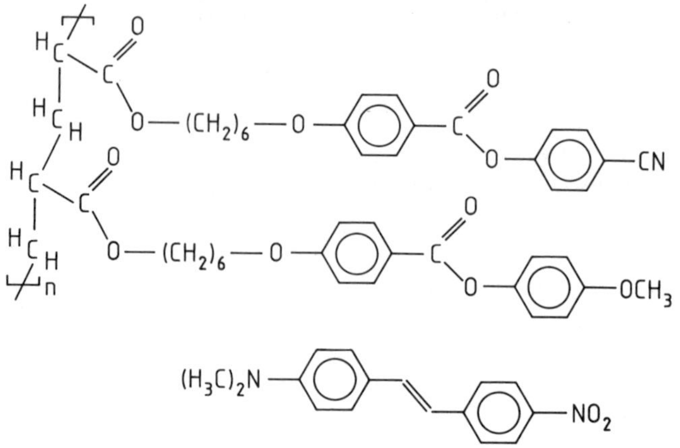

Figure 12.5 Solid solution of the organic molecule DANS in a mesogenic acrylate copolymer host, for improved polar ordering.

undesired light scattering. Moreover, the T_g values of the above-mentioned liquid crystalline copolymer hosts are currently rather low (18.5°C) indicating that such structure must be kept at fairly low temperatures after poling in order to prevent relaxation of the induced polar ordering; polymers with higher T_g values are desirable. In the case of isotropic PMMA solid solutions, T_g is around 100°C.

12.6.3 Side chain polymers

In order to increase the concentration of optically nonlinear molecules in polymer matrices and simultaneously to make use of liquid crystal assisted induced polar ordering, copolymers of mesogenic side chain moieties (esters) and of optically nonlinear side groups (stilbenes) have been synthesized (Le Barny *et al.*, 1986). The structure of such a polymer is shown in Figure 12.6.

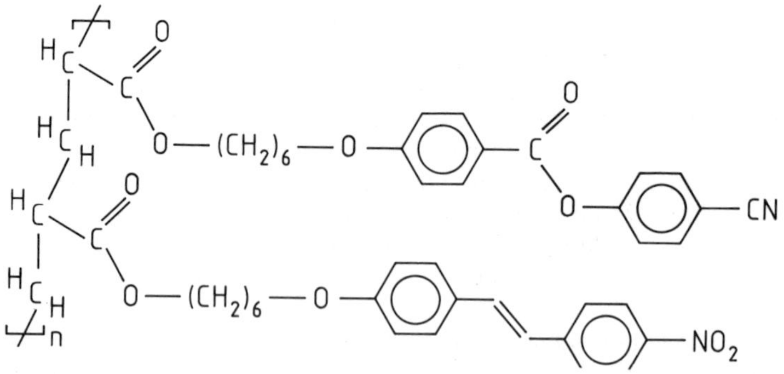

Figure 12.6 Copolymers, containing mesogenic and optically nonlinear side groups, for improved polar ordering.

The T_g of this system is still rather low (33°C). Electric field poling has been carried out with a field of $2 \times 10^6 \, \text{V m}^{-1}$. The experimentally obtained harmonic generation coefficient for this polymer, using 1064 nm incoming light, was found to be six times that of the inorganic crystal lithium niobate.

Besides the problem of the low T_g, severely limiting the application of the above-mentioned polymers in devices, a high concentration of the nonlinear molecule will influence the liquid crystalline properties of the copolymer.

12.7 Electric field poling of polymers

Many hyperpolarizable molecules (such as DANS) exhibit a long conjugated π-electron system (Pi), asymmetrically substituted by an electron donor (D) and an electron acceptor (A). Such APiD molecules show a fairly large β-component, almost coinciding with the direction of the permanent dipole moment μ_0. The permanent dipole moment, in its turn, almost coincides with the direction from A to D, which is the direction of the long (z-)axis of the molecule.

On one hand, this permanent dipole moment is a drawback because the dipole–dipole interaction energy counteracts a state of ideal polar alignment. On the other hand, the permanent dipole moment provides the driving force for ordering ('poling') of the molecules by an external electric field.

If idealized 'one-dimensional' and dispersion-free APiD molecules (β_{zzz} only) are sandwiched in a layer between two plane parallel electrodes (corresponding symmetry is $C_{\infty v}$ or ∞mm), the relation between the nonzero components of $\chi^{(2)}$ and β can be simply written as (Williams, 1984; Meredith $et\ al.$, 1982, 1983):

$$\chi^{(2)}_{ZZZ} = N \times F \times \beta_{zzz} \times \langle \cos^3 \theta \rangle \tag{12.10}$$

$$\chi^{(2)}_{XXZ} = \chi^{(2)}_{YYZ} = \chi^{(2)}_{XZX} = \chi^{(2)}_{YZY} = \chi^{(2)}_{ZXX}$$
$$= \chi^{(2)}_{ZYY} = N \times F \times \beta_{zzz} \times \langle \cos \theta \cdot \sin^2 \theta \rangle / 2 \tag{12.11}$$

where θ is the only significant angular variable, representing the angle between the permanent dipole moment (z direction of the molecular frame) and the electric poling field (Z direction in the macroscopic frame). The factor N is the number density of the APiD molecules and F is the local field correction factor. The expressions between brackets are the polar ordering parameters. Upper case indices refer to the macroscopic systems; lower case characters to the molecular frame.

To achieve maximum values for $\chi^{(2)}$-components, the factors N, F, β_{zzz} and the corresponding polar ordering parameters have to be optimized. Concerning the optimization of the polar order parameters, it will be shown hereafter that $\langle \cos^3 \theta \rangle$ can be made larger than $\langle \cos \theta \cdot \sin^2 \theta \rangle / 2$. If the same pair of electrodes is used for poling as well as for electro-optic switching after poling (based on $r_{33} = -2 \cdot \chi^{(2)}_{ZZZ} / n_z^4$), the factor $\chi^{(2)}_{ZZZ}$ is the most important tensor component. Axial ordering, already spontaneously present in liquid crystalline systems, enhances the field-induced polar ordering with respect to $\langle \cos^3 \theta \rangle$, while simultaneously decreasing $\langle \cos \theta \cdot \sin^2 \theta \rangle / 2$. Therefore, liquid crystalline polymers offer some advantage over isotropically ordered systems with respect to induced polar ordering and nonlinear properties finally obtained (Meredith $et\ al.$, 1982, 1983; Le Barny, 1986; Singer $et\ al.$, 1986, 1987; Van der Vorst and Picken, 1987; Stamatoff $et\ al.$, 1986; Griffin $et\ al.$, 1986).

12.8 Theoretical models for electric field poling

In the literature, four molecular statistical models for electric field poling of hyperpolarizable molecules have been described, yielding $\langle \cos^3 \theta \rangle$-values as a function of poling field strength and poling temperature. These four models are: the isotropic model; the Ising model (Meredith *et al.*, 1982, 1983); the model of Singer, Kuzyk and Sohn (SKS model, Singer *et al.*, 1986); and the extended self-consistent Maier–Saupe model by Van der Vorst and Picken (MSVP, Van der Vorst and Picken, 1987). These models do not describe the dynamics of polar ordering during the poling process, but yield the degree of ordering if sufficient time is allowed to complete the poling by reaching thermodynamic equilibrium.

If poling can only be achieved at or above T_g, while below T_g no orientational relaxation occurs, the net polar ordering at room temperature would be the same as that obtained at T_g. For experimental verification of the polar ordering theories, by comparing theoretically and experimentally derived $\chi^{(2)}$ values, such relaxations must be absent or suppressed. In order to diminish the influence of relaxation as much as possible, $\chi^{(2)}$ experiments should be performed with the poling field kept on at T_g or at room temperature. After switching off the poling field, relaxation studies can be carried out separately.

In the above-mentioned models, orientational-averaged values of any function $A(\theta)$ are calculated, using the orientational distribution function $G(\theta)$ as a weight factor:

$$\langle A \rangle = \left[\int_{-1}^{+1} \mathrm{d}(\cos \theta) \cdot A(\theta) \cdot G(\theta) \right] \bigg/ \left[\int_{-1}^{+1} \mathrm{d}(\cos \theta) \cdot G(\theta) \right] \tag{12.12}$$

where the fraction of the molecules with a polar angle in the range θ to $\theta + \mathrm{d}\theta$, is proportional to $G(\theta) \cdot \sin \theta \cdot \mathrm{d}\theta$. In Boltzmann statistics:

$$G(\theta) = \exp[-U(\theta)/kT] \tag{12.13}$$

in which $U(\theta)$ is the molecular energy in the presence of an electric field and can be written as the sum of the terms:

$$U(\theta) = U_0(\theta) + U_1(\theta) + U_2(\theta) \tag{12.14}$$

where $U_i(\theta)$, with $i = 0, 1, 2$, is proportional to the ith power of the electric field strength E. The expressions for $U_i(\theta)$ in the four different models can be found in Table 12.4. The models have identical expressions for $U_1(\theta)$, representing the energy term for the

Table 12.4 Energy expressions in the four different models

Term	Isotropic model	Ising model	SKS model	MSVP model
$U_0(\theta)$	0	0 for $\theta = 0, \pi$ ∞ for $\theta \neq 0, \pi$	No analytical formula*	$-\xi \langle P_2 \rangle P_2(\cos \theta)$
$U_1(\theta)$	$-\mu_0 E \cos \theta$	$-\mu_0 E \cos \theta$	$-\mu_0 E \cos \theta$	$-\mu_0 E \cos \theta$
$U_2(\theta)$	0	0	0	$-\Delta \alpha E^2 P_2(\cos \theta)/3$

*In the SKS model, the product $A(\theta) \cdot \exp[-U_0(\theta)/kT]$, in which $A(\theta) = \cos^3 \theta$ and $(\cos \theta \cdot \sin^2 \theta)/2$, is expanded in a series of Legendre polynomials; the corresponding coefficients are the microscopic axial order parameters of the liquid crystal host. These values have to be determined experimentally.

permanent dipole moment in the electric field; they differ in their choice of $U_0(\theta)$ and of $U_2(\theta)$.

The term $U_2(\theta)$ is the θ-dependent part of the energy term, connected with the linearly induced dipole moment ($-\alpha_{zz}\cdot E^2/2$); this term is currently taken into account in the MSVP model only; it is ignored in the other three models. The θ-dependence is caused by the strong anisotropy of α and is inherent to long conjugated π-electron systems. If we define for the 'one-dimensional' (= cylindrical symmetric) APiD molecules: $\alpha_\parallel \equiv \alpha_{zz}, \alpha_\perp \equiv (\alpha_{xx} + \alpha_{yy})/2 \cong \alpha_{xx} \cong \alpha_{yy}; \; \bar{\alpha} \equiv (\alpha_{xx} + \alpha_{yy} + \alpha_{zz})/3 = (\alpha_\parallel + 2\cdot\alpha_\perp)/3$ and $\Delta\alpha \equiv \alpha_\parallel = \alpha_\perp > 0$, it follows that:

$$-\alpha_{zz}\cdot E^2/2 = -(\alpha_\parallel\cdot\cos^2\theta + \alpha_\perp\cdot\sin^2\theta)\cdot E^2/2 \tag{12.15}$$

which can be written as

$$-\bar{\alpha}\cdot E^2/2 - (\Delta\alpha\cdot E^2\cdot P_2(\cos\theta))/3 \tag{12.16}$$

in which $P_2(\cos\theta)$ is the second-order Legendre polynomial of $\cos\theta$; $P_2(\cos\theta) = (3\cdot\cos^2\theta - 1)/2$. Since $U_2(\theta)$ is minimal for $\theta = 0$ or π, the term $U_2(\theta)$ by itself leads to field-induced axial ordering. Such axial order corresponds to an elongation of the distribution function along the Z-axis, while keeping the distribution function centrosymmetrical. Axial ordering is represented by the axial order parameter $\langle P_2 \rangle$.

In the literature describing the electric field poling of APiD molecules, it has been already recognized that field-induced polar ordering is enhanced by axial ordering. Such axial ordering is already spontaneously present (without a field) in a liquid crystal. This enhancement is supported theoretically by the fact that $\langle\cos^3\theta\rangle$ calculated via the Ising model (assuming perfect axial ordering with $\langle P_2 \rangle = 1$), is five times higher than in the isotropic case in which axial ordering is completely absent, $\langle P_2 \rangle = 0$ (Meredith et al., 1982, 1983):

$$\langle\cos^3\theta\rangle_{\text{Ising}} = a \tag{12.17}$$

$$\langle\cos^3\theta\rangle_{\text{isotropic}} = a/5 \tag{12.18}$$

Here, $a = \mu_0 E/(kT)$; we have seen the approximate solutions of the Ising and isotropic models, eqns (12.17) and (12.18), before in the form of eqn (12.9). Exact solutions have been published in the literature (Van der Vorst and Picken, 1987). The enhancement can be understood qualitatively by comparison of the Ising model with the isotropic case. In the Ising model, two states are allowed ($\theta = 0, \pi$), each with a maximum magnitude of $\cos^3\theta$, but with opposite signs (± 1). At zero field strength, both states are equally populated, so their large, but opposite, contributions to $\langle\cos^3\theta\rangle$ completely cancel. However, if an electric field is present, the corresponding energy levels are split up and are 'far' apart. So the lower state with $\theta = 0$, is more densely populated than the higher lying one with $\theta = \pi$, giving a certain non-zero value for $\langle\cos^3\theta\rangle$. In the isotropic model, many states of intermediate energy are also allowed. For every state θ, there exists a state $\pi - \theta$ with opposite sign for $\cos^3\theta$. However, owing to the fact that these states are not very far apart in energy, the difference in population is smaller. Moreover, the magnitude of their (opposite) $\cos^3\theta$ is also smaller; therefore, $\langle\cos^3\theta\rangle$ is smaller.

To exploit the axial ordering for enhancement of electric field-induced poling, liquid crystalline poling media (e.g. LC-side chain polymers) doped with APiD-molecules are being studied (Meredith et al., 1982, 1983; Singer et al., 1986, 1987). Also, liquid crystalline side chain copolymers, built up from hyperpolarizable and mesogenic side groups, are under investigation; the mesogenic moieties in the resulting polymer taking care of the liquid crystallinity of the entire system (Le Barny, 1986). In this case it is hoped that most of the cooperative axial order of the mesogenic groups is 'transferred' to the hyperpolarizable groups, through alignment of the APiD molecules along the mesogenic moieties. One can imagine that such an alignment occurs through the mechanism of anisotropic repulsion between the APiD and the liquid crystalline side chain molecules; both types of molecules are rod-shaped, exhibiting large anisotropies. Also, anisotropic attractions between the mutually induced dipole moments may contribute to the ordering. When one realizes that it is these very interactions between liquid crystal molecules that confer on them their mesogenic properties, and when one notices the resemblance of APiD molecules to molecules able to form a liquid crystalline phase, it is obvious that APiD molecules themselves should also possess some mesogenity. In other words, APiD molecules in concentrated 'single-rod systems' (with rod–rod interactions), like side chain homopolymers with APiD side groups, should also experience a certain tendency towards mutual axial alignment (Van der Vorst and Picken, 1987; Griffin *et al.*, 1986).

If the potentially mesogenic interactions in the 'mixed-rod systems' (copolymers) as well as in the 'single-rod systems' (homopolymers) below T_c are large enough to overcome the thermal counteraction, the system will exhibit a liquid crystalline phase. In this phase, spontaneous cooperative axial order will exist below T_c, not having taken into account other interfering phase transitions.

The Maier–Saupe mean field theory, as extended by Van der Vorst and Picken (MSVP theory), describes the electric field poling of mesogenic APiD molecules in such concentrated single-rod systems. Depending on the model parameters, the system can be isotropic or liquid crystalline at zero field strength. The tendency towards mutual alignment can be expressed by the effective single particle energy $U_0(\theta)$ of the form:

$$U_0(\theta) = -\xi \cdot S(T) \cdot P_2(\cos \theta) \tag{12.19}$$

This expression was originally used by Maier and Saupe (1958, 1959, 1960) in their theory for (nematic) liquid crystals. An APiD molecule, placed in an environment of APiD neighbours already axially ordered (director along the Z-axis), will experience a mean field tending to orient it along that director. The θ-dependence of the corresponding potential energy can be approximated by the second order Legendre polynomial of $\cos \theta$. The constant ξ is the absolute 'strength' of the potential and depends on the types of the interacting species. Since in the present case only APiD–APiD interactions are considered, ξ is more or less a materials constant.

The Maier–Saupe order parameter S describes the dependence of the relative strength of the orienting potential on the degree of axial order of the environment, which is temperature-dependent. For example, above T_c, thermal fluctuations are so large that spontaneous axial ordering does not take place. So if no axial order in the environment is present, the APiD molecule under study will not experience any orienting potential, resulting in S equal to zero above T_c.

The usual choice for S is

$$S = \langle P_2 \rangle \tag{12.20}$$

since $\langle P_2 \rangle$ is a measure for axial order. A self-consistent value of $\langle P_2 \rangle$, satisfying eqns (12.12)–(12.16), (12.19) and (12.20), can be found by iterative numerical calculus for a particular set of parameters (μ_0, $\Delta\alpha$, ξ, T, E). After substituting the $\langle P_2 \rangle$ obtained in the expression for $U_0(\theta)$, other averages can be calculated, e.g. $\langle \cos^3 \theta \rangle$.

The input parameters for the MSVP model can be derived as follows. The molecular parameters μ_0 and $\Delta\alpha$ can be calculated theoretically or determined experimentally. However, since local field effects are not considered explicitly in the model, the input parameters must in principle be corrected for these effects; μ_0 and $\Delta\alpha$ should be 'dressed' quantities. Alternatively, E itself should be corrected for local field effects.

Electric field poling is assumed to be carried out at the polymers' glass transition temperature (T_g), a temperature that can be determined by Differential Scanning Calorimetry (DSC), for example. The Maier–Saupe theory itself provides a way of determining ξ. This theory reveals a relationship between ξ and T_c (at zero field strength), where the first-order nematic–isotropic phase transition takes place if $T_c > T_g$, or would take place if $T_c < T_g$:

$$T_c \approx 0.22\xi/k \qquad (12.21)$$

For polymers, showing liquid crystallinity at $T_g(\xi > 4.5kT_g)$, ξ can be determined from observation of the nematic–isotropic transition at $T_c > T_g$, applying a polarization microscope. This observation is not possible for polymers which are isotropic at $T_g(\xi < 4.5kT_g)$, owing to the glassy nature of the polymer which prevents the nematic–isotropic transition taking place at $T_c < T_g$. In this case, however, T_c can be estimated from the temperature dependence of field-induced birefringence (Gramsbergen et al., 1986).

Since analytical expressions for the order parameters in terms of the model parameters cannot be derived, self-consistent numerical methods must be applied for each set of parameters to calculate the corresponding S. The results so obtained can be presented using the following reduced (non-redundant) set of model parameters:

$$a \equiv \mu_0 E/(kT_g)$$
$$b \equiv \Delta\alpha E^2/3kT_g$$
$$c \equiv \xi/kT_g = (0.22\tau)^{-1}$$
$$\text{or } a, b, \tau = T_g/T_c$$

Some calculated results of the MSVP model in terms of the model parameters a, b and τ are presented in Figure 12.7, showing plots of $\langle P_2 \rangle$ values as a function of a for different values of b and τ.

Note that in the isotropic cases (solid line and $\tau > 1$), considerable axial order can be induced by an electric field, especially when the system is pulled through the nematic–isotropic transition by the electric field (e.g. around $a = 0.55$, $b = 0.0$, $\tau = 1.1$). We will return later to this phenomenon of field-induced phase transitions. Such electric field-induced axial ordering is not only due to the energy term U_2, but results also from the combined effect of the terms U_0 and U_1, as also can be seen in Figure 12.7. To show the combined ($U_0 + U_1 + U_2$) field dependence of the axial order, the reduced parameter set a, b, τ is not very suitable, since both a and b contain the field strength E. In this case the full parameter set is more advantageous.

Whereas in the initially isotropic cases (solid lines and $\tau > 1$), considerable axial order can be induced by the electric field, in systems initially already liquid crystalline ($\tau < 1$), the axial order is only slightly enhanced by the field. Therefore, the degrees of

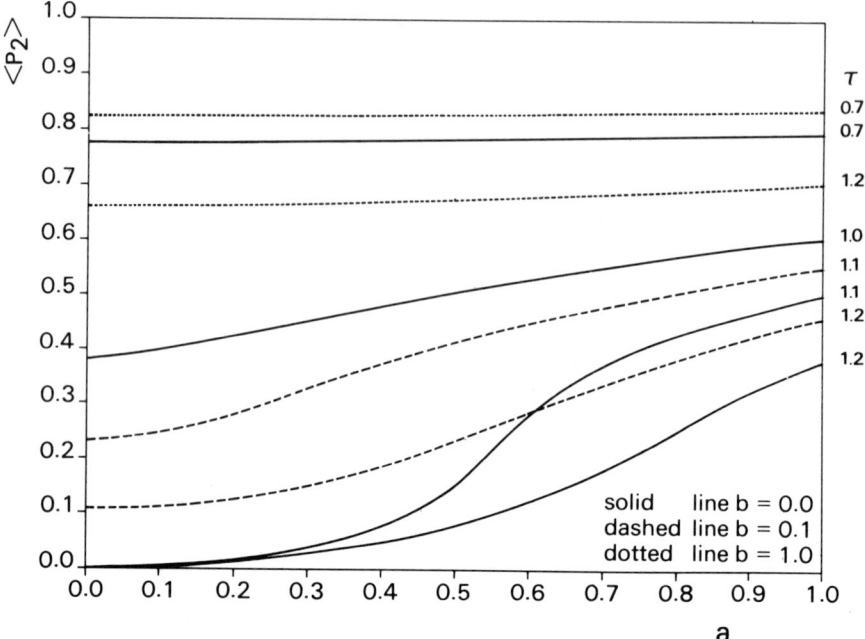

Figure 12.7 Calculated axial order in the MSVP model as a function of the reduced dipole energy $a = \mu_0 E / kT$.

axial ordering of initially isotropic and liquid crystalline samples, respectively, approach each other for increasing field strength.

In Figure 12.8 the polar order term $\langle \cos^3 \theta \rangle$ is shown as a function of a for different values of b and τ.

For comparison, the solutions for the isotropic and Ising models have also been added. For low field strengths ($a < 0.2$, $b \approx 0$), the initially isotropic cases ($\tau > 1$) behave according to the isotropic model. However, the initially liquid crystalline cases ($\tau < 1$) do show enhanced polar ordering, although the enhancement is not by the factor of five corresponding to the Ising model.

Interestingly, at higher field strengths (e.g. $a > 0.2$ and $b > 0$), the field-induced polar ordering becomes enhanced owing to field-induced axial ordering; this enhancement is more pronounced in initially isotropic than in initially liquid crystalline samples. In fact, in the initially isotropic case ($\tau > 1$), the term $\langle \cos^3 \theta \rangle$ shows a superproportional behaviour as a function of a (for $b = 0$, 0.1), this enhancement being due to the axial order induced through U_0 and U_1. However, in the initially liquid crystalline cases ($\tau < 1$), this behaviour is subproportional, owing to saturation effects.

The same remark concerning the electric field dependence of $\langle \cos^3 \theta \rangle$ can be made as was made previously for $\langle P_2 \rangle$; since both a and b contain the field strength E, the full, instead of the reduced, parameter set should be used. Concluding, one may say that in initially isotropic, and to a lesser extent in liquid crystalline systems, the polar ordering term $\langle \cos^3 \theta \rangle$ is significantly enhanced by field-induced axial ordering.

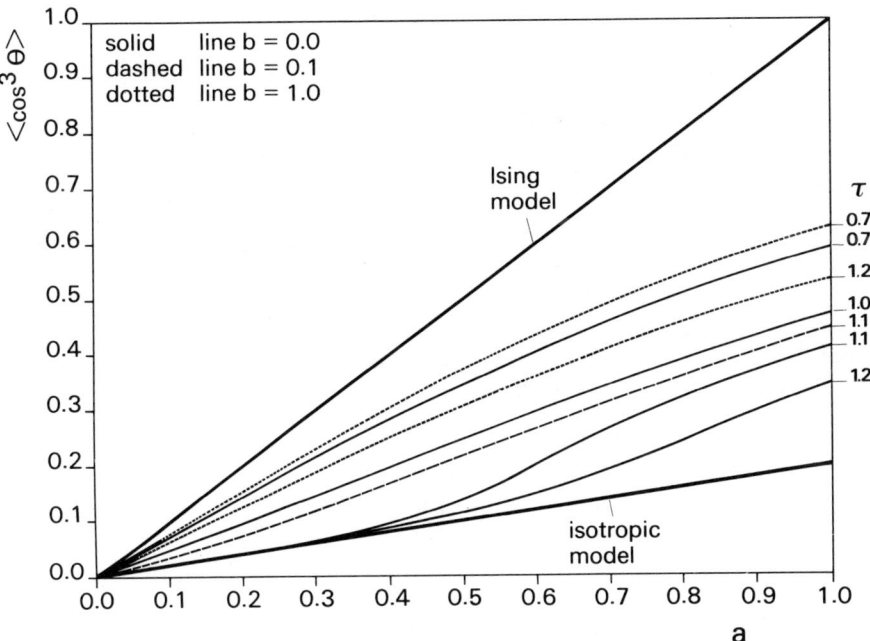

Figure 12.8 Calculated polar order in the MSVP model as a function of the reduced dipole energy $a = \mu_0 E/kT$.

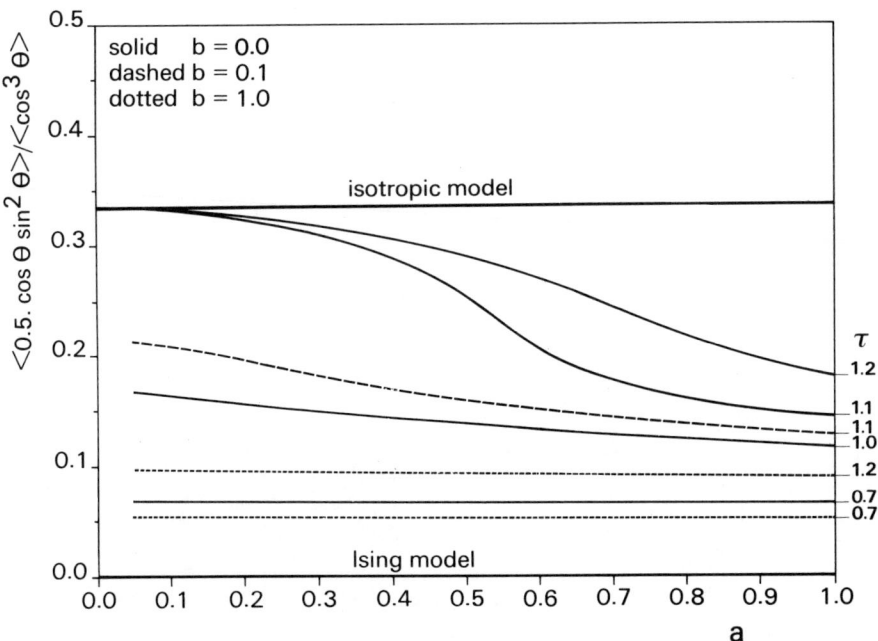

Figure 12.9 Calculated ratio of the two polar order parameters in the MSVP model as a function of the reduced dipole energy $a = \mu_0 E/kT$.

In Figure 12.9 the ratio $\langle 0.5 \cos \theta \sin^2 \theta \rangle / \langle \cos^3 \theta \rangle$ is shown as a function of a, b and τ, using the MSVP model. The initially isotropic model cases ($\tau > 1$) behave at 'low field strength' ($a < 0.2$; $b = 0$) according to the isotropic model, where the ratio is $1/3$. In all other cases, the ratio is smaller than $1/3$, owing to axial order, either initially present or field-induced. In the Ising model with complete axial order ($\langle P_2 \rangle = 1$), the ratio is zero because $\langle 0.5 \cos \theta \sin^2 \theta \rangle$ is zero.

In Figure 12.10 the correlation between polar and axial order is depicted graphically. The polar order parameters $\langle \cos^3 \theta \rangle$ and $\langle 0.5 \cos \theta \sin^2 \theta \rangle$ are always induced by a field and more or less proportional to E, whereas axial order can either be spontaneously present or induced (enhanced) by a field. In Figure 12.10 not the polar order parameters themselves but the values relative to the corresponding isotropic model solutions ($\propto E$) are shown as a function of the calculated $\langle P_2 \rangle$ values for different sets of model parameters a, b, τ. It is seen that the field-induced polar order term $\langle \cos^3 \theta \rangle / (a/5)$ is enhanced by the presence of axial order. The large bandwidth is due to saturation effects already mentioned.

It is also seen that the other polar order term $\langle 0.5 \cos \theta \sin^2 \theta \rangle / (a/15)$ decreases for $\langle P_2 \rangle$ values larger than about 0.3; in the Ising limit this term even becomes zero. It is clear from this picture that the polar order term $\langle \cos^3 \theta \rangle$ term is dominating, and finally the only one when $\langle P_2 \rangle$ approaches to 1.

Whereas the reduced parameter set a, b, c (or τ) is useful for plotting and tabulation purposes, it is inappropriate for showing the total effect (through $U_0 + U_1 + U_2$) of the poling field or the poling temperature. For that purpose, the total parameter set must

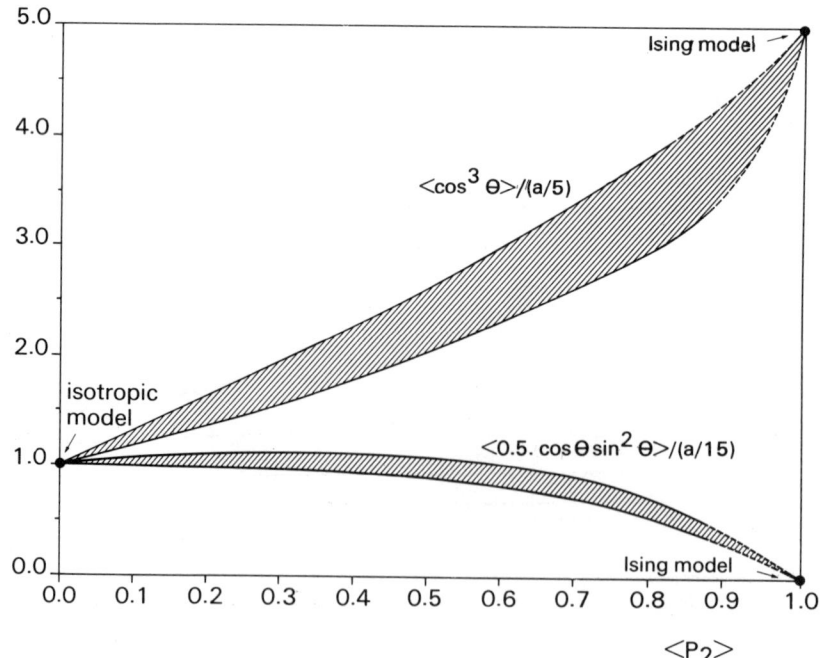

Figure 12.10 Correlation between axial and polar order in the MSVP model.

be used. Hereafter, some calculated results of the MSVP model are given for a specific model case with the parameter settings $\mu_0 = 7D$, $\Delta\alpha = 47 \text{ Å}^3$ and $T_g = 380 \text{ K}$. As shown by Van der Vorst and Picken (1987), both for $\langle P_2 \rangle$ and $\langle P_4 \rangle$, considerable axial ordering can be induced by the field for initially isotropic samples ($T_c < T_g$). In the case where liquid crystallinity is already present at $T_g (T_c > T_g)$, the electric field-induced axial ordering is rather small as compared with the value already present. In Figure 12.11 $\langle \cos^3 \theta \rangle$ is shown as a function of the field strength for two different clearing temperatures. Also, here liquid crystallinity (axial order) initially present helps to achieve polar ordering at low field strengths. At the higher field strength, the two curves corresponding to the different T_c values approach each other. The limiting isotropic and Ising models are also given.

Conclusively, one may say that liquid crystallinity is advantageous at the lower field strength in achieving considerable enhancement of polar ordering. At higher field strength, the field-induced axial ordering in initially isotropic samples approaches that of initially liquid crystalline samples, reducing the need for initial liquid crystallinity at high field strengths.

In Figure 12.12, the MSVP model results of Figure 12.11 are compared with those of the model used by Singer et al. (1986); in their (SKS) model $\langle \cos^3 \theta \rangle$ is expressed as follows:

$$\langle \cos^3 \theta \rangle^{SKS} = a(1/5 + 4\langle P_2 \rangle/7 + 8\langle P_4 \rangle/35) \qquad (12.22)$$

The SKS model is meant for APiD molecules in liquid crystalline mixed-rod systems,

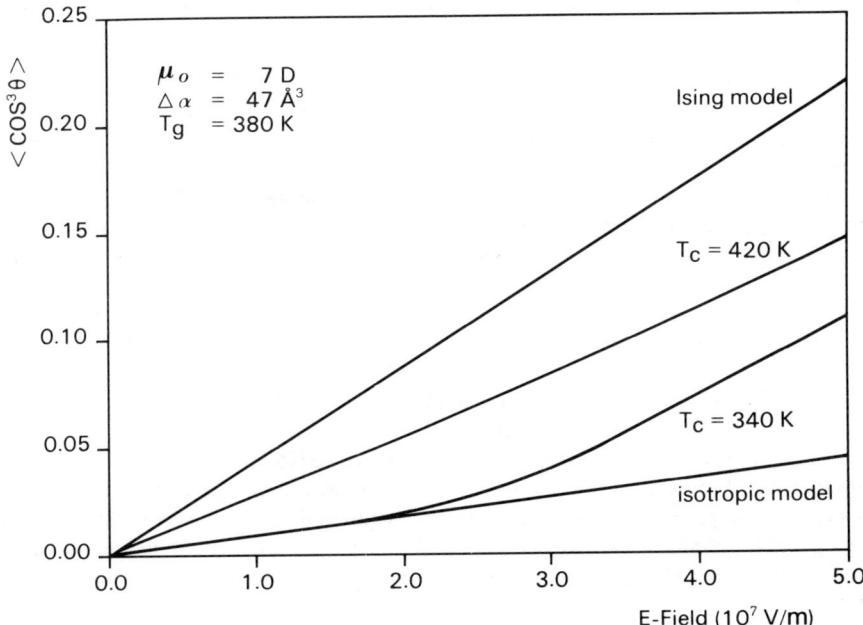

Figure 12.11 Calculated polar order in the MSVP model as a function of the poling field strength E.

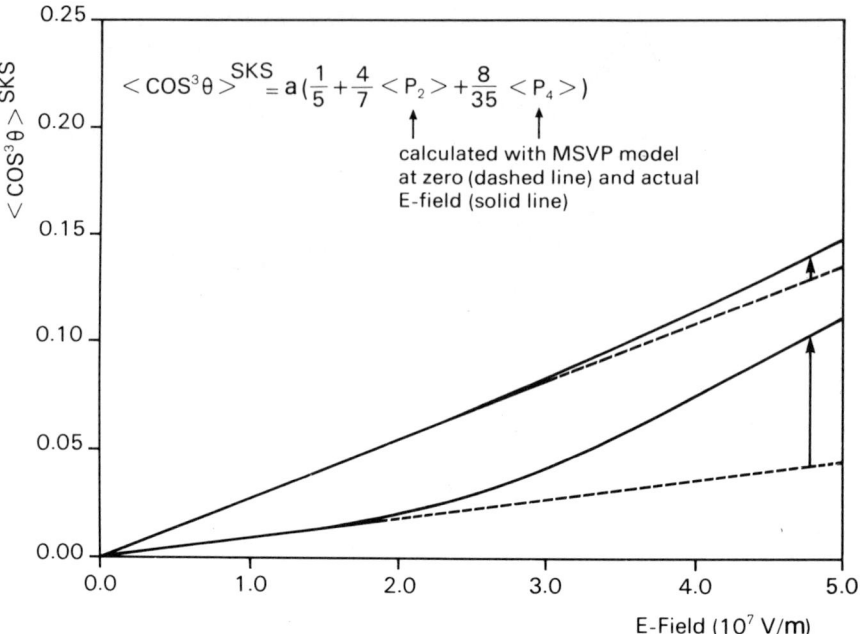

Figure 12.12 Calculated polar order in the SKS model as a function of the poling field strength E. The arrows indicate the enhancement of the polar order by the field-induced axial order.

characterized by the axial order parameters $\langle P_2 \rangle$ and $\langle P_4 \rangle$. Here, $\langle P_4 \rangle$ is the averaged fourth-order Legendre polynomial of $\cos \theta$. The terms $\langle P_2 \rangle$ and $\langle P_4 \rangle$ should be determined experimentally and they are assumed to be rather insensitive to the electric field strength, where $\langle \cos^3 \theta \rangle^{SKS}$ is a linear function of E (only through a).

In order to compare the MSVP and SKS models, the SKS model is also applied for the case of single-type rod-like molecules exhibiting both mesogenic and hyperpolarizable properties, forming a system with axial order parameters $\langle P_2 \rangle$ and $\langle P_4 \rangle$. To avoid the experimental determination of $\langle P_2 \rangle$ and $\langle P_4 \rangle$, the corresponding values have been calculated using the MSVP model at $E = 0$ and the parameters of the above-mentioned model case (broken lines).

As can be seen in Figure 12.12, $\langle \cos^3 \theta \rangle^{SKS}$ for $T_c = 340$ K behaves completely according to the isotropic model, whereas in the case with $T_c = 420$ K, the enhancement is by their initial ($E = 0$) axial order only.

Taking the actual field strengths into account, the SKS model yields the solid lines indicated in Figure 12.12 for the two T_c cases. Now, the difference between the two models is negligible and only due to the termination of the power series in the SKS case (Singer et al., 1986). So, the MSVP and SKS models are capable of reaching the same results; the main difference being the way in which the field-induced axial order is taken into account.

In the MSVP model, the field-dependent axial order parameters can be calculated, omitting the necessity for measuring these values in the presence of the electric field. For concentrated single-rod systems like side chain homopolymers, the calculation is

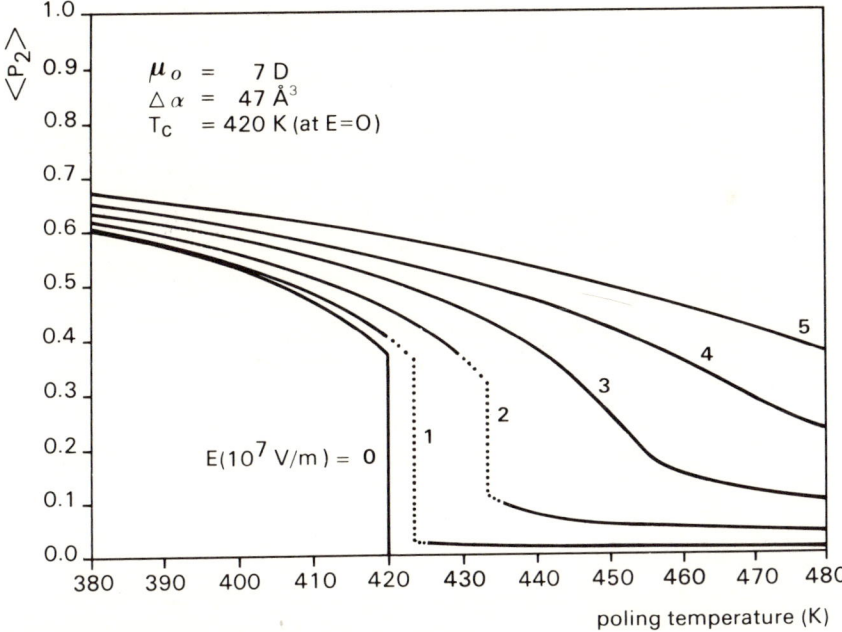

Figure 12.13 Calculated axial order in the MSVP model as a function of temperature for several poling field strengths E (in units of $10^7\,\mathrm{V\,m^{-1}}$).

straightforward. In the case of mixed-rod systems, like APiD dopes in liquid crystals or side chain copolymers with mixed APiD and mesogenic side groups, the current MSVP model needs to be extended further.

Throughout this chapter we have defined T_c as the temperature at which the discontinuous nematic–isotropic phase transition takes place (if $T_c > T_g$) or would take place (if $T_c < T_g$) in the *absence* of an electric field. The importance of this zero-field clearing temperature is that it is related to ξ through the relation $T_c \approx 0.22\xi/k$. The *actual* clearing temperature *in the presence of a field* will be larger than T_c, and it will increase with increasing E. In other words, during poling above T_c, a system can be 'pulled through' the phase transition, permitting the actual clearing temperature to become equal to or larger than the poling temperature. Examples of this have been mentioned before. Also, the discontinuous jump of $\langle P_2 \rangle$ at the phase transition (≈ 0.4 at $E = 0$) will decrease with increasing field, until it becomes zero. From there on, the phase transition will be continuous. The above findings are demonstrated in Figure 12.13, where $\langle P_2 \rangle$ has been plotted against the poling temperature, for several values of E and for fixed $\mu_0 = 7D$, $\Delta\alpha = 47\,\text{Å}^3$ and $T_c = 420\,\text{K}$. Owing to the finite temperature steps in the calculations, the actual clearing temperature has not been determined very accurately, as is indicated by the broken lines.

12.9 Application of optically nonlinear polymers in devices

Optically nonlinear materials must exhibit suitable properties in order to become applicable in electro-optic and related devices. Some of the required materials properties are: sufficiently, large optically nonlinear effect, small (electric or optical) driving powers, fast response, ease of processing, compatibility with a variety of substrates, low temperature processing, no cross-contamination, etc.

Are these requirements fulfilled by organic (polymeric) materials? Organic compounds have shown nonlinear coefficients, larger than found in some inorganic crystals. Especially in the case of frequency doubling, organic materials appeared to be more effective than the inorganic crystals, owing to the purely electronic character of the nonlinear effect in organic materials permitting sizeable effects in the optical frequency range. It must be mentioned, however, that inorganic crystals exist (e.g. barium titanate) exhibiting very large electro-optic coefficients, but coupled to a very large dielectric constant, making the materials response to applied fields rather slow.

Inorganic as well as organic crystals are difficult to grow and not easy to process (by cutting and, polishing) for device purposes. The realization of optical waveguides in crystals is an especially laborious process; in lithium niobate, waveguides are made by titanium diffusion at 1100°C, or by a proton exchange process. Such waveguides are mostly confined to regions close to the surface of the crystals, excluding complex 3D-waveguiding structures.

Langmuir–Blodgett films, consisting of organic nonlinear molecules, have shown only second-order optical nonlinearity in monolayer film or at most in a few consecutive layers. Apparently, the non-centrosymmetry necessary for the effect cannot be maintained over many deposited layers. For guided wave elements, films consisting of more than 500 layers are needed to achieve a waveguide thickness of the order of one micrometre. Also, the optical quality of many such deposited L–B layers is not yet good enough; much work needs to be done, including polymerization after deposition, to improve the mechanical properties.

Optically nonlinear side chain polymers are believed to be very attractive for applications in electro-optic and related devices. By attaching sufficient nonlinear side groups to the polymer backbone, sizeable nonlinear coefficients will be achieved if the material can be properly poled by an electric field. Polymers permit the coating at relatively low temperatures of large areas by spin coating, dipping, doctor blading, etc., on a variety of substrates; this may permit even the integration of these polymers with electronic ICs. The electronic nature of the nonlinearity combined with the low dielectric constant enables very fast electro-optic circuits to be realized. The tailorability of the molecular structures permits applications-specific development of polymeric material; properties such as colour, hardness and T_g can be controlled to some extent. Also, the etching, metallization and photolithographic processing of polymeric structures seem to be very promising. Finally, by repeated deposition of polymeric thin films, 3D-structures may be realized, enabling complex optical processing (switching) to be carried out.

However, many problems still must be solved before optically nonlinear polymers can become a routinely applied material. The electric field poling of the material is a difficult step; contaminants (dust, solvents, etc.) may cause problems in the poling of polymers contained between plane parallel electrodes (relatively large conducting surface at a small distance). Often, the thermal stability of polymers is questioned: this is an area that requires detailed study in order to establish the thermal properties of such materials, providing the basis for improvements. However, polymeric materials (such as polyimides) are known that can withstand temperatures up to about 400°C. The chemistry of polymers may ultimately result in the realization of optically nonlinear polymers that are stable at such high temperatures.

Generally, one may say that there exists no ideal material, organic or inorganic, fulfilling all possible design requirements. It is believed, however, that side chain polymers have sufficient potential to become a material with very attractive properties for application in optically nonlinear devices.

Although device-making using organic polymeric materials is still in its infancy, some results have already been achieved with respect to the realization of optical waveguides, or in the demonstration of the nonlinear effects. Electro-optic modulators in poled polymeric slab waveguides have been reported by Lytel *et al.* (1987); the electro-optic coefficient was measured as 2.8 pm V^{-1}.

It is expected that a wealth of electro-optic devices such as optically-addressed spatial light modulators, tunable notch filters, optical amplifiers and laser beam deflectors, based on nonlinear polymers, will be developed in research laboratories (Lytel *et al.*, 1987).

A general structure for the development of many devices can be envisaged as follows: it basically consists of a multilayer structure incorporating a substrate, electrodes for poling and (later) switching, an electro-optically active layer in which the light propagates, and buffer layers preventing the undesired interaction of the propagating light beam with the electrodes. A schematic representation of such a multilayer structure is shown in Figure 12.14.

Besides guided wave applications, alternative architectures exist in which the light propagates perpendicular to the multilayer structure. For this latter, 'free-space' type of device, (e.g. spatial light modulators, dynamic gratings, etc.), some or all of the layers in the structure should be transparent to the wavelength to be used.

To gain some insight about the optical nonlinearities that can be achieved with the currently existing polymers, the following example is worked out. The field-induced change in refractive index (dn) through the Pockels effect for an electro-optically active material can be expressed as follows:

$$dn = \chi^{(2)} \cdot E/n \qquad (12.23)$$

in which E is the switching electric field strength and n the refractive index. By inserting the values $\chi^{(2)} = 41$ pm V^{-1} ($= 0.1 \times$ lithium niobate), $E = 10^{+7}$ V m^{-1} ($= 10$ V μm^{-1}) and $n = 1.65$, the change in refractive index equals 2.5×10^{-4}. In order to achieve a phase shift of π for light with a wavelength of 820 nm, an optical pathlength equal to $820 \times 10^{-9}/(2 \times 2.5 \times 10^{-4}) = 1.64 \times 10^{-3}$ m ($= 1.64$ mm) is necessary. This is then the required length of the active part of an interferometer, for instance.

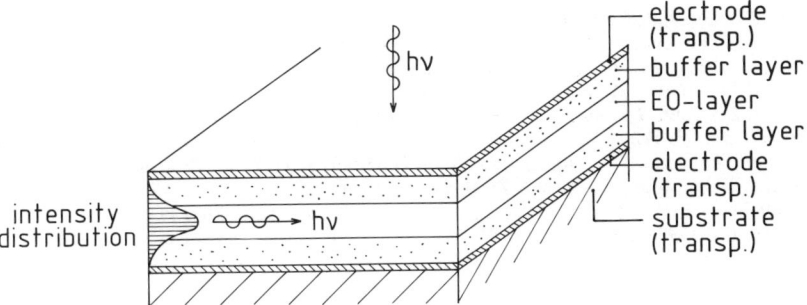

Figure 12.14 Schematic representation of a multilayer structure for the realization of optically nonlinear devices.

Considering the dispersive behaviour of the $\chi^{(2)}$-value, it can be expected that for selected molecules at particular wavelengths, $\chi^{(2)}$-values about an order of magnitude larger than those of lithium niobate can be found. Then, for the same applied field strength as above, an optical path length of only 17 μm suffices to achieve a phase shift equal to π. The inevitably corresponding larger optical attenuation caused by the increased absorption in the polymer (applied wavelength closer to the resonant transition) can be allowed over the optical pathlength of only 17 μm; an optical loss of several dBs may then be experienced.

For guided wave structures, incorporating complex and relatively lengthy optical pathways, the increased absorption may become problematic. In such cases a balance must be found between the maximum allowable optical loss and the desired nonlinear coefficient.

Acknowledgements

The authors thank Mr D.F. van Embden for his contribution to the artwork, and Dr S.J. Picken for performing MSVP calculations.

References

Chemla, D.S. and Zyss, J. (eds.) (1987) *Nonlinear Optical Properties of Organic Molecules and Crystals*, Academic Press, Orlando.

Garito, A.F., Singer, K.D. and Teng, C.C. (1983) in *Molecular Optical Properties of Organic and Polymeric Materials*, ed. Williams, D.J., ACS Symp. Ser. **233**, American Chemical Society, Washington DC.

Gramsbergen, E.E., Longa, L. and de Jeu, W.H. (1986) *Phys. Rept.* **135(4)**, 195.

Griffin, A.C., Bhatti, A.M. and Hung, R.S.L. (1986) *SPIE Conf. Proc.* **862**, 65.

Halbout, J.-M. and Tang, C.L. (1987) in *Nonlinear Optical Properties of Organic Molecules and Crystals*, eds. Chemla, D.S. and Zyss, J. Academic Press, Orlando.

Handbook of Lasers, The Chemical Rubber Company, Cleveland, (1971), 447.

Le Barny, P. (1986) *SPIE Conf. Proc.* **682**, 56.

Lytel, R., Lipscomb, G.F., Thackara, J., Altman, J., Elizondo, P., Stiller, M. and Sullivan, B. (1987) *SPIE Conf. Proc.* **824**, 152.

Maier, W. and Saupe, A. (1958) *Z. Naturforsch.* **13A**, 564.

Maier, W. and Saupe, A. (1959) *Z. Naturforsch.* **14A**, 882.

Maier, W. and Saupe, A. (1960) *Z. Naturforsch.* **15A**, 287.

Meredith, G.R., Van Dusen, J.G. and Williams, D.J. (1982) *Macromolecules* **15**, 1385.

Meredith, G.R., Van Dusen, J.G. and Williams, D.J. (1983) in *Molecular Optical Properties of Organic and Polymeric Materials*, ed. Williams, D.J., ACS Symp. Ser. **233**, American Chemical Society, Washington DC, 109.

Oudar, J.L. (1977) *J. Chem. Phys.* **67(2)**, 446.

Oudar, J.L. and Chemla, D.S. (1977) *J. Chem. Phys.* **66(6)**, 2664.

Shen, Y.R. (1984) *The Principals of Nonlinear Optics*, Wiley, New York.

Singer, K.D., Kuzyk, M.G. and Sohn, J.E. (1987) *J. Opt. Soc. Amer.* **B4**, 968.

Singer, K.D., Sohn, J.E. and Lalama, S.J. (1986) *Appl. Phys. Lett.* **49**, 248.

Stamatoff, J.B., Buckley, A., Calundann, G., Choe, E., De Martino, R., Khanarian, G., Leslie, T., Nelson, G., Stuetz, C., Teng, C. and Yoon, H. (1986) *SPIE Conf. Proc.* **682**, 85.

Tabiryan, N.V., Sukhov, A.V. and Zel'dovich, B.Ya. (1986) *Mol. Cryst.* **136**, 1.

Van der Vorst C.P.J.M. and Picken, S.J. (1987) *SPIE Conf. Proc.* **866**, 99.

Williams, D.J. (1984) *Angew. Chem.* **96**, 637.

13 The application of side chain liquid crystal polymers in optical data storage

C.B. McARDLE,* GEC Research, Hirst Research Centre, East Lane, Wembley, Middlesex HA9 7PP, UK.

13.1 Introduction

The demonstration of optical storage in (side chain) liquid crystal polymers (LCPs) can be performed very simply. All that is required is a relatively low-powered laser, (e.g. 10 mW HeNe), an appropriately dyed LCP on a substrate, a simple lens and perhaps a translation stage. It is even unnecessary to initialize the sample to a particular aligned condition, as bulk annealing of the sample produces one optical texture, whilst local laser heating and subsequent quenching produces a distinctly different one. The coexistence of the textures thus produced generates the requisite optical contrast for a storage element in the laboratory demonstrator. The transformation of such experiments, even when greatly refined, from the laboratory into commercially interesting products, is however, a much more complex procedure. The rate of this transformation is influenced by a great many factors, including the value of the potential markets which might be associated with the new medium and by the rate of progress made in directly competitive materials.

At present no LCP optical storage products are commercially available and many competitive media exist. An analysis of the open literature reveals that relatively few systematic studies on optical storage in LCPs have been conducted and that few specific storage device applications have been identified. Whilst it would be náive, particularly from an industrialist's standpoint, to put too much emphasis on what has been published pertaining to device outlets, the lack of detailed studies on optical storage using LCPs is still a little surprising when one contrast the situation for low-molar-mass (LMM) LC storage media. In this case numerous systematic studies have been documented and many reviews exist (Sincerbox, 1977; Kahn, 1973; Taylor et al., 1974; Hareng et al., 1975, 1976; Dewey et al., 1977; Dewey, 1980; Urabe et al., 1983; Dewey, 1984; Saski, 1986). Furthermore, many specific applications have been identified and demonstrated for the latter media, a non-exhaustive list of which includes artwork generators (Heinz et al., 1977; 1983; Smith et al., 1979; Kaneko, 1987; Kahn, 1987a) erasable optical disks (Kahn et al., 1984; Birecki et al., 1983; Higashiyama, 1985), full colour 'electronic slides' for CAD work (Kahn et al., 1987b) and light modulators for colour electronic imaging (Lu et al., 1987).

Evidently, therefore, much further work is required before LCP optical storage products can be realized, and it is fortunate that such a wealth of data already exists for

*Present address: Loctite (Ireland) Ltd, Research and Development Laboratories, Whitestown Industrial Estate, Tallaght, Co. Dublin, Ireland.

the LMMLC media which can be applied to the LCP case. Rapid progress is being made in the development of sophisticated polymeric liquid crystalline materials (see Chapters 3 and 4) and when this is combined with polymer processing technologies, storage applications for LCPs, which complement rather than compete against those offered by equifunctional LMMLC media, should follow naturally.

The objectives of the present chapter are to provide a condensed summary of the state of the art relating to LCP optical storage applications, to detail basic proof of concept experiments on simple LC copolysiloxane stores, and finally to present a case for the application of a hypothetical LCP product.

13.2 The role of polymers in optical storage applications

The role of polymeric materials in general optical recording applications has recently been reviewed by Smith (1981), Kaemf (1985, 1987) and Kuder (1986). Polymers can function in both passive and active roles in this application. In the passive roles, polymers form materials for disk substrates (Takeshima *et al.*, 1984; Kato *et al.*, 1986), subbing layers, protective coatings for imaging layers and dielectric spacers in antireflection trilayer devices, for example. In active roles, the polymer can form all, or part of, the imaging layer (s), e.g. it can function as a film forming binder for the imaging film, alternatively, it forms the actual imaging layer and thus usually possesses some inherent photo-, thermo- or electroresponsive property or properties.

Both write-once and erasable polymeric recording media have been described in the literature. The write-once media are usually (but not exclusively, see Gambogi *et al.*, 1987, for example) absorber hybrids, although photopolymers and polymers capable of thermoplastic deformation are well known in both holographic and digital optical recording (Booth, 1977; Sheriden, 1972; Ralston, 1980; Elmasry, 1986; Credelle *et al.*, 1972), thermoplastic polymers are in principle erasable types but are not always used as such. An interesting new class of photopolymers for near-infrared holographic recording has been described recently by Pinsl *et al.* (1986, 1987*b*) (see also Gerbig *et al.*, 1983).

In the dye/polymer types, ablation or laser-induced dye diffusion recording mechanisms may be used, and these have been reviewed by Alexandru *et al.* (1984). The metal particle/polymer types also employ ablation (Drexler, 1981) or sintering of metal particles (Reich *et al.*, 1987). Polymeric matrices have also been used for materials which exhibit persistent spectral hole-burning in the very interesting area of frequency domain optical storage (Moerner, 1985; Moerner *et al.*, 1987; Itoh *et al.*, 1987). Erasable polymer-based recording media are not as common as the write-once types. These include systems capable of undergoing temperature-induced phase separation of included components (Dabisch, 1979; Dabisch *et al.*, 1980), thinly and thickly overcoated dye/polymer layers (Gupta *et al.*, 1986; 1985) and dye/polymer bilayers which require preferential heating of selected layers during the act of writing and erasing (Hartman *et al.*, 1987). Side chain liquid crystalline polymeric recording materials, the subject of the present chapter, also fall into this class, and a condensed review of the international research activity centred upon this class of macromolecular recording materials now follows.

13.2.1 *Overview of side chain LCPs in optical storage applications*

Write-once and erasable, digital, analogue and holographic optical storage in nematic, cholesteric and smectic LCPs of different classes have been demonstrated since 1983. That such a variety of techniques and samples have been utilized might imply that the application of LCPs in optical storage was a well-understood device area which is backed up by many detailed studies. As mentioned already, examination of the open literature reveals that full reports dedicated to the actual laser addressing of LCPs are confined to half-a-dozen or so research papers or letters, and few of these are from industrial laboratories. The purpose of the present section attempts to provide a condensed summary and analysis of what has been published.

Several mechanisms can be used for thermo-optical recording on LCPs. The majority of these are relevant to LMMLC storage media, and since these have recently been documented by Sasaki (1986), they will not be considered further here. One of the most popular systems requires the initial condition of the LCP to be the homeotropic state. Thus, for example, a smectic LCP of positive dielectric anisotropy confined in a traditional LCD may be transformed from the bulk scattering state by heating to the isotropic phase and subsequently cooling in the presence of an electric field. On removing the field, the clear uniaxially aligned condition remains stored either in a mesophase or in the glass phase of the polymer. The sample in this condition is ready to receive laser-written information. The laser energy is usually coupled into the recording medium by resonance absorption, so that inclusion of an appropriate dye in the LCP host is common practice. Optical contrast results when the laser locally heats the LCP, usually to the isotropic phase, and is then switched or moved off, so that subsequent quench cooling results. The addressed bit cools back to an unoriented polydomain state which strongly scatters light. This represents positive contrast writing, the scattering features appearing dark against a bright background. Selective erasure is provided by reversing this process, but permitting cooling of the re-addressed bit under the influence of a field in order to locally reinstate the homeotropic condition. The field-assisted thermal mode may also be used to provide negative contrast writing, if the initial condition of the sample is scattering.

Apart from the example cited above, other variants exist which employ different initial recording conditions. For example, Schmidt (1984a, b) has used an inherently coloured LC copolyacrylate of negative dielectric anisotropy which possessed a nematic phase as a recording medium. The initial sample condition was the homogeneous state in this case. This arrangement has the potential advantage of improved recording sensitivity to a linearly polarized laser source due to efficient coupling of laser energy. Data pertaining to selective erasure of this sample are not available.

It is known also that aligning solely by interaction with agents coated on to cell walls is much more difficult for LCPs than for LMMLC material, and usually requires annealing processes to maintain reproducibly good quality samples. It is likely, then, that in the above case careful annealing of the sample after bulk erasure would be necessary, thus presenting some operation inconvenience. The analogue of this material with positive dielectric anisotropy has not been prepared, presumably due to the difficulties associated with radical-initiated polymerization of cyano-terminated vinyl monomers (Griffin et al., 1986). Such a material, however, would be a very attractive candidate for the normal positive contrast recording mode.

Initial bulk scattering LCP samples have been proposed as recording media which would employ the so-called 'scattering on scattering' laser writing mechanism (Coles, 1986; Coles et al., 1985a). This mechanism suffers the disadvantage that selective erasure cannot be implemented. This is so, because a low scattering density texture is usually produced by slow cooling or annealing the LCP sample, whilst the high scattering density or birefringent feature (cf. Coles, 1985a) is produced by quench cooling from the laser addressed zones. It is thus difficult to envisage how laser treatment can be used to locally restore the initial annealed texture.

The recent interest in thermo-optical recording on side chain LCPs was probably stimulated by the first reported observation of the phenomenon in a nematic system by Shibaev *et al.* (1983, 1985). Although of a preliminary nature, these results were significant in that durable storage in a glass phase, underlying the nematic phase, was demonstrated, as were thermal management and bulk erasure in a truly high molecular weight LCP. Various other workers have subsequently demonstrated the phenomenon (Schmidt, 1984a, b; Coles *et al.*, 1985a–c; Ueno *et al.*, 1986; McArdle *et al.*, 1987; Pinsl *et al.*, 1987).

Coles *et al.* (1985a–c) studied fully substituted smectic LC copolysiloxanes. Several points have arisen from their studies which are important to examine because of their implications for device application. The first point relates to the level of achievable contrast with LCPs. From Coles *et al.* (1985a) the implication appears to be that the particular polysiloxane studied, known as PG296, provides an order of magnitude at least of improvement in contrast over LMMLC storage systems. However, Hughes and Daley (1987) have shown this not to be the case when both types of LC storage media are assessed under equivalent experimental conditions. Additionally, the latter authors describe the important effect of thermal history on the condition of PG296, and indicate that fast cooling produces poor scattering whilst slow cooling produces superior scattering in this material. Consideration of these results raise questions as to exactly what achievable contrast might be expected when PG296 is laser-addressed, bearing in mind that quench rates from the heated bit may be $\geqslant 1000°C\,s^{-1}$. A second important point relates to recording sensitivity. Sensitivity may be defined as the minimum amount of energy required to produce a *stable* detectable change in the recording medium. For PG296, a sensitivity of $4\,nJ\,\mu m^{-2}$ has been quoted (Coles, 1985b). From the information given and assuming a circular bit, we may also estimate the required laser power at the sample (i.e. after insertion losses) to be about 15 mW. This sensitivity figure is perfectly reasonable in the light of other studies on LCPs (McArdle *et al.*, 1987; Pinsl *et al.*, 1987a). It is not apparent, however, how the sensitivity, resolution and power requirements, of $<1\,nJ\,\mu m^{-2}, 0.5\,\mu m$ and 0.5 mW (calculated) respectively are arrived at in a subsequent publication for what appears to be the same material (Coles, 1986). Laser-addressed PG296 gives interesting birefringent textures, however, a further point which requires clarification regarding this material is what role the polarization plane of the laser light plays, if any, in this particular type of realignment process (Coles, 1985b; Coles *et al.*, 1985a). Clearly, thorough systematic studies are required for LCP recording media before a sufficiently detailed knowledge can be attained and subsequently applied to device application.

Our group has chosen to study LC copolysiloxanes with less than 100% mesogen substitution, and an evaluation of these materials, aimed simply at demonstrating basic device functions, is given in the next section. These materials have certain advantages and are convenient to handle; however, they suffer the disadvantage of having sub-ambient T_g's so that we could not exploit the full range of properties available to polymeric liquid crystals. Storage in the highly viscous smectic phase is adequate for laboratory work but may not be so for practical devices; this depends very much on the intended application. In real-life operation, for example, data on a LC optical store may be continuously subjected to 'nonselective' erasure due to cumulative effects of field application incurred during the act of editing (Sincerbox, 1977). This inevitably occurs during readout, so the sample is simultaneously being warmed by absorption of projection light. Thus the accepted specification of a 10-year lifetime continuous working device may best be met by storage in a glass phase, particularly if mechanical stresses (flexing, etc.) are superimposed on the problems already identified, and easily-accessed glassy phases are available to polymers anyway.

A novel study of erasable optical storage in side chain LCPs which exploits glass-phase storage has been described by Eich and co-workers (1986, 1987a–d). These authors have employed a photochromic LC copolyester as an erasable holographic storage medium, the initial recording condition being either the homeotropic (Eich et al., 1986, 1987a) or the homogeneous (Eich et al., 1987c) state. Eich et al. (1987a) have tentatively suggested that the laser-induced refractive index changes in this material result from a combination of photo- and thermal effects, and further work continues (Eich, 1987d). Thus, photogenerated cis-forms of the azobenzene side chains in the copolymer locally perturb mesogenic side chain ordering when the polymer is simultaneously locally heated from a smectic to a nematic mesophase. On recooling, the azobenzene moieties relax to their more stable trans forms, but the perturbation suffered by the mesogenic side chains results in significant changes in birefringence (0.01).

Impressive phase gratings with high diffraction efficiency can be stored in this way, and are exceptionally interesting from the standpoints of both optical mass storage and holographic optical element formation.

It is interesting to note that the combination of photo- and thermal effects has also been used as an imaging technique for LMMLCs. Thus, colour changes resulting from reversible (Sackmann, 1971; Godberg et al., 1973) or irreversible (Haas et al., 1969; Adams et al., 1971) changes in cholesteric helical pitch have been reported. The latter effect has also been demonstrated in LCPs (Pinsl et al., 1987a). Furthermore, reversible photo-induced changes in mesophase transition temperature in LMM smectic LCs have been used in novel imaging applications (Haas et al., 1977; Ogura et al., 1983; Leier et al., 1979; Tazuke et al., 1987). Various classes of photochromic LCPs are known to be under investigation at the present time (see Chapter 3 and 4).

Non-linear optical self-diffraction in side chain LCPs has been reported (Eich et al., 1985; Shibaev et al., 1985). It is worth noting in passing that similar effects have been utilized for dynamic recording media, e.g. in thin oil films (Da Costa, 1986; Cormier et al., 1978).

Pinsl et al. (1987a) have provided an elegant study of write-once optical storage on cholesteric copolysiloxanes. Their materials were doped with carbon black or benzophenone; alternatively, benzophenone moieties were inherent in the LC copolymeric structure. In either case, laser-induced photochemistry of the additive led to irreversible destruction of the helical pitch in the LCP and hence destroyed its ability to reflect light completely in the addressed zones ($\geqslant 200 \,\mu m$ diameter). In this way 80% changes in reflectivity could be produced in about 250 ns following a 10-ns pulse from a Nd:YAG pumped dye laser (680 nm). Although the system is write-once, by employing energies below the saturation energy, only surface marking occurs. Subsequent bulk heating of the sample above its transition temperature redistributes the damaged material. Implementation of this self-healing process offers the prospect of some form of erasability or extended working life. It is interesting to note that the redistribution of damaged material in the smectic LCPs we have studied also leads to extended device life after bulk erases (see section 13.2.2).

Japanese scientists have been very active in LCP optical storage device application (e.g. Takayanagi et al., 1987), in particular in cholesteric LCPs (Ueno et al., 1987a–c). Ueno et al. (1986), for example, have described an erasable system (without electrical access) addressed by a thermal head rather than by laser for large area display applications. This work is a good example of the exploitation of both polymeric and liquid crystalline properties available in the cholesteric polysiloxane they have studied.

13.2.2 An experimental approach to the evaluation of LCPs for erasable optical storage

The foregoing section has briefly reviewed the laser addressing of LCPs since 1983. The purpose of the present section is to detail some erasable optical storage 'proof of concept' experimental work, conducted exclusively on smectic A siloxane-based LC copolymers. These, together with nematic, cholesteric and chiral smectic C homo- and copolymeric siloxanes, have been undergoing macroscopic and device evaluation in our laboratories since about 1985. Some interesting optical storage results performed on some of the other mesomorphic types have been omitted from the present discussion but will appear in future publications.

The principal goal of our work was to demonstrate all the functions necessary in a

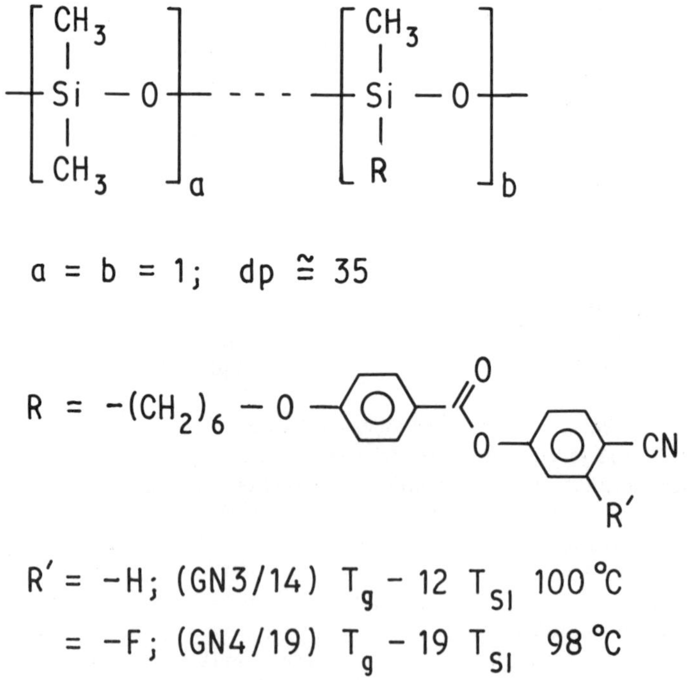

$$a = b = 1; \quad dp \cong 35$$

$$R = -(CH_2)_6 - O - \langle O \rangle - C(=O) - O - \langle O \rangle - CN$$

with R'

$$R' = -H; \ (GN3/14) \quad T_g - 12 \quad T_{SI} \quad 100\ ^\circ C$$
$$= -F; \ (GN4/19) \quad T_g - 19 \quad T_{SI} \quad 98\ ^\circ C$$

Figure 13.1 Structure and phase behaviour of LC copolysiloxane storage materials.

fatigue-resistant optical memory based upon a polymeric liquid crystal. The evaluation was performed using a simple system which could, in principle, be incorporated at low cost in a practical system design.

LC copolysiloxanes less than 100% side chain substituted were the materials of choice because of their ease of handling in sample preparation, and also because their relatively unhindered structures and highly flexible backbones implied a high degree of responsiveness to electric field induced poling and erasure. The structures and phase behaviour of two key materials coded GN3/14 and GN4/19 are shown in Figure 13.1. The methyl and mesogenic groups in these structures are randomly distributed and the polydispersity based upon the unsubstituted backbone and compared against polystyrene standards was 1.5–1.9.

These particular materials were prepared by Nestor from the liquid crystal group at Hull University, UK. Their quality was monitored by atomic absorption spectroscopy, FTIR and electron impact solid probe pyrolysis mass spectroscopy. These analyses checked for colloidal Pt remaining from the hydrosilylation reaction, residual Si-H and alkeneic monomer respectively. Unacceptably high levels of any of these cause problems in device application due to electrical shortage, cross-linking leading to variable phase behaviour and spherulitic crystallization. TGA revealed a 1% weight loss at about 380°C for GN3/14. This method of analysis is more accurate for assessing thermal stability than mass spectroscopy. The stability previously quoted (McArdle et al., 1987) is thus a little optimistic at 450°C. These analyses are important, since thermal modelling laser heating indicates that central zones in addressed bits reach temperatures of these magnitudes.

As part of a macroscopic evaluation programme pertinent to optical storage applications, data on the thermal conductivity, specific heat capacity, density and viscosity of key materials were obtained, some data for GN3/14 are set out below.

	70°C	110°C
Thermal conductivity ($Wk^{-1} cm^{-1} K$)	1.4×10^{-3}	1.6×10^{-3}
Specific heat capacity, c_v ($J cm^{-3} K$)	7.2	9.15
Viscosity (cp) at 76°C.	~1200	190

The basic optical storage devices we employed worked exclusively in the transmissive mode and used glass substrates. The cells were of very simple design, and no attempt was made to optimize their thermal characteristics at this proof-of-concept stage. Subsequently, however, the devices have been thermally modelled (Evans et al., 1987) and some work on plastic substrates has been started. The LCPs were confined between 1 mm thick sodalime glass plates coated with indium tin oxide (ITO) with a 700 Å SiO_2 barrier layer between the glass and the electrode material in a traditional LCD sandwich-type arrangement. The ITO sheet resistance ranged from 50–100 Ω/square. The cell cavity spacing was controlled by chopped optical fibre spacers (8 or 10 μm) deposited by the usual techniques over all but the working electrode area (~1 cm^2) of the cell. No surface treatments were used on cell walls, principally because these were largely ineffective for unannealed LCP samples. Anisotropically conductive seals were used to provide 'through cell' connections so that connections for both electrodes (for the AC field) could be on the same (base) plate, thus some etching was requied on this plate. The top plate of the optical storage device was unetched and its uniform ITO layer served to provide a pathway for the passage of DC heating currents which were used to stabilize and regulate the temperature at which the LCP was studied. A value of $50\,\Omega$/square sheet resistance was chosen to accommodate the resistive heating and feedback circuitry. The accuracy of such a system is about $\pm 1°$, which was adequate for these experiments. Improved accuracy would have necessitated thermal lagging, with concomitant restrictions on input and output optical apertures. Figure 13.2 shows how the LCP devices were optically and electrically accessed on the optical bench. This arrangement was mounted on a horizontal translation stage and proved very convenient for repositioning a given cell to a new sampling area, as well as for the interchange of cells without disturbing the optics. Cell filling was performed from the molten dyed LCPs *in vacuo*. Typical fill times were about 40 minutes for a 3–$5\,\text{cm}^2$ cell, depending on LCP type. This included the sample outgassing period.

The initial recording condition in our positive contrast writing mode work was exclusively the homeotropic state. This was produced by electric field aligning the samples as they cooled from the melt, and required $\sim 75\,\text{V}_{\text{rms}}$ at $3\,\text{kHz}$ (sinewave).

The instrumentation used for studying the laser addressing of LCP optical stores is shown in Figure 13.3.

The equipment shown in Figure 13.3 uses a single beam approach (cf. Lee *et al.*, 1982) to address the test medium and gather information on write, read, selective and bulk erase, grey scale modes and fatigue characteristics. The source laser operating cw at $632.8\,\text{nm}$ is modulated by an analogue-driven acousto-optic modulator (working in first order), about the writing threshold for the sample.

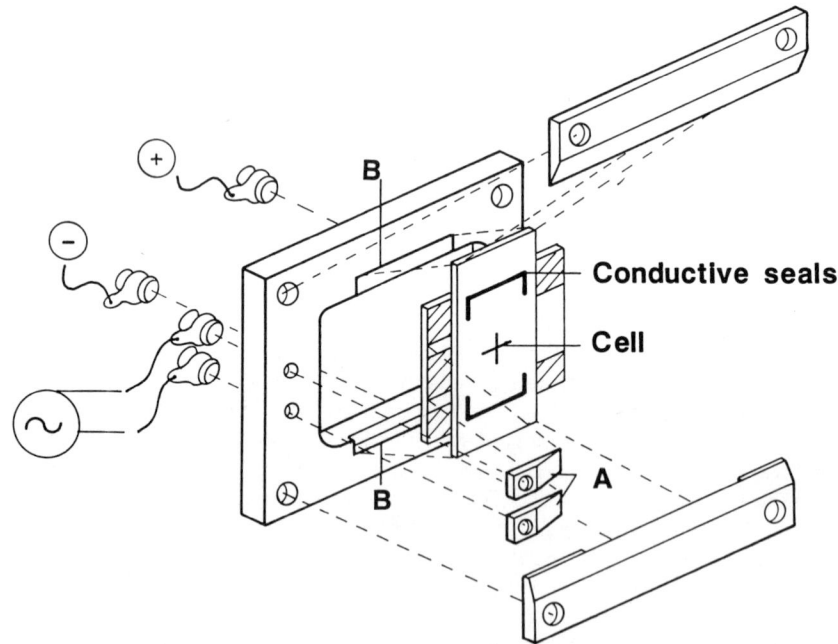

Figure 13.2 Exploded view of LC test cells with optical and electrical access for both resistive heating (DC current) and erase fields (AC field). The DC current is applied via copper inserts *B* in the perspex holder, whilst the AC field is applied via two Be–Cu spring contacts. *A*.

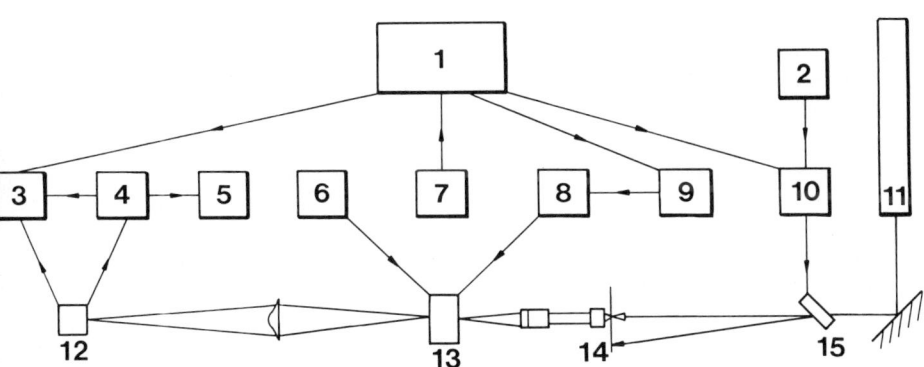

Figure 13.3 Experimental optical set-up for evaluation of LC optical stores: (1) control electronics; (2) pulse generator; (3) XY recorder; (4) digital storage oscilloscope; (5) microcomputer; (6) temperature controller; (7) function generator; (8) auxiliary oscilloscope; (9) high-voltage amplifier; (10) RF driver; (11) laser; (12) photodetector; (13) LC samples; (14) spatial filter; (15) acousto-optic modulator.

The subthreshold condition is used as a monitor or read level. The system was adjusted so that the laser power ratio between the read and write phases was about 1:10. The writing lens used was a × 40 Nikon extra-long-working-distance objective with 0.5 NA and a 6 mm aperture. It was found necessary to purify the output beam from the modulator by spatial filtering before filling the aperture of the objective. Laser power at the LCP was always ⩽ 10 mW. A simple aspheric lens was used to collect the output light cone from the sample, integrating over its entire aperture and imaging on to a red extended photomultiplier detector.

Before describing experiments on the dynamics of laser-addressing LCPs, it is noteworthy to mention some simple experiments relating to the formation of scattering textures and the effects of cooling rates. In this connection we found LC copolysiloxanes with less than 100% full mesogen substitution to behave differently to some of the homo- and copolymeric LC siloxanes with full substitution reported in the literature. We could, for example, cool our copolymers at rates of up to $1 \, \text{K s}^{-1}$ under an electric field and still produce the homeotropic state. This, however, is not the case with some LC polysiloxanes (Attard *et al.*, 1986). Heating then fast cooling our samples in the absence of a field produced dense scattering; the faster the rate, the denser the scattering textures. On the other hand, slow cooling the LC copolysiloxanes (e.g. GN3/14) produced clear samples which were composed of large focal conics which do not scatter light efficiently. Hughes and Daley (1987) reported that the fully substituted LC copolysiloxanes known as PG296 produced only very weak scattering when rapidly cooled in the absence of a field, but produced the densest scattering textures when slowly cooled over a two-hour period.

Notwithstanding that the quench rate from laser heating in this type of application may be ⩾ $1000°C \, \text{s}^{-1}$ (Armitage, 1981), it would appear from these simple experiments that the semisubstituted LC copolysiloxanes are superior candidate materials for studying the dynamics of laser addressing. It is obvious that for positive contrast (scattering on clear) mode LCP optical stores, the production of dense scattering in the absence of a field and the restoration of the homeotropic state in the presence of a field on quenching from the melt, are essential prerequisites in any material proposed for selectively erasable thermo-optical recording.

The processes of bit formation and erasure impress modulation on the read beam in laser addressing experiments using the apparatus in Figure 13.3 because, in positive contrast writing, the written bit is composed of dense smectic focal conic units which cause direct attenuation of transmission by scattering and some absorption of the probing beam level. This can be appreciated from Figure 13.4, which is a photograph of the read channel output cone intercepted by a screen at the position where the aspheric collecting lens would normally reside (cf. Figure 13.3). The situation before and after writing is shown in Figure 13.4a and b respectively. In our laser evaluation experiments, the inter-event period between write and erase (etc.) was a minimum of 750 ms, and contrast ratios of $\geqslant 50\%$ were normal. We have elected to use low data rates, i.e., relatively long inter-event periods, to test and ensure bit stability. It is known, for example, that bit formation is not instantaneous but requires development (Birecki et al., 1983) and experience tells us that small written features may not be stable but may contract and even disappear some time after first formation. The latter phenomenon, known as spot or line shrinkage, has been identified as a problem with some LMMLC materials, and presumably arises from a combination of their tendency to realign under the influence of both the surrounding well orientated bulk matrix and the cell walls.

Figure 13.5 is a typical read channel output signal over four write/erase cycles obtained for polymer GN4/19 using the experimental arrangement of Figure 13.3. The 200μ s write pulse comes on at point A, and bit formation is seen as an attenuated signal as described previously. Another write pulse of equal duration (point B) combined with application of an electric field (60 V_{rms}, 3 kHz sinewave) causes bit erasure and

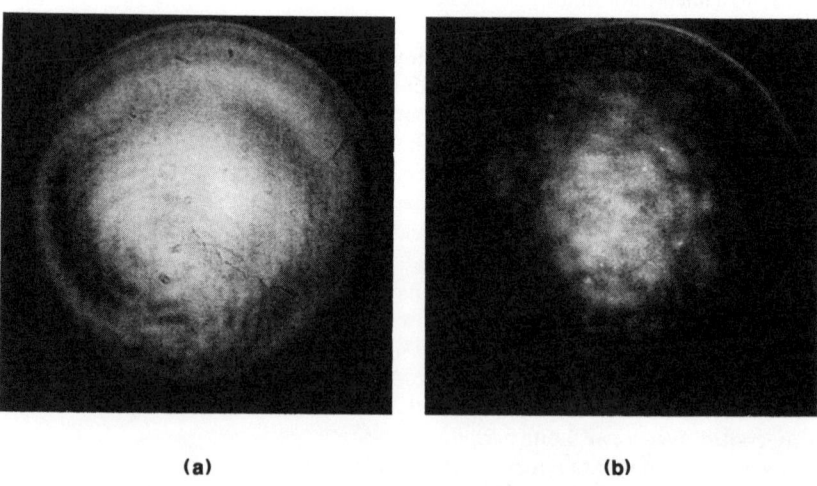

(a)   **(b)**

Figure 13.4 Photographs of the output light cone in the read channel at a position equivalent to the full aperture of the aspheric collecting lens in Figure 13.3. (a) Situation before writing, read beam passing through homeotropic blue dyed GN4/19; (b) Situation after writing, read beam attenuated by scattering spot. The actual intensities recorded in the photographs are misleading because a 40-s exposure was used in both cases (see text).

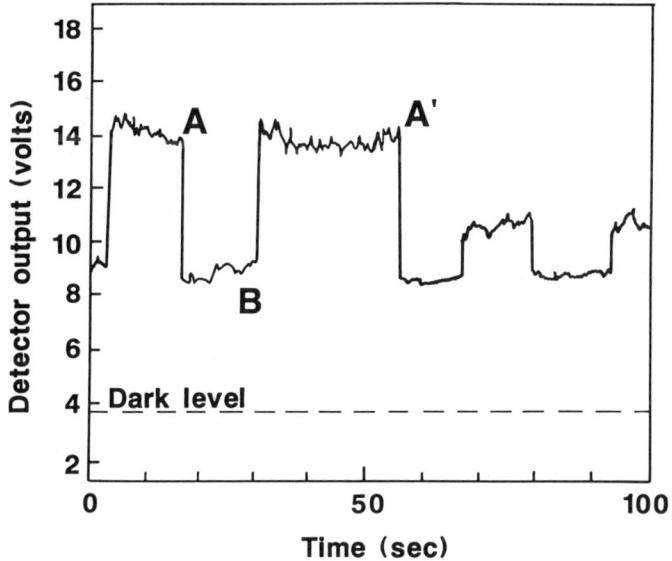

Figure 13.5 Typical output read channel for GN4/19. A 200 μs write pulse was applied at point A. An identical pulse was applied at point B together with a 60 V(RMS) 3 kHz sinewave. Point A' represents the start of a second cycle. The third and fourth cycles were obtained as above, except the erase field had half the magnitude. Experiment performed at 55°C.

restoration of the original level. In the third and fourth cycles, the magnitude of the electric field was halved during erasure to produce a grey scale effect.

On the basis of the resuts in Figure 13.5, it occurred to us that it would be of interest to study the reproducibility of the write and erase cycles, generated in a random fashion, by superimposition of read channel signals in time. Thus by observing read beam attenuation on bit formation (binary 0) and its restoration on bit erasure (binary 1) it is possible to generate a pseudo-random bit stream, which upon superimposition in time generates a digital 'eye' pattern describing the material's response to *all possible* combinations of binary logic sequences. The digital eye diagram is a well-established concept for evaluation of read channel quality in both WORM optical disk technology (Bouwhuis *et al.*, 1985; Heemskerk *et al.*, 1984) and in digital communications systems (Feher, 1981). In the former case, eye patterns are used to evaluate the playback channel quality by observing the detector output on an oscilloscope synchronized with the clock of the bit stream. In the latter case, where temporal rather than spatial constraints prevail, eye diagrams are used with smoothing and encoding techniques to monitor optimization of data bit rates with a minimum of crosstalk in the digital channel.

The nature of our experiment is slightly different. In our case the dynamic nature of the experiment originates from repeatedly writing and erasing a data bit at the same location on a static sample, in constrast to WORM systems where the information itself is static (e.g. pits or bumps), but the action of rotating the sample (disk) impresses modulation on the probe beam generating the playback signal.

Figure 13.5 can be broken down into three variable binary bits corresponding to three periods, e.g. (i) before write, (ii) after write/before erase and (iii) after erase. Following the convention used in optical signal processing, based on variable light intensities (Elion *et al.*, 1984), the point to A' in Figure 13.5 would correspond to the 101 sequence.

In order to describe all the possible arguments sets for three variable binary bits, the eight-row table, shown in Table 13.1, is required. Table 13.1 also shows schematically the outputs from a photodetector receiving the modulated read beam. On continuously repeating these sequences we statistically average to equal numbers of write and erase events for a large number of cycles. Superimposition of the detector outputs then generates an 'eye' pattern with three pupils (cf. idealized case in Table 13.1) which can be used to highlight several features.

We have designed some control circuitry which allows automated writing and erasing on to LCPs in a pseudo-random fashion. At the heart of this circuit is an 8 kbyte EPROM which contains all the data necessary to control and synchronize erase field parameters, laser pulses and enable superimposition of output read channel data in the eye diagram format. A full description of the function of individual controllers is described elsewhere (McArdle *et al.*, 1989).

The results of an eye diagram experiment conducted on LCP GN4/19 at 55°C using a 200 μs pulse for writing and erasing are shown in Figure 13.6 for 500 repeat write/erase cycles. It can be seen immediately that the written in and erased out states do not overlap exactly on any one curve under these rigorous test conditions. This test is

Table 13.1 List of argument sets for three variable binary bits with schematic detector outputs. 1 corresponds to bright (erased) condition, 0 corresponds to dark (written) condition

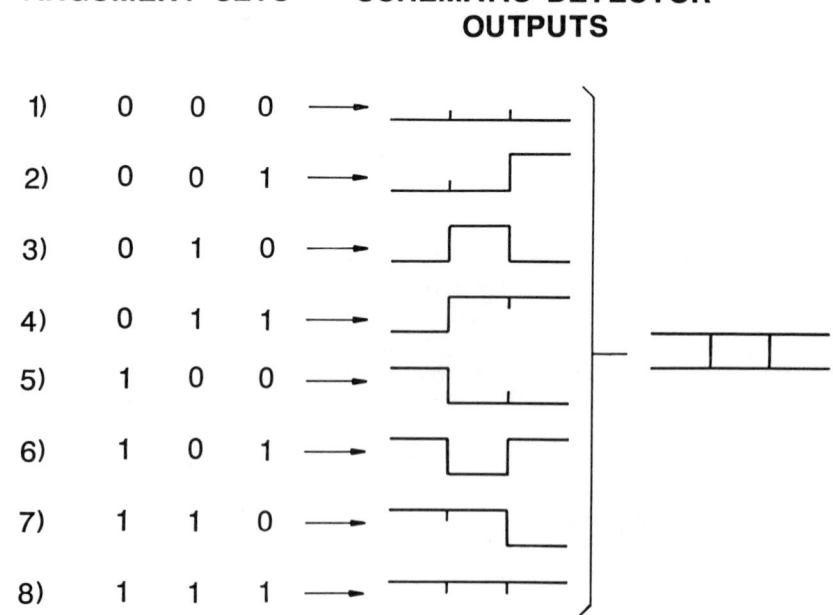

representative of what might be expected of these media in real-life operation, whether for digital or analogue applications, since it examines all possible combinations of binary logic with the previous history of the location under test changing continuously throughout the experiment.

The eye diagram format shown in Figure 13.6 also highlights several other features. For example, it shows up the best and worst cases for written and erased bits and indicates that the kinetics are comparable for both processes, erasure being slightly faster than writing. This is not the case for many optical storage media as will be discussed later. Under the conditions described for an LCP, the change from 10 to 90% levels in the erase cycle takes approximately 200 ms. The magnitude of the read signal is very large, typically 8 V for the erased state compared with a written state magnitude of under 2 V. The eye diagram thus also gives an indication of where decision levels might be placed within this range and show clearly that differentiation between two states could be performed on much shorter timescales using a fraction of the full-scale detector output signal. It can also be seen that true equilibria for written and erased states are established within the 750-ms sampling period we have chosen. Attempts to perform the experiments without ensuring equilibrium give meaningless results, and care should be taken to adjust the sampling period to suit cell parameters, measuring temperatures, etc. It is also important to ensure that the LC experiences the electric field to the maximum extent on cooling during the erase cycles. In this connection we have experimented with different length electric field pulses gated just before the high intensity laser pulse comes on and continuing for some time after laser pulse cessation. We are thus confident that the LCs are 'seeing' the field to the full extent on cooling.

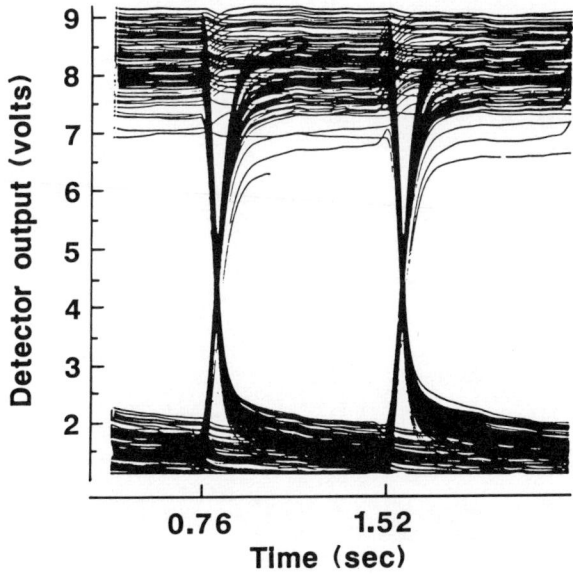

Figure 13.6 Digital eye diagram for LCP GN4/19 dyed to 3% wt/wt recorded after 500 write–erase cycles. Conditions employed were: write pulse = erase pulse = 200 μs; temp = 55°C. Erase field 60 V_{rms}, 3 kHz sinewave applied just before laser pulse and continued for approx. 300 ms thereafter.

As the eye diagram experiment progressed, we noticed a contraction of the digital pupil height and have interpreted this as fatiguing of the material. We shall discuss this fatigue and its probable mechanism later.

Perhaps unique amongst erasable optical recording media is the ability of LCs to show grey scale (Hareng *et al.*, 1975). This may be achieved in either the writing or erasing cycle with appropriate control. The reproducibility of grey scale can also be texted using our eye diagram analysis. A single intermediate grey level of, say, half the voltage required for complete (selective) erasure will generate a third logic level, thereby producing 3^3 possible argument sets, following the principles outlined already. Table 13.2 indicates the additional arguments sets required when testing for one extra

Table 13.2 List of additional argument sets required to describe three variable binary bits with an extra logic level and the resulting detector outputs. 1 corresponds to bright (erased) conditions, 2 corresponds to grey (intermediate) condition and 0 corresponds to dark (written) condition

ARGUMENT SETS   SCHEMATIC DETECTOR OUTPUTS

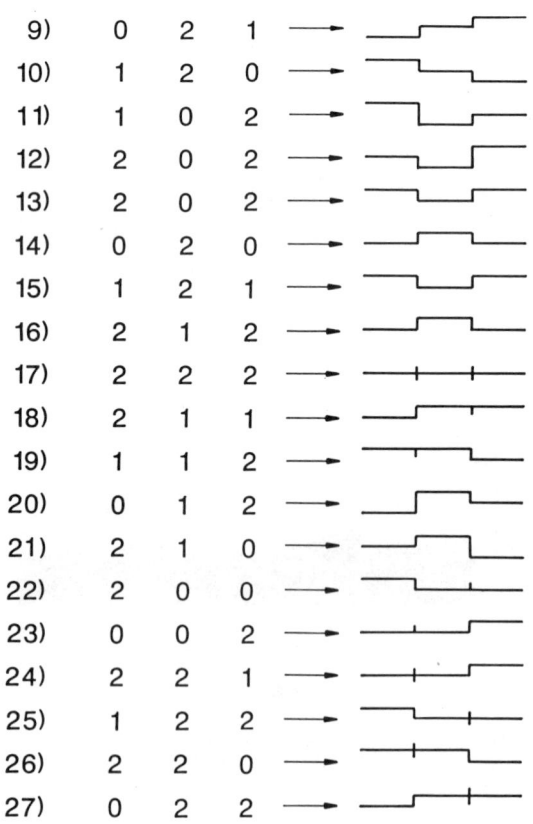

9)	0	2	1
10)	1	2	0
11)	1	0	2
12)	2	0	2
13)	2	0	2
14)	0	2	0
15)	1	2	1
16)	2	1	2
17)	2	2	2
18)	2	1	1
19)	1	1	2
20)	0	1	2
21)	2	1	0
22)	2	0	0
23)	0	0	2
24)	2	2	1
25)	1	2	2
26)	2	2	0
27)	0	2	2

level. A combination of Tables 13.1 and 13.2 then describes how all possible states may be randomly sampled. Superimposition of read channel data in this case generates a six-pupil eye diagram. As the number of logic levels to be tested increases, we impose increasing demands on electronic memory. In this connection it is worth pointing out that the m-sequencing technique commonly used by microwave engineers would prove

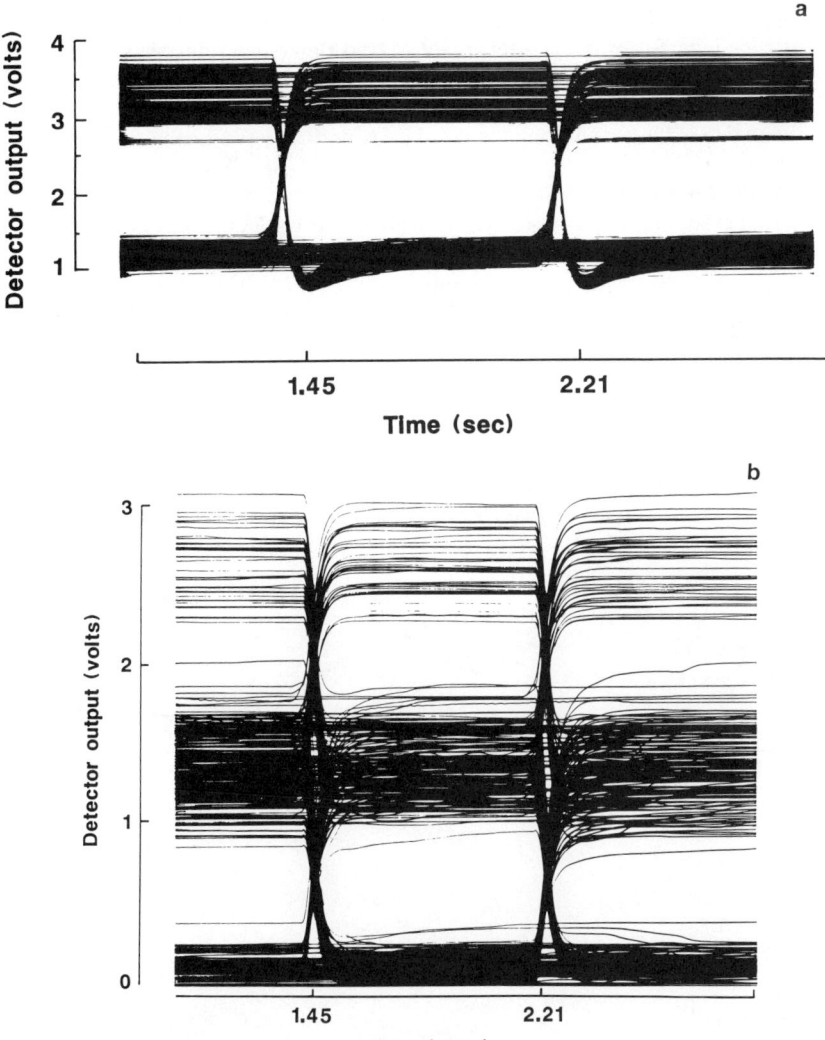

Figure 13.7 Digital eye diagrams for LMM S2 containing 3% wt/wt of dye. (*a*) Three-pupil eye diagram after 1000 write-erase cycles; (*b*) Six-pupil eye diagram after approx. 400 write–erase cycles. Conditions employed were: write pulse = erase pulse 100 μs; temperature 24°C. Erase field = 80 V_{rms} full erase and 30 V_{rms} at 3 kHz in a sinewave (intermediate erase). Field applied just before laser pulse and continued for $\sim$ 300 ms thereafter. Note that in this experiment the intermediate level has been sampled more frequently than the upper one.

useful. This technique makes use of simple linear feedback shift registers to examine 2^{n-1} argument sets, where n can be large.

Figure 13.7 shows eye diagrams for the LMMLC optical storage medium S2 (BDH, Poole, UK). Figure 13.7b shows a result with one grey level. The conditions used were 100 μ s pulses to cause events with 80 V_{rms} for full and 30 V_{rms} for intermediate erasure, both using a 3 kHz sinewave. The six pupil eye diagram is a novel result for erasable optical storage media, although multilevel eye diagrams are well known in digital communications, for example with modified and unmodified duobinary techniques (Feher, 1981). For LCP GN4/19, clear distinction between the three levels was more difficult to sustain on prolonged testing, and much more work is required to gain an understanding of these results and on how best to control reproducibility of grey scales in LCPs.

Regarding the further interpretation of eye diagrams, we may look for parallels in the literature pertaining to digital optical recording and CDROM technology to assess what further information might be extracted. Eye height as a function of time is commonly used to evaluate read channel quality in such applications whilst eye breadth gives information about bit edge acuity (Bouwhuis *et al.*, 1985). In our case, we have only considered the eye height as a parameter, since our samples are static. We have plotted eye height as a function of cycle number to highlight material fatigue for both S2 and GN4/19; the results are shown in Figure 13.8.

At first sight it would appear that the polymer was more resilient than S2, since the eye pupil remains open for longer. However, the sensitivity of the S2 sample necessitated adjustment of experimental parameters with the result that the power densities were unequal in the two cases (15.3 kW cm^{-2} for S2 and 11.8 kW cm^{-2} for

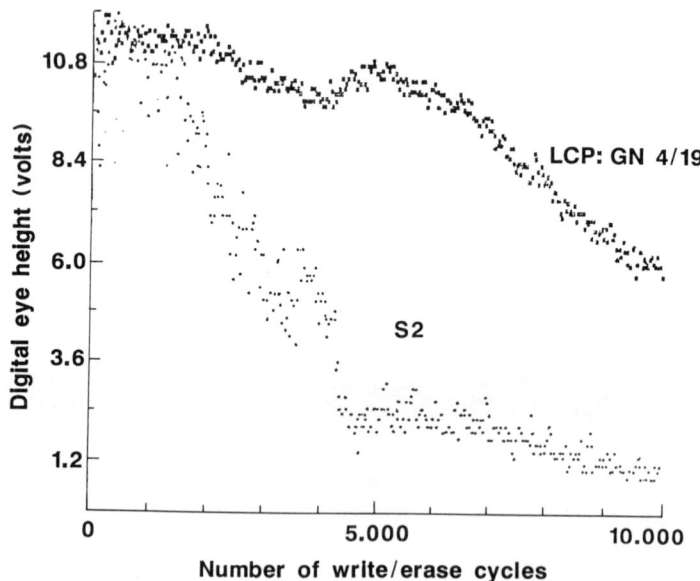

Figure 13.8 Plot of digital eye height against number of write–erase cycles for GN4/19 and S2. Conditions employed are described in the text.

GN4/19) and hence a direct comparison cannot be made. Nevertheless, Figure 13.8 shows that both LC materials fatigue when dyed with a highly absorbing pleochrome ($\varepsilon_0^{633} \sim 5 \times 10^4 \, l \, cm^{-1} \, mol^{-1}$) and addressed at 633 nm. We have conducted follow-up experiments using the eye diagram technique to investigate systematically the effect of changing power densities, absorption coefficients, etc., for S2 and found that no fatigue at all was observed after 2×10^4 cycles on S2 at 23°C when it contained a dye with a molar decadic absorption coefficient approximately five times lower than that described above. If dyes are to be used, it is clear that a judicious selection should be made, and that other factors in addition to absorption coefficient may also be important (Sporer, 1987). Birecki *et al.* (1983) have shown that coupling laser energy into an undyed smectic LC similar to S2 via a metal absorber electrode in thermal contact with the matrix, produces no fatigue after 9×10^3 cycles. All this evidence suggests that the observed fatigue is due to resonant absorption into the LC matrix. This is hardly surprising when one considers that we have both scattering and absorption in these matrices, and are using average power densities of $13 \, kW \, cm^{-2}$, together with large electric fields (approximately $8 \, MV \, m^{-1}$) for up to 50% of the time.

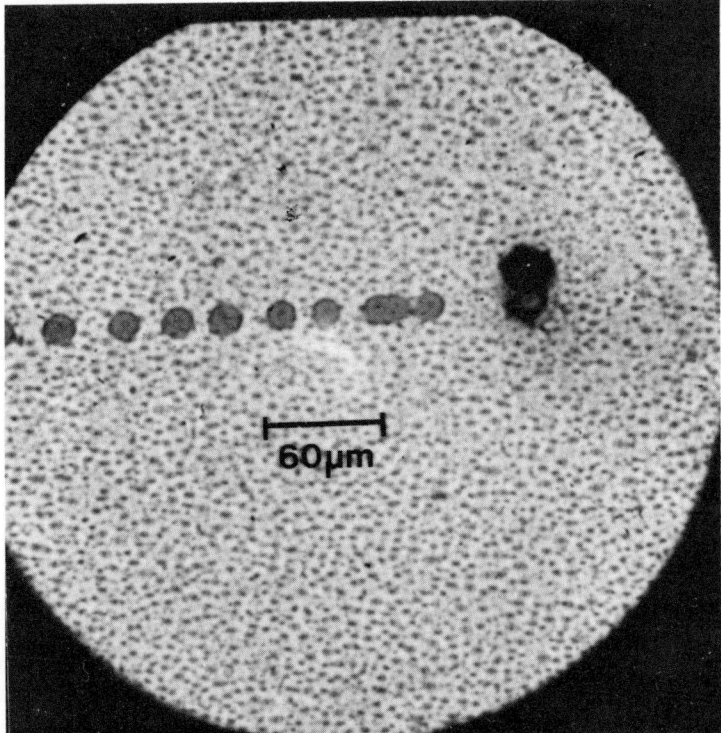

Figure 13.9 Photograph of once-written data on GN4/19 for positive contrast writing compared to the result for a multiply-written and erased bit at the extreme right of the track. Conditions employed as detailed in caption to Figure 13.6. Background speckles are defects (1–2 μm) in the homeotropic sample.

It should be noted that the laser damage thresholds for optical grade polymers for single-shot experiments are reported to be in the order of $170\,kW\,cm^{-2}$ for $300\,\mu s$ pulse widths, whilst multishot testing may reduce this level to 10% of the original value (O'Connell et al., 1983). Furthermore, resonant absorption and scattering in polymers is known to initiate fatigue (Aldoshin et al., 1979; O'Connell et al., 1983). Figure 13.9 gives a direct comparison of the condition of once-written data and a damaged bit written and erased on LCP GN4/19 some 12 000 times. For the sake of completion, we note that the energy damage threshold for ITO electrode materials is $2.2–3.1\,J/cm^2$ (Pawlewicz et al., 1979).

It should be noted that the fatigue effect described applies when the medium is written and selectively erased in the same location many times without any bulk erasure. Bulk erasure always has a restorative effect on a fatigued location, sometimes reinstating it to a completely unfatigued condition. However, further systematic studies of this effect are required (cf. self-healing process alluded to by Pinsl et al., 1987a).

From the foregoing results, we conclude that we have demonstrated most of the functions necessary for a relatively fatigue-resistant erasable optical memory based on LCPs. The demonstration of stable long-term high resolution memory on the LC copolysiloxanes has also been shown, and will be discussed briefly at the end of this section. The eye diagram technique we have used has proved useful as a method for studying reproducibility and fatigue in LC media in general, as well as providing information valuable for systems aspects. In principle, the method could be extended to other erasable media, with appropriate modifications. In particular, the elegant static tester proposed by the IBM group (Chen et al., 1985) could implement this method. A summary of the potential uses of the method applied to LCPs and LMMLCs as we see them is given in Table 13.3.

With regard to fatigue, bit stability, contrast ratio and kinetic characteristics, liquid crystalline media compare well with other, more widely advocated, erasable media. Table 13.4 summarizes some important characteristics, taken from the literature (Chen et al., 1985; Clemens, 1983; Barton et al., 1986) for other media, mostly based upon amorphous–crystalline phase change mechanisms. An ideal medium requires high fatigue resistance ($\geqslant 10^6$ cycles), long-term bit stability, comparable write and erase characteristics and high contrast amongst its major virtues. Unprotected Te of course is useless as a storage medium. However, arsenic doping up to a point improves bit stability and fatigue characteristics somewhat (Clemens, 1983). Newer Te alloys have good fatigue resistance but poor stability (Chen et al., 1985). Sb_2Se phase change media are quite impressive, apparently offering good cyclability, stability and impressive kinetic performance. However, the better kinetic characteristics correlate with decreased cyclability (Barton et al., 1986). It is a great advantage if write and erase rates are comparable, since in this case one laser operation is possible without degrading data rates during erasure. It is the slow rate of recrystallization which is largely responsible for the lack of progress with many of the Te-based erasable media. Of course magneto-optic media offer a combination of useful properties and continue to dominate as contenders for erasable optical disks, in spite of their relatively slow data rates and low signal-to-noise ratio; the prospects for improving the data rate for these media, however, seem good (Shieh and Kryder, 1987).

With regard to testing high-information-content analogue storage on the LC copolysiloxanes (GN3/14 and GN4/19), our collaborators, LaserScan Laboratories, UK, produced the data shown in Figure 13.10 from digitized data which was used to

Table 13.3 Summary of the usefulness of the eye diagram technique as it relates to LC erasable optical storage media

Digital 'eye' patterns can be used to highlight the performance of W.R.E.M. (Write, Read, Erase, Many times) LC media in a number of ways.

They can:

1. Indicate the best and worst cases of write and erase

2. Highlight the difference between write and erase kinetics

3. Offer information on the placement of decision levels and indicate the ease of discrimination between 'n' logic levels where $n >= 2$

4. Provide the basic information to systems engineers for development of encoding techniques

5. Establish systems operational characteristics relative to cell parameters, erase field condition(s), tolerable probe beam levels, etc.

6. Provide information on optical system arrangements, ie., use of reflective or transmissive cells, collection of scattered or attenuation of transmitted light, value of Schlieren techniques, etc.

7. Highlight settling times and hence provide conclusions on optical data verification techniques such as D.R.A.W.

8. Enable positive and negative contrast modes to be evaluated and directly compared

9. Offer some conclusions on the effectiveness or otherwise of various surface treatments on cell walls.

Table 13.4 Comparison of some important characteristics for selected erasable optical storage media and LC materials

Medium	Fatigue after: (Write/erase cycles)	Bit Stability	Write pulse / Erase pulse	Contrast (%)
1. $Te_{100}As_0Ge_0$	50	7sec	–	–
2. $Te_{98}As_{1.9}Ge_{0.1}$	–	3hrs	–	–
3. $Te_{96.8}As_3Ge_{0.2}$	20uncapped 5000capped	3wks.	2×10^{-3}	50
4. $Te_{88}As_5Ge_7$	20	Long Term	0.1	–
5. $Te_{87}Ge_8Sn_5$	1×10^6	20wks	1.6×10^{-3}	33
6. Sb_2Se	1×10^6	Long Term	5×10^{-3}	10
Sb_2Se	1×10^4	Long Term	0.25	10
7. Magneto-Optic	1×10^6	Long Term	1	(low)
8. Dyed LCP (GN4/19)	1×10^4	Long Term	1	>80
9. Dyed LC S2	5×10^3	Long Term	1	>80
10. Undyed LC	tested to 9×10^3			

Contrast $= (A_1 - A_0)/A_1$

where $A_1 =$ Signal amplitude of erased state

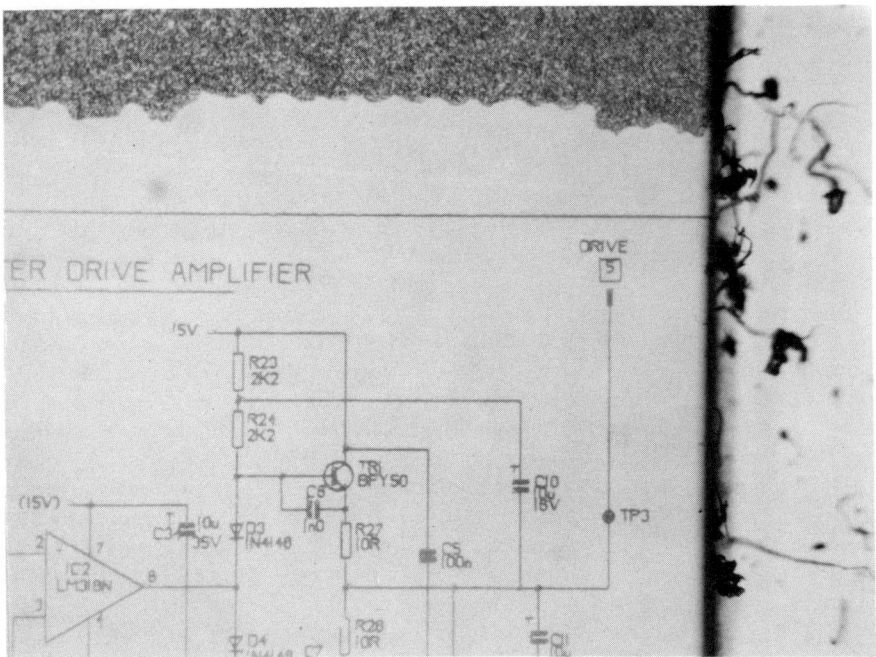

Figure 13.10 Part of a circuit diagram laser-written on to homeotropically aligned and dyed GN3/14. The upper part of the photograph shows the off-electrode region which is unaligned. The cell was filled from the right-hand edge where traces of polymer remain. Note also difference in transmission through homeotropically aligned LCP and ITO coated glass overhang at right edge. This is due to dye absorption in the cell which reduces contrast ratio. (Written data © Laser-Scan Laboratories. Cambridge, U.K., reproduced with permission).

drive a deflection unit to scan our optical stores (see also McArdle *et al.*, 1987). The oldest sample at the time of writing (February 1988) has retained its details for about two years (stored in ambient conditions) and may still be read and used to produce good quality hardcopy images from standard microfiche reader-printers. Graphical data with grey scale achieved by dither have also been produced on GN4/19 samples. In Figure 13.10 the sample was scanned at 24°C at about 60 mm s^{-1}, and the linewidth is about 14 μm^{-2} at 24°C (75% incident power absorbed) with a 2% dye loading (ε_0^{633} for dye $\cong 5 \times 10^4$ l cm^{-1} mol^{-1}), but significant improvements in this figure have been obtained subsequently with different samples and still without optimization of the devices.

These results are encouraging and lead us to the opinion that a likely first LCP product would make further use of the polymeric nature of these materials and be targeted at the erasable analogue storage and display markets where high data rates may not be so important. Evidence to support this assumption is already apparent from the report by Ueno *et al.* (1986), who have used a cholesteric side chain polysiloxane in the form of a recording sheet.

Having detailed the proof of concept experiments pertaining to erasable optical storage on some LCPs, the next stage involved in developing potential applications of these media requires more consideration of how the LCPs might be packaged and addressed in a system. A case for the application of hypothetical LCP products in optical storage is presented in the next section.

13.3 A case for the application of LCPs as optical storage media

In this section the intention is to discuss a strategic approach to the application of LCPs as storage media. Basically five questions will be addressed:

1. What are the virtues of LCPs?
2. How can these be incorporated in product design?
3. What potential has a specific LCP product in the marketplace and are identified markets growing or in decline?
4. What systems constraints will be imposed upon our hypothetical LCP product?
5. What competition exists and what future improvements are required in the LCP case?

13.3.1 *Materials properties available to LCPs*

Liquid crystal polymers are heralded in the literature as unique materials, combining the properties of macromolecules with those of conventional electro-responsive LMMLCs. Furthermore, the ease of polymer processability is often alluded to as an additional materials advantage for LCPs. Whilst the former statement is certainly true, it makes little commercial sense to simply utilize these relatively new media, as mere substitutes for LMMLCs, in applications where the latter offer more than adequate performance. Regarding polymer processability, it should be stressed that the preparation of the high-quality thin polymer films, free from impurities and pinholes, demanded by the electronics industry, is by no means an easy task. Few examples appear in the literature describing attempts to actually exploit polymeric properties in the preparation of LCP samples (Bouligand *et al.*, 1974; Ueno *et al.*, 1986; Pinsl *et al.*, 1987a; Griffin *et al.*, 1988), and in these cases no electrical access to the sample was required. In the case where electrical access is desired, examination of the literature reveals that working electrode areas of $> 1 \text{ cm}^2$ have not been reported, and this is for good reason (cf. Chapter 7, p. 217).

Apparently therefore, with little exception, there are few true examples where the unique combination of properties available to polymeric liquid crystals have been exploited to the full. The notable exceptions to date involve some, but not all, aspects of erasable optical storage, particularly with regard to holography (Eich et al., 1986, 1987a–d), the promising experiments in non-linear optics (cf. Chapter 12) and the conception of a completely new class of material, the LC elastomers (cf. Chapter 10) which have no LMMLC equivalent. With regard to erasable (non-holographic) optical storage, this situation provides the necessity for a strategic approach to product design and device application for currently available LCP materials. In what follows an attempt is made to identify niche areas of application which exploit both the polymeric and liquid crystalline properties of these media to the full.

13.3.2 Incorporation of LCP-specific properties in product design

Side chain LCPs, as we have seen from Chapters 3 and 4, are made up to a large extent from classical polymer backbones, e.g. polysiloxanes, polyacrylates and polyesters. The latter types in particular may have high T_gs and good mechanical properties. Such materials are compatible with the coating technologies developed for the paper and textile industries. Good-quality cholesteric LCP films coated on paper have, for example, been used as lecture demonstrations by Finkelmann. In order to produce an embodiment with both optical and electrical access, however, it is necessary to confine the LCP, either between two optically transparent electrodes (OTEs), or between one metallic electrode and one OTE giving the option of transmissive or reflective device geometry. The OTE may be on a glass or rigid plastic substrate, or perhaps, more interestingly for our application, the OTE may be on flexible (non-birefringent if necessary) plastic foils. Such substrates and their metallized analogues are commercially available, are of high quality, and are less expensive than their glass counterparts.

The confinement of LCPs in all-plastic embodiment raises the obvious possibility of potentially novel recording film applications. In principle we can draw upon the wealth of literature that already exists on the laboratory and reel-to-reel production of both simple and complex plastic LCDs and apply it to LCP product design (Penz et al., 1985; Culley et al., 1980; Umeda et al., 1982; Takahashi et al., 1981; Hirai et al., 1987). Furthermore, the results of recent experiments (Parker et al., 1987) have indicated that the competitive LMMLC storage media cannot usefully be employed in such embodiments. Unlike polymers, these are not sufficiently viscous to prevent serious data corruption when plastic cells containing them are deformed. In this context, correctly designed LCPs offer distinct materials advantages.

At the outset of any product design, sight should never be lost of the compliance of the test media with the official standards imposed upon directly competing media for the same or similar applications. It can thus be instructive to consult relevant standards (e.g. for microfilm) and consider them in LCP product design (Parker, 1987; BS 4210: Parts 1–3, 1977). With regard to optical storage devices confined in plastic foils, the properties of LCPs such as high viscosity, tailorable phase behaviour, compatibility with coating techniques and low solvent power for the plastic substrates make their chances of successful compliance to standards much greater than for LMMLC storage media.

Having identified film-based recording products as a specific area which is best suited to LCPs and which is potentially exploitable, we need next consider what market(s) might exist for such hypothetical LCP devices and attempt to gain some information on the systems required to handle these media. To this end we continue our discussion by considering the Information and Image Processing (I&IP) industry and some aspects of laser scanning technology.

13.3.3 *Potential of film-based LCP recording media*

The basic film-based LCP media we are considering to exploit are addressed exclusively by laser via thermal as opposed to photonic mechanisms. Ideal diffraction-limited operation in a system ensures high recording densities, and hence such media are well suited to micrographics applications in general.

Considered purely as a subset of optical mass storage, traditional micrographics has many advantages, including highly compact storage (1×10^6 documents occupy less than 1ft^3), efficient retrieval mechanisms, human readable forms, low cost, simple back-up techniques, etc. A major disadvantage however, is the lack of electronic output and limited access. These factors have been recognized, and the micrographics industry is currently undergoing change. The convergence of computing, communications, electronics, optics and printing techniques with micrographic storage is an emerging technological area. This has been described in different ways by Scribbins (1985) and the Association for Information and Image Management (AIIM). AIIM have named this broader discipline Information and Image Processing or Management (I&IP or I&IM) (1987). Several groups of technology support I&IM, one of which is Computer Output to Microfilm (COM). COM utilizes data-processing peripherals to permit recording of computer-generated data and/or graphics on to microfilm at high speeds. COM recorders consist of several basic components, of which deflection and display sections and film processing sections are relevant to our discussion.

AIIM market surveys have found not to be true the implicit assumption that microfilm will be made obsolete by the currently emerging Electronic Image Management (EIM) technology based around optical disk systems. Microfilm/microforms have relevant cost and functional advantages which will keep them competitive in many I&IM applications well into the 1990s. The most likely scenario is that microfilm, optical disk and other media will coexist for some time. The cost per Megabyte for optical disk technology will only approach that for micrographic media by 1995. Currently microform masters at $\times$ 48 reduction are about 14 times less expensive, and microform duplicates some 110 times cheaper, than optical disk technology (Coopers and Lybrand Report, 1987).

Table 13.5 has been constructed to highlight the potential attributes of a system using a hypothetical LCP erasable optical recording film as compared with existing technology. From the foregoing discussion it is evident that such a product is compatible with what appears to be a growing high technology market. Furthermore, the LCP film offers a number of advantages and has a higher degree of functionality than other film-based recording media. The hypothetical LCP product is a direct imaging read-write and verify medium. It requires no subsequent processing and can be handled in ambient lighting conditions. The thermotropic LCP film can be integrated into systems utilizing digital laser address techniques such as COM. Consequently, needs such as multiple and/or remote access to image information can be met

System Performance Rating

Attributes of Systems	Optical Based (WORM)	Microform (Conventional)	Hypothetical LCP film
1. Permits high volume mass input	Poor	Good	Poor
2. Media easily duplicated	Poor	Good	Good
3. Established standards	Poor	Good	Poor
4. Legal acceptability of records	Poor	Good	Poor
5. Archival life of record	Adequate	Good	Adequate
6. Images resolution	Good	Good	Good
7. Compaction of Image Information	Good	Good	Good
8. Speed and ease of retrieval	Good	Good	Good
9. Communications capablities	Good	Poor	Good
10. Immediate availability of information	Good	Poor	Poor
11. Ease of updating records	Good	Poor	Good
12. Edit function	Poor	Poor	Good
13. Low cost	Poor	Good	Good

Table 13.5 Comparison of the attributes of hypothetical LCP media and conventional microfilm with WORM optical disk technology (cf. Coopers and Lybrand Report, 1987)

instantaneously on-line through the use of standard, and probably already existing, data communication networks, protocols and terminals. The LCP medium has an edit function, unique in a film-based product, and additionally can offer bulk erasability and the capability of grey scaling (although the last point requires much further experimental evaluation (McArdle *et al.*, 1989; Simon *et al.*, 1986)).

It is important to keep in perspective where such functionality can be fully exploited and in what circumstances it may not be required. Selective erasability is highly desirable in the case when the user wishes to modify records, e.g. in government statistical data, military personnel records, real estate, etc. The possibility of doing this simultaneously and remotely is attractive from the standpoint of convenience and the assurance of proper record updating and is also important for the sale of requisite workstation hardware. For other I&IM applications, however, the edit function may be detrimental. For example, in business and engineering document systems, the re-recording of images to show each modification may be highly desirable as it creates an audit trail or history of design changes. Erasability also raises obvious issues pertaining to legal acceptability of data. The LCP embodied between plastic OTEs is an unusually versatile medium. It may have dual mode functionality offering editability only, or editability then subsequent updatable archivability, since in the latter mode of operation the user need only cut off the electrical access contact pads. This mode circumvents the less stringent legal acceptability issues and simultaneously achieves the maximum degree of flexibility of usage of the LCP film.

In addition to the above utilities, the liquid crystalline nature of the active material in the film, opens up interesting possibilities of novel duplication processes and duplicate functionality by analogy to LC light valve and spatial light modulator technology (Clark and McArdle, 1987; cf. Efron *et al.*, 1983).

Before leaving our discussion on possible LCP outlets in the micrographics industry, it is noteworthy to mention that holographic LCP storage media could also make an impact in this area.

The reader is referred to works of Jamberdino (1981) and Yuan (1987) for an insight as to how holography can be employed in micrographics.

13.3.4 *Systems and media aspects*

Confined in the fashion suggested in section 13.3.2, the hypothetical LCP recording product might be considered as being similar to other recording films, except that it has extra functionality and does not require film processors in its systems architecture. In principle then, an insight as to what might be required in a system based around this product can be gained by considering systems already developed for high resolution imagery (Pierson, 1979; Ulmer, 1979), flying spot laser printer technology (Stark-weather, 1980; Kenville, 1981; Beiser, 1987) and COM systems (Miyauchi *et al.*, 1983; Hauser 1983). This exercise is quite instructive as it also serves to illustrate just how competitive the performance of the LCP medium must be.

Figure 13.11 shows a hybrid block diagram of a laser-based micrographic image generator, according to Mohr (1984) and Gillespie *et al.* (1979), which can be used as a basis for discussion.

As with LMMLC storage media, LCPs can be doped with near-infrared absorbing dyes (cf. Urabe *et al.*, 1984); alternatively and uniquely they may be molecularly designed to selectively absorb energy in this spectral region. Laser diodes thus makes a

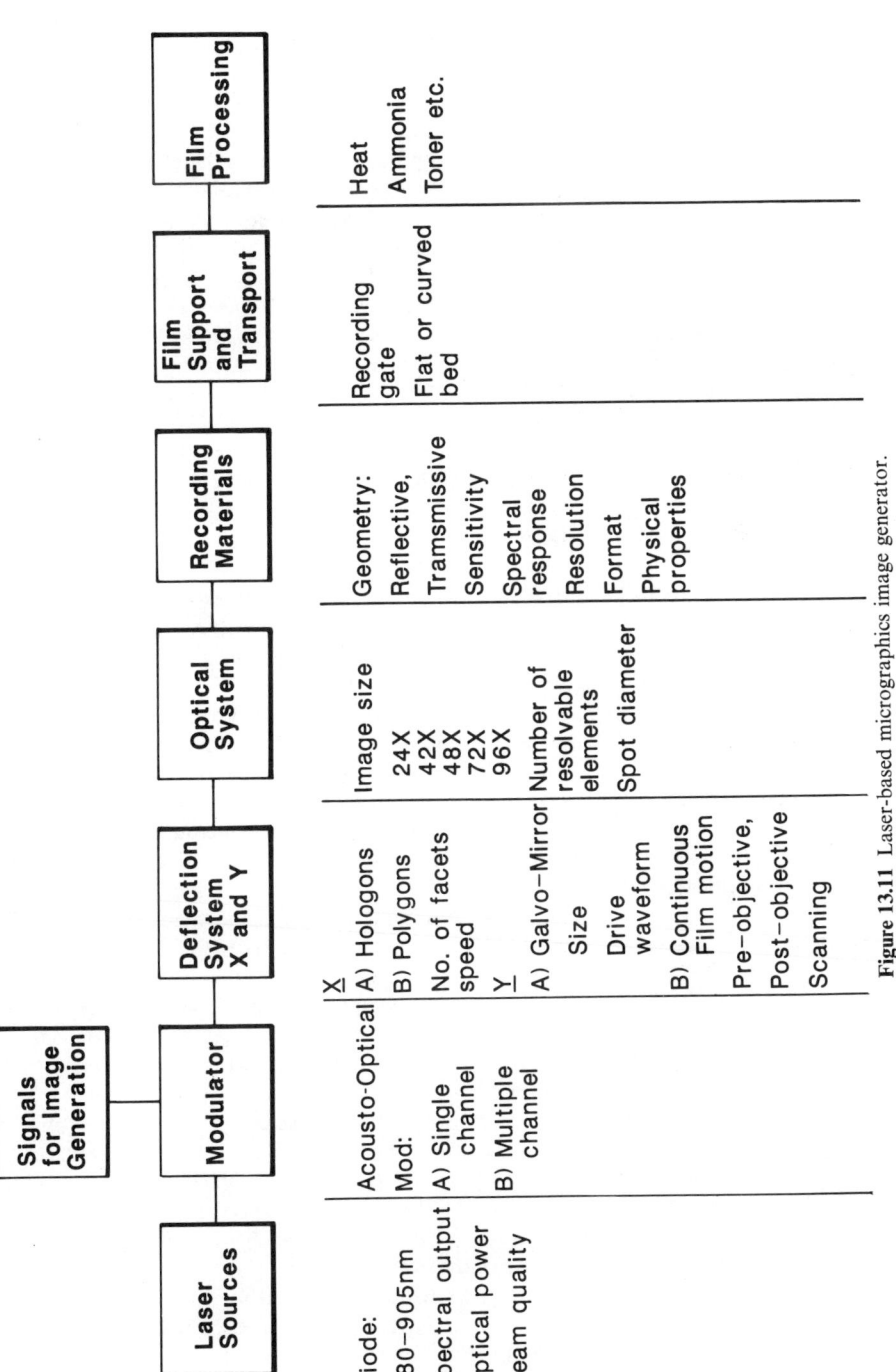

Figure 13.11 Laser-based micrographics image generator.

sensible choice of addressing source. These lasers have many advantages, including direct modulation (hence dispensing with the external modulator in Figure 13.11), low cost and small size (Finck et al., 1980). Recently there have been major advances in diode laser design which are highly significant for erasable optical recording in general. These include improvements in power output, operating wavelength and reliability (Urita, 1986, 1987; Botez 1987; Ueda, 1987) and the combination of two or three lasers on the same chip. Such configurations have been designed for easily alignable optical heads with write-read and write-read-erase capabilities respectively (Hattori et al., 1987; Yamaguchi, 1987). Laser diodes may also be used in multichannel operation to address the recording medium (Dewey, 1983; Carlin, 1984). Laser diodes (20 mW cw), complete with beam-shaping optics, are now commercially available at low cost as spin-off products from the highly successful optical disk technology pioneered by Philips, and high-power products (1 W cw for 805–840 nm) have recently been announced by Sony.

A wide variety of laser beam deflection systems have been described in the literature and comprehensively reviewed by many authors (Marshall, 1987; Herzog, 1987; Beiser, 1984, 1983, 1981, 1974a, b; Starkweather, 1980; Beiser and Marshall, 1976; Jamberdino, 1980, 1979; Dunn, 1980). The discussion which follows is limited to only a few directly relevant examples.

Carlson et al. (1969) have described a very elegant scanning/digitizing system for recording microimages which also comprises a projection display. A system such as this might well be suited to scanning LCP media and merits a detailed analysis. A schematic of the system is given in Figure 13.12.

This configuration employs pre-objective scanning via a rotating polygon mirror. The polygon rotates simultaneously around its C axis and oscillates in the direction indicated by arrow B to produce 1 2-D row-by-row scan of the refocused spot at the field lens (9). The limits of the scan are provided by photodiodes (10, 11, 27 and 28). The scanning pattern at lens (9) is transferred to recording medium (16) via beam splitter (14) and the high numerical aperture recording lens (15). The field lens (9) is designed and located so as to form an image of spot S which fills the entrance pupil of the recording lens, thereby ensuring minimum spot size and high power density. Scanning lens (24) magnifies the 2-D scan at lens (9), imaging the pattern on to a full sized document, in this way the document scanning beam and the recording beam are automatically synchronized. The photodiode (25) is disposed with respect to the document to receive scattered light reflected from the page and its output is subsequently used to drive Pockels cell modulator (4). Rotational polarization variations are only converted to amplitude modulation after passing analyser (13).

In this system, substantially the entire output laser power is converted into a tightly focused spot of 2 μm diameter at the recording medium. The recording field in this case is flat and has dimensions of 3 mm × 3 mm, constituting 2.25 Mbits of information. The total line length per frame amounts to 4.5 m and the scan speed per frame is < 2.5 s. From the data available (Carlson et al., 1969) and assuming 30 mW laser power into a circular spot of $\pi \mu m^2$, the sensitivity of the examplar dye in polymer medium cited by Carlson et al. (1969) is around 10 nJ μm^{-2}, which is comparable with the unoptimized sensitivity of some LCPs (McArdle et al., 1987). Clearly then, even early reported scanning systems are capable of producing impressive performance applicable to LCP media. Furthermore, even relatively insensitive materials are capable of faster imaging times than some currently available micrographic products (Wolf, 1981; Hill, 1987).

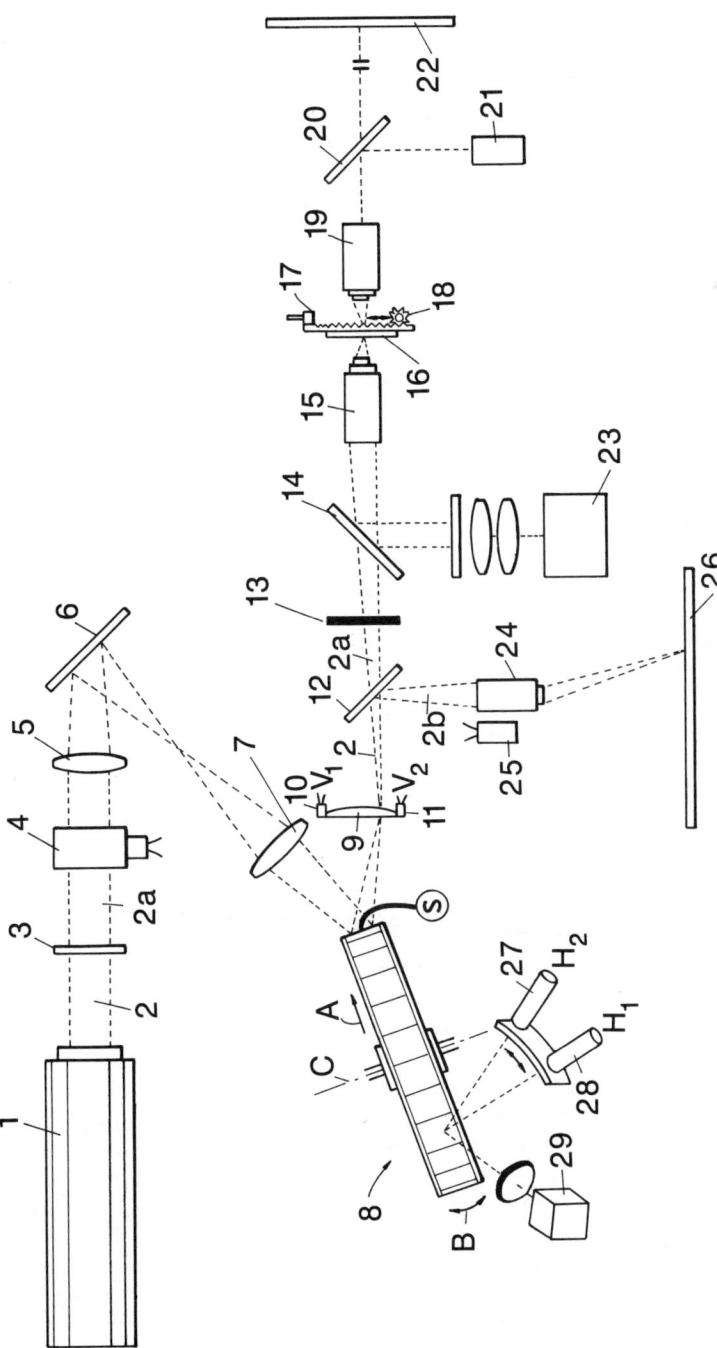

Figure 13.12 Two-dimensional scanning/digitizing system for microimaging (redrawn from Carlson *et al.*, 1969).

The time gains for conventional media due to photographic exposure are offset somewhat by subsequent development and processing; thus thermo-optic media such as LCPs can be competitive, given good system design.

Having illustrated the potential of one well-established system design relevant to LCP application, it is important at this point to examine more closely what standard format requirements are employed in general micrographic products. This is useful, since it enables rough calculations to be performed to provide estimates pertaining to prospective media resolution sensitivity and energy requirements. Conventional roll film, for example, is a continuous length of microfilm on reel in either 16, 35 or 105 mm widths. COM data images may be recorded in several different orientations on roll film in order to accommodate different retrieval systems (Hauser, 1983). These include the $0°$ and $-90°$ rotation modes and the so called 2-up or 4-up 'comic strip' modes. Microfiche is yet another possibility where data is recorded in rows and columns. Irrespective of format, the information recorded on microfilm will have a specific frame size which will be a reduced version of the original document. The standard reduction ratios used are $\times 24$, $\times 42$ and $\times 48$, and a future goal is $\times 96$.

On the basis of the foregoing information we can establish some spatial and temporal requirements of currently available products, then enquire as to whether LCP materials could fulfill similar roles. Table 13.6 describes some requirements which the reader can easily calculate from existing (or anticipated) micrographic, product formats.

Several points arise from Table 13.6. Firstly, we note high data capacities achievable through small bit sizes. The laser spot size for $\times 24$ reduction is easily possible, but the $\times 96$ poses more severe systems constraints and implies higher hardware costs. It is perhaps helpful to provide an idea of what these bit sizes correspond to in real character terms at this stage. The character font used in optical character recognition (OCR) is usually displayed in a bit map comprising of 35 bits disposed in a matrix of 7 rows and 5 columns. Thus characters in the $\times 24$ and $\times 96$ cases occupy about 68 and 42 μm^2 respectively. In reality, the scan densities (and data capacities) quoted in Table 13.6 are overestimates, since a finite space is required between successive blocks of 7 rows. The tolerable size of this gap is important when one considers selective erase implementation and system addressability.

The linear raster scan speed has been arbitrarily selected at $10 \, m \, s^{-1}$, and retrace time, etc. has not been allowed for. It should be noted that much faster speeds have been reported for thermo-optic but non-LCP recording films (Noguchi, 1982). At such rates, the bit dwell time is obviously short, requiring high sensitivity, high available laser power, or both, in the media and system. Again it is noteworthy to point out that high data-rate recording has been reported for photo-optical recording on LCPs (Pinsl et al., 1987a). The Y-advance rate quoted in Table 13.6 relates to film motion or a linear vertical scan. In this system the horizontal raster is provided by a flying spot scan from a spinning polygon which does not simultaneously oscillate about its vertical axis, a much more practical concept (cf. Carlson et al., 1969). Three possible options for provision of a vertical scan are (i) a precision translation stage for film advance; (ii) an additional pre-objective scanner; and (iii) an additional post-objective scanner. The penultimate option would require further complications in flat field optics design. The last option may not be realistic when the working distance of the objective required to achieve such high resolution (e.g. in $\times 96$ reduction) is taken into consideration. The

Table 13.6 Simplified spatial and temporal data for micrographics products under $\times 24$ and $\times 96$ image reduction

Spatial and Temporal Data	Image Reduction Factors	
	x24	x96
A6[a] Format (row x column)	7x14	28x57
Image count	98	1596
Frame size (length x breadth) (mm)	15x10.6	3.75x2.6
Frame area (mm^2)	159	9.75
Fiche data capacity[b] (Bytes)	98x10^6	1.6x10^9
Bit size[c] (μm)	4.45	1.1
Bit density/frame (horizontal)	2.3x10^3	2.3x10^3
Scan density/frame (vertical)	3.4x10^3	3.4x10^3
Total line length/frame (m)	35.72	8.86
Total time/frame[d] (sec)	3.57	0.88
Dwell time/bit (μsec)	0.387	0.110
Data rate (Mbits/sec)	~2.6	~9
Y−Advance rate (mm/sec)	4	4

(a) Standard A6 format = 148mm x 105mm

(b) Assuming 1 frame ≡ 1 A4 page ≡1M Byte (cf Starkweather 1980)

(c) For square rather than circular bit. When square side length = circle diameter, the area of the square bit is 22% greater than for the circular bit. This influences total data capacity.

(d) Assuming a 10m/sec linear raster scan and excluding electronics set up time and retrace.

estimated depth of focus for, say, 850 nm radiation in the × 24 and × 96 cases is about 30 μm and 2 μm respectively (Schreiber, 1976; Hubby, 1983).

Energy and power requirements may also be calculated from Table 13.6 and assuming 10 nJ μm^{-2} as a typical unoptimized LCP sensitivity. To expose an entire frame in the × 24 and × 96 cases requires 1.6 J and 98 mJ respectively which correspond to approximately 445 mW and 111 mW incident powers with the imaging times as indicated in Table 13.6. These high values are in accordance with other thermo-optic recording films such as the Fuji Photo Film product known as LDF (Miyauchi et al., 1983). The LDF product has sensitivity $\leqslant 0.8$ nJ μm^{-2} and requires laser powers P of ~ 120 mW $< P < 1W$ to record high resolution data at 0.25 s page^{-1} (Miyauchi et al., 1983; Noguchi, 1982). We should be encouraged in the knowledge that breakthroughs in GaAlAs lasers have recently been announced and that one-watt cw laser diodes are now commercially available (Laser and Optronics, December 1987) and gains in LCP sensitivity can be made; additionally, thermal management may also be exploited.

Unlike passive recording films, the LCP film is electroresponsive and possesses an edit function. In practical terms, individual bit erasure cannot be realistically considered and the most simple mode of selective erase implementation would likely edit lines at a time, e.g. in 7-line blocks. This would necessitate a rescan operation in the presence of an electric field. The positional control of the vertical deflector is of critical importance with regard to edit implementation, since the system requires high addressability, i.e. the ratio between line width and the number of points per width must be small. Thus for 10 addressable points per line in our example we would require repositional accuracies of 0.44 μm (for × 24) and 0.11 μm (for × 96). Given that the data in Table 13.6 are overestimated for simplicity, such tolerances would be minimal. In any case these accuracies are achievable; however, the 0.11 μm value presents significant system design challenges and increased sophistication in a system is commensurate with increased cost. For horizontal positional control, auxiliary beam methods may usefully be employed (Toyen, 1976; Miyauchi et al., 1983).

Apart from spinning polygonal scanners, other systems have been used to scan film. For example the Kodak Komstar system architecture for COM utilises two galvo-driven mirrors disposed orthogonally to one another to produce line and page scans. Datagraphix Inc. use a hologon for line scans and a low inertia scanner for page scans (Montagu, 1987).

In different applications flexible recording media can offer some system design simplifications. Thus flexible films may be fed through curved recording gates which match the scan locus of a post-objective scanned laser beam. In this respect correctly designed LCP media could be used in applications which are not suited to the equifunctional LMMLC optical storage media. Pre-objective scanning on to a flat bed requires either complex field flattening ($f\theta$) lenses or dynamic focus control (Hartfield, 1973; Montagu, 1987). Figure 13.13 illustrates these points schematically.

It should be reiterated at this point that, apart from raster scanning, laser scanning in the graphic arts frequently employs vectorial techniques. These techniques often utilize straightforward or novel combinations of galvo-driven mirrors. The data presented in Figure 13.10 were produced by vectorial scanning on to LCP samples by LaserScan Laboratories, UK.

Another product design feature which has ramifications in system design is whether reflective or transmissive geometries should be employed. Reflective embodiments offer

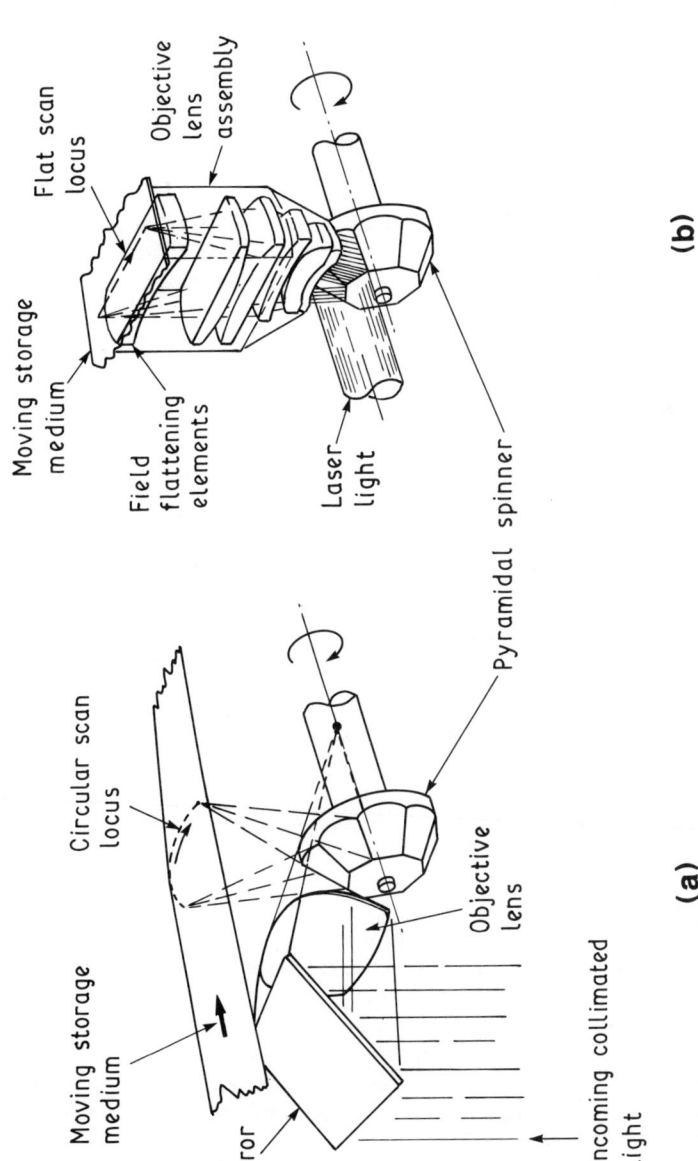

Figure 13.13(*a*) Post-objective scanning on to a curved field; (*b*) pre-objective scanning on to a flat field (redrawn from Biesier, 1974).

Flat scan locus

Objective lens assembly

Moving storage medium

Field flattening elements

Laser light

Pyramidal spinner

(b)

Circular scan locus

Moving storage medium

Objective lens

Mirror

Incoming collimated light

(a)

several advantages and have been favoured by many workers experimenting with LMM S$_A$ LC media (Dewey, 1980; Birecki *et al.*, 1983; Kahn *et al.*,· 1987). The advantages include efficient and wavelength-independent laser coupling into the metal absorber electrode, and less optical constraints on large area display systems with regard to objective working distance. However, contrast ratio may be slightly poorer than for transmissive embodiments unless antireflection coatings are applied, and resolution suffers slightly due to thermal spread in the absorber. Furthermore, complex off-axis projection systems are required (Dewey, 1984). Transmissive geometries invariable require the presence of selective absorbing pleochromes in the LC matrix. Since background absorption degrades contrast ratio, and since too high an initial dye content can give problems with crystallization and disruption of host ordering perfection, there is a practical limit as to the initial pleochrome content. Judiciously designed LCPs may obviate some of these problems, however, since they can be inherently coloured (Ringsdorf *et al.*, 1987).

13.3.5 *Competitive technology and scope for the future*

LCPs are undoubtedly an interesting class of materials which show promise in optical data storage applications, whether as film-based products or media for optical disks (cf. Pinsl *et al.*, 1987a). Their outlets in erasable holographic optical storage look particularly interesting and they can be used to produce holographic optical elements (Eich *et al.*, 1986; 1987a–d).

Obviously LMMLC storage media offer serious competition to LCPs. The former have equivalent functionality, have been extensively researched and have recently reached the marketplace through Greyhawk Systems Inc. (Kahn *et al.*, 1987b; see also *Lasers and Optronics*, June 1987, p. 42; Owyama, 1987). Whilst it is true that LCPs are more amenable to molecular tailoring than their LMMLC counterparts, so that phase behaviour, inherent coloration and other properties can be directly controlled, it is perhaps the disparity of mechanical properties as compared to LMMLCs which account for the most important LCP asset. In principle this leads to film-based applications for LCPs which complement those already existing and adequately covered by LMMLCs, e.g., large area projection display. In film format, LCP media offer much functionality and their integrity for stable image storage is excellent in ambient conditions; this could not be said for a photochromic recording film, for example.

Elsewhere in the LC field, there exists other potential competition for LCPs. Demus *et al.*, (1987a, b; 1985) have recently announced an exciting new class of glassy nematic LMMLCs for optical storage applications. Unlike LCPs, the orientation of these materials can readily be controlled by cell walls, they are monodisperse media and operate with moderate voltages. Some further refinements in their phase behaviour are still required, however. To date (February 1988) the mechanical properties of these media have not been reported. Various workers have experimented with dispersed or encapsulated LMM (nematic) LCs in polymeric binders which are subsequently fashioned into large area flexible displays (e.g. Drziac, 1986). Laser-addressed versions of such displays have been reported (Fergason *et al.*, 1986); however, there may be problems with particle size in the media for high-resolution applications. Vaz *et al.* (1986; 1987a, b) have developed similar technology recently but with submicron nematic, cholesteric or smectic LC droplets dispersed in uv or thermally curable polymers. The dyed smectic versions are obviously interesting for thermo-optical recording.

Laser-addressed high-quality LCP films may find outlets in high-resolution recording, in graphic arts and in COM and CAR (Computer Aided Retrieval) applications. In the latter type of application it should also be noted that the technological competition currently existing in (large) office automation and document image processing systems is fierce. A full description of this technology is beyond the scope of this chapter (see however, Hendley, 1988; Cawkell, 1988; Adrian *et al.*, 1987; Krönert, 1987). Nevertheless, for the small- and medium-sized user, COM/CAR systems offer a cost-effective method of optical data storage, and improvements in media and their functionality are always viewed favourably in such niche areas. Refinements in LCP recording sensitivity, chemical purity and mechanical properties coupled with more systematic studies on the processing of these materials will further the case for application of LCPs storage media considerably.

Acknowledgements

The present work summarizes a part of the GEC Research contribution to a national research initiative between Hull University, LaserScan Laboratories (Cambridge) and GEC, which was partly funded by the UK Department of Trade and Industry. The author is indebted to his many colleagues in all three institutions, in particular Professor G.W. Gray FRS, Gary Nestor and co-workers at Hull, Andy Parker at LaserScan Labs and Mike Clark, Nick Porter, Christine Haws, Anita Russ and Bryn Evans at GEC. Andy Parker is thanked especially for critically reading the MS and for helpful suggestions.

References

Adams, J. and Haas, W. (1971) *J. Electrochem. Soc.* **118**, 2026.
Adrian, R., Davis, C., Russell, T. and Burns, W. (1987) *J. Photogr. Sci.* **35**, 97.
Adloshin, J., Manekov, A., Nechitailo, V. and Pogonin, V. (1979) *Sov. Phys. Tech. Phys.* **49(11)**, 1412.
Alexandru, L., Hopper, M., Loutfy, R., Sharp, J. and Vincett, P. (1984) in *Materials for Microlithography*, eds. Thompson, L., Willson, C. and Frechet, J., ACS Symp. Ser. **266**, American Chemical Society, Washington DC.
Armitage, D. (1981) *J. Appl. Phys.* **52(7)**, 4843.
Attard, G. and Williams, G. (1986) *Polym. Commun.* **27**, 66.
Barton, R., Davis, C., Rubin, K. and Lim, G. (1986) *Appl. Phys. Letts.* **48(19)**, 1255.
Beiser, L. (1974a) in *Laser Applications*, ed. Ross, M., vol. 2, Academic Press, New York, 53.
Beiser, L. (1974b) *Proc. SPIE* **53**.
Beiser, L. and Marshall, G., (1976) *Proc. SPIE* **84**.
Beiser, L. (1981) *Proc. SPIE* **299**.
Beiser, L. (1983) *Proc. SPIE* **390**.
Beiser, L. (1984) *Proc. SPIE* **498**.
Beiser, L. (1987) *Information Display*, **2**, 17.
Birecki, H., Naberhuis, S. and Kahn, F. (1983) *Proc. SPIE* **420**, 194.
Booth, B. (1977) *J. Appl. Photogr. Eng.* **3**, 24.
Botez, D. (1987) *Laser Focus*, March, 68.
Bouligand, Y., Cladis, P., Liebert, L. and Strzelecki, L. (1974) *Mol. Cryst. Liq. Cryst.* **25**, 233.
Bouwhuis, G., Braat, J., Huijser, A., Pasman, J., Van Rosmalen, G. and Schouhamer Immink, K. (1985) *Principles of Optical Disc Systems*, Adam Hilger, Bristol.
Carlin, D., Bednarz, J., Kaiser, C., Connolly, J. and Harvey, M. (1984) *Appl. Optics*, **23(24)**, 4613.
Carlson, C., Bernstein, H. and Stone, E. (1969) US Patent 3 465 352.
Cawkell, A. (1988) *Inf. Media and Technol.* **21(1)**, 19.
Chen, M., Rubin, K., Marrello, V., Gerber, U. and Jipson, V. (1985) *Appl. Phys. Letts.* **46(8)**, 734.

Clark, M. and McArdle, C. UK Patent Appl. 8 723 394.
Clemens, P. (1983) *Appl. Optics* **22(20)**, 3165.
Coles, H. and Simon, R. (1985*a*) *Polymer* **26**, 1801.
Coles, H. (1985*b*) *Faraday Disc. Chem. Soc.* **79**, 201.
Coles, H. and Simon, R. (1985*c*) UK Patent 2 146 787A.
Coles, H. (1986) in *Fine Chemicals for Electronics Industry*, ed. Bamfield, E., Spec. vol. **60**, Royal Society of Chemistry, London, 97.
Coopers and Lybrand Report (1987) *Information and Image Management: The Industry and the Technologies*, (available from Association for Information and Image Management, Silver Spring, Maryland 20910, USA; AIIM Cat. No D016).
Cormier, M., Blanchard, M., Rioux, M. and Beaulieu, R. (1978) *Appl. Optics* **17(22)**, 3622.
Credelle, T. and Spong, F. (1972) *RCA Review* **33(1)**, 206.
Culley, B. and Surtani, K. (1980) US Patent 4 228 574.
Dabisch, W. (1979) European Patent 0 000 868 Al.
Dabisch, W., Kuhn, P., Müller, S. and Narayanan, K. (1980) European Patent 0 014 826 A2.
Da Costa, G. (1986) *Opt. Engr.* **25(9)**, 1058.
Demus, D. and Pelzl, G. (1985) DD 242 624 A1.
Demus, D., Pelzl, G. and Wedler, W. (1987*a*) *Eurodisplay '87 Proc.*, 71.
Demus, D., Weissflog, W., Pelzl, G., Wedler, W., Dehne, J., Andreas, R. and Weigeleben, A. (1987*b*) DE 3 635 584 A1.
Dewey, A. (1980) in *The Physics and Chemistry of Liquid Crystal Devices*, ed. Sprokel, G., Plenum, New York, 219.
Dewey, A. (1984) *Opt. Engr.* **23(3)**, 230.
Dewey, A., Jacbos, J., Huth, B., Sincerbox, G., Sprokel, G., Juliana, A. and Koepcke, R. (1977) *SID Symp. Dig. '77*, 108.
Dewey, A., Anderson, S., Cheroff, G., Feng, J., Handen, C., Johnson, H., Leff, J., Lynch, R., Marinelli, C. and Schmiedeskamp, R. (1983) *SID Symp. Dig. '83*, 36.
Drexler, J. (1981) *J. Vac. Sci. Technol.* **18(1)**, 87.
Drzaic, P. (1986) *J. Appl. Phys.* **60(6)**. 2142.
Dunn, S. (1980) *Proc. SPIE* **223**.
Dunn, S. (1987) *Proc. SPIE* **759**.
Efron, U., Braatz, P., Little, M., Schwartz, R. and Grinberg, J. (1983) *Opt. Engr.* **22(6)**, 682.
Eich, M. (1987*d*) Unpublished PhD thesis Technische Hochschule, Darmstadt.
Eich, M. and Wendorff, J. (1985) *Makromol. Chem.* **186**, 2639.
Eich, M. and Wendorff, J. (1987*a*) *Makromol. Chem., Rapid Commun.* **8**, 467.
Eich, M. and Wendorff, J. (1987*b*) DE 3 603 267 A1.
Eich, M., Reck, B., Ringsdorf, H. and Wendorff, J. (1986) *Proc. SPIE* **682**, 93.
Eich, M., Wendorff, J., Reck, B. and Ringsdorf, H. (1987*c*) *Makromol. Chem., Rapid Commun.* **8**, 59.
Elion, H.J. and Morozov, V. (1984) *Optoelectronic Switching Systems in Telecommunications and Computers*, Marcel Dekker, New York.
Elmasry, M. (1986) EPA 0 194 747 A2.
Feher, K. (1981) *Digital Communications: Microwave Applications*, Prentice-Hall, New Jersey.
Fergason, J. and Fang, N. (1986) *Japan Display '86 Proc.* post deadline paper PD8.
Finck, J., van der Lakk, J. and Schrama, J. (1980) *Philips Tech. Rev.* **39(2)**, 37.
Gambogi, W., Traynor, L. and Buhks, E. (1987) *Topical Optical Data Storage*, 151, March 11–13, Nevada.
Gillespie, R. and Lee, S. (1979) *Proc. SPIE* **169**, 116.
Goldberg, N. and Fergason, J. (1973) US Patent 3 720 658.
Griffin, A., Bhatti, A. and Hung, R. (1986) *Proc. SPIE* **682**, 65.
Griffin, A., Hall, C., Hoyle, C., Gross, J., Venataram, K. and McArdle, C. (1988) *Makromol. Chem. Rapid Commun.* **9**, 463.
Gupta, M. and Strome, F. (1985) *Topical Meeting on Optical Data Storage*, WBB1-1, Washington DC.
Hass, W., Adams, J. and Wysocki (1969) *J. Mol. Cryst. Liq. Cryst.* **7**, 371.
Haas, W., Nelson, K., Adams, J. and Dir, G. (1974) *J. Electrochem. Soc.* **121(12)**, 1667.
Hareng, M. and Le Berre, S. (1975) *Electron. Lett.* **11**, 73.
Hareng, M. and Le Berre, S. (1976) *J. de Physique* C3, 135.

Hartifield, E. (1973) *Laser Focus*, April, 47.

Hartman, J. and Lind, M. (1987) *Topical Meeting on Optical Data Storage*, 155, March 11–13, Nevada.

Hattori, R., Yagi, T., Yamashita, K., Kagawa, H., Ishii, M., Takamiya, S. and Mitsui, S. (1987) *Proc. SPIE* **740**, 12.

Hauser, C. (1983) *J. Appl. Photog. Engr.* **9(1)**, 47.

Heemskerk, J. and Schouhamer Immink, K. (1984) *Philips Tech. Rev.* **40**, 157.

Heinz, R. and Oehole, R. (1977) *Western Elec. Engr.* **21**, 2.

Hendley, A. (1988) *Inf. Media and Technol.* **21(1)**, 17.

Herzog, K. (1987) *Proc. SPIE* **759**.

Higashiyama, T. (1985) JP 60 236 132 [through *CA Selects: Liquid Crystals* **10**, 104 (1986), entry 177945w].

Hill, K. (1987) *Microfilm and Imaging Systems*, **9**, Jan./Feb.

Hirai, H., Sekiguchi, H., Miyata, S. and Kobayashi, S. (1987) *Appl. Phys. Lett.* **50(13)**, 818.

Hubby, L. (1983) *Proc. SPIE* **390**, 79.

Hughes, A., Daley, R. (1987) *Mol. Cryst. Liq. Cryst.* **148**, 163.

Itoh, U. and Tani, T. (1987) *Topical Meeting on Optical Data Storage*, 147, March 11–13, Nevada.

Jamberdino, A. (1979) *Proc. SPIE* **200**.

Jamberdino, A. (1980) *Proc. SPIE* **222**.

Jamberdino, A. (1981) *Opt. Engr.* **20(3)**, 387.

Kaemf, G. (1985) *Ber. Bunsenges. Phys. Chem.* **89**, 1179.

Kaemf, G. (1987) *Polymer J.* **19(2)**, 257.

Kahn, F. (1973) *J. Appl. Phys. Letts.* **22**, 111.

Kahn, F. (1987) *Proc. SPIE* **760**.

Kahn, F., Birecki, H. and Burmeister, R. (1984) GB 2 090 673 B.

Kahn, F., Kendrick, P., Leff, J., Livoni, L., Loucks, B. and Stepner, D. (1987) *SID Symp. Dig. '87*, 247.

Kaneko, E. (1987) *Liquid Crystal TV Displays*, KTK Scientific Publishers, Tokyo, Chap. 8.

Kato, Y., Imai, S., Isobe, M., Manabe, K. and Nakarai, T. (1986) *Proc. SPIE* **695**, 38.

Kenville, R. (1981) *Opt. Engr.* **20(3)**, 330.

Krönert, G. (1987) in *ESPRIT '87 Achievements and Impact, Part 2*, ed. Commission of European Community Directorate General, New Holland, Amsterdam, 1367.

Kuder, J. (1986) *J. Imaging Technol.* **12(3)**, 140.

Lee, N., Kohanazadch, Y., Shepherd, P. and Chao, S. (1982) *Proc. SPIE* **329**, 186.

Leier, C. and Pelzl, G. (1979) *J. Prakt. Chem. Leipzig* **321**, 197.

Lu, S. and Hochbaum, A. (1987) *SID Symp. Dig. '87*, 367.

Marshall, G. (1987) *Laser Beam Scanning*, Marcel Dekker, New York.

McArdle, C., Clark, M., Haws, C., Wiltshire, M., Parker, A., Nestor, G., Gray, G., Lacey, D. and Toyne, K. (1987) *Liquid Crystals* **2**, 573.

McArdle, C., Clark, M., Evans, W., Porter, N. and Haws, C. (1989) *J. Appl. Phys.* (submitted).

Miyauchi, A., Ohnishi, A., Noguchi, M. and Washizawa, Y. (1983) *J. Appl. Photogr. Engr.* **9(1)**,7.

Moerner, W. (1985) *J. Molec. Electron.* **1**, 55.

Moerner, W., Macfarlane, R. and Lenth, W. *Topical Meeting on Optical Data Storage*, March 11–13, Nevada, 151.

Mohr, S. (1984) *Proc. SPIE* **489**, 148.

Montagu, G. (1987) in *Laser Beam Scanning*, ed. Marshall, G., Marcel Dekker, New York, Chap. 5.

Noguchi, M. (1982) *Appl. Optics* **21(15)**, 2665.

O'Connell, R. and Saito, T. (1983) *Opt. Engr.* **22(4)**, 393.

Ogura, K., Hirabayashi, A., Ueijima, A. and Nakamura, K. (1982) *Jpn. J. Appl. Phys.* **21**, 969.

Owyama, Y. (1987) *Nikkei Electronics* **10(5)** 98.

Parker, A. (1987) *Internal Rept. Laser-Scan Labs., Cambridge, UK*, June [see also BS 4210: Parts 1–3, 1977 and DD27: April 1973 (British Standards Institution, London].

Pawlewicz, W., Mann, I., Lowdermilk, W. and Milam, D. (1979) *Appl. Phys. Lett.*, **34(3)**, 196.

Penz, P., Sampsell, J. and Collins, D. (1985) *IEEE Trans. Electron Devices* **ED32** (11), 2206.

Pierson, P. (1979) *Proc. SPIE* **200**, 100.

Pinsl, J., Bräuchle, Chr. and Kreuzer, F. (1987a) *J. Molec. Electron.* **3**, 9.

Pinsl, J., Deeg, F. and Braüchle, Chr. (1987*b*) *J. Amer. Chem. Soc.* **109**, 6479.
Ralston, L. (1980) *Proc. SPIE* **218**, 54.
Reich, S., Mazur, S., Avakian, P. and Wilson, F. (1987) *J. Appl. Phys.* **62(1)**, 287.
Ringsdorf, H., Schmidt, H-W., Eilingsfeld, H. and Etzbach, K-H.. (1987) *Makromol. Chem.* **188**, 1355.
Sackmann, E. (1971) *J. Amer. Chem. Soc,* **93**, 7088.
Sasaki, A. (1986) *Mol. Cryst. Liq. Cryst.* **139**, 103.
Schmidt, H.-W. (1984*a*) Dissertation Doktor der Naturwissenschaften, Universität Mainz, 72.
Schmidt, H.-W. (1984*b*) DE 3 429 438.
Schreiber, W. (1976) *Proc. SPIE* **84**, 21.
Scribbins, P. (1985) *IEE Colloq. on Optical Mass Data Storage, Dig.* 1985/58, May [Institution of Electrical Engineers, London].
Sheriden, N. (1972) *IEEE Electron Devices* **19**, 1003.
Shibaev, V., Krostromin, S., Platé, N., Ivanov, S., Yu Vetrov, V. and Yakovlev, I. (1983) *Polym. Commun.* **24**, 364.
Shibaev, V., Krostromin, S., Platé, N., Ivanov, S., Yu Vetrov, V. and Yakovlev, I. (1985) *Polym. Sci. Technol.* **28**, 345.
Shieh, H. and Kryder, M. (1987) *J. Appl. Phys.* **61(3)**, 1108.
Simon, R. and Coles, H. (1986) *Liquid Crystals* **1(3)**, 281.
Sincerbox, G. (1977) *IBM Tech. Repts.* RJ1969, RJ1973.
Sincerbox, G. (1987) in *Laser Beam Scanning*, ed. Marshall, G., Marcel Dekker, New York.
Smith, M., Burns, R. and Tsai, R. (1979) *Proc. SPIE* **200**, 171.
Smith, T. (1981) *J. Vac. Sci. Technol.* **18(1)**, 100.
Sporer, A. (1987) *Appl. Optics* **26(7)**, 1240.
Starkweather, G. (1980) in *Laser Applications*, eds. Goodman, J. and Ross, M., Vol. 4, Academic Press, New York, 125.
Takashasi, S., Shimokawa, O., Inoue, H., Uehara, K., Horose, T. and Kikuyama, A. (1981) *SID Symp. Dig.* 86.
Takayanagi, T. and Kawaguchi, H. (1987) JP 62 170 396 [through *CA Selects: Liquid Crystals* 13954n, (1988)].
Takeshima, M. (1984) *Proc. SID* **25(3)**, 219.
Taylor, G. and Kahn, F. (1973) *J. Appl. Phys.* **22**, 111.
Tazyke, S., Kurihara, S. and Ikeda, T. (1987) *Chem. Lett.* 911.
Toyen, G. (1976) *Proc. SPIE* **84**, 138.
Ueda, Y. (1987) *Jpn. J. Electr. Eng.* 51.
Ueno, T., Nakamura, T. and Tani, C. (1986) *Proc. Japan Display '86*, 290.
Ueno, T., Nakamura, T., Tani, C., Hoshino, H., Yoshio, K., Takada, K. and Samura, H. (1987*a*) JP 62 154 340 [through *CA Selects: Liquid Crystals*, 2, 29583b (1988)].
Ueno, T., Nakamura, T., Tani, C., Hoshino, H., Yoshio, K., Takada, K. and Samura, H. (1987*b*) JP 62 157 341 [through *CA Selects: Liquid Crystals*, 3, 46951/n (1988)].
Ueno, T., Nakamura, T., Tani, C., Hoshino, H., Yoshio, K., Takada, K. and Samura, H. (1987*c*) JP 62 157 342 [through *CA Selects: Liquid Crystals*, 3, 29587f (1988)].
Ulmer, D. (1979) *Proc. SPIE* **200**, 114.
Umeda, T., Igawa, Y., Simazaki, T., Miyashita, T. and Nakauo, F. (1982) USP 4 362 771.
Urabe, T., Arai, K. and Ohkoshi, A. (1983) *J. Appl. Phys.* **54(3)** 1552.
Urabe, T., Usai, H., Arai, K. and Ohkoshi, A. (1984) *Proc. SID* **25(4)**, 299.
Urita, K. (1987*a*) *Jpn. Semicond. Tech. Repts.* **2(4)**, 36.
Urita, K. (1987*b*) *Jpn. J. Electr. Eng.* **31**,
Vaz, N. and Smith, G. (1986) EP 0 205 261 A2.
Vaz, N., Smith, G. and Montgomery, G. (1987*a*) *Mol. Cryst. Liq. Cryst.* **146**, 1.
Vaz, N., Smith, G. and Montgomery, G. (1987*b*) *Mol. Cryst. Liq. Cryst.* **146**, 17.
Wolf, D. (1981) *J. Micrographics* 23, August.
Yamaguchi, T. (1987) *Jpn. J. Electr. Eng.* 54.
Yuan, W. (1987) *Proc. SPIE* **747**, 135.

14 Physicochemical studies and analytical applications of mesomorphic polysiloxane (MEPSIL) solvents by gas–liquid chromatography

G.M. JANINI, Department of Chemistry, University of Kuwait, Kuwait City 13060 Kuwait, R.J. LAUB*, Department of Chemistry, San Diego State University, San Diego, California 92182, USA; J.H. PURNELL, Department of Chemistry, University College of Swansea, Swansea, Wales SA2 8PP, UK and O.S. TYAGI[†] Regional Research Laboratory (Council of Scientific and Industrial Research), Hyderabad 500 007, India.

14.1 Introduction

14.1.1 *Historical overview*

From its inception (Martin and James, 1952), gas–liquid chromatography (GLC) has developed into a sophisticated tool that currently enjoys widespread use both in physicochemical studies (Laub and Pecsok, 1978) and in chemical analysis (Novak and Leclercq, 1988). Morevoer, the relevant theory, techniques, and instrumentation are all advanced to the point that, today, gas-chromatographic measurements are carried out routinely that even a few years ago were considered unmanageably complex (Laub and Purnell, 1988).

In particular, the kinds of physicochemical investigations to which GLC has been applied include the assessment of equilibrium partition coefficients of volatile solutes distributed between liquid (stationary, S) and gaseous (mobile, M) phases, together with the associated thermodynamic properties of solution; various transport properties, such as diffusion coefficients and interfacial fluxes; kinetic aspects of on-column reactions, including rate constants and Arrhenius parameters; and various intrinsic molecular properties, such as solute vapour pressures, boiling points, and heats of vaporization. In addition, GLC has also been widely applied in recent years to the study and characterization of the properties of polymers via the technique that has come to be labelled, somewhat misleadingly, as 'inverse' gas chromatography (Aspler, 1985).

At the same time, advances made in analytical instrumentation and techniques have ensured the continued development and expansion of applications of GLC in separations, where the theoretical limits of maximal resolution in minimal time of analysis, coupled with maximal sample throughput, have today largely been reached. This is in no small part a consequence of the ready availability of modern detection

---

*To whom correspondence should be addressed.
[†]On leave to San Diego State University.

systems that are highly sensitive as well as selective, relatively inexpensive, and simple to operate and maintain, e.g., desk-top mass spectrometers. In addition, the advent of open tubular (capillary) columns fabricated particularly from flexible fused-silica materials has revolutionized the efficiency of GC systems, to the point that separations requiring 100 000 or more theoretical plates are regarded at present as commonplace (Lambert et al., 1984).

It is also fair to say at this point that the somewhat empirical aspects of analytical GLC, namely the selection of an appropriate stationary phase and set of conditions for a particular separation, have, at least in principle, been overcome with the plenary window-diagram optimization strategy originated by Laub and Purnell (1975). Their procedures make use of simple graphical techniques for the systematic deduction of global sets of optimal conditions for a particular analysis, as opposed to single optima yielded by other methods (e.g. SIMPLEX) which often give only local minima or maxima. The window-diagram methodology has thereby largely supplanted what formerly amounted to the selection of phases and conditions by an analyst virtually entirely on the basis of his experience, as augmented by what literature reports there might be of the separation to hand. The window strategy has also been extended and amplified over the years such that it is today routinely employed not only in gas and liquid chromatography (Laub and Purnell, 1978), but in addition in spectroscopy (Laub et al., 1979) and electrochemistry (Anderson and Laub, 1981).

14.1.2 Physicochemical basis of GLC separations

The physicochemical basis of analytical GLC separations is conveniently represented by the expression:

$$\alpha_{i/j} = (\gamma_A^\infty f_A^0)_{\text{solute } j}/(\gamma_A^\infty f_A^0)_{\text{solute } i} \tag{14.1}$$

where $\alpha_{i/j}$ is the so-called relative volatility (separation) of solutes i and j, which are identified so as to maintain $\alpha_{i/j} \geqslant 1$; γ_A^∞ is the Raoult's law mole-fraction-based solute activity coefficient at infinite dilution in the stationary phase; and f_A^0 is the bulk-solute fugacity, which is commonly replaced with little associated error by the vapour pressure p_A^0, the two quantities being related at temperature T through the solute second-interaction virial coefficient B_{AA} and the gas constant R:

$$f_A^0 = p_A^0 \exp p_A^0 (B_{AA} - \bar{V}_A)/RT \tag{14.2}$$

The solute activity coefficient is related in turn to the concentration-based liquid–gas partition coefficient $K_R^0 = C_A^S/C_A^M$ via the equation:

$$K_R^0 = RT/\gamma_A^\infty f_A^0 \bar{V}_S \tag{14.3}$$

where C_A^i are the solute molar concentrations in phases S and M, $\bar{V}_S$ is the molar volume of the solvent, and the superscript zero on K_R indicates that the full virial correction for gas phase nonideality has been applied (Laub, 1984) (the effects are negligible with hydrogen and helium carriers).

Since measurement of partition coefficients entails assessment of the column stationary-phase volume V_S and density ρ_S, it is rather more convenient, particularly with polymeric materials, to express eqn (14.3) in terms of what are known as specific retention volumes V_g^T:

$$V_g^T = K_R^0/\rho_S = RT/\gamma_A^\infty f_A^0 M_S \tag{14.4}$$

where M_S is the solvent molecular weight. Further, since, V_g^T in fact corresponds to the volume of carrier (corrected for gas-phase compressibility and the system void space) required to elute the solute peak maximum through the column, it is common practice to convert it to standard temperature:

$$V_g^0 = (273.15/T)V_g^T = 273.15R/\gamma_A^\infty f_A^0 M_S \tag{14.5}$$

In practice, specific retention volumes are calculated from solute retentions, the carrier flow rate, column pressure drop, and so forth, via the expression:

$$V_g^0 = (3/2)(t_R - t_A)F_0(273.15/T_0)[(p_0 - p_w)/p_0]$$
$$\times \{[(p_i/p_0)^2 - 1]/[(p_i/p_0)^3 - 1]\}/w_S \tag{14.6}$$

where t_R and t_A are the retentions of a solute and a nonretained substance (e.g. air); F_0 is the mobile-phase flow rate measured at the column outlet at temperature T_0 and outlet pressure p_0; p_i is the inlet pressure; p_w is the vapour pressure of water (included if a soap-bubble flowmeter is employed): and w_S is the column weight of stationary phase.

Within the approximation that the activities of infinitely-dilute solutes can be replaced by concentrations, the partition coefficient is a true thermodynamic equilibrium constant and, hence, the fundamental equations of equilibrium thermodynamics apply. In the case of GLC the process in question is simply one of a vapour-phase solute dissolving in a liquid solvent, and so

$$\Delta\bar{G}_S/RT = -\ln K_R^0 \tag{14.7}$$

$$\Delta\bar{H}_S/RT = -\ln K_R^0 + C_1 \tag{14.8}$$

$$\Delta\bar{S}_S = (\Delta\bar{G}_S - \Delta\bar{H}_S)/T \tag{14.9}$$

where $\Delta\bar{G}_S$, $\Delta\bar{H}_S$, and $\Delta\bar{S}_S$ are respectively the Gibbs' free energy, enthalpy, and entropy of solution.

In the particular case of polymer solvents it is more convenient to measure V_g^0 rather than K_R^0 as mentioned earlier, for which some modification of eqns (14.7–14.9) is required (Littlewood, 1964):

$$\Delta\bar{G}_S' = \Delta\bar{G}_S - RT\ln(T\rho_S/273.15) \tag{14.10}$$

$$\Delta\bar{H}_S' = \Delta\bar{H}_S + RT - RT^2\eta_S \tag{14.11}$$

$$\Delta\bar{S}_S' = \Delta\bar{S}_S + R - RT\eta_S + R\ln(T\rho_S/273.15) \tag{14.12}$$

where the C_i are constants, and η_S is the coefficient of expansion of the solvent:

$$\eta_S = -(1/\rho_S)(\partial\rho_S/\partial T)_p \tag{14.13}$$

Since, for polymer liquids, η_S is small (c. 1×10^{-3} deg^{-1}), the respective pairs of quantities $(\Delta\bar{G}_S, \Delta\bar{G}_S')$, $(\Delta\bar{H}_S, \Delta\bar{H}_S')$, and $(\Delta\bar{S}_S, \Delta\bar{S}_S')$ are usually found to be identical to within $\pm 3\%$.

The excess properties of the solute–solvent solution are easily derived from the temperature-dependence of γ_A^∞:

$$g^E/RT = -\ln\gamma_A^\infty \tag{14.14}$$

$$h^E/RT = -\ln\gamma_A^\infty + C \tag{14.15}$$

$$s^E = (h^E - g^E)/T \tag{14.16}$$

A conceptual difficulty arises in the use of eqn (14.5) or (14.6) with polymeric stationary phases, namely, since the solute retentions stay finite γ_A^∞ must approach zero as M_S tends to infinity. For example, those for n-hexane and benzene solutes at 30°C with a PDMS of $c.$ 10^5 Da are of the order of 10^{-3} (Ashworth et al., 1984). Also, there is generally some uncertainty associated with the measurement of M_S, as well as the inevitable degree of polydispersity of polymers. Accordingly, and in an effort to circumvent at least the first of these inconveniences, Patterson, Guillet, and their co-workers (Patterson et al., 1971) recommended the use of weight-fraction-based solute activity coefficients $^w\gamma_A^\infty$, which are related to their mole-fraction counterparts $^x\gamma_A^\infty$ through the expression:

$$^w\gamma_A^\infty = {}^x\gamma_A^\infty M_S/M_A \qquad (14.17)$$

Eqn (14.5) now takes the form

$$V_g^0 = 273.15R/^w\gamma_A^\infty f_A^0 M_A \qquad (14.18)$$

The latter relation has the practical advantage that the solute molecular weight M_A has replaced that of the solvent in the denominator; moreover, it has been shown that the values of $^w\gamma_A^\infty$ for a given solute do indeed approach constancy as M_S becomes large. However, the disadvantage is then incurred that $^w\gamma_A^\infty$ can assume virtually any positive value, even in the instance of ideal solutions, depending upon the ratio of the solute and solvent molecular weights. That is, the significance of an activity coefficient of unity is thereby lost.

It is also possible to show that the partition coefficient and specific retention volume are related to what is known as the solute capacity factor $k' = (t_R - t_A)/t_A$:

$$k' = (w_S/V_M)V_g^T = (T/273.15)(w_S/V_M)V_g^0 = (V_S/V_M)K_R^0 \qquad (14.19)$$

where V_M is the column volume of carrier gas, and where the quotient V_M/V_S is known as the phase ratio β. Eqn (14.19), which in various forms is applicable both to GC and LC, is often called the 'fundamental' equation of chromatography insofar as it relates the solute distribution between phases S and M, viz., the equilibrium constant K_R^0, to the practical experimental makeup of the system, namely, the phase ratio. Moreover, separations can be expressed equally in terms of k', V_g^0, or K_R^0:

$$\alpha_{i/j} = k_i'/k_j' = V_{g,i}^0/V_{g,j}^0 = K_{R,i}^0/K_{R,j}^0 \qquad (14.20)$$

where the use of capacity factors has the clear advantage that the requisite data, t_R and t_A, can be taken directly from strip-chart tracings.

Two additional expressions of analytical separations are also common, these being the resolution R_s and the separation factor S_f, the latter having been formulated relatively recently by Jones and Wellington (1981):

$$R_s = (1/4)[(\alpha - 1)/\alpha][k'/(k' + 1)]N^{1/2} \qquad (14.21)$$

$$S_f = (t_{R,i} - t_{R,j})/(t_{R,i} + t_{R,j}) = 2R_s/N^{1/2} \qquad (14.22)$$

where N, the number of theoretical plates exhibited by the system, is a form of expression of the kinetic bandspreading of solutes as they pass through the column.

14.1.3 Blended stationary phases

The above relations bring out clearly that GLC separations are a function of solute vapour pressures p_A^0 as well as the extent of their interaction with the stationary phase, i.e., values of γ_A^∞. Moreover, separations could easily be predicted if the respective activity coefficients could somehow be forecast. Unfortunately, and despite considerable progress made in this area in recent years, it is still impossible to calculate, *a priori*, the γ_A^∞ for systems other than those comprised of *n*-alkanes to better than *c*. $\pm 5\%$. In contrast, an efficient GLC column can resolve two peaks of α of as little as 1.015, that is, γ_A^∞ data must be accurate to *c*. $\pm 1\%$ to be of value in predicting (thence optimizing) gas-chromatographic resolutions.

Despite the limited success in predicting GC separations with pure solvents, there have been noteworthy gains made in describing solute elution behaviour with blended stationary phases. For example, conventional theories of solutions yield an expression of the following form (Harbison *et al.*, 1979):

$$\ln K_{R(M)}^0 = \phi_B \ln K_{R(B)}^0 + \phi_C \ln K_{R(C)}^0 + \phi_B \phi_C (\bar{V}_A / \bar{V}_C) \chi_{C(B)} \tag{14.23}$$

where the solute partition coefficient with a binary solvent M ($= B + C$) is said to be related logarithmically to the volume-fraction (ϕ_i) weighted average of those with each of the pure phases; and where $\chi_{C(B)}$ is a Flory-type interaction parameter, which is often found to be a function of the solvent make-up. Eqn (14.23) can also be cast in terms of activity coefficients:

$$\ln \gamma_{A(M)}^\infty = \phi_B \ln \gamma_{A(B)}^\infty + \phi_C \ln \gamma_{A(C)}^\infty + \phi_C \ln (\bar{V}_C / \bar{V}_B)$$
$$+ \ln (\bar{V}_B / \bar{V}_M) + \phi_B \phi_C (\bar{V}_A / \bar{V}_C) \chi_{C(B)} \tag{14.24}$$

Laub, Purnell and Vargas de Andrade (Purnell and Vargas de Andrade, 1975; Laub and Purnell, 1976) have alternatively shown that very many ternary systems (with the solute at infinite dilution) obey a surprisingly simple relation that describes $K_{R(M)}^0$ as linear in solution composition:

$$K_{R(M)}^0 = \phi_B K_{R(B)}^0 + \phi_C K_{R(C)}^0 = \phi_C \Delta K_R^0 + K_{R(B)}^0 \tag{14.25}$$

where $\phi_B = 1 - \phi_C$, and where $\Delta K_R^0 = K_{R(C)}^0 - K_{R(B)}^0$. Apparently, the data for higher-component mixtures can then be represented by a summation:

$$K_{R(M)}^0 = \sum_{i=1}^{n} \phi_i K_{R(i)}^0 \tag{14.26}$$

Simple derivation also shows that the relevant form of eqn (14.25) in terms of specific retention volumes is given by

$$V_{g(M)}^0 = w_B V_{g(B)}^0 + w_C V_{g(C)}^0 = w_C \Delta V_g^0 + V_{g(B)}^0 \tag{14.27}$$

where the solute V_g^0 with a mixed phase M is a solvent-component weight-fraction w_i ($i = B$ or C) average of those with the two individual phases, as first deduced by Primavesi (1959). Further, for columns of equal phase ratio per unit length, the relation can be put in terms of capacity factors (Hildebrand and Reilley, 1964):

$$k_{(M)}' = w_B k_{(B)}' + w_C k_{(C)}' \tag{14.28}$$

Some modification of eqn (14.28) is required when the capacity-factor data are taken with packings of unequal liquid loading and/or particle size (Laub and Purnell, 1978b; Laub *et al.*, 1978a).

Eqns (14.25)–(14.28) can alternatively be cast in terms of activity coefficients; the appropriate form for two-component solvents is given by

$$1/\gamma^{\infty}_{A(M)} = x_B/\gamma^{\infty}_{A(B)} + x_C/\gamma^{\infty}_{A(C)} \qquad (14.29)$$

In addition, Laub (1986) has claimed recently that deviations of eqn (14.29) from what amounts to ideal behaviur of binary mixtures of B + C can be taken into account by use of what he dubbed a mean-solvent activity coefficient $\bar{\gamma}_M$:

$$\ln \bar{\gamma}_M = g^E_M/2RT = (1/2)[x_B \ln \gamma_{B(M)} + x_C \ln \gamma_{C(M)}] \qquad (14.30)$$

where g^E_M is the excess free energy of mixing of B + C, x_i are mole fractions, and $\gamma_{i(M)}(i = B$ or C) are the finite-concentration activity coefficients of the solvent-components. Eqn (14.29) is then modified to the form

$$1/(\gamma^{\infty}_{A(M)}\bar{\gamma}_M) = x_B/\gamma^{\infty}_{A(B)} + x_C/\gamma^{\infty}_{A(C)} \qquad (14.31)$$

Negative deviations of B and C from ideality thus give curvature that is concave from the linearity predicted by eqn (14.29), while positive deviations give rise to convex curves. (This result has brought about some controversy; however, given the log form of eqn (14.30) and composition-dependence of $\chi_{C(B)}$, eqns (14.24) and (14.31) are in fact not altogether dissimilar.)

In any event, whichever form of $\gamma^{\infty}_{A(M)} = f(M)$ may ultimately prove to be the more realistic, clearly one need only substitute the function into eqn (14.1) in order to obtain a complete description of the separation of pairs of solutes across the complete compositional range, pure B to pure C. For example, in instances in which eqn (14.25) is found to hold, the separation of solutes i and j is given in terms of ϕ_C by the expression

$$\phi_C = [K^0_{R(C)i} - \alpha_{i/j}K^0_{R(B)j}]/[\alpha_{i/j}\Delta K^0_{R,j} - \Delta K^0_{R,i}] \qquad (14.32)$$

Further, plot of $\alpha_{i/j}$ against the stationary-phase composition (window diagrams) reveal upon inspection all values of ϕ_C at which the separation of all pairs of solutes is feasible to within the capacity of some user-specified level of column efficiency, or, if beyond the limit of the existing system, the efficiency that must be attained in order to achieve the desired extent of resolution.

Expressions of the form of eqn (14.32) also provide insight into the spectrum of separations that can be achieved with combinations of solvents, where blends of phases (the compositions of which can be deduced quantitatively via the window-diagram strategy) provide access to intermediate regions of selectivities that are otherwise unattainable with any neat solvent. Moreover, appropriate mixtures of B and C can be used to replicate (hence eliminate the need for) other pure stationary liquids; the series of methylphenylsilicone OV phases provides an example of the success of such an approach in winnowing the number of solvents from the 800 or so that are currently available commercially to a manageable few (Chien et al., 1980, 1984). Indeed, it is entirely conceivable that all GC separations could be carried out with a mere half-dozen or so standard stationary phases, used either neat or in admixture as required.

The primary criterion for the choice of such a set of standard solvents is that each must provide separations on a unique basis. For example, if a particular poly(dimethylsiloxane) (PDMS) phase, say, SE-30, failed to yield the desired resolution for the solutes to hand then a similar material, e.g. OV-1, would probably also prove to be unsatisfactory. That is, blends of SE-30 with OV-1 would yield selectivities that would not differ much from those offered by either pure phase. In contrast, blends of OV-1 with a polysiloxane of some phenyl content could be expected to give relative retentions, and even orders of elution, that would differ substantially from either pure solvent.

A number of secondary criteria can also be broadly defined regarding the ideal physical properties of phases irrespective of their chemical nature. These include ready availability in reasonable quantity; narrow polydispersity if polymeric; negligible vapour pressure at well in excess of 300°C; appropriate viscosity and surface tension as a room temperature liquid so as to give a uniform film when coated on to inert support materials; and provision of crosslinking (preferably carried out *in situ*) as well as chemical attachment to the walls of capillary tubes. These attributes clearly point to a polymeric material, for which the additional capability of ready functionalization of the backbone should then also obtain.

All of the above characteristics are of course exhibited by polysiloxanes and, on this basis, Laub and co-workers (Chien *et al.*, 1981, 1983; Laub, 1987; Laub and Purnell, 1988) selected as the first of their bank of standard fluids a PDMS. Materials such as OV-1, OV-101, SE-30, and so forth, are widely available, inexpensive, can readily be refined, are thermally stable to in excess of 350°C, and give high column efficiency even when used in thick films with packed columns. Phases of these types also give separations that are founded primarily on solute vapour pressure, which is advantageous at the outset since only rarely are mixtures encountered for which at least some resolution cannot be realized on the basis of relative volatility. The second of their phases, chosen so as to be substantially different from the first, was OV-17, a poly(methylphenylsiloxane) (Chien *et al.*, 1984). This solvent gives separations even of hydrocarbons that differ markedly from those found with PDMS (Chien *et al.*, 1983). However, the studies of both materials completed thus far can hardly be claimed at the present time to be comprehensive, and much remains to be done in evaluating their suitability overall as 'standard' GC solvents.

14.1.4 *Small-molecule liquid-crystalline stationary phases*

Small-molecule liquid-crystalline stationary phases were considered at one time to be a very promising class of materials from which to draw a third standard solvent. The materials give GC separations that are very different from those that can be obtained with any other phase and, as a result, have been employed over the years in a wide variety of analytical applications. In addition, several noteworthy studies of their physicochemical properties have been carried out by gas chromatography, as reviewed by Janini (1979).

Among the first such solvents to be employed as GC phases were chiral-nematic cholesteryl esters. These compounds are easily synthesized and purified, and, of equal importance in analytical GLC, exhibit a wide range of transition temperatures, an indication of which is provided in Table 14.1. All of the materials shown are useful at least to some extent as mesomorphic GC solvents, particularly for some chiral separations, albeit each is liquid-crystalline over a more or less narrow temperature span. However, broadly speaking, their volatility becomes unacceptably high as the column temperature is made to approach the point at which each fluid turns isotropic.

By far the more numerous have been reports of smectic and nematic GLC solvents, which date back to the mid-1950s (Kelker, 1963, 1978; Schroeder, 1974; Janini, 1979). These materials give separations on the basis of solute shape, one of the more

Table 14.1 Transition temperatures $T(°C)$ of some small-molecule liquid crystals used as GLC stationary phases[a,b]

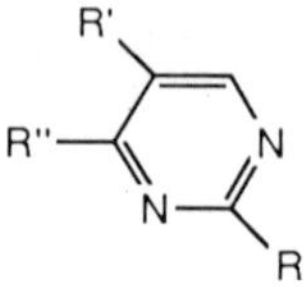

R	R'	R''	
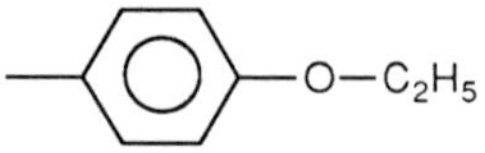 $-\bigcirc-O-C_2H_5$	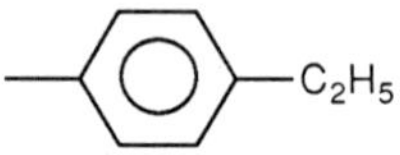 $-\bigcirc-C_2H_5$	—H	k 153 n 214 i
	—R		k 196 n 270 i
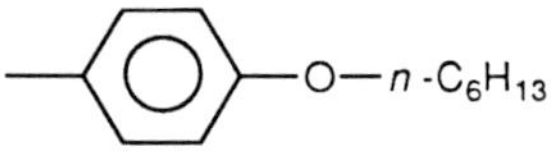 $-\bigcirc-O-n\text{-}C_6H_{13}$	—R	—H	k 104 s 185 i
		—OH	k 201 s 288 i
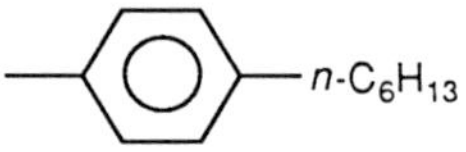 $-\bigcirc-n\text{-}C_6H_{13}$	—R	—H	k 104 s 185 i
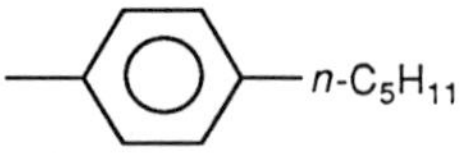 $-\bigcirc-n\text{-}C_5H_{11}$	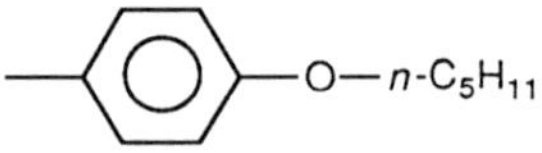 $-\bigcirc-O-n\text{-}C_5H_{11}$	—H	k 80 s 211 i
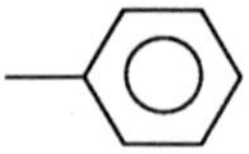 $-\bigcirc$	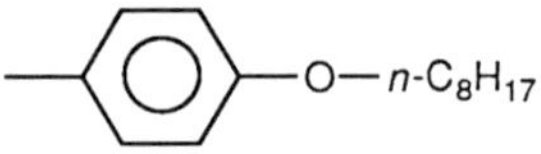 $-\bigcirc-O-n\text{-}C_8H_{17}$	—H	k 84 s 162 i

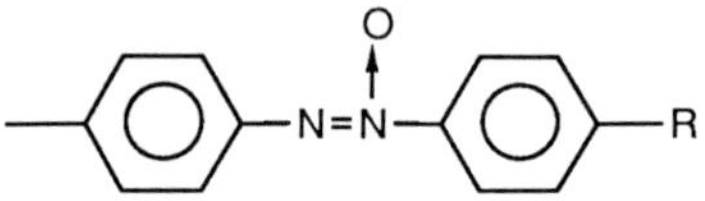

R =	$-O-CH_3$	k 119 n 135 i
	$-O-C_2H_5$	k 139 n 169 i
R =	$-O-n\text{-}C_4H_9$	k 105 n 136 i
	$-O-n\text{-}C_6H_{13}$	k 81 n 128 i
	$-\overset{O}{\underset{\parallel}{C}}-O-C_2H_5$	k 114 n 120 i

Table 14.1 (Contd.)

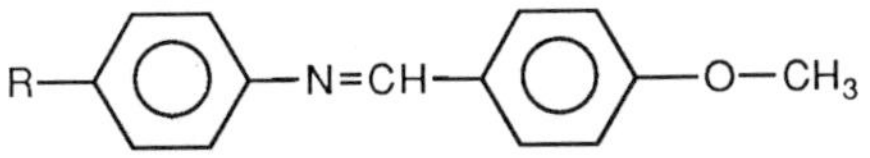

$$R-\text{◯}-N=CH-\text{◯}-O-CH_3$$

R =		
—CN		k 106 n 117 i
—n-C_4H_9		k 20 n 48 i
—O—C_2H_5		k 83 n 107 i
—$\overset{\overset{O}{\|\|}}{C}$—$O$—$CH_3$		k 79 n 102 i

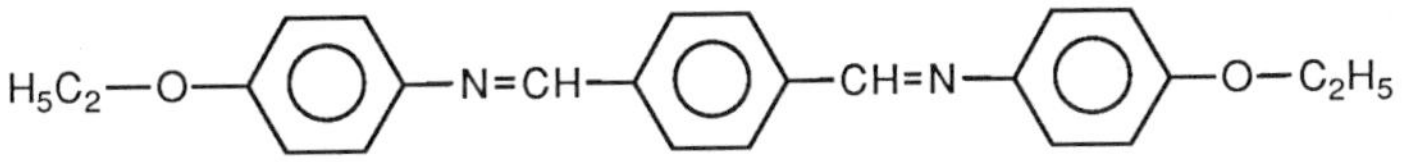

$$H_5C_2-O-\text{◯}-N=CH-\text{◯}-CH=N-\text{◯}-O-C_2H_5$$

k 200 n 320 i

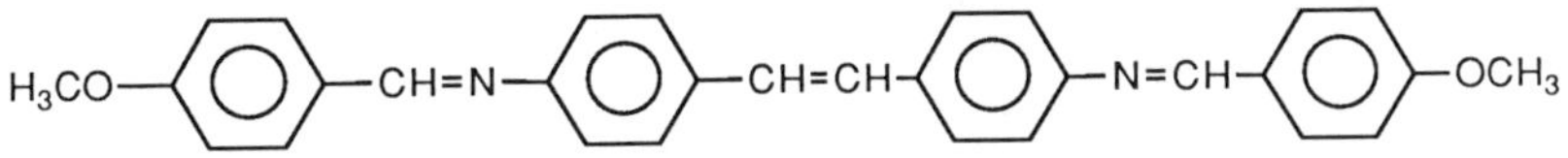

$$H_3CO-\text{◯}-CH=N-\text{◯}-CH=CH-\text{◯}-N=CH-\text{◯}-OCH_3$$

k 274 n 340 i

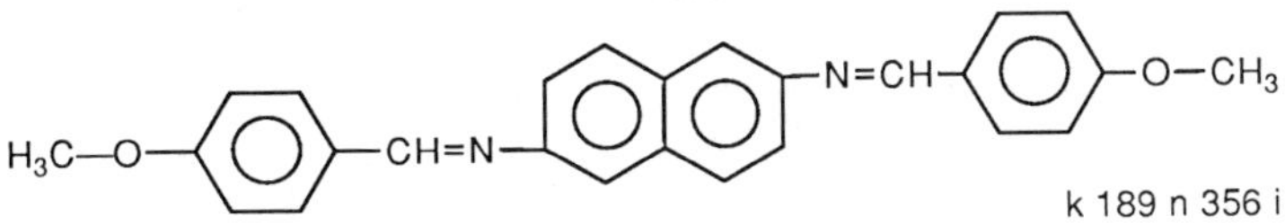

$$H_3C-O-\text{◯}-CH=N-\text{◯◯}-N=CH-\text{◯}-O-CH_3$$

k 189 n 356 i

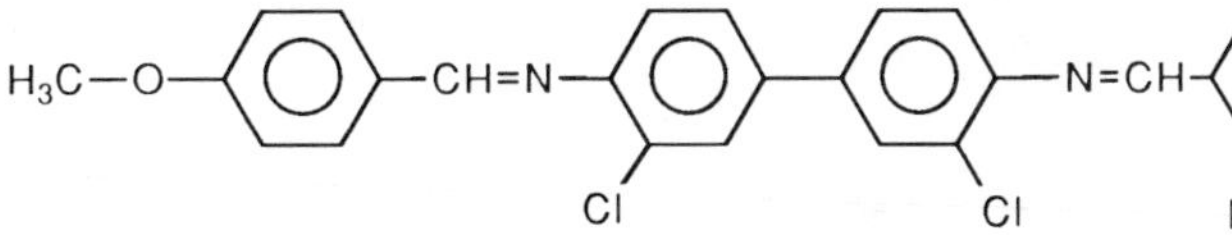

$$H_3C-O-\text{◯}-CH=N-\underset{Cl}{\text{◯}}-\underset{Cl}{\text{◯}}-N=CH-\text{◯}-O-CH_3$$

k 154 n 344 (dec) i

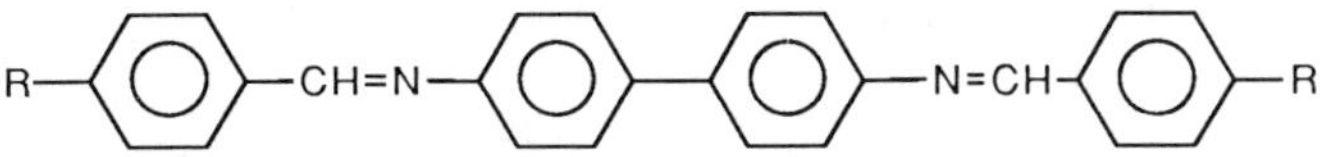

$$R-\text{◯}-CH=N-\text{◯}-\text{◯}-N=CH-\text{◯}-R$$

R =		
—H		k 239 n 265 i
—O—CH_3		k 266 n 390 i

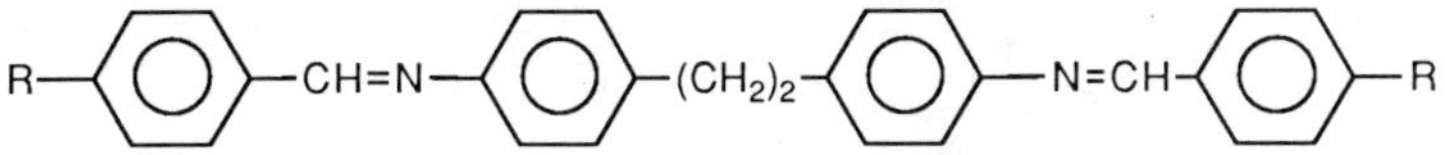

$$R-\text{◯}-CH=N-\text{◯}-(CH_2)_2-\text{◯}-N=CH-\text{◯}-R$$

R =		
—CH_3		k 197 n 287 i
—Cl		k 232 n 318 i

Table 14.1 (Contd.)

—CN	k 227 n 367 i
—O—CH$_3$	k 181 n 337 i
—O—n-C$_4$H$_9$	k 159 s 188 n 303 i
—O—n-C$_6$H$_{13}$	k 127 s 229 n 276 i

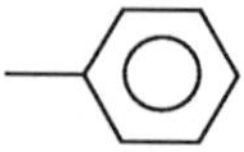

k 257 n 403 i

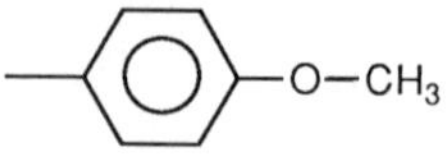

k 253 n 270 (dec) i

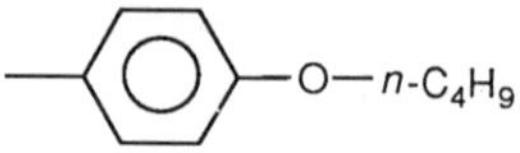

k 232 n 331 i

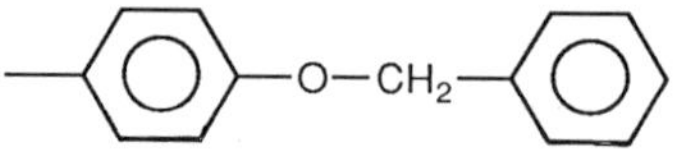

k 270 n 346 i

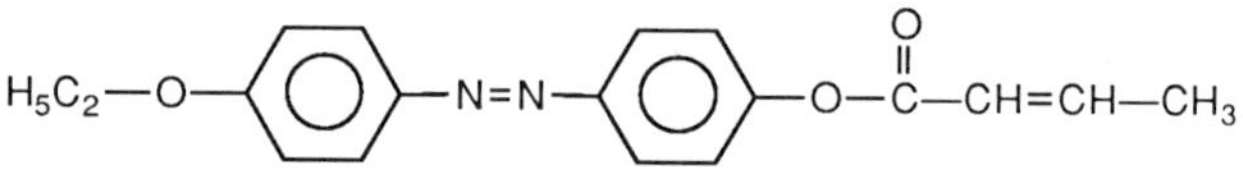

k 110 n 197 i

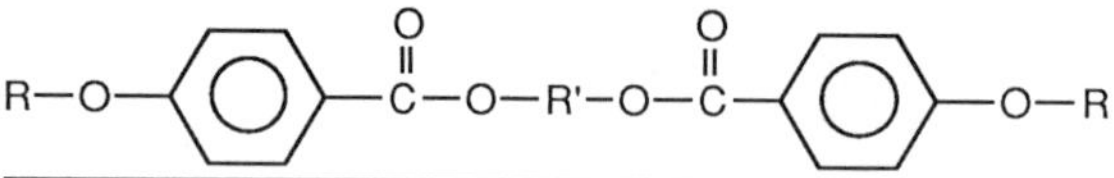

R =	R' =	
—n-C$_4$H$_9$	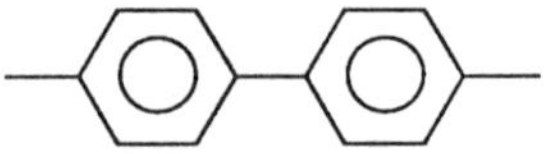	k 171 s 184 n 358(dec) i
—n-C$_7$H$_{15}$		k 83 s 125 n 206 i
	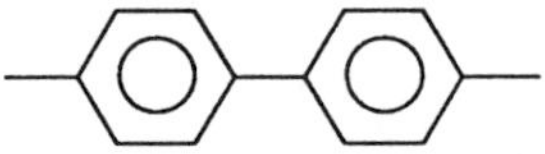	k 150 s 211 n 316 i

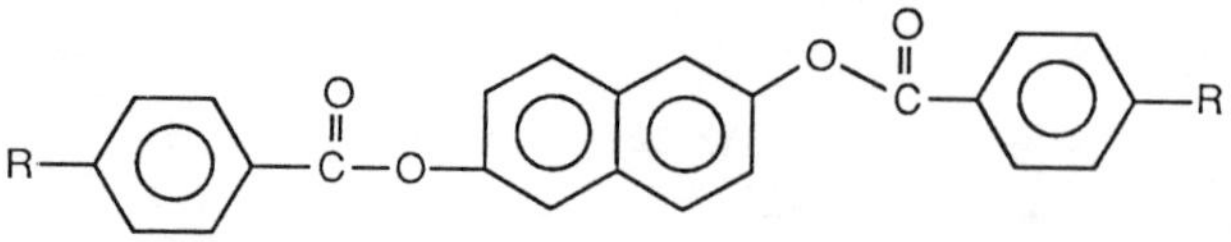

R =	
—CH$_3$	k 207 n 318 i
—O—CH$_3$	k 220 n 350 i
—O—n-C$_5$H$_{11}$	k 145 n 272 i
—O—n-C$_7$H$_{15}$	k 153 n 245 i
—O—n-C$_9$H$_{19}$	k 144 n 227 i

Table 14.1 (Contd.)

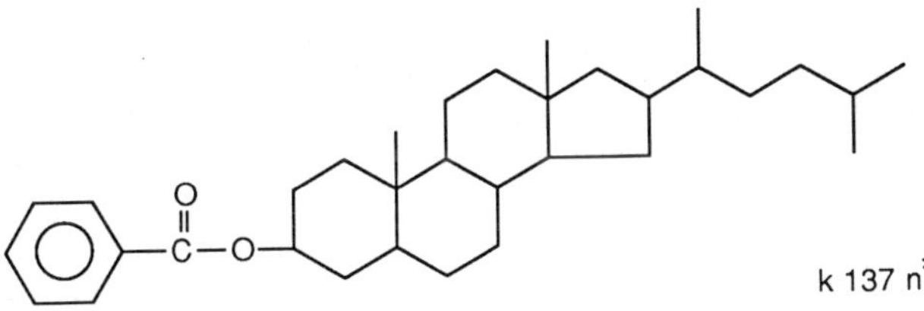

k 137 n*155 i

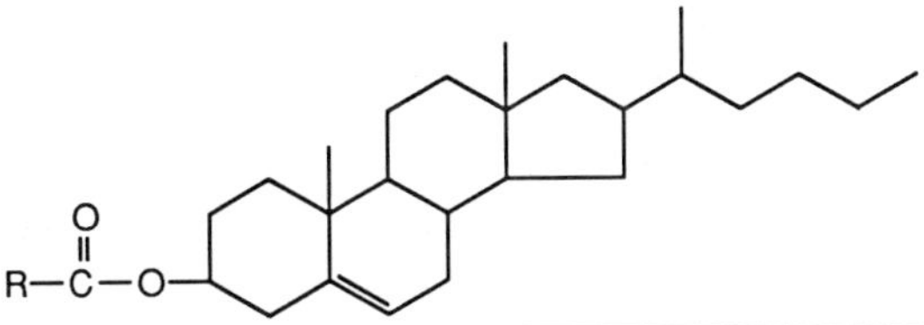

R =		
—Cl	k 118 n* 125 i	
—CH_3	k 95 n* 117 i	
—C_2H_5	k 97 n* 114 i	
—n-C_6H_{13}	k 96 n* 112 i	
—n-C_7H_{15}	k 97 n* 110 i	
—n-C_8H_{17}	k 78 s*81 n* 92 i	
—$CH(C_2H_5)(CH_2)_3CH_3$	k 30 n* 50 i	
—n-$C_{13}H_{27}$	k 71 s* 79 n*83 i	
—n-$C_{14}H_{29}$	k 77 s*79 n* 83 i	
—n-$C_{16}H_{33}$	k 76 s*80 n*83 i	
—$(CH_2)_7(CH=CH—CH_2)(CH_2)_6CH_3$	k 39 s* 44 n* 49 i	
—$(CH_2)_7(CH=CH—CH_2)_2(CH_2)_3CH_3$	k 20 s* 44 n* 49 i	
—$(CH_2)_7(CH=CH—CH_2)_3CH_3$	k 35 s* 45 n* 48 i	
—O—$(CH_2)_8(CH=CH—CH_2)(CH_2)_6CH_3$	k-10 s* 18 n*31 i	
—O—$(CH_2)_2$—O—$(CH_2)_2$—O—C_2H_5	k-2 n*15 i	

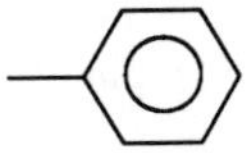

k 150 n* 178 i

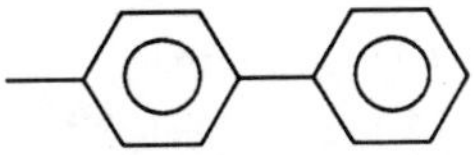

k 178 n*290 i

---

[a]Kelker (1978); Janini (1979). Virtually none of these materials have been used as MEPSIL sidechains.
[b]k, crystalline; s, smectic; n, nematic; i, isotropic; *chiral.

important examples of which is the resolution of benzo[a]pyrene (a carcinogen) from benzo[e]pyrene. A number of types of small-molecule mesomorphic GLC phases have been reported, a few representative examples of which are shown in Table 14.1. The most widely used at present are the nematic toluidine solvents BMBT, BBBT, and BPhBT that were pioneered by Janini and co-workers (Janini et $al.$, 1975, 1976a, 1976b, 1980; Zielinski Jr., et $al.$, 1981; Haky and Muschik, 1982; Witkiewicz, 1982). The nematic state is particularly favoured because of the appreciably greater column efficiency that can be achieved with it relative to smectics. (This must evidently also be the case generally, because of more efficient mass transfer into and out of one- as opposed to two-dimensional ordered fluids.)

However, the small-molecule liquid crystals employed to date as GC solvents have proved in practice to have a number of drawbacks. For example, the BiBT series of compounds fail to wet inert GC supports such as Chromosorbs, which results in very uneven coatings that give rise in turn to poor efficiency with packed-column systems. Nor does the use of capillary columns result in any substantial improvement, roughly 300 plates per foot being about as good as can be hoped for (Laub et $al.$, 1980; Laub, 1981)—approximately one-fifth that which can be obtained with PDMS phases (Lambert et $al.$, 1984). Many of the compounds also tend to be unacceptably volatile at as little as 10°C above their melting points; most have in any event somewhat narrow mesomorphic ranges (cf. Table 14.1).

Laub and co-workers (Laub et $al.$, 1980; Laub, 1981) attempted to overcome some of the above difficulties by blending, for example BBBT with a methylphenylsiloxane, with, however, only some success. For example, while the efficiency of the mixed solvents was improved substantially over that of pure BBBT, and while the volatility of the mesomorphic compounds could also be reduced in approximate accord with Raoult's law, its transition temperatures could not thereby be altered. Indeed, the mixed phases behaved as if diachoric (Laub and Purnell, 1976), and followed eqn (14.28) precisely.

At this point Laub and colleagues turned to mesomorphic polysiloxane stationary phases, which at that time had just recently been introduced, and which have since proved to be very useful indeed for GLC separations. In addition, these materials are also intrinsically interesting from the standpoint of their physicochemical properties, for which high-precision GLC (Laub and Pecsok, 1978; Laub et $al.$, 1978; Laub and Purnell, 1988) is ideally suited as a characterization technique. Each of these aspects of side chain siloxane polymers is therefore considered in turn in the sections that follow.

14.2 Selected aspects of the synthesis and characterization of MEPSIL phases

Hydrosilation is of course a well-known method of forming silicon–carbon bonds via the insertion of Si-H hydride hydrogen into vinyl $CH_2=CH-R$ in the presence of a catalyst (Speier, 1979), as reviewed elsewhere in this text (cf. Chapter 4). However, there are a number of subtleties involved both in carrying through the reaction sequence and in the product work-up, several pitfalls in the practical aspects of each of which have led in some cases to irreproducibility in transition temperatures of as much as $\pm 30°C$ (Jones et $al.$, 1984; Markides et $al.$, 1985). We therefore briefly recount here our own observations (Finkelmann et $al.$, 1981, 1982; Apfel et $al.$, 1985; Janini et $al.$, 1985, 1987, 1988; Laub, 1988) so as to help newcomers to this field, as well as encourage and promote active study and consideration in general of the synthesis of MEPSIL polymers.

14.2.1 General reaction scheme

The overall synthetic scheme is shown in Figure 14.1: poly(hydrogenmethylsiloxane) (PHMS) (I) under a nitrogen blanket is reacted (refluxing benzene or tetrahydrofuran; 24 h) with monomer (II) containing a vinyl group whereupon, in the presence of a platinum catalyst (III), hydrosilation takes place yielding the respective side chain polysiloxane (IV). The extent of reaction can be monitored during its course by removing samples for infrared (IR) and nuclear magnetic resonance (NMR) spec-

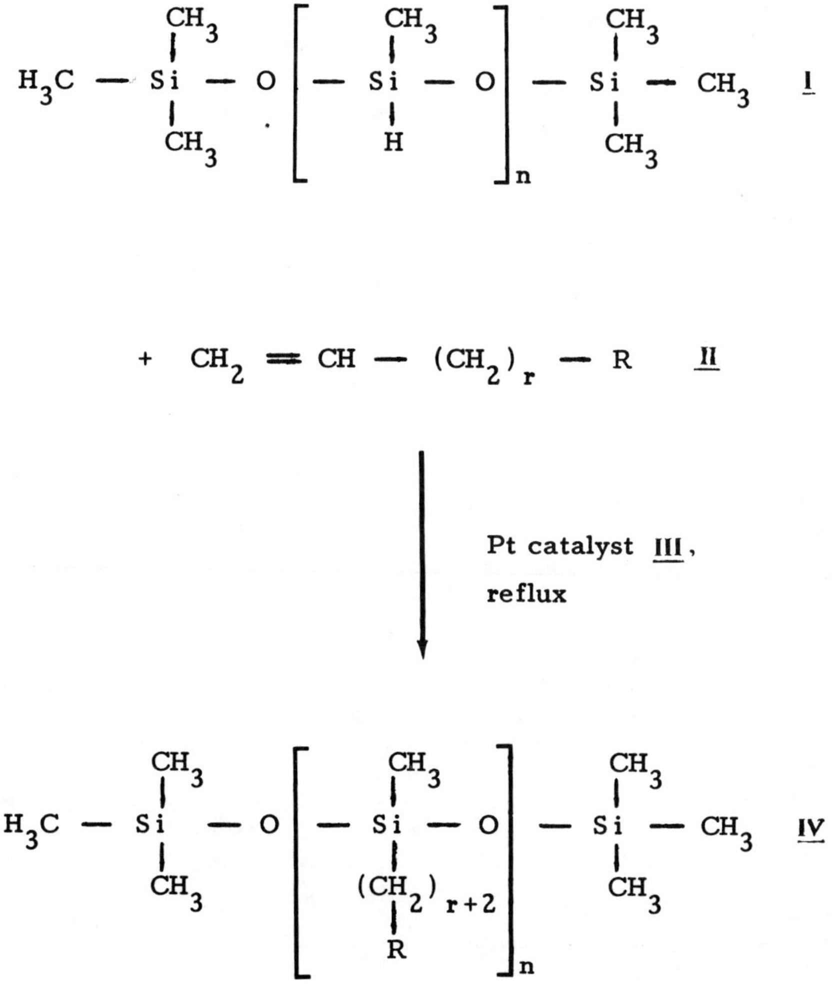

Figure 14.1 Reaction scheme for synthesis of MEPSIL polymers.

troscopic analysis. Product work-up is carried out by dissolution from toluene with dry methanol, followed by centrifugation; the mother liquor is then decanted and the pellet of crude product washed with several additional volumes of methanol. The dissolution process is repeated until the transition temperature becomes constant, three or four repetitions usually being sufficient (see also later).

14.2.2 *Polydispersity of poly(hydrogenmethylsiloxane)* (*I*)

We have obtained consistent results with Silar, Petrarch, and Wacker versions of PHMS, which are available in lots of average degree of polymerization of *c.* 30 to 60. However, all PHMS as received are of course polydisperse and so, require some clean-up prior to reaction in order that the final product be reasonably homogeneous, hence exhibit reproducible properties. Preparative-scale gel-permeation chromatography (GPC) is one such means; however, we found several years ago (Apfel *et al.*, 1985) that fractional dissolution is also a facile technique for bulk-scale isolation of starting reactant polymer. Thus, 20 cm^3 of PHMS is dissolved in 80 cm^3 of HPLC-grade dry benzene in a water-jacketed 250 cm^3 separatory funnel thermostated at 20°C and equipped with a stirrer, to which is added gradually with mild agitation 140 cm^3 of dry methanol. Since methanol reacts with hydride hydrogen even at moderate tempera-tures to produce molecular hydrogen and poly(methoxymethylsiloxane) (evidenced by bubble formation in the mixture, which becomes increasingly pronounced when heated further), the solution must not be allowed to warm beyond 25°C. Upon standing for 2 h, two phases separate. The lower layer consists of *c.* 80% v/v of the starting PHMS and is the higher-molecular-weight fraction. This material is drawn off and refractionated with 80 cm^3 benzene and 100 cm^3 methanol. The recovered polymer is placed under a vacuum of *c.* 10^{-3} Torr at room temperature for 12 h so as to remove residual solvent, and then stored under nitrogen. Recoveries amount routinely to 40%, i.e. 8 cm^3 of PHMS. Polydispersity can be checked by analytical-scale GPC, for which we employ high-performance TSK 3000 H columns with tetrahydrofuran (THF) mobile phase and refractive-index or uv (220 nm) detection.

Some comment is in order at this point regarding the importance of the polymer molecular mass to the properties, particularly the transitions, of the resultant MEPSIL phases. First, in GLC, the lower glass- or crystal/mesomorphic temperature is important when the column is operated in that region. However, as discussed later in section 14.3, the column efficiency deteriorates rapidly as MEPSILs are cooled to within 20°C or so of their lower transitions. Thus, since, as a result, MEPSIL columns are routinely operated at well in excess of the solvent melt point, the lower transition temperature generally is not of concern. (This is not to say, however, that a solid–mesomorphic transition temperature as low as possible, preferably at or below room temperature, is not desirable.) In contrast, the upper transition of a MEPSIL is crucial in determining not only the useful mesomorphic range of the solvent, but also the kinds of separations to which it can be applied. For example, five-ring PAH are intractable below *c.* 250°C regardless of the type of GC column or phase employed, and so, in order for the mesomorphic properties of a MEPSIL solvent to be taken advantage of in the gas chromatography of such samples, it must remain in the nematic state to at least 300°C.

As a result, Laub and co-workers (Apfel *et al.*, 1985; Janini *et al.*, 1985), among others, have evaluated at least in passing the dependence of transition temperatures on the reactant-polymer degree of polymerization DP. They found, not surprisingly, that the lower transition was affected, but not the upper nematic–isotropic clearing temperature, over a relatively narrow range of DP. For example, PHMS of DP 38 and 58 yielded respective meltpoints of 91° and 119°C for a methoxybiphenyl MEPSIL, but nematic–isotropic transitions of 317° and 319°C, i.e., to within experimental error, identical t_{n-i}. These findings were rationalized as a consequence of the dependence of melting on precise close packing of molecules in the solid state, for which polymers of different DP will pack differently. In contrast, the only ordering extant in the liquid-crystalline state is an average parallel orientation of the long axis of the side chains, which with spacers of at least three methylene units are free to cluster regardless of the length of the backbone to which they are attached. Thus, the upper mesomorphic–isotropic transitions hardly appear to be affected by the reactant-polymer molecular weight above DP of about 20. This is highly favourable insofar as the important properties of the resultant product-polymers are independent of the starting reactant PHMS.

Gray and co-workers (Nestor *et al.*, 1987) have recently presented corroborative evidence of the above findings (see also Chapter 4 of this text). First, a careful study of the precision of preparation of a particular cyanobiphenyl MEPSIL (5-carbon spacer) showed a spread of 9°C on the upper smectic–isotropic transition after three dissolution cycles (dichloromethane solvent; methanol precipitant), which was hardly improved to a spread of 7°C after 8–10 dissolution steps. In contrast, and as expected on the basis of the argument presented above, variation of the lower glass–smectic transition temperature was narrowed from 7° to 1°C by repeated dissolution. Further, it seems unlikely that the improvement could be entirely due simply to the removal of unreacted monomer, since we have obtained essentially the same results in our own work, and since we have also found by high-performance gel-permeation and column-liquid chromatography (HPLC) that all detectable monomer can effectively be removed by dissolution from warm toluene after only 3–4 cycles (see later). Gray *et al.* have in addition examined a series of polymers of DP 40–107, and found that the clearing point of the resultant (cyanobiphenyl) MEPSIL, while drifting higher as DP was increased, in fact varied only from 128° to 145°C. If we allow for an experimental variation of ± 7°C, as seems reasonable in light of the data cited earlier, these clearing points fall to within 3° of being identical.

14.2.3 *Synthesis of mesomorphic monomers (II): 4-((4-allyloxy)benzoyl)oxy-4, 4'-methoxybiphenyl*

Many of the monomers employed to date to produce MEPSIL polymers are of the form indicated in Figure 13.1, and comprise a spacer group of methylene units a central aromatic linkage, and a liquid-crystalline end-group. Spacers of $n > 4$ yield smectic phases, and so we have routinely made use of allyl units, i.e. $n = 3$. Also, the aromatic linkage that has provided many useful materials is the 4-oxybenzoate moiety, although others are also under active study by us (Janini *et al.*, 1985, 1987, 1988) as well as others. A variety of end-groups have been evaluated, as reported in this text and elsewhere (Finkelmann, 1984); to date the most useful from the standpoint of homopolymer nematic GLC solvents has been found to be 4-methoxybiphenyl, a corresponding MEPSIL of which gives a crystal–nematic transition at 91°C and a nematic–isotropic clearing point of 317°C, the broadest nematic range yet produced.

We therefore present the synthesis of this monomer in some detail as an example of procedures that are applicable generally to the fabrication of materials of these kinds.

4-(allyloxy)benzoic acid is first prepared as follows. Into a 1-dm^3 three-neck round-bottom flask equipped with a reflux condenser, an addition funnel, a heating bath, and a magnetic stirrer, is placed 90 g (0.65 mol) of 4-hydroxybenzoic acid. To this is added 400 cm^3 of methanol at 25°C, which results in a brown solution. Next, added dropwise is 105 g (1.9 mol) of potassium hydroxide in 125 cm^3 of distilled water. The reaction mixture is heated to reflux and 90 g (0.74 mol) of allyl bromide is then added over a 1-h period. The reaction mixture is allowed to reflux for an additional 8 h, at which time 250 cm^3 of methanol is removed by single-stage distillation. The remainder of the reaction mixture is cooled to 25°C and added to 1 dm^3 of distilled water. The aqueous solution is extracted three times with 200 cm^3 of diethyl ether to remove organic impurities and the ether extracts discarded. The aqueous phase is heated to 40°C in a fume hood and neutralized with 300 cm^3 of 20% aqueous HCl, which causes the desired product to precipitate. The crude solid is recovered by suction filtration and then purified by crystallization from absolute ethanol, yielding 107 g (60%), m.p. 165°C.

4-(allyloxy)benzoyl chloride is prepared next, as follows. To a 100-cm^3 round-bottom flask equipped with a reflux condenser, a drying tube, a magnetic stirrer, and a heating bath, is added 50 cm^3 (0.42 mol) of freshly-distilled thionyl chloride, into which is dissolved 50 g (0.28 mol) of 4-(allyloxy)benzoic acid with continuous stirring. Ten drops of dimethylformamide are added, the reaction mixture is protected from moisture, and allowed to stir overnight at room temperature. It is then heated to 60°C for an additional 2 h to ensure complete reaction. The excess thionyl chloride is removed *in vacuo*, and the crude product vacuum-distilled at 100°C/1 Torr, yielding 50 g (90%) of a pale yellow liquid which is stored in a tightly-sealed glass container.

To 400 cm^3 of 10% aqueous sodium hydroxide cooled separately to 0°C in an ice bath is added 75 g (0.40 mol) of 4, 4'-dihydroxybiphenyl with vigorous stirring. Fifty grams (0.40 mol) of dimethyl sulphate is then added to the reaction mixture over a period of 1 h, again with vigorous stirring, while the system is kept scrupulously under nitrogen. After the addition is complete, the resultant precipitate is removed by filtration with a sintered glass filter and then redissolved in 400 cm^3 of 10% aqueous sodium hydroxide. The solution is heated briefly to boiling, allowed to cool, and the solid filtered off. The solid mass is then placed in 3 dm^3 of distilled water, the solution is heated to boiling, and then filtered while still hot. The filtrate is next reheated to 70°C and acidified with 20% HCl. The desired product thereupon precipitates as white crystals, which are purified by crystallizing twice from ethanol, yielding 24 g (30%) of 4-hydroxy-4'-methoxybiphenyl.

Into a 250-cm^3 three-neck round-bottom flask equipped with a dropping funnel, a reflux condenser, and a magnetic stirrer is next placed under a nitrogen blanket 12 g (0.66 mol) of 4-hydroxy-4'-methoxybiphenyl prepared as above, dissolved in 100 cm^3 of dry pyridine. To this solution is added over a 1-h period with stirring 12 g (0.06 mol) of 4-(allyloxy)benzoyl chloride dissolved in 20 cm^3 of pyridine. The reaction mixture is stirred for an additional 3 h at room temperature, and then heated to 60°C for 2 h. It is cooled again to room temperature, and then added to 400 cm^3 of distilled water, following which the mixture is acidified with 20% aqueous HCl. The resultant precipitate is removed by suction filtration, and then washed with 400 cm^3 of saturated aqueous sodium bicarbonate, followed by an equal volume of distilled water. The crude monomer is crystallized twice from acetone, yielding 15 g (70%) of 4-((4-allyloxy)benzoyl)oxy-4,4'-methoxybiphenyl, m.p. 147–148.5°C.

14.2.4 Synthesis of dicyclopentadienylplatinum(II) chloride catalyst (III)

This aspect of the synthesis of MEPSIL phases is arguably the most crucial, from several points of view. First, we and others have found that hexachloroplatinic acid, even when freshly prepared (indeed, particularly when freshly prepared) gives some decompositon of the reactant siloxane backbone, leading to cross-linking as well as the formation of oligomers and some cyclization. Also, the catalyst decomposes nearly completely under the conditions of the hydrosilation reaction to colloidal platinum, which cannot be removed entirely from solution even after many cycles of solution/dissolution of the product. This results in turn, *inter alia*, in residual undesirable dielectric properties of the polymer. We have in contrast made use in our work over the years of a milder-acting reagent, dicyclopentadienylplatinum(II) chloride. However, as noted elsewhere (Apfel *et al.*, 1985), the fabrication of this material does not proceed as described by some. We have therefore developed an alternative synthetic scheme, presented here, that is simple and straightforward and that reproducibly gives good yields of readily-purified product. Thus, to 6 cm^3 of glacial acetic acid in a 25-cm^3 round-bottom flask is added 2.75 g of the hydrate of hexachloroplatinic acid. The solution is next diluted with 10 cm^3 of distilled water and then heated to 70°C, whereupon 2 cm^3 of dicyclopentadiene is added. The vigorously-stirred mixture is kept at this temperature for 24 h, following which the off-white precipitate is collected by suction filtration, decolorized with charcoal, and crystallized twice from THF, yielding 1.2 g (70%), m.p. 218°C. The structure as well as purity of the white crystalline product is easily confirmed by NMR as well as mass spectrometry (MS); a cluster of molecular ions (due to the naturally-occurring platinum isotopes) as well as a simple fragmentation pattern are obtained with conventional 70 eV electron impact ionization. The catalyst is stable in air, does not require any extraordinary handling procedures, and retains its activity over at least three years of ordinary shelf storage.

14.2.5 Synthesis of MEPSIL polymers (IV)

Using the materials synthesized as described above, the 3-carbon spacer methoxybiph-enyl MEPSIL is made as follows. Into a 25-cm^3 round-bottom flask modified with a stopcock and an injection port is placed 1.0 g (2.3 × 10^{-3} mol) of the methoxybiphenyl monomer dissolved in 100 cm^3 of dry and freshly-distilled benzene or THF solvents. Into this is pipetted 0.15 g (4.3 × 10^{-5} mol; 60.1 g equiv^{-1}) of starting polymer, i.e. a 10% excess of monomer over a mole ratio of 1:1 based upon siloxane hydrogen. The flask is next equipped with a reflux condenser and the contents are protected from air and moisture with a stream of nitrogen. The reaction mixture is heated to boiling, and 100 μg of catalyst per gram of polymer is then injected with a syringe as a solution of 1 mg catalyst per cm^3 of dichloromethane solvent. The mixture is then refluxed under nitrogen for 12 h, whereupon a second addition of catalyst is made, followed by refluxing for a further 12 h. The solution is then cooled and approximately three-fourths of the polymer is precipitated by the addition of methanol. The resultant solid is centrifuged, collected, and reprecipitated three or four times from warm toluene, again by the addition of methanol. The wet gel-like product is partially dried under a stream of nitrogen, followed by complete drying in a vacuum oven at 110°C for 24 h. The cream-to-paper-white material is stored in the dark.

Our experience in using these procedures over the previous five years is as follows. Benzene, THF, and toluene reaction media give identical results, so long as the solvents are dry. In this regard the aromatic materials are slightly favoured insofar as they are less hygroscopic than THF. Drying can in any event be carried out in the usual way by refluxing over sodium, where a small amount of benzophenone added to the solution provides an indication of the reduction of the moisture content to a few parts per million (the solution turns from straw yellow to deep blue). If the catalyst described here is employed, there is nothing gained by carrying out more than 4 dissolutions of the product, since little colloidal platinum is produced in the initial reaction. In addition, the use of warm toluene as the 'good' solvent effectively removes virtually all unreacted monomer after 3 cycles of the clean-up procedure, as we have verified by 'high-performance' GPC and column liquid chromatography (LC). Also, maintaining the solution temperature at $c.$ 40°C while effecting dissolution with methanol reacts all residual hydride hydrogen with the formation of $\equiv$Si—O—CH$_3$ and molecular hydrogen. There is therefore no need to add small-molecule 1-alkenes to the reaction mixture prior to the first precipitation step, as has been recommended by some as an alternative means of removing residual $\equiv$Si—H. A further aid in eliminating platinum from the final product is precipitation of less than the full amount of MEPSIL polymer from the initial reaction solution. That is, some of the material is sacrificed at this point, which is well worth the gain realized in the purity of the final product. The exact amount of methanol required must be determined by trial and error, however, since the solubility of MEPSIL polymers varies. The addition of methanol should also be carried out slowly both to avoid platinum forming large particles (thence coming out of solution) as well as to minimize the metal being carried down with rapidly precipitated polymer. (Unfortunately, ultrafiltration of toluene solutions of MEPSIL appears to be of limited utility in removing colloidal platinum.)

A selection of the MEPSIL phases that we have synthesized over the years, including a number of copolymers, is shown in Table 14.2.

14.2.6 Extent of reaction

There have been some reports of evaluation of the extent of reaction by grating IR, where the Si–H bond at 2180 cm^{-1} is monitored. We found this method to be insensitive above $c.$ 75% completion. However, modern Fourier-transform IR spectrometers are capable of much higher levels of sensitivity, and can easily detect 1% residual unreacted hydride hydrogen. An alternative is NMR, where the Si–H proton band is $c.$ 4–5 ppm removed from the signal for tetramethylsilane (TMS). Our experience is that 220 MHz FT-NMR provides about the same sensitivity for Si–H as does FT-IR, and we now routinely make use of both (Janini et al., 1987).

In contrast, elemental analysis is insufficiently accurate to reflect differences of better than $\pm$ 10% extent of reaction. For example, for the methoxybiphenyl MEPSIL of degree of polymerization 58 the molecular formula is $Si_{60}C_{1398}H_{1410}O_{291}$. The theoretical elemental weight composition is therefore $Si_{6.86}C_{68.39}H_{5.79}O_{18.96}$; whereas, in an evaluation of elemental analysis, triplicate runs of what gave all indication of being a fully-reacted material (FT-IR, FT-NMR) yielded $Si_{6.96}C_{67.76}H_{5.81}O_{19.47}$ (Apfel et al., 1985; Janini et al., 1987).

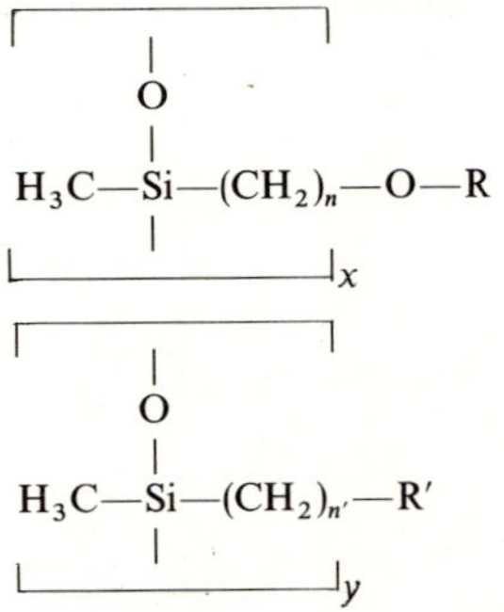

$$\left[-O-\underset{\underset{H_3C}{|}}{Si}-(CH_2)_n-O-R \right]_x \qquad \left[-O-\underset{\underset{H_3C}{|}}{Si}-(CH_2)_{n'}-R' \right]_y$$

R	R'	n	n'	x	y	Monomer	Polymer
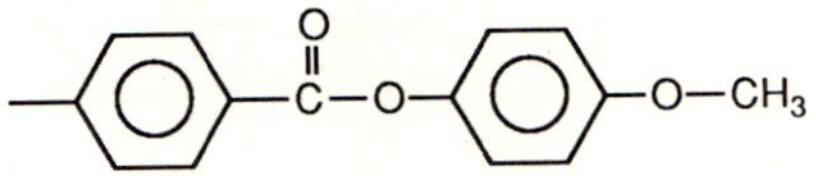 $O=C-\!\!\!\bigcirc\!\!\!-\!\!\!-O-\!\!\!\bigcirc\!\!\!-O-CH_3$	H	3	1	58	0	k 90 i	g 15 n 61 i
		4				k 87 i	g 15 n 91 i
		6				k 63 i	g 5 s 46 n 112 i
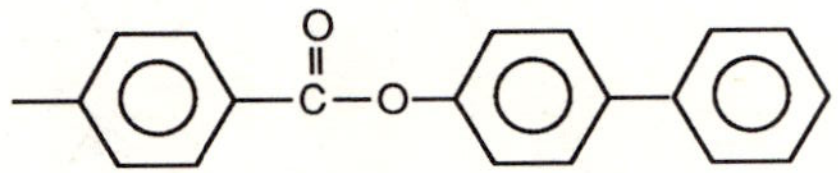 $O=C-\!\!\!\bigcirc\!\!\!-\!\!\!-O-\!\!\!\bigcirc\!\!\!-\!\!\!\bigcirc$	H	3	1	58	0	k 139 i	k 95 n 125 i
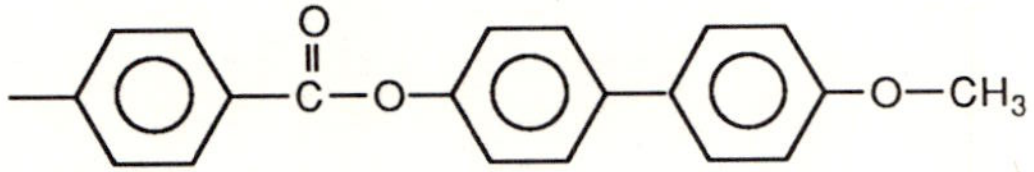 $O=C-\!\!\!\bigcirc\!\!\!-\!\!\!-O-\!\!\!\bigcirc\!\!\!-\!\!\!\bigcirc\!\!\!-O-CH_3$	H	3	1	58	0	k 147 n 249 i	k 119 n 319 i
				36	0		k 91 n 317 i
				29	7		k 134 n 300 i
				28	8		k 131 n 300 i
				26	10		k 132 n 300 i
				20	16		k 95 n 247 i
				12	24		k 88 s 115 n 169 i
		4	1	58	0		k 89 s 99 n 304 i
				12	24		k 88 s 124 n 170 i
	c	3	3	27	9		k 81 n 237 i
				18	18		k 63 n 175 i
				9	27		k 36 n 121 i

Table 14.2 (Contd.)

R	R'	n	n'	x	y	Monomer	Polymer
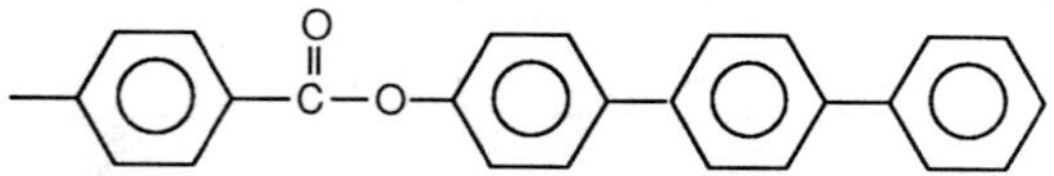	H	3	1	58	0	k > 350	k 200 n 360 (dec) i
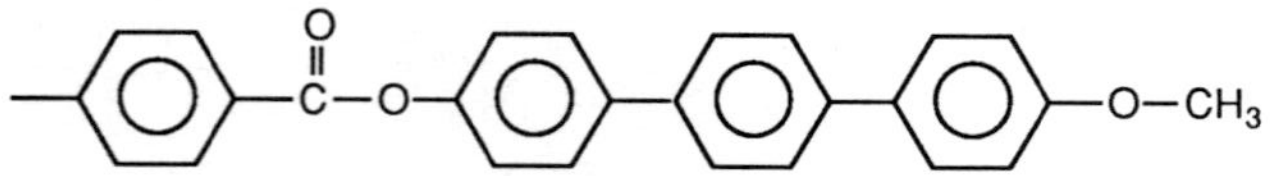	H	3	1	58	0	k > 350	k 200 n 360 (dec) i
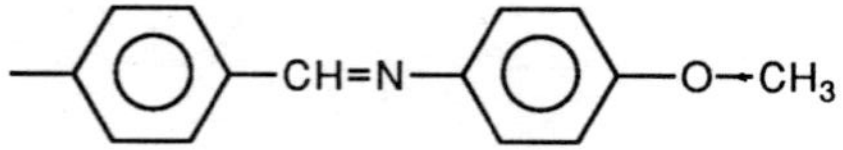	H	3	1	58	0	k 111 i	k 100 n 123 i
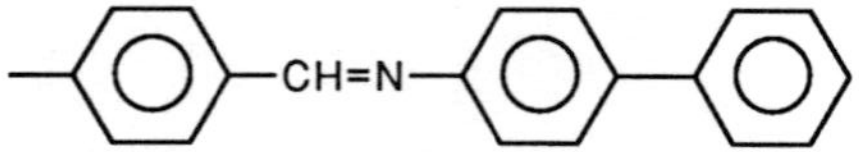	H	3	1	58	0	k 152 n 166 i	k 183 s 262 i
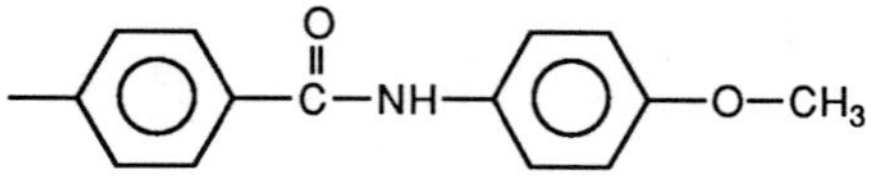	H	3	1	58	0	k 170 i	k > 350 (dec)

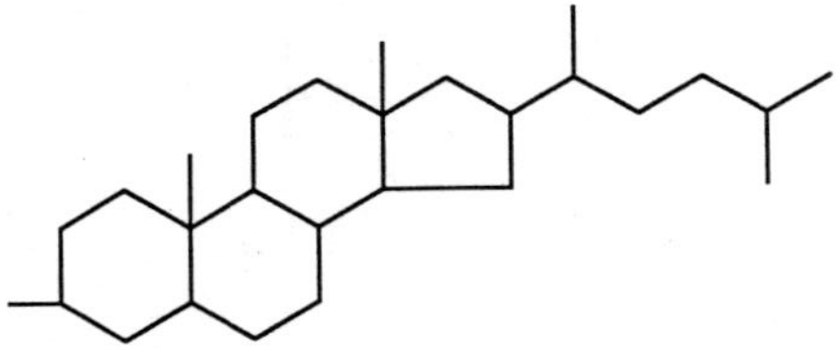 H	3	1	58	0	k 33 s 62 i	g 59 s 174 i
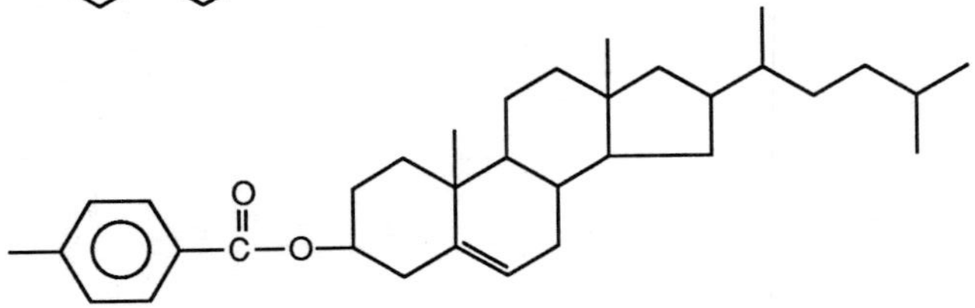 H	3	1	58	0	k 120 n* 243 i	g 130 s* 287 i

[a] Apfel et al. (1985); Janini et al. (1985, 1987).
[b] DSC, hot-stage microscopy.
[c] R' =

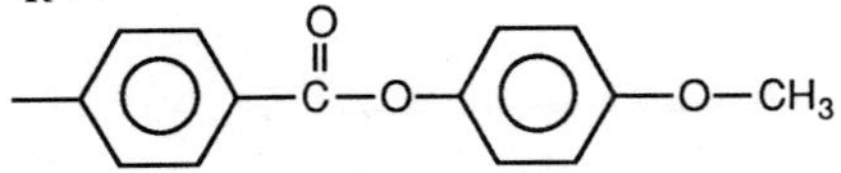

14.3 Applications of MEPSIL phases in GLC analysis

14.3.1 *Practical operating temperature limits*

Van't Hoff plots of $\log(t_R - t_A)$ against reciprocal absolute temperature for a number of polycyclic aromatic hydrocarbon test-solutes with the above-described methoxybiphenyl MEPSIL were found by Laub and co-workers (Apfel *et al.*, 1985) to be linear from 300° to 100°C, which indicates that there is no substantial change in the physical state of the melt on cooling to the latter limit. MEPSIL phases can also be supercooled to well below their melting temperatures. However, by way of contrast, Figure 14.2 presents plots of the relative separation α and resolution R_S of anthracene and phenanthrene as functions of temperature that are typical of those observed generally with MEPSIL solvents. While the separation factor increases steadily as the temperature is decreased, the resolution, which takes account of column efficiency (N) as well as selectivity (α), gives a maximum at *c.* 150°C, that is, there is substantial peak broadening as the nominal melting point is approached. The lower practical operating limit of this MEPSIL is therefore roughly 140°C. (A more quantitative guideline might be specified as, say, the temperature at which the column efficiency exhibited by a given solute decreases by 10% from that observed at the resolution curve-maximum.)

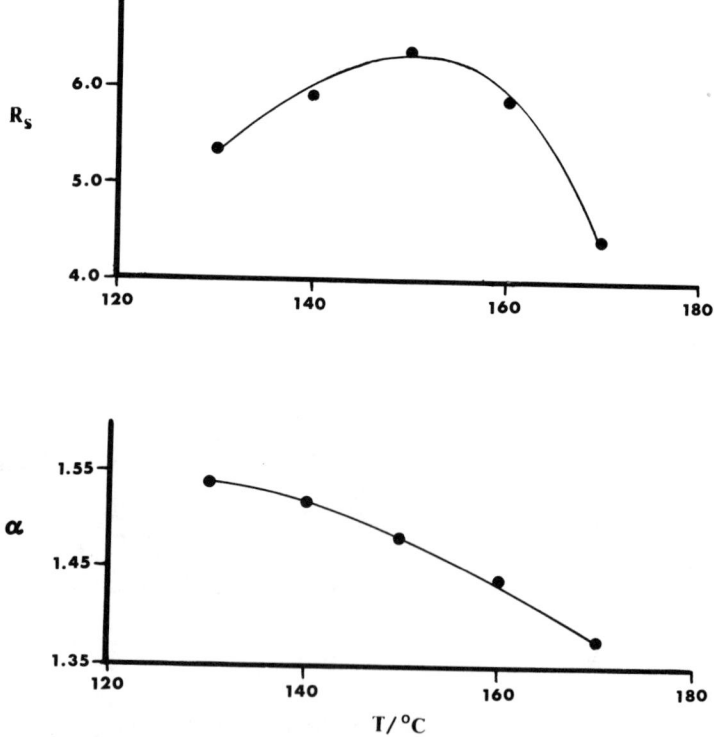

Figure 14.2 Plots of α and R_S against temperature for anthracene and phenanthrene solutes with a methoxybiphenyl MEPSIL (3-carbon spacer) stationary phase (Apfel *et al.*, 1985).

Regardless, the lower temperature limits of the most useful MEPSILs are far from satisfactory at the present time, values much closer to room temperature being desirable as mentioned earlier.

The upper practical operating limit of a MEPSIL phase is governed largely by two factors, namely, the mesomorphic–isotropic transition temperature (beyond which, generally, the selectivity of these phases is substantially reduced; however, see below), and the thermal stability of the materials. For example, the nematic–isotropic transition of the methoxybiphenyl MEPSIL occurs at 319°C as mentioned previously, and as illustrated in Figure 14.3 by the van't Hoff plots of benzo[a]pyrene (upper curve) and benzo[e]pyrene (the vertical dashed line gives the DSC-determined t_{n-i}). That is, a discontinuity occurs in the retentions that is coincident with the DSC-measured phase transition of the solvent. (In this instance, however, interestingly, the selectivity is very largely retained, as evidenced by the notable separation between the two solute lines to the left of the transition. There may well therefore be some residual ordering extant with this material, perhaps induced as a result of being coated onto the walls of a capillary tube as a thin film, since the solute lines for benzo[a]pyrene and benzo[e]pyrene would otherwise be expected to coalesce with isotropic MEPSIL.) In any event, the thermal

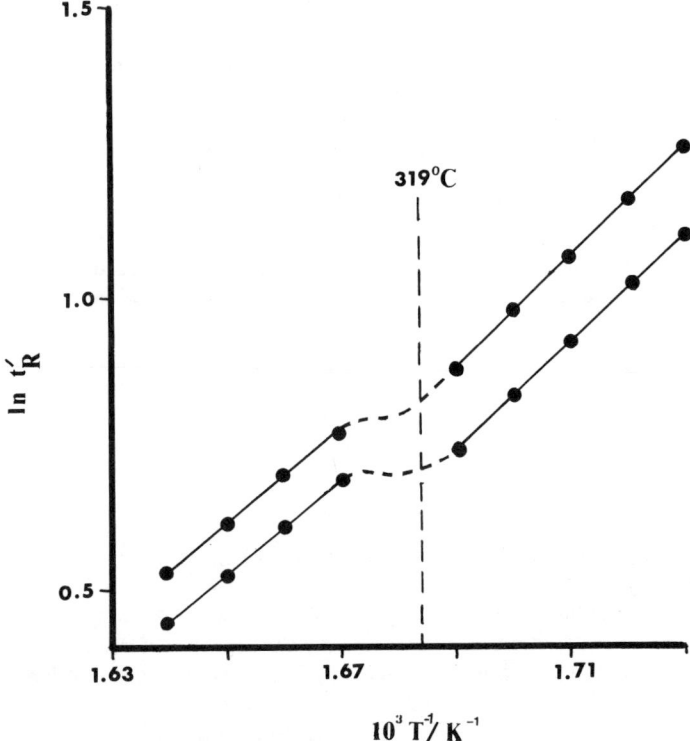

Figure 14.3 Plots of log $t_{R'}$ against reciprocal absolute temperature for benzo[a]pyrene (upper curve) and benzo[e]pyrene solutes with MEPSIL stationary phase of Figure 14.2. Vertical dashed line indicates DSC-determined nematic–isotropic transition temperature (319°C) (Apfel et al., 1985).

stability of this and other MEPSILs in inert atmospheres is limited at present to c. 280°C, at which point cleavage occurs at the ester function; thermogravimetric analysis (TGA) has also shown that the weight loss at this temperature approaches 1%. However, while stationary-phase 'bleed' is never desirable, from the standpoint of GLC analysis this level of solvent volatility is not overtly serious. For example, both N and R_S for benzo[a]pyrene and benzo[e]pyrene with several MEPSIL phases were found to be constant to within $\pm 5\%$ over one week of continuous operation at 280°C (Apfel et al., 1985). Moreover, only gradual deterioration of methoxybiphenyl MEPSIL was observed when columns containing this solvent were operated continuously at or above 300°C. The MEPSIL hence can be used at least for brief periods (as in temperature-programming) at upwards of 300°C without seriously affecting column performance. The practical operating temperature limits of the phase therefore encompass a large portion of its mesomorphic range.

Since, as noted above, MEPSILs of the form of that described decompose thermally at the ester linkage, Laub and colleagues undertook some time ago the synthesis of side chain polysiloxanes with other central linkages, including amide and Schiff's base functionalities (Janini et al., 1985). Unfortunately, these proved to be far too rigid when employed with methoxybiphenyl end-groups, as indicated by the high melting points shown in Table 14.2. The tactic may yet prove useful, however, as for example with end-groups comprised of multiple-substituted aromatic units designed to limit (or perhaps even eliminate altogether) crystal formation of the resultant polymer. On the other hand, chemical bonding and/or cross-linking of MEPSIL GC phases will obviously be of little value in this regard, since the point of decomposition is in the side chain and not along the polymer backbone. (Indeed, PDMS is thermally stable to in excess of 350°C both in bulk and thin-film states.)

A series of copolymer MEPSIL solvents was also synthesized several years ago by Laub and co-workers from reactant poly(hydrogenmethyl-dimethylsiloxane) (PHMDMS) polymers (Petrarch, Inc., USA) in an alternative attempt to lower the melting point of such phases while largely preserving the high clearing point (Janini et al., 1985):

$$\begin{bmatrix} CH_3 \\ | \\ Si-O \\ | \\ H \end{bmatrix} \begin{bmatrix} CH_3 \\ | \\ Si-O \\ | \\ CH_3 \end{bmatrix}$$

Unfortunately, however, the upper transition temperature for a given mesomorphic side chain was found to fall geometrically as $x \to 0$ (cf. Table 14.2). For example, for x/y of 1/3 with the methoxybiphenyl end-group, the nematic–isotropic transition falls fully 150° to 169°C. In addition, a smectic region appears that extends from the melting point, 80°, to 150°C. Thus, although the lower solid–mesomorphic transition temperature is indeed reduced in copolymers of this type, the upper nematic–isotropic transition is dramatically affected as well. Copolymers of these kinds have yet to be explored in any real detail, however, and may yet be found to hold some advantage in the quantitative adjustment both of the lower and upper transition temperatures of MEPSIL stationary phases (Janini et al., 1987).

14.3.2 *Column efficiency*

The unique flexibility of the polysiloxane backbone as well as its low surface tension ensures quite good column efficiency for MEPSIL phases. For example, glass-capillary columns containing the methoxybiphenyl version exhibit roughly 2500 plates per metre (chrysene solute, $k' > 10$; 230°C), which compares very favourably with c. 3000–3500 N m^{-1} obtained for columns of identical dimensions that contain SE-30 or SE-52 phases. (Blending the MEPSIL with SE-30 yields still further improvement, as described later in this chapter.)

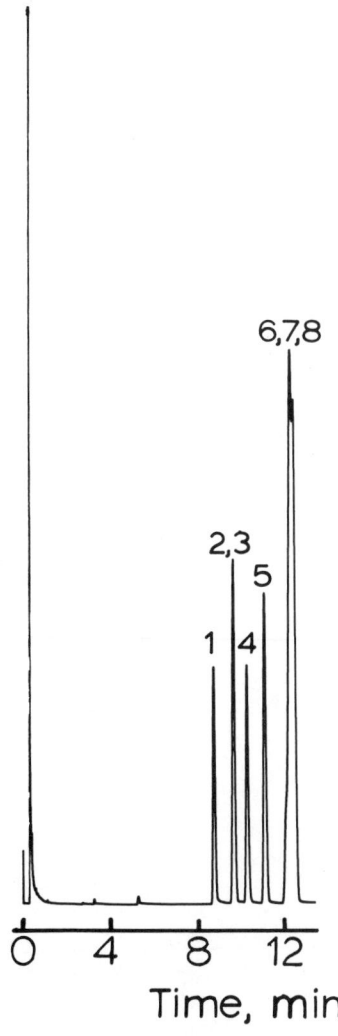

Figure 14.4 Separation of alkyl-substituted aromatic hydrocarbons of Table 14.3 with SE-30 PDMS at 85°C (Finkelmann *et al.*, 1981).

14.3.3 *Selectivity*

MEPSIL phases have thus far been applied to the GLC separation of only a few classes of solutes, primarily because, until quite recently, the materials were commercially available only in bulk form and not as packed or coated capillary columns. Even so, the kinds of samples that have been resolved are sufficiently impressive to warrant their evaluation in a wide range of separations problems. A few applications from the work of Laub and co-workers are presented as examples.

(i) *Polycyclic aromatic hydrocarbons (PAHs).* The first demonstration of the quite remarkable gas-chromatographic selectivity of nematic MEPSIL phases is shown in Figures 14.4 and 14.5 (Finkelmann *et al.*, 1981, 1982). The first chromatogram, Figure 14.4, gives the separation of several alkyl-substituted aromatic hydrocarbons with SE-30 (PDMS) solvent at 85°C, where several overlaps are clearly evident. The next illustration, Figure 14.5a, provides the chromatogram obtained with a methoxy-phenyl MEPSIL (4-carbon spacer) at 100°C, at which temperature the solvent is isotropic. The elution order is obviously quite different in this tracing from that

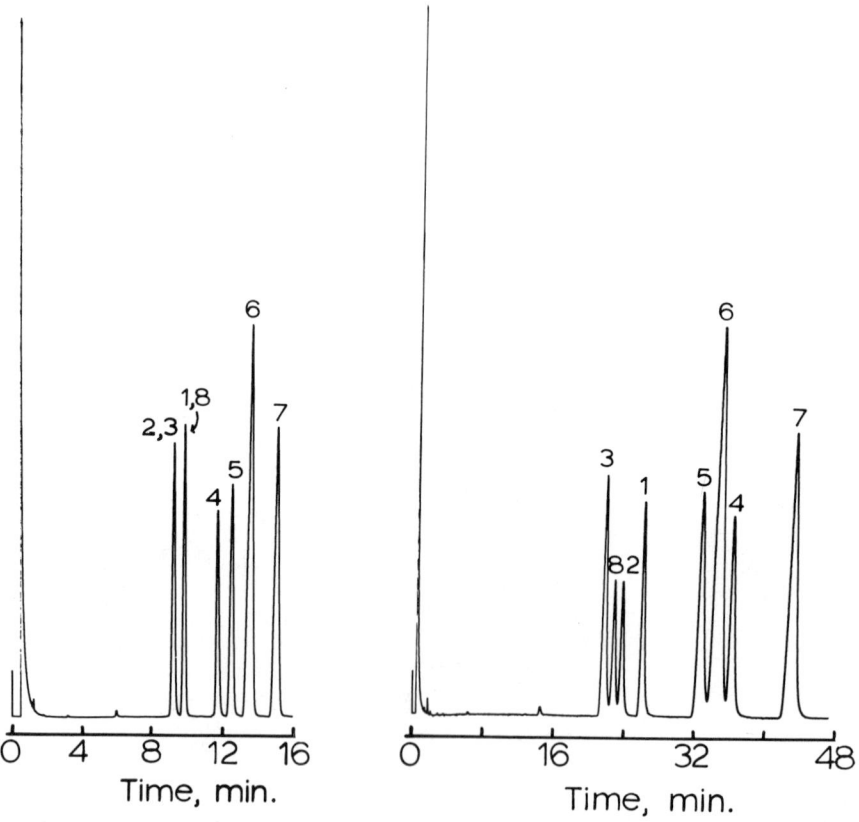

Figure 14.5 As in Figure 14.4; methoxyphenyl MEPSIL stationary phase (4-carbon spacer) at 100°C (isotropic) and 70°C (nematic).

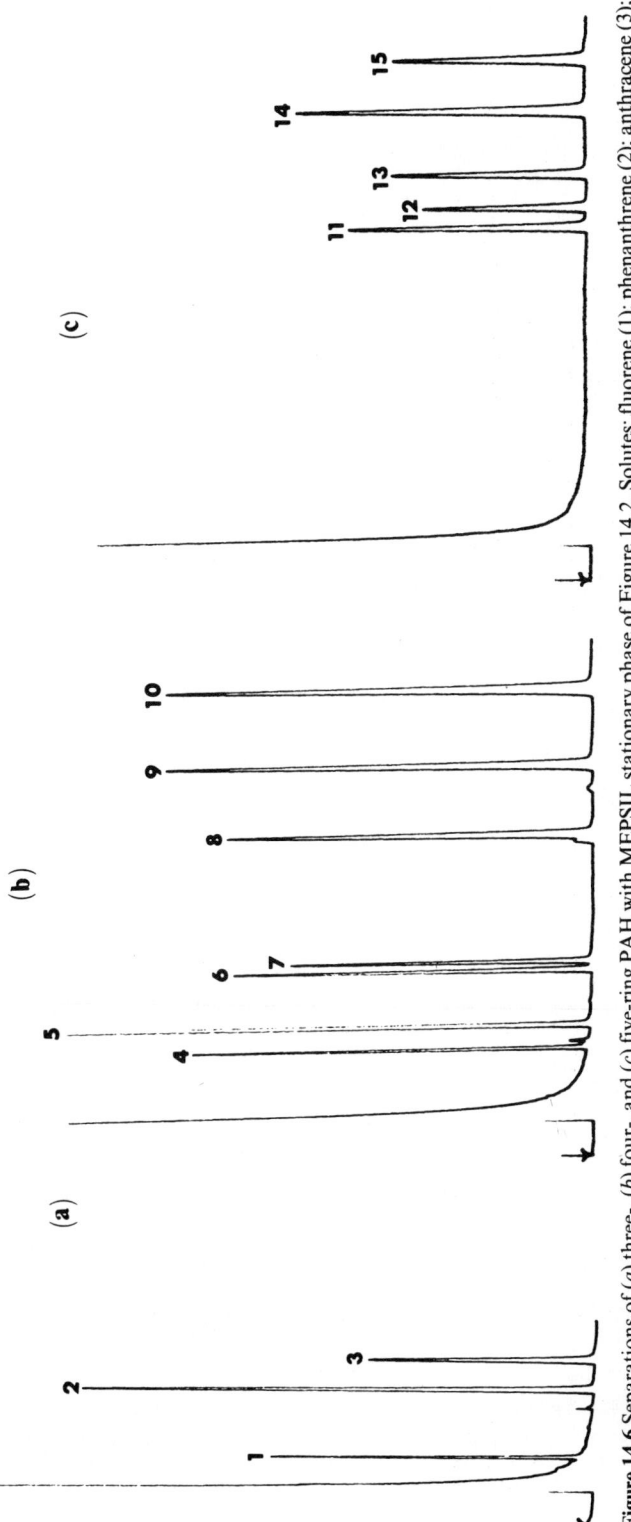

Figure 14.6 Separations of (*a*) three-, (*b*) four-, and (*c*) five-ring PAH with MEPSIL stationary phase of Figure 14.2. Solutes: fluorene (1); phenanthrene (2); anthracene (3); fluoranthene (4); pyrene (5); 1, 2-benzofluorene (6); 2, 3-benzofluorene (7); triphenylene (8); benz[*a*]anthracene (9); chrysene (10); benzo[*b*]fluoranthene (11); benzo[*k*]fluoranthene (12); benzo[*e*]pyrene (13); perylene (14); and benzo[*a*]pyrene (15). Column temperatures: (*a*), 180°; (*b*), 235°; (*c*), 275°C isothermal (Apfel *et al.*, 1985).

preceding; however, the solute mixture has still to be fully resolved. The third chromatogram, Figure 14.5*b*, presents the result obtained with the same phase at 70°C, where it is in the nematic state, and where all solutes have been fully separated. Moreover, the elution order is substantially different from those extant either in (*a*) or (*b*), and follows roughly what would be expected on the basis of the aspect (length-to-breadth) ratio of the solutes. For example, 2, 3-dimethylnaphthalene (no. 7) is more rodlike than, say, 1, 3-dimethylnaphthalene (no. 5) and so, the former elutes well after the latter.

Figure 14.6 presents the isothermal separation of some three-, four-, and five-ring PAHs with a methoxybiphenyl MEPSIL (Apfel *et al.*, 1985; Janini *et al.*, 1988; see also Kong *et al.*, 1982; Bradshaw *et al.*, 1985; Markides *et al.*, 1985; Nishioka *et al.*, 1986; Bayona *et al.*, 1987). One of the more interesting aspects of the latter chromatogram is that benzo[*a*]pyrene is separated from benzo[*e*]pyrene to such an extent that at least four other peaks could be placed between the two, yet these compounds are otherwise difficult to resolve with most other GC phases (the elution order with PDMS is benzo[*e*]pyrene + benzo[*a*]pyrene, followed by perylene). Figure 14.7 then gives the programmed-temperature separation of a synthetic mixture of 17 of the three- to five-ring PAH encountered frequently *inter alia* in environmental samples.

(ii) *Polychlorinated biphenyls (PCBs)*. Programmed-temperature separations of Aroclor 1254 and Aroclor 1260 respectively are presented in Figures 14.8 and 14.9 (Janini *et al.*, 1988), where the Ballschmiter–Zell number (1980) and substitution

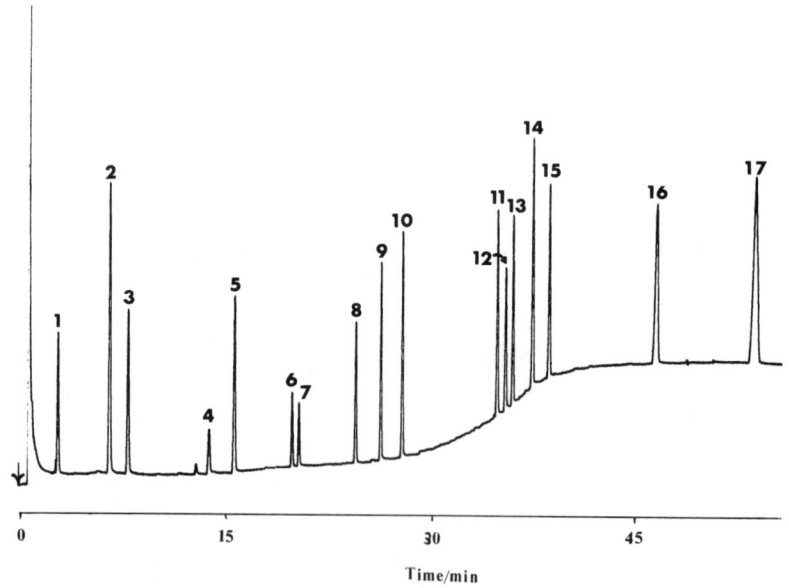

Figure 14.7 Programmed-temperature separation of common PAH with MEPSIL phase of Figure 14.2. Solutes numbered as in Fig 14.6. Additional solutes: 1, 2, 3, 4-dibenzanthracene (16); benzo[*ghi*]perylene (17). Column temperature: 140°C 3min⁻¹; 140–280°C at 4°C min⁻¹ (Apfel *et al.*, 1985).

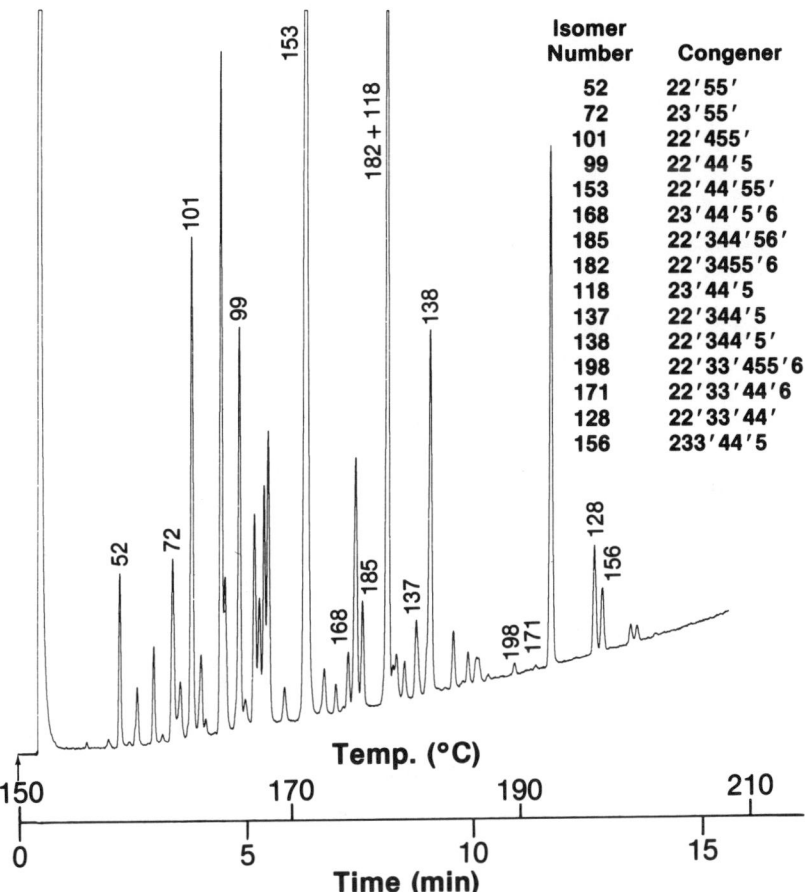

Isomer Number	Congener
52	22'55'
72	23'55'
101	22'455'
99	22'44'5
153	22'44'55'
168	23'44'5'6
185	22'344'56'
182	22'3455'6
118	23'44'5
137	22'344'5
138	22'344'5'
198	22'33'455'6
171	22'33'44'6
128	22'33'44'
156	233'44'5

Figure 14.8 Programmed temperature open-tubular column separation of Aroclor 1254 with MEPSIL phase of Figure 14.2: Temperature 150°C 1 min^{-1}; 150–210°C at 4°C min^{-1}. Numbered peaks identified by injection of standards (Janini *et al.*, 1988).

pattern are given for the peaks that were identified with standards. There are 209 discrete PCBs, called congeners, that are comprised of 1 to 10 chlorine atoms attached to biphenyl; as a result, commercial materials as well as environmental samples that contain these compounds generally give quite complex GLC patterns. The shape selectivity provided by MEPSIL phases may therefore not necessarily provide any real advantage over the use of high-resolution capillary columns containing other phases (Zielinski, Jr. *et al.*, 1986). Even so, liquid-crystalline solvents yield distinctly different elution orders of PCB from any other phases, that is, will provide complementary information that should enhance efforts at the identification of congeners in complex mixtures. The potential advantages of MEPSIL phases for difficult separations of large numbers of solutes with multiple-column switching systems, just now becoming available commercially, should also not be overlooked.

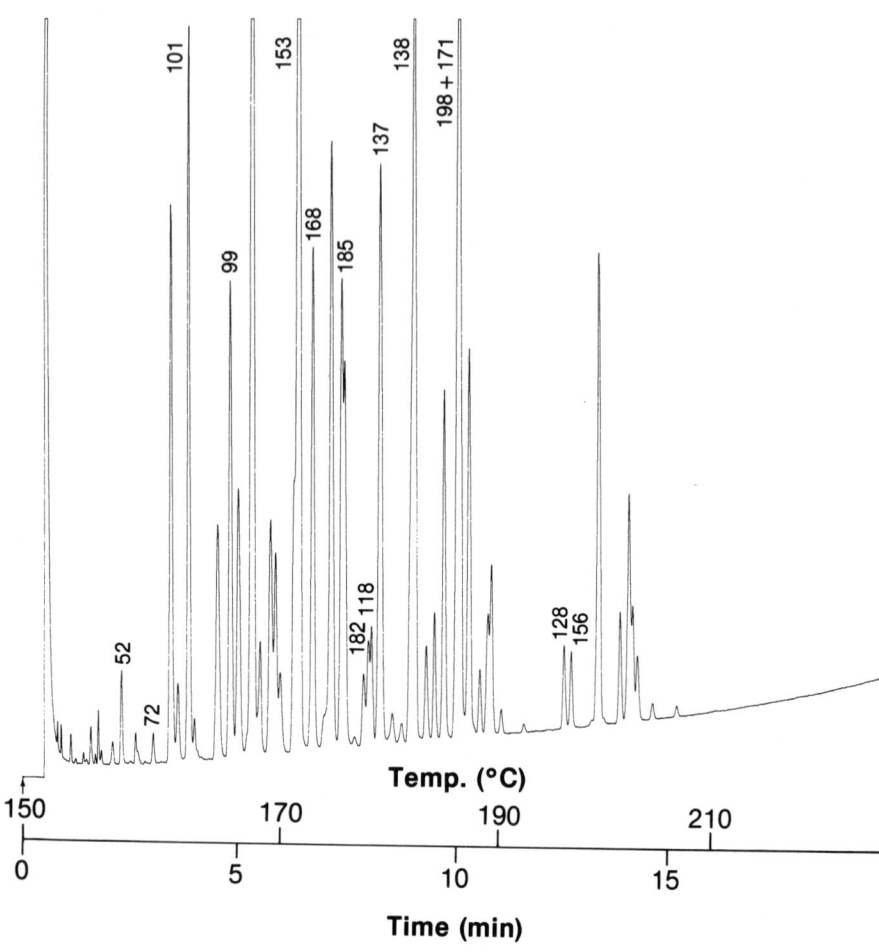

Figure 14.9 As in Figure 14.8: Aroclor 1260.

(iii) *Fatty acid methyl esters (FAMEs)*. Figure 14.10 shows the isothermal (150°C) separation of some *cis* and *trans* fatty acid methyl ester (FAME) solutes with the methoxybiphenyl MEPSIL solvent. The order of elution is quite remarkable when compared with that generally observed with conventional phases. First, all of the *cis/trans* isomer pairs are fully resolved in less than 10 minutes. Second, the *trans* isomers are retained longer than the respective *cis* compounds of the same carbon number, which is the reverse of that observed with most other solvents commonly employed for FAME analysis. The temperature-programmed separation of a range of FAMEs is shown next in Figure 14.11, where elution is in order of increasing carbon number. In addition, within groups of compounds of the same carbon number, retentions increase with increasing saturation. That is, the last-eluting compound of each group is the fully-saturated isomer. This marks another sharp departure from the elution order found with conventional phases, with which, almost without exception,

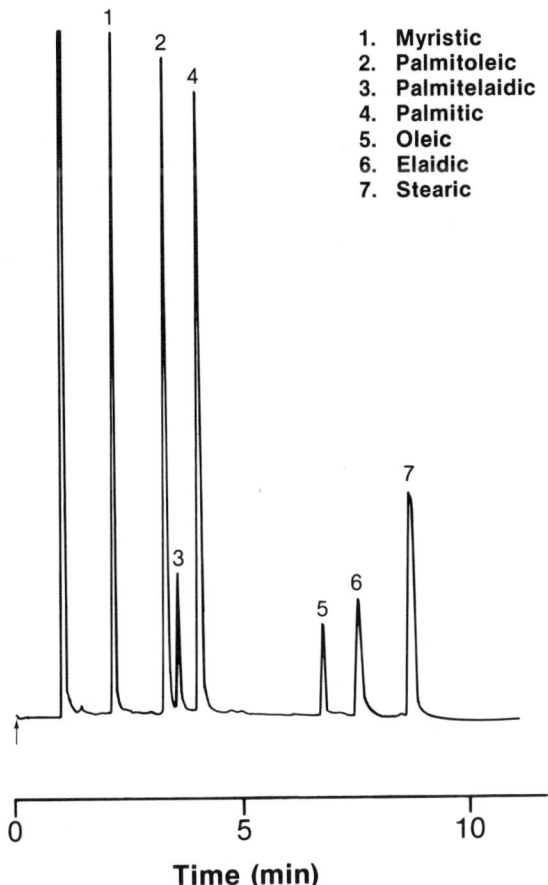

1. **Myristic**
2. **Palmitoleic**
3. **Palmitelaidic**
4. **Palmitic**
5. **Oleic**
6. **Elaidic**
7. **Stearic**

Figure 14.10 Isothermal open-tubular column separation of *cis* and *trans* fatty acid methyl esters at 150°C with MEPSIL phase of Figure 14.2 (Janini *et al.*, 1988).

the saturated isomer elutes first and the retentions of related isomers then increase with increasing unsaturation.

14.3.4 *MEPSIL solvents in admixture with PDMS phases*

In order to realize the advantages of blended phases in analytical GLC separations, one or other versions of the relations presented in section 14.1.3 must of course apply, such that retentions with neat phases can be used to predict those with mixed solvents. However, it might be supposed that blending a MEPSIL phase with, say, a PDMS could well reduce the upper transition (hence narrow the mesomorphic span) to such an extent that the selectivity requisite for the separation to hand might be lost. Fortunately, the use of blended phases in GLC is not so problematic. First, in the instance of packed columns, the simplest as well as most practicable means of realizing mixed phases is mechanically blending appropriate amounts of support materials that

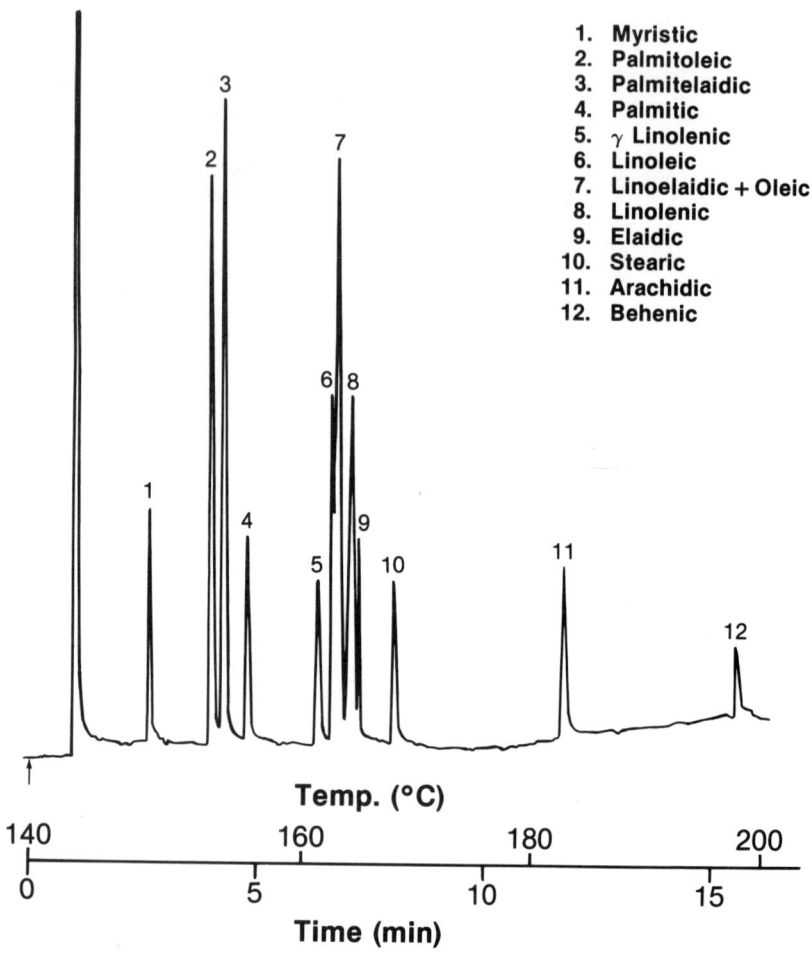

Figure 14.11 Programmed-temperature open-tubular column separation of fatty acid methyl esters with MEPSIL phase of Figure 14.2. Temperature: 140°C 1 min^{-1}; 140–200°C at 4°C min^{-1} (Janini *et al.*, 1988).

have been coated separately with pure phases. Moreover, since mixtures of pure-phase packings in effect constitute immiscible liquids, eqns (14.25–14.29) (hence (14.32)) must apply exactly (Chien *et al.*, 1980). Further, essentially the same situation pertains to serial-connected pure-phase capillary columns, so long as the pressure drop across each is correctly taken account of (Purnell and Williams, 1983, 1984, 1985*a,b*; Purnell *et al.*, 1985, 1986, 1987). Even so, it is conceivable that there might be some advantage to the use of intimate blends comprised in part of MEPSILs, for example, in order to improve column efficiency. The matter is therefore of some interest in analytical GLC, and we consider below some recent results obtained with mixtures of methoxybiphenyl MEPSIL + PDMS.

Shown in Figure 14.12 are plots of $\log t'_R$ against T^{-1} for benzo[*a*]pyrene (upper curve in each) and benzo[e]pyrene test-solutes with 50/50 and 25/75 w/w intimate blends of a methoxybiphenyl MEPSIL (3-carbon spacer) with SE-30. While the van't Hoff plot with the pure MEPSIL exhibits a discontinuity at 319°C (see Figure 14.3), no such discontinuity is evident with either blend. Instead there is some curvature in the plots over the temperature ranges that encompass the DSC-measured t_{n-i} (vertical dashed line in each figure). Also, the upper transition temperature of the MEPSIL was barely affected when blended with SE-30. In fact, t_{n-i} was depressed by only 11°C even when the phase was diluted by as much as 3:1. The chromatographic selectivity of blended MEPSILs is therefore left intact at temperatures in excess of 300°C.

Plots of $\log t'_R$ against T^{-1} for various probes with the above blended MEPSIL

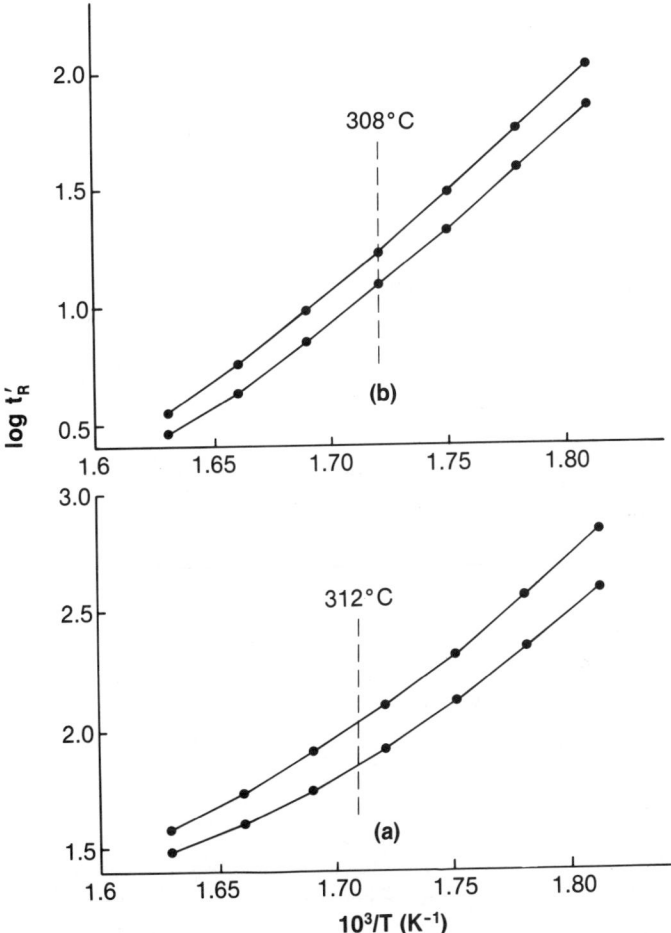

Figure 14.12 Plots of log t'_R against reciprocal absolute temperature $10^3 \; T^{-1} K^{-1}$ for benzo[a]pyrene (upper curve) and benzo[e]pyrene solutes with (a) 50/50 MEPSIL/SE-30 and (b) 25/75 MEPSIL/SE-30 admixed stationary phases. DSC transition temperatures indicated by vertical dashed lines (Janini et al., 1988).

solvents were also found to be linear from 170° to 80°C. Moreover, DSC scans of several mixtures showed no isotherms on cooling the melts from 250°C to ambient. It might therefore be thought that the lower practical operating temperature of MEPSILs could be lowered when admixed. In order to gauge to what extent this might be so, the performance of the methoxybiphenyl MEPSIL blended 1:3, 1:1, and 3:1 with SE-30 was assessed in terms of plots of α and R_s against temperature for 1,3- and 2,3-dimethylnaphthalene solutes, which have reasonable retentions at moderate column temperatures. The results are presented in Figure 14.13, and show that operation near the lower temperature limits of melts of the admixed MEPSIL gives greatly diminished column efficiency. In fact, while the separation factor increases as the temperature of each column is lowered, the resolution plots pass through maxima that are reminiscent of the behaviour of the neat MEPSIL shown earlier in Figure 14.2.

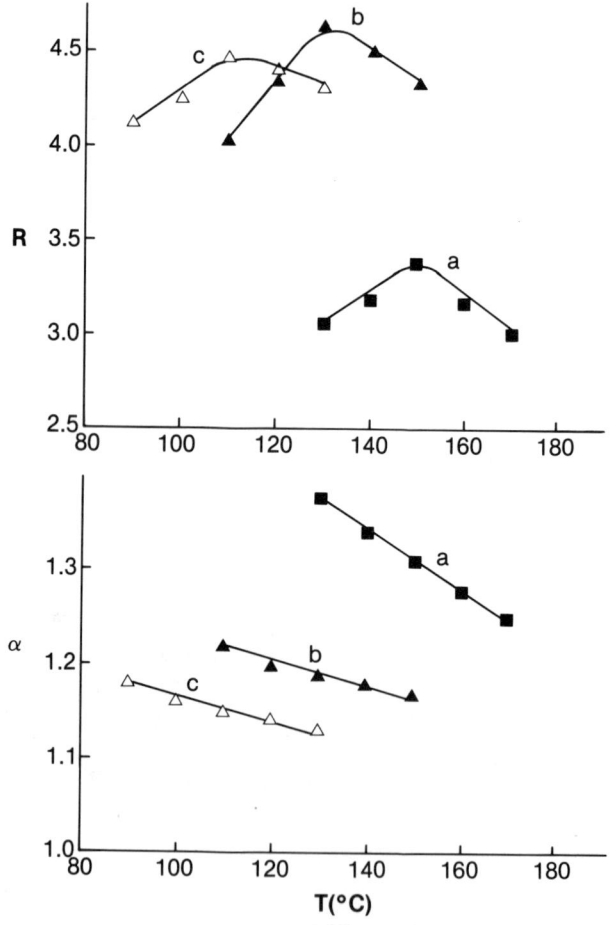

Figure 14.13 Plots of α and resolution R_s against temperature $t(°C)$ for the solute pair 2,3-/1,3-dimethylnaphthalene with (a) MEPSIL; (b) 50/50 MEPSIL/SE-30; and (c) 25/75 MEPSIL/SE-30 admixed stationary phases (Janini *et al.*, 1988).

In contrast, while blended-MEPSIL columns routinely yield on average only c 1000 N m^{-1} at lower temperatures, their efficiency rises dramatically as the temperature is raised. For example, the 1:1 MEPSIL/SE-30 column gave 3200 N m^{-1} with benz[a]anthracene solute at 240°C, a notable improvement over that provided by the pure MEPSIL. (However, no further increase in efficiency was observed upon increasing the extent of dilution to 1:3.) One consequence of this is that, while the α values shown in Figure 14.13 are largest with the neat phase, the higher column efficiency of the blended solvents actually results in resolutions obtained with them being the superior.

Figure 14.14 illustrates next the effects of dilution of the MEPSIL phase on the separations of the three- and four-ring benzenoid isomers anthracene/phenanthrene at 200°C, and chrysene/benz[a]anthracene at 235°C (each pair coelutes with neat SE-30 at these temperatures). Not suprisingly, the α values were found to be greatest with pure MEPSIL; however, as shown, the selectivity of the mixed phases is not related linearly to the extent of dilution. In fact, the alphas were only slightly diminished for blends of up to 1:1 and, further, the loss of selectivity was more than compensated for by the

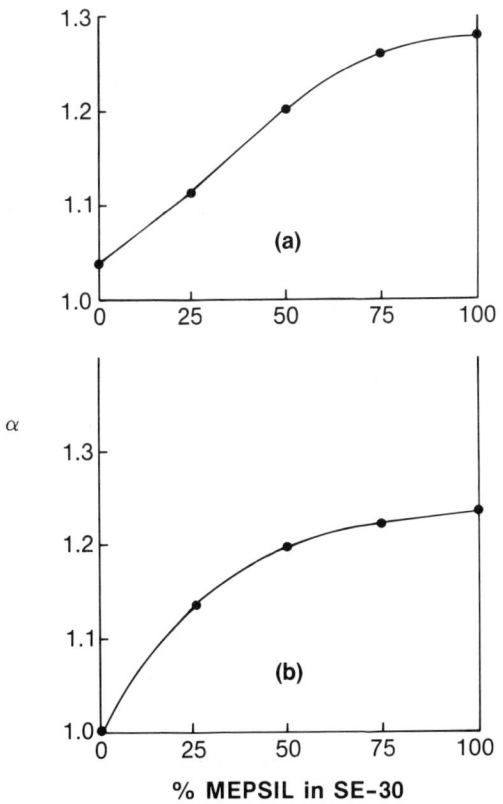

Figure 14.14 Plots of alpha for PAH as a function of stationary-phase composition. (a) Anthracene/phenanthrene (200°C); (b) chrysene/benz[a]anthracene (235°C) (Janini *et al.*, 1988).

gains realized in efficiency. (There would thus appear to be advantage in instances such as these in constructing window diagrams from plots of R_s against column composition, as opposed to α, as pointed out some time ago by Laub, 1983.)

The practical utility of blended MEPSIL phases is further illustrated in Figures 14.15–14.17. The first of these shows the programmed-temperature separation of a synthetic mixture of common PAHs (most of which are priority pollutants) with a 1:1 MEPSIL/SE-30 column; while the separations of some isomeric methylbenz[*a*]anthracene and methylbenzo[*a*]pyrene solutes are then given in the

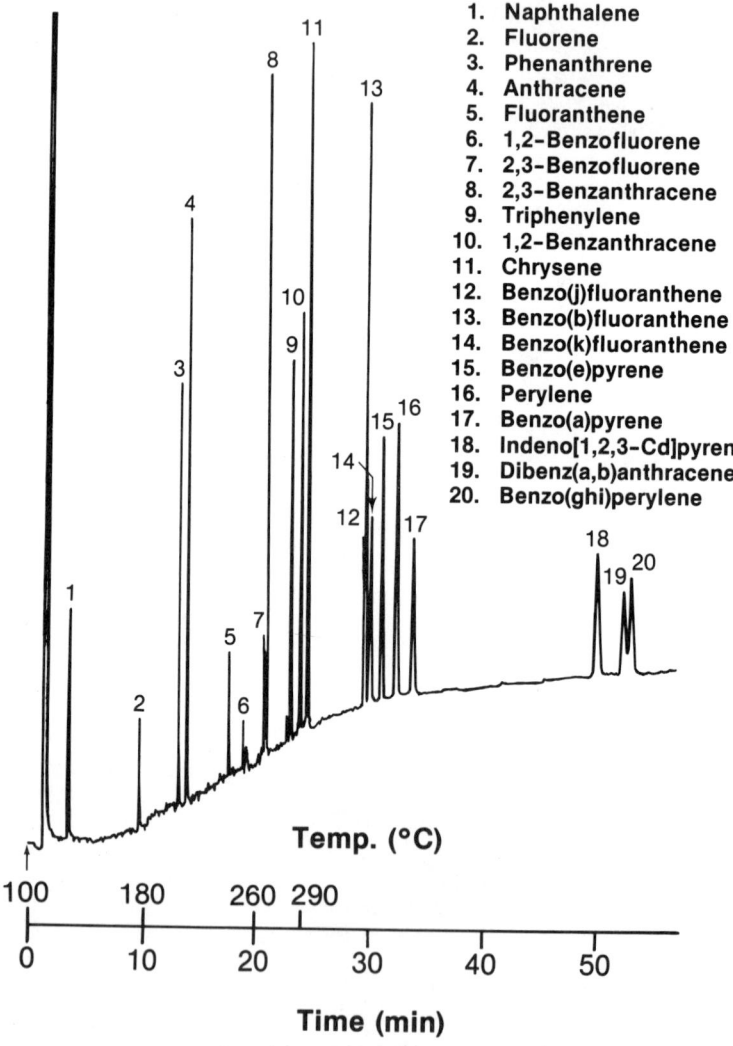

1. Naphthalene
2. Fluorene
3. Phenanthrene
4. Anthracene
5. Fluoranthene
6. 1,2–Benzofluorene
7. 2,3–Benzofluorene
8. 2,3–Benzanthracene
9. Triphenylene
10. 1,2–Benzanthracene
11. Chrysene
12. Benzo(j)fluoranthene
13. Benzo(b)fluoranthene
14. Benzo(k)fluoranthene
15. Benzo(e)pyrene
16. Perylene
17. Benzo(a)pyrene
18. Indeno[1,2,3–Cd]pyrene
19. Dibenz(a,b)anthracene
20. Benzo(ghi)perylene

Figure 14.15 Programmed-temperature open-tubular column separation of common PAH solutes with a 50/50 MEPSIL/SE-30 stationary phase. Temperature 100°C 2 min^{-1}; 100–290°C at 8°C min^{-1} (Janini *et al.*, 1988).

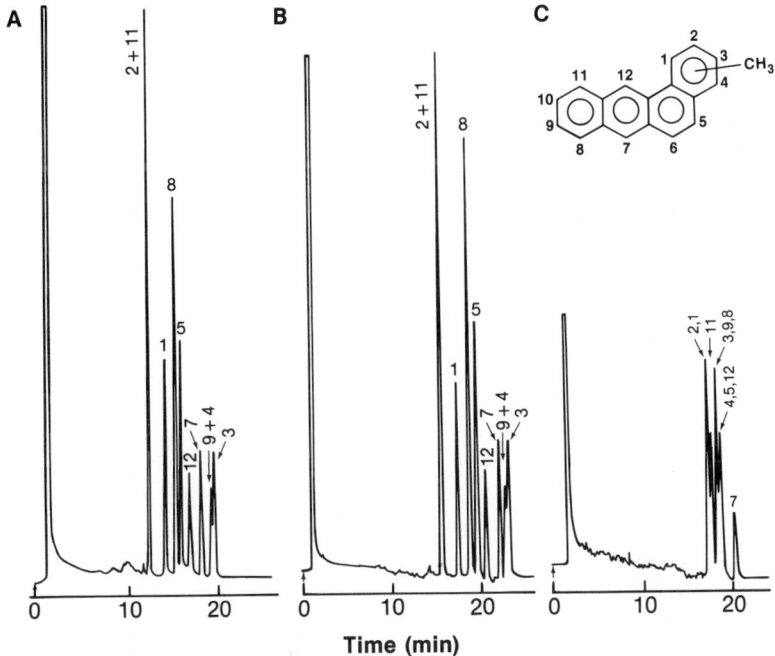

Figure 14.16 Isothermal open-tubular column separation of indicated isomeric methyl-substituted benz[a]anthracene solutes at 240°C with (a) MEPSIL, (b) 50/50 MEPSIL/SE-30, and (c) SE-30 stationary phases (Janini et al., 1988).

figures following. Interest in the analysis of the latter compounds stems from their biological activity as among the most potent aromatic alkylating agents found in the environment; various isomers are also carcinogenic (Hecht et al., 1974). We note in passing that the order of elution of the PAHs in Figure 14.15 is identical to that obtained with neat MEPSIL, Figure 14.7. In addition, the blended phase gives a measurable adjusted retention time for naphthalene, whereas this solute elutes with (or very close to) the solvent front with pure MEPSIL. Also, the separation of isomers of methyl-substituted PAH is virtually impossible with SE-30; moreover, attempts at resolving the methylbenz[a]anthracenes with SE-54 by GC (Garrigues et al., 1987) as well as by HPLC (Wise et al., 1981) have by and large not proved fruitful. Some separations of these compounds with GC capillary columns containing a smectic MEPSIL have been reported (Bradshaw et al., 1986), although, given the imprecision of the transition temperatures of the material (Jones et al., 1984; Markides et al., 1985), it is not clear at present to what extent these results can be reproduced. In any event, and, whereas the SE-30 column, (c) in Figure 14.16, separated only 5 peaks, the MEPSIL columns resolved 8 of the 10 methylbenz[a]anthracene isomers injected at 240°C. Further, the 1:1 blended MEPSIL phase showed a substantial improvement over neat MEPSIL in the separation of the methylbenzo[a]pyrene solutes, Figure 14.17. Indeed, it appears likely that careful window-diagram optimization of the blended phases could well result in full resolution of all 12 isomers of each of these families of PAH.

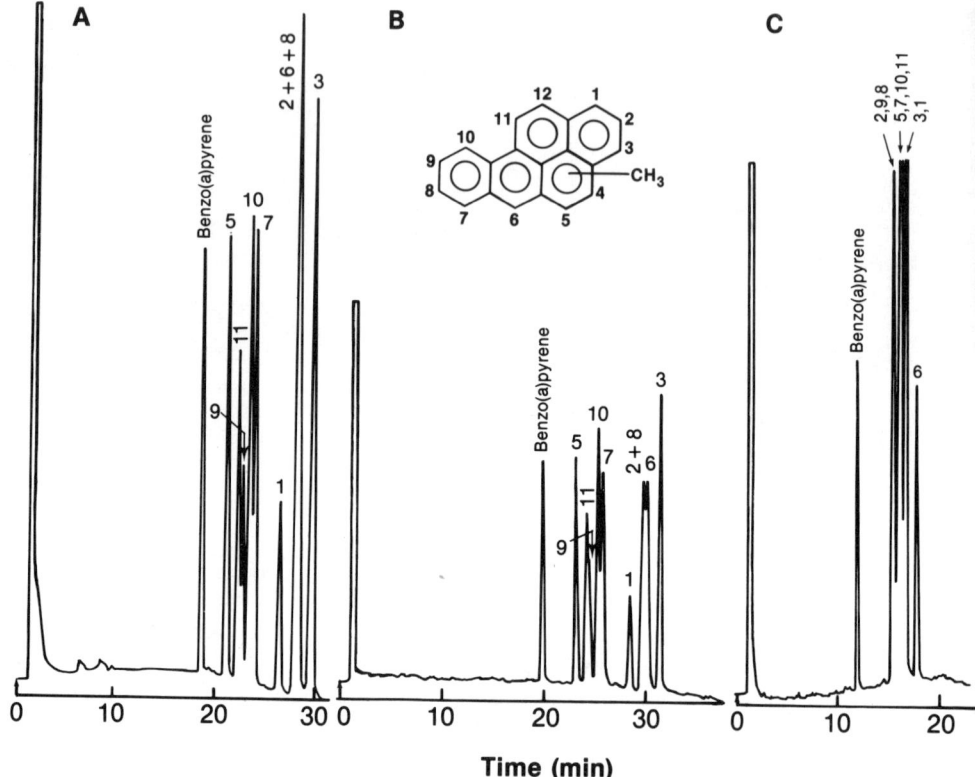

Figure 14.17 As in Figure 14.16; isomeric methyl-substituted benzo[*a*]pyrene solutes at 250°C.

14.3.5 *Comparison with mesomorphic polyacrylate (MEPCRYL) phases*

Liquid-crystalline acrylate monomers readily undergo radical polymerization with 2, 2'-azoisobutyronitrile (AIBN) initiator to yield mesomorphic polyacrylate (MEPCRYL) polymers, the details of some of which are presented elsewhere in this text. Rokushika, Naikwadi, and co-workers (Rokushika *et al.*, 1985; Naikwadi *et al.*, 1986; Jadhav *et al.*, 1987) appear to have been the first to employ nematic MEPCRYLs as GLC solvents; one of the most useful has a glass–nematic transition t_{g-n} of 85°C, and a nematic–isotropic transition temperature of 291°C:

$$\begin{array}{c} | \\ CH_2 \\ | \\ HC-CO_2-(CH_2)_2-O-Ph-CO_2-Ph-N=N-Ph-n\text{-}C_4H_9 \\ | \end{array}$$

As far as can be ascertained from published chromatograms, separations of PAH for example with MEPCRYLs are at least the equal of those observed with MEPSIL phases. Of particular interest as well is the resolution of isomeric insect pheromones with a MEPCRYL phase, where near-baseline separation of E, E-, Z, Z-, Z, E-, and

E, Z-9, 12-tetradecadienyl acetate solutes was achieved in a little over an hour's time with a packed column (Rokushika *et al.*, 1985). However, other chromatograms illustrated in the same work gave indication of considerable peak-tailing. In addition, the upper temperature limit of these materials is subject to the same drawback as governs the most successful of the nematic MEPSIL phases, namely, decomposition at the side chain ester linkage(s). It is therefore not clear at present what advantages, if any, MEPCRYL phases might eventually be shown to have over their MEPSIL counterparts. Nevertheless, there are as yet only a few data available upon which to base such comparisons, and any additional observations on the matter hence must await further and comprehensive study and investigation of both of these classes of GLC solvents.

14.3.6 *Chiral MEPSIL phases*

As mentioned at the outset of this chapter, chloresteryl esters were one of the earliest types of liquid-crystalline phases to be employed in GLC. When a cholesterol end-group monomer is attached to a polysiloxane backbone (see Apfel *et al.*, 1985; Janini *et al.*, 1985; Adams *et al.*, 1987, for recent examples) the result is invariably a twisted smectic-C phase, where the twist occurs between two-dimensional ordered sheets of side chain. However, each cholesterol end-group of course still contains a chiral centre, and so, the resultant MEPSIL polymer exhibits optical activity.

Chiral monomers that are not liquid-crystalline can also be attached as side chains to siloxane polymers, giving chiral (yet nonmesomorphic) polysiloxanes. The best-known compounds of these types are those containing amino acid residues, which were first synthesized by Bayer and Frank (1979). The phases have since proved useful for the GLC separation of a wide range of enantiomers, including all protein amino acids (Bayer and Frank, 1980, 1981; Bayer *et al.*, 1980; Frank *et al.*, 1977, 1978a,b; Koppenhoefer and Bayer, 1984).

However, bar cholesterics, true chiral–nematic mesomorphic siloxane polymers were synthesized only relatively recently (Finkelmann, 1984). The materials are of the usual form of aromatic end-group MEPSILs, except that the *p*-alkoxy group is replaced with one that is optically active, such as 2-methyl-1-butoxy or 3-methyl-2-butoxy. The rationale for doing so is that a single stationary phase might thereby be made to exhibit selectivity on the basis both of solute shape and chirality. Unfortunately, chiral recognition with such materials has thus far proved to be weak. For example, Aggarwal *et al.* (1987) and Bradshaw *et al.* (1987) have recently synthesized a series of chiral polysiloxanes, the structure of the most successful of which as a chiral selective GLC phase for *N*-pentafluoropropionyl isopropyl esters of amino acids is shown below:

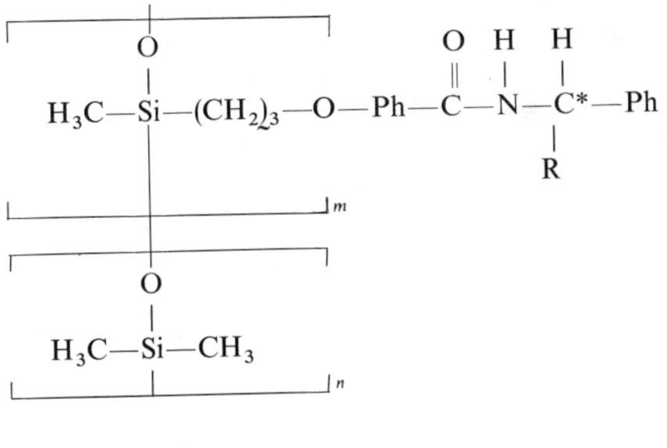

$$
\left[H_3C-\underset{\underset{O}{|}}{Si}-(CH_2)_3-O-Ph-\overset{\overset{O}{||}}{C}-\overset{\overset{H}{|}}{N}-\overset{\overset{H}{|}}{\underset{\underset{R}{|}}{C^*}}-Ph \right]_m \left[H_3C-\underset{\underset{O}{|}}{Si}-CH_3 \right]_n
$$

where $m{:}n \approx 1{:}13$. It was observed that there must be a very considerable gap between the chiral side chains in order that enantiomer resolution take place, as had previously been found also by Bayer and Frank (1980). The explanation for this is that solutes such as amino acids must be able to interact simultaneously with two adjacent side chains, that is, there must be sufficient space, but just sufficient, for the solute to fit between. The separation of derivatized amino acid enantiomers appears to require that $m{:}n$ be about 1:6, as employed by Bayer and co-workers; further, the separation worsens substantially if m is made either larger or smaller. The resolution of amino acids achieved by Bayer *et al.*, therefore remains, indisputably, the best that has yet been achieved by GLC with chiral polysiloxane stationary phases. On the other hand, the requisite spacing is so large as to preclude mesomorphic ordering. Whether a suitable compromise that exhibits both chiral and nematic selectivity can eventually be realized therefore remains at present open to speculation.

14.4 GLC studies of the physicochemical properties of MEPSIL phases

14.4.1 *Transition temperatures*

It is usually the case that gas-chromatographic retentions exhibit discontinuities, or at least distortions, at solvent phase transitions. Illustration of this was provided earlier in Figure 14.3, where the plots of $\log t'_R$ against T^{-1} were discontinuous at the nematic–isotropic transition temperature of a methoxybiphenyl MEPSIL. Thus, van't Hoff plots, which are otherwise invariably linear, can be used to detect changes in the mesomorphic state of the solvent melt. (Unfortunately, this is usually not so for stationary phases in the solid state.) The technique thus complements DSC and hot-stage light-polarized microscopy. GLC is also useful for the investigation of transitions of mesomorphs both in bulk (viz., thick-film packed-column systems) and in thin-film coatings on a variety of (capillary-tube) surfaces, analogous to studies carried out with coated microscope cover slips. There is, however, the additional advantage that, once assembled, the GLC system can then be employed to determine a wide range of thermodynamic properties of probe-solute/solvent interactions over comprehensive ranges of temperature quickly, easily, and accurately, as mentioned at the outset of this chapter (see also below). On the other hand, it must be noted that the GLC method of determining phase transitions has the drawback that retentions in such regions are indistinct, and there is therefore some spread on the values thereby derived. That in Figure 14.3, for example, extends over 8°C (dotted-line range). Nor does it appear that the situation can be improved much beyond that shown, insofar as the closer a transition temperature is approached the more unstable is the state of the liquid phase. Thus, smaller and smaller amounts of probe-solute must be injected to avoid masking (or, indeed, causing) a change in the properties of the solvent. The best that can be achieved in GLC is, therefore, transition temperatures that must be inferred from extrapolation, the accuracy of which will be limited by the sensitivity of the detection system, and which in any event should probably be taken as no better than $\pm 1°C$.

It is worth noting at this point that, barring catastrophic conditions, since bulk vapour pressures are not discontinuous, the sharp changes observed in van't Hoff plots at phase transitions must be due to abrupt swings in the respective solute infinite-dilution activity coefficients.

14.4.2 *Thermodynamic properties of solution*

As mentioned in section 14.1.2, the usual thermodynamic and excess properties of solution can be derived from the temperature dependencies of V_g^0 and γ_A^∞. Studies of these kinds with MEPSIL phases are currently under way in several research groups, and swift progress can be anticipated now that many of the details of MEPSIL synthesis have been clarified. In the meantime, Laub (1988) has recently reported the initial results obtained with a more expedient means of assessing, via 'family' plots, some of the physicochemical properties of MEPSILs.

Recall that the Clausius–Clapeyron equation relates the solute vapour pressure to temperature in the manner:

$$\log p_A^0 = -\Delta \bar{H}_v/2.3\,RT + C \tag{14.33}$$

where $\Delta \bar{H}_v$ is the solute molar heat of vaporization, and C is a constant of integration. Substituting this relation into the log-form of eqn (14.5), and assuming that fugacity effects are negligible,

$$\log V_g^0 = -a \log p_A^0 + C \tag{14.34}$$

where $a = \Delta \bar{H}_s / \Delta \bar{H}_v$. Thus, a 'family' plot of log V_g^0 against log p_A^0 for a give solute over a range of temperatures should be linear and of slope -1 for ideal solutions, as first pointed out some time ago by Hoare and Purnell (Hoare and Purnell, 1955, 1956; Purnell, 1957, 1962). Further, since $\Delta \bar{H}_s + \Delta \bar{H}_v = h^E$, it must also be the case that $a = (h^E/\Delta \bar{H}_v) - 1$, that is, the excess enthalpy can also be determined from the slopes of the plots.

Laub and Tyagi (1988) have recently tested the above equations with normal and branched alkane solutes with *n*-alkane solvents; and with a large number of aliphatic, cyclic, olefinic, and aromatic solutes with the OV series of poly(methylphenylsiloxane) solvents. They found some interesting effects of the solute molar volume on the apparent nonideality of the solute–solvent systems, as well as novel graphical means of displaying their results. Moreover, they pointed out that, since the solute capacity factor k' is related to V_g^0 simply through constants, plots of log k' against log p_A^0 should also be effective in gauging solution ideality. Further, there is then realized the considerable advantage of obviating measurement of the column content of stationary phase (whether by volume or mass), all of the requisite data being available directly from strip-chart tracings. This mode of data reduction should therefore prove useful as an alternative method of evaluating solute–MEPSIL interactions; two examples are presented below.

The Antoine constants A, B, and C of the relation (Dreisbach, 1955; Boublik *et al.*, 1973):

$$\log p_A^0 = A - B/(t + C) \tag{14.35}$$

(t in °C) as well as the capacity factors k' for a number of PAH solutes with a methoxyphenyl MEPSIL (4-carbon spacer) at 70°C (nematic) and 100°C (isotropic) (Finkelmann *et al.*, 1982) are provided in Table 14.3. Also shown for comparison are the capacity factors at 85°C with SE-30 PDMS. Figure 14.18 then illustrates the family regression found with the MEPSIL solvent when in both the nematic (left-hand side) and isotropic states. There is clearly evident, first, a substantial discontinuity in the regression that corresponds with the nematic–isotropic transition of the phase at 95°C.

Table 14.3 Antoine constants[a] and capacity factors[b] for listed solutes with a methoxyphenyl MEPSIL (4-carbon spacer) and SE-30 PDMS at indicated temperatures $T(°C)$

| | | | | | | k' | |
| | | Eqn (14.35) | | | MEPSIL | | SE-30 |
No.	Solute	A	B	C	70°C	100°C	85°C
1	Biphenyl	7.24541	1998.725	202.733	65.0	34.0	30.6
2	2-Ethylnaphthalene	7.4350	2131.1	212.2	59.0	31.9	33.9
3	1-Ethylnaphthalene	7.30532	2024.2	201.1	54.2	31.9	33.9
4	2,6-Dimethylnaphthalene	7.3968	2080.3	200.8	90.6	41.0	36.1
5	1,3-Dimethylnaphthalene	7.3978	2085.0	200.9	82.2	43.9	39.1
6	1,4-Dimethylnaphthalene	7.4040	2112.0	201.0	87.8	48.1	43.7
7	2,3-Dimethylnaphthalene	7.40396	2111.9	201.1	109.0	53.3	43.7
8	Diphenylmethane	7.51935	2197.1	211.6	56.8	34.0	43.4

[a]Dreisbach (1955); Boublik et al. (1973). [b]Laub (1988).

Secondly, leaving aside solutes 1 and 8 as not closely related to the 'family' set of alkyl-substituted naphthalenes, a reasonable straight line can indeed be constructed through the data for solutes 2–7 with the isotropic form of the MEPSIL, the slope a and linear least-squares correlation coefficient r being -0.97 and 0.98, respectively. (The regression might in fact be much better than this, since the retentions were taken with an analytical gas chromatograph, that is, none of the usual precautions associated with high-precision GC (Laub et al., 1978) were observed as it was the intent of the work at that time simply to test whether the phase exhibited shape selectivity.) In contrast, the solute retentions with the MEPSIL in the nematic state (left-hand side of Figure 14.18) exhibit considerable scatter, as reflected by a decrease of the correlation coefficient to 0.92 for the naphthalenic compounds, and to a mere 0.75 (i.e., no statistical connection) when

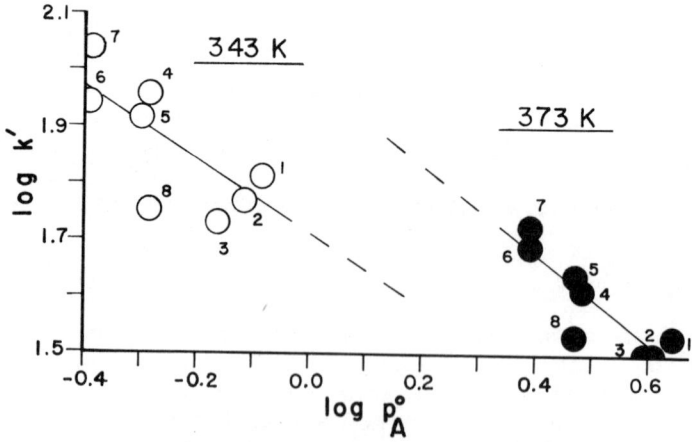

Figure 14.18 Plots of $\log k'$ against $\log p_A^0$ for solutes of Table 14.3 with a methoxyphenyl MEPSIL solvent at 70°C (nematic) and 100°C (isotropic). Symbol size reflects approximate experimental uncertainty on k' (Laub, 1988).

the data for nos. 1 and 8 are included. Thus, correlations of retentions drawn solely on the basis of vapour pressure appear to be inappropriate for mesomorphic phases; presumably better suited would be some or other parameter based on shape, or even size, such as the solute aspect ratio or molar volume. Even so, it must also be said that the overall patterns of the data points are not altogether dissimilar on passing from the nematic state to the isotropic, that is, the solute vapour pressure continues to play a substantial role. There can therefore be expected at least some residual family behaviour of homologous series of solutes irrespective of the state of the solvent, although how prominent this might be with smectic or chiral phases, for example, remains at present open to speculation.

Table 14.4 next presents the relevant vapour-pressure and capacity-factor data for some three- and four-ring PAHs with a methoxybiphenyl MEPSIL (3-carbon spacing) at 180°C and 235°C, both temperatures being within the nematic range of this phase. The 'family' regression is then shown in Figure 14.19, and is reasonably good despite the use of an analytical gas chromatograph, there being evident some obvious experimental scatter as commented upon above. However, that there should be any correlation at all for these quite disparate compounds with a nematic liquid crystal, and then at two temperatures more than 50°C apart, is intriguing. The solute vapour pressures must therefore exert considerable influence on the retentions, despite the evident strong shape selectivity of the mesomorphic stationary phase. Nevertheless, few data of the relevant form are at present available for further testing, and the regressions of many more fused-ring solutes comprising several well-defined families are required before any additional generalities can be drawn.

We note in passing that plots such as those shown in Figures 14.18 and 14.19 are important from a purely analytical standpoint, since they can be used to predict retentions (hence, separations) of other family-member or related compounds. Also, homologous-series plots of high accuracy can usually be derived from regressions of the retentions of no more than four or five compounds. In addition, the figures tend to imply that relatively small differences between solute families, say alkyl-substituted naphthalenes v. biphenyls, are sufficient to give rise to separate retention lines. This suggests in turn the possibility of qualititive identifications being carried out in terms of family-regression slopes and intercepts. Moreover, the data for homologous series of solutes at a common temperature often give rise to grid-like patterns, which could then provide a kind of two-dimensional confirmation of at least the chemical nature of an

Table 14.4 Antoine constants and capacity factors[a] for listed solutes with a methoxybiphenyl MEPSIL (3-carbon spacer) at indicated temperatures $T(°C)$

		Eqn (14.35)			k'	
No.	Solute	A	B	C	180°C	235°C
1	Anthracene	7.37609	2518.815	218.509	1.07	
2	Phenanthrene[b]				3.23	
3	Fluorene	7.76176	2637.095	243.190	4.05	
4	Fluoranthene	6.37310	1756.355	118.428		2.04
5	Pyrene	5.61838	1122.025	15.158		2.73

[a]Laub (1988). [b]$\log p_A^0 = (29.5477 - 4743.0)/(T - 6.7893 \log T)$.

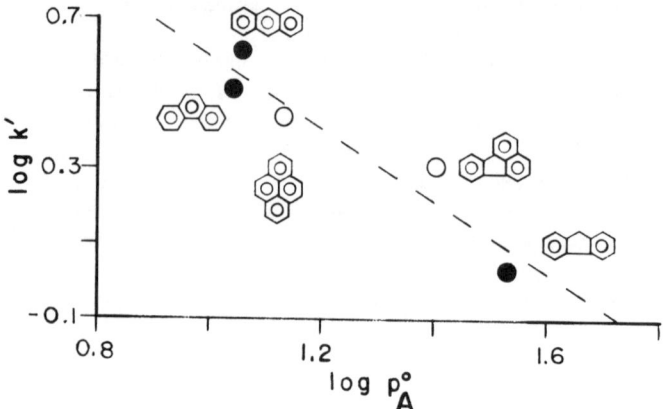

Figure 14.19 Plots of $\log k'$ against $\log p_A^0$ for solutes of Table 14.4 with a methoxybiphenyl MEPSIL (nematic) stationary phase at 180°C (filled circles) and 235°C (open circles) (Laub, 1988).

unknown (including any chirality, provided that appropriate standards were available) and, in favourable cases, its identity as well.

From the standpoint of assessment of physicochemical properties, plots such as those shown in Figures 14.18 and 14.19 are useful for the determination of vapour pressures and, recalling eqn 14.33, heats of vaporization. The GLC method of determining these quantities in this way will be particularly valuable in instances where the solutes are available only in minute quantities, or are substantially impure. Indeed, it should be possible to determine vapour pressures of quite high accuracy for many compounds over extensive ranges of temperature from the retentions obtained from only a few chromatographic runs. A recent example has in fact been reported by Heath and Tumlinson (1986), who employed log retention plots to determine the vapour pressures of trace acetate ester constituents of pheromones. An interesting feature of their work was that good correlations were obtained with a chiral–nematic stationary phase, cholesteryl p-chlorocinnamate.

Acknowledgements

We gratefully acknowledge support received for this work in part from the US Department of Energy Office of Basic Energy Sciences (analytical and polymer-synthetic work) and from the National Science Foundation (physicochemical studies). RJL also thanks A. Pelter and the Department of Chemistry, University College of Swansea, Wales, for the gracious hospitality extended him during the course of manuscript preparation. Figures 14.2, 14.3, 14.6, 14.7, 14.8, 14.9, 14.10, 14.11, 14.12, 14.13, 14.14, 14.15, 14.16 and 14.17 are reproduced by permission of ACS, Washington, DC; Figures 14.4 and 14.5 by permission of Batelle Press, Columbus, Ohio; Figures 14.18 and 14.19, and Tables 14.3 and 14.4 by permission of Gordon and Breach Sci. Publ. SA, NY.

References

Adams, N.W., Bradshaw, J.S., Bayona, J.-M., Markides, K.E. and Lee, M.L. (1987) *Mol. Cryst. Liq. Cryst.* **147**, 43.

Aggarwal, S.K., Bradshaw, J.S., Eguchi, M., Parry, S., Rossiter, B.E., Markides, K.E., and Lee, M.L. (1987) *Tetrahedron* **43**, 451.

Anderson, L.B. and Laub, R.J. (1981) *J. Electroanal. Chem.* **122**, 359.
Apfel, M.A., Finkelmann, H., Janini, G.M., Laub, R.J., Luhmann, B.-H., Price, A., Roberts, W.L., Shaw, T.J., and Smith, C.A. (1985) *Anal. Chem* **57**, 651.
Ashworth, A.J., Chien, C.-F., Furio, D.L., Hooker, D.M., Kopecni, M.M., Laub, R.J. and Price, G.J. (1984) *Macromolecules* **17**, 1090.
Aspler, J.S. (1985) in *Pyrolysis and GC in Polymer Analysis*, eds. Liebman, S.A. and Levy, E.J., Marcel Dekker, New York, Chap. 9.
Ballschmiter, K. and Zell, M. (1980) *Fresenius' Z. Anal. Chem.* **302**, 20.
Bayer, E. and Frank, H. (1979) German Patent 2 740 019, March 1979.
Bayer, E. and Frank, H. (1980) in *Modified Polymers*, ACS Symp. Ser. **121**, American Chemical Society, Washington DC, 341.
Bayer, E. and Frank, H. (1981) German Patent 3 005 024, August 1981.
Bayer, E., Frank, H. and Nicholson, G.J. (1980) German Patent 2 838 760, March 1980.
Bayona, J.M., Tarbet, B.J., Chang, H.C., Schregenberger. C.M., Nishioka, M., Markides, K.E., Bradshaw, J.S. and Lee, M.L. (1987) *Int. J. Environ. Anal. Chem.* **28**, 263.
Boublik, T., Fried, V., and Hala, E. (1973) *The Vapour Pressures of Pure Substances*, Elsevier, Amsterdam.
Bradshaw, J.S., Adams, N.W., Johnson, R.S., Tarbet, B.J., Schregenberger, C.M., Pulsipher, M.A., Andrus, M.B., Markides, K.E., and Lee, M.L. (1985) *HRC CC, J. High Resolut. Chromatogr. Chromatogr. Commun.*, **8**, 678.
Bradshaw, J.S., Schregenberger, C.M., Chang, K.H.C., Markides, K.E. and Lee, M.L. (1986) *J. Chromatogr.*, **358**, 95.
Bradshaw, J.S., Aggarwal, S.K., Rouse, C.A., Tarbet, B.J., Markides, K.E. and Lee, M.L. (1987) *J. Chromatogr.*, **405**, 169.
Chien, C.-F., Kopecni, M.M. and Laub, R.J. (1980) *Anal. Chem.* **52**, 1402, 1407.
Chien, C.-F., Kopecni, M.M. and Laub, R.J. (1981) *HRC CC, J. High Resolut. Chromatogr. Chromatogr. Commun.*, **4**, 539.
Chien, C.-F., Furio, D.L., Kopecni, M.M. and Laub, R.J. (1983a) *HRC CC, J. High Resolut. Chromatogr. Chromatogr. Commun.*, **6**, 577.
Chien, C.-F., Furio, D.L., Kopecni, M.M. and Laub, R.J. (1983b) *HRC CC, J. High Resolut. Chromatogr. Chromatogr. Commun.*, **6**, 669.
Chien, C.-F., Kopecni, M.M. and Laub, R.J. (1984) *J. Chromatogr. Sci.* **22**, 1.
Dreisbach, R.R. (1955) *Physical Properties of Chemical Compounds*, Vol. I, American Chemical Society, Washington, DC.
Finkelmann, H. (1984) *Adv. Polymer Sci.*, **60/61**, 99.
Finkelmann, H., Laub, R.J., Roberts, W.L. and Smith, C.A. (1981a) Paper presented at the Sixth International Symposium on Polynuclear Aromatic Hydrocarbons, Columbus, Ohio, October 1981.
Finkelmann, H., Laub, R.J., Roberts, W.L. and Smith, C.A. (1981b) in *Polynuclear Aromatic Hydrocarbons: Physical and Biological Chemistry*, eds. Cooke, M.W., Dennis, A.J. and Fisher, G.L., Battelle Press, Columbus, 275.
Frank, H., Nicholson, G.J. and Bayer, E. (1977) *J. Chromatogr. Sci.* **15**, 174.
Frank, H., Nicholson, G.J. and Bayer, E. (1978a) *Angew. Chem. Int. Edn. Engl.*, **17**, 363.
Frank, H., Nicholson, G.J. and Bayer, E. (1978b) *J. Chromatogr.* **146**, 197.
Garriguas, P., Marniesse, M.-P., Wise. S.A., Bellocq, J. and Ewald M. (1987) *Anal. Chem.* **59**, 1695.
Haky, J.E. and Muschik, G.M. (1981a) *J. Chromatogr.* **214**, 161.
Haky, J.E. and Muschik, G.M. (1981b) *J. Chromatogr.* **238**, 367.
Harbison, M.W.P., Laub, R.J., Martire, D.E., Purnell, J.H. and Williams, P.S. (1979) *J. Phys. Chem.* **83**, 1262.
Heath, R.R. and Tumlinson, J.H. (1986) *J. Chem. Ecol.* **12**, 2081.
Hecht, S.S., Bondinell, W.E. and Hoffman, D.J. (1974) *J. Natl. Cancer Inst.* **153**, 1121.
Hildebrand, G.P. and Reilley, C.N. (1964) *Anal. Chem.* **36**, 47.
Hoare, M.R. and Purnell, J.H. (1955a) *Research* **8**, S41.
Hoare, M.R. and Purnell, J.H. (1955b) *Trans. Farady Soc.* **52**, 222.
Jadhav, A.L., Naikwadi, K.P., Rokushika, S., Hatano, H. and Ohshima, M. (1987) *HRC CC, J. High Resolut. Chromatogr. Chromatogr. Commun.* **10**, 77.
Janini, G.M. (1979) *Adv. Chromatogr.* **17**, 231.
Janini, G.M., Johnston, K., Zielinski, W.L. Jr. (1975) *Anal. Chem.* **47**, 670.

Janini, G.M., Muschik, G.M., Schroer, J.A. and Zielinski, W.L. Jr. (1976a) *Anal Chem.* **48,** 1879.
Janini, G.M., Muschik, G.M. and Zielinski, W.L. Jr. (1976b) *Anal. Chem.* **48,** 809.
Janini, G.M., Sato, R.I. and Muschik, G.M. (1980) *Anal. Chem.* **52,** 2417.
Janini, G.M., Laub, R.J. and Shaw, T.J. (1985) *Macromol. Chem., Rapid Commun.* **6,** 57.
Janini, G.M., Laub, R.J., Pluyter, J.G.L. and Shaw, T.J. (1987) *Mol. Cryst. Liq. Cryst.* **153,** 479.
Janini, G.M., Muschik, G.M., Issaq, H.J. and Laub, R.J. (1988) *Anal Chem.* **60** (in press).
Jones, B.A., Bradshaw, J.S., Nishioka, M. and Lee, M.L. (1984) *J. Org. Chem.* **49,** 4947.
Jones, P., and Wellington, C.A. (1981) *J. Chromatogr.* **213,** 357.
Kelker, H. (1963) *Ber Bunsenges. Phys. Chem.* **67,** 698.
Kelker, H. (1978) *Adv. Liq. Cryst.* **3,** 237.
Kong, R.C., Lee, M.L., Tominaga, V., Pratap, R., Iwao, M. and Castle, R.N. (1982) *Anal. Chem.* **59,** 1802.
Koppenhoefer, B. and Bayer, E. (1984) *Chromatographia,* **19,** 123.
Lambert, B.J., Laub, R.J., Roberts, W.L. and Smith, C.A. (1984) in *Ultra-High Resolution Chromatography,* ed. Ahuja, S., ACS Symp. Ser. **250,** American Chemical Society, Washington DC, 49.
Laub, R.J. (1981) in *Chromatography, Equilibria, and Kinetics: Proc. Faraday Soc. Symp. (No. 15),* ed. Young, D.A., Royal Society of Chemistry, London, 179.
Laub, R.J. (1983) in *Physical Methods in Modern Chemical Analysis,* Vol. 3, ed. Kuwana, T., Academic Press, New York, 250 (*see also* Purnell *et al.* (1981) below).
Laub, R.J. (1984) *Anal. Chem.* **56,** 2110, 2115.
Laub, R.J. (1986) in *Chromatography and Separations Chemistry,* ed. Ahuja, S., ACS Symp. Ser. **297,** American Chemical Society, Washington DC, 6.
Laub, R.J. (1987) *HRC CC, J. High Resolut. Chromatogr. Chromatogr. Commun.* **10,** 565.
Laub, R.J. (1988) *Mol. Cryst. Liq. Cryst.* (in press.)
Laub, R.J. and Pecsok, R.L. (1978) *Physicochemical Applications of Gas Chromatography,* Wiley-Interscience, New York, (1978).
Laub, R.J. and Purnell, J.H. (1975) *J. Chromatogr.* **112,** 71.
Laub, R.J. and Purnell, J.H. (1976) *J. Amer. Chem. Soc.* **98,** 30, 35.
Laub, R.J. and Purnell, J.H. (1978a) *J. Chromatogr.* **161,** 49.
Laub, R.J. and Purnell, J.H. (1978b) *J. Chromatogr.* **161,** 59.
Laub, R.J. and Purnell, J.H. (1988) *HRC CC, J. High Resolut. Chromatogr. Chromatogr. Commun.,* submitted for review.
Laub, R.J., Purnell, J.H., Summers, D.M. and Williams, P.S. (1978a) *J. Chromatogr.* **155,** 1.
Laub, R.J., Purnell, J.H., Williams, P.S., Harbison, M.W.P. and Martire, D.E. (1978b) *J. Chromatogr.* **155,** 233.
Laub, R.J., Pelter, A. and Purnell, J.H. (1979) *Anal. Chem.* **51,** 1878.
Laub, R.J., Roberts, W.L. and Smith, C.A. (1980) *HRC CC, J. High Resolut. Chromatogr. Chromatogr. Commun.* **3,** 355.
Littlewood, A.B. (1964) *Anal. Chem.* **36,** 1441.
Markides, K.E., Chang, H.-C., Schregenberger, C.M., Tarbet, B.J., Bradshaw, J.S. and Lee, M.L. (1985a) *HRC CC, J. High Resolut. Chromatogr. Chromatogr. Commun.* **8,** 516.
Markides, K.E., Nishioka, M., Tarbet, B.J., Bradshaw, J.S. and Lee, N.L. (1985b) *Anal. Chem.* **57,** 1296.
Martin, A.J.P. and James, A.T. (1952) *Biochem. J.* **50,** 679.
Naikwadi, K.P., Jadav, A.L., Rokushika, S., Hatano, H. and Ohshima, M. (1986) *Makromol. Chem.* **187,** 1407.
Nestor, G., White, M.S., Gray, G.W., Lacey, D. and Toyne, K.J. (1987) *Makromol. Chem.* **188,** 2759.
Nishioka, M., Jones, B.A., Tarbet, B.J., Bradshaw, J.S. and Lee, M.L. (1986) *J. Chromatogr.* **357,** 79.
Novak, J. and Leclercq, P.A. (1988) *Quantitative Analysis by Gas Chromatography,* 2nd edn., Marcel Dekker, New York.
Patterson, D., Tewari, Y.B., Schreiber, H.P. and Guillet, J.E. (1971) *Macromolecules* **4,** 356.
Primavesi, G.R. (1959) *Nature (London)* **184,** 210.
Purnell, J.H. (1962) *Gas Chromatography,* Wiley, London, Chap. 10.
Purnell, J.H. (1975) in *Gas Chromatography 1956,* ed. Desty, D.H., Butterworth, London, 52.
Purnell, J.H. and Vargas de Andrade, J.M. (1975) *J. Amer. Chem. Soc.* **97,** 3585, 3590.

Purnell, J.H. and Williams, P.S. (1983) *HRC CC, J. High Resolut. Chromatogr. Chromatogr. Commun.* **6**, 569.

Purnell, J.H. and Williams, P.S. (1984) *J. Chromatogr.* **292**, 197.

Purnell, J.H. and Williams, P.S. (1985a) *J. Chromatogr.* **321**, 289.

Purnell, J.H. and Williams, P.S. (1985b) *J. Chromatogr.* **325**, 1.

Purnell, J.H., Williams, P.S. and Zabierek, G.A. (1981) in *Proc. 4th Int. Symp. Capillary Chromatogr.*, ed. Kaiser, R.E., Bad-Duerkheim, FRG, 573.

Purnell, J.H., Rodriguez, M. and Williams, P.S. (1985) *J. Chromatogr.* **323**, 402.

Purnell, J.H., Rodriguez, M. and Williams, P.S. (1986) *J. Chromatogr.* **358**, 39.

Purnell, J.H., Jones, J.R. and Wattan, M.H. (1987) *J. Chromatogr.* **399**, 99.

Rokushika, S., Naikwadi, K.P., Jadhav, A.L. and Hatano, H. (1985) *HRC CC, J. High Resolut. Chromatogr. Chromatogr. Commun.* **8**, 480.

Schroeder, J.P. (1974) in *Liquid Crystals and Plastic Crystals*, Vol. 1, eds. Gray, G.W. and Winsor, P.A., Ellis Horwood, Chichester, 356.

Speer, J.L. (1979) *Adv. Organometal. Chem.* **17**, 407.

Wise, S.A., Bonnett, W.J., Guenther, F.R. and May, W.E. (1981) *J. Chromatogr. Sci.* **19**, 457.

Witkiewicz, Z. (1982) *J. Chromatogr.* **251**, 311.

Zielinski, W.L. Jr., Scanlan, R.A. and Biller, M.M. (1981) *J. Chromatogr.* **209**, 87.

Zielinski, W.L. Jr., Miller, M.M. Ulma, G. and Wasik. S.P. (1986) *Anal. Chem.* **58**, 2692.

Index

443